WORLD HEALTH ORGANIZATION

INTERNATIONAL AGENCY FOR RESEARCH ON CANCER

IARC MONOGRAPHS
ON THE
EVALUATION OF CARCINOGENIC RISKS TO HUMANS

Chlorinated Drinking-water; Chlorination By-products; Some Other Halogenated Compounds; Cobalt and Cobalt Compounds

VOLUME 52

This publication represents the views and expert opinions
of an IARC Working Group on the
Evaluation of Carcinogenic Risks to Humans,
which met in Lyon,

12–19 June 1990

1991

IARC MONOGRAPHS

In 1969, the International Agency for Research on Cancer (IARC) initiated a programme on the evaluation of the carcinogenic risk of chemicals to humans involving the production of critically evaluated monographs on individual chemicals. In 1980, the programme was expanded to include the evaluation of the carcinogenic risk associated with exposures to complex mixtures.

The objective of the programme is to elaborate and publish in the form of monographs critical reviews of data on carcinogenicity for chemicals and complex mixtures to which humans are known to be exposed, and on specific exposures, to evaluate these data in terms of human risk with the help of international working groups of experts in chemical carcinogenesis and related fields, and to indicate where additional research efforts are needed.

This project is supported by PHS Grant No. 5-UO1 CA33193-08 awarded by the US National Cancer Institute, Department of Health and Human Services. Additional support has been provided by the Commission of the European Communities since 1986.

©International Agency for Research on Cancer 1991

ISBN 92 832 1252 5

ISSN 0250-9555

All rights reserved. Application for rights of reproduction or translation, in part or *in toto*, should be made to the International Agency for Research on Cancer.

Distributed for the International Agency for Research on Cancer
by the Secretariat of the World Health Organization

PRINTED IN THE UK

CONTENTS

NOTE TO THE READER 7

LIST OF PARTICIPANTS 9

PREAMBLE

    Background 15
    Objective and Scope 15
    Selection of Topics for Monographs 16
    Data for Monographs 17
    The Working Group 17
    Working Procedures 18
    Exposure Data 18
    Biological Data Relevant to the Evaluation of Carcinogenicity to
        Humans 20
    Evidence for Carcinogenicity in Experimental Animals 21
    Other Relevant Data in Experimental Systems and Humans 23
    Evidence for Carcinogenicity in Humans 25
    Summary of Data Reported 28
    Evaluation 29
    References 33

GENERAL REMARKS 37

THE MONOGRAPHS

    Chlorinated drinking-water

       1. Description of the process 45
          1.1 History of chlorination of drinking-water 45
          1.2 Overview of the addition of chlorine during drinking-water
              treatment 46
          1.3 Impurities in chlorine gas and liquid 53
          1.4 Alternative disinfectants for drinking-water 54

CONTENTS

2. Occurrence and analysis of compounds formed by the chlorination of drinking-water 55
   2.1 Occurrence 55
   2.2 Analytical methods 68
3. Biological data relevant to the evaluation of carcinogenic risk to humans 72
   3.1 Carcinogenicity studies in animals 72
   3.2 Other relevant data 77
   3.3 Epidemiological studies of carcinogenicity in humans 107
4. Summary of data reported and evaluation 122
   4.1 Exposure data 122
   4.2 Experimental carcinogenicity data 125
   4.3 Human carcinogenicity data 125
   4.4 Other relevant data 127
   4.5 Evaluation 128
5. References 128

Some chemicals used in the chlorination of drinking-water

Sodium chlorite 145
Hypochlorite salts 159

Chlorination by-products

Bromodichloromethane 179
Bromoform 213
Chlorodibromomethane 243
Halogenated acetonitriles 269
    Bromochloroacetonitrile
    Chloroacetonitrile
    Dibromoacetonitrile
    Dichloroacetonitrile
    Trichloroacetonitrile

Some other halogenated chemicals

Bromoethane 299
Chloroethane 315
1,1,2-Trichloroethane 337

Cobalt and cobalt compounds 363

CONTENTS

SUMMARY OF FINAL EVALUATIONS 473

APPENDIX 1. SUMMARY TABLES OF GENETIC AND RELATED EFFECTS 477

APPENDIX 2. ACTIVITY PROFILES FOR GENETIC AND RELATED EFFECTS 491

SUPPLEMENTARY CORRIGENDA TO VOLUMES 1-51 513

CUMULATIVE INDEX TO THE *MONOGRAPHS* SERIES 515

NOTE TO THE READER

The term 'carcinogenic risk' in the *IARC Monographs* series is taken to mean the probability that exposure to an agent will lead to cancer in humans.

Inclusion of an agent in the *Monographs* does not imply that it is a carcinogen, only that the published data have been examined. Equally, the fact that an agent has not yet been evaluated in a monograph does not mean that it is not carcinogenic.

The evaluations of carcinogenic risk are made by international working groups of independent scientists and are qualitative in nature. No recommendation is given for regulation or legislation.

Anyone who is aware of published data that may alter the evaluation of the carcinogenic risk of an agent to humans is encouraged to make this information available to the Unit of Carcinogen Identification and Evaluation, International Agency for Research on Cancer, 150 cours Albert Thomas, 69372 Lyon Cedex 08, France, in order that the agent may be considered for re-evaluation by a future working group.

Although every effort is made to prepare the monographs as accurately as possible, mistakes may occur. Readers are requested to communicate any errors to the Unit of Carcinogen Identification and Evaluation, so that corrections can be reported in future volumes.

IARC WORKING GROUP ON THE EVALUATION OF CARCINOGENIC RISKS TO HUMANS: CHLORINATED DRINKING-WATER; CHLORINATION BY-PRODUCTS; SOME OTHER HALOGENATED COMPOUNDS; COBALT AND COBALT COMPOUNDS

Lyon, 12–19 June 1990

LIST OF PARTICIPANTS

Members

M.W. Anders, Department of Pharmacology, The University of Rochester Medical Center, 601 Elmwood Avenue, Rochester, NY 14642, USA

B.K. Armstrong[1], Commissioner of Health for Western Australia, Health Department of Western Australia, PO Box 8172, Sterling Street, Perth, WA 6000, Australia (*Chairman*)

A. Astrup Jensen, Danish Institute of Technology, Department of Environmental Technology, Section of Environmental and Occupational Toxicology, PO Box 141, 2630 Taastrup, Denmark

D. Beyersmann, Division 2-Biochemistry, University of Bremen, 2800 Bremen 33, Germany

R. Bull, College of Pharmacy, Washington State University, Pullman, WA 99164-6510, USA

K. Cantor, Environmental Epidemiology Branch, National Cancer Institute, 443 Executive Plaza North, Bethesda, MD 20892, USA

C.E.D. Chilvers, Department of Public Health Medicine and Epidemiology, The University of Nottingham Medical School, Queen's Medical Centre, Nottingham NG7 2UH, UK

---

[1]Present address: International Agency for Research on Cancer, 150 cours Albert Thomas, 69372 Lyon Cedex 08, France

C.-G. Elinder, Department of Medicine, Section for Renal Medicine, Karolinska Institute and Hospital, 104 01 Stockholm, Sweden

S.L. Eustis, National Toxicology Program, National Institute of Environmental Health Sciences, PO Box 12233, Research Triangle Park, NC 27709, USA

M. Fielding, Water Research Centre, Medmenham Laboratory, Henley Road, Medmenham, PO Box 16, Marlow, Buckinghamshire SL7 2HD, UK

M. Ikeda, Department of Public Health, Kyoto University, Faculty of Medicine, Kyoto 606, Japan (*Vice-Chairman*)

H. Kappus, Institute for Toxicology, Society for Research on Radiation and the Environment, Ingolstädter Landstrasse 1, 8042 Neuherberg, Germany

N.K. Mottet, Department of Pathology, SM-30, University of Washington, School of Medicine, Seattle, WA 98195, USA

H. Norppa, Institute of Occupational Health, Topeliuksenkatu 41 a A, 00250 Helsinki, Finland

A. Pinter, Department of Morphology, National Institute of Hygiene, Gyali ut 2-6, 1966 Budapest, Hungary

K.-H. Schaller, Institute for Occupational and Social Medicine, University of Erlangen-Nürnberg, Schillerstrasse 25, 8520 Erlangen, Germany

B. Schwetz, National Toxicology Program, National Institute of Environmental Health Sciences, PO Box 12233, Research Triangle Park, NC 27709, USA

J. Siemiatycki, Institut Armand-Frappier, Québec University, 531 boulevard des Prairies, Case Postale 100, Laval, Québec H7N 4Z3, Canada

F.W. Sunderman, Jr, Departments of Laboratory Medicine and Pharmacology, University of Connecticut, School of Medicine, 263 Farmington Avenue, Farmington, CT 06032, USA

S. Swierenga, Drug Identification Division, Banting Building, Tunney's Pasture, Ottawa, Ontario K1A 0L2, Canada

B. Teichmann, Academy of Sciences, Central Institute for Molecular Biology, Robert-Rössle-Strasse 10, 1115 Berlin Buch, Germany

Representatives/Observers

Representative of Tracor Technology Resources, Inc.

S. Olin, Tracor Technology Resources, Inc., 1601 Research Boulevard, Rockville, MD 20850, USA

Representative of the Chemical Manufacturers' Association

N. Krivanek, E.I. du Pont de Nemours & Co., Haskell Laboratories, PO Box 50, Newark, DE 19714, USA

Representative of the Cobalt Development Institute

B. Swennen, Metallurgie Hoboken-Overpelt, Division of SA ACEG-Union Minière NV, Leemanslaan 36, 2430 Olen, Belgium

Representatives of the Commission of European Communities

M. Draper, 10 West Mayfield, Edinburgh EH9 1TQ, UK

E. Krug, Commission of the European Communities, Health and Safety Directorate, Bâtiment Jean Monnet, BP 1907, 2920 Luxembourg, Grand Duchy of Luxembourg

Representative of the European Chemical Industry, Ecology and Toxicology Centre

D.G. Farrar, ICI Chemicals and Polymers Ltd, Occupational Health, PO Box 13, The Heath, Runcorn, Cheshire WA7 4QF, UK

Representative of the International Life Sciences Institute

L. Fishbein, ILSI-Risk Science Institute, 1126 Sixteenth Street NW, Suite 111, Washington DC 20007, USA

Representative of the US Environmental Protection Agency

J. Orme, Health Effects Assessment, Section II, Office of Drinking Water, Environmental Protection Agency, Waterside East Building, Room WH-550D, 401 M Street SW, Washington DC 20460, USA

Secretariat

H. Bartsch, Unit of Environmental Carcinogenesis and Host Factors

P. Boffetta, Unit of Analytical Epidemiology

G. Burin, Promotion of Chemical Safety, Division of Environmental Health, World Health Organization, Geneva, Switzerland

E. Cardis, Unit of Biostatistics Research and Informatics

M. Coleman, Unit of Descriptive Epidemiology

M.-J. Ghess, Unit of Carcinogen Identification and Evaluation

E. Heseltine, Lajarthe, Montignac, France

V. Krutovskikh, Unit of Mechanisms of Carcinogenesis

M. Marselos[1], Unit of Carcinogen Identification and Evaluation

D. McGregor, Unit of Carcinogen Identification and Evaluation

---

[1]Present address: Department of Pharmacology, Medical School, University of Ioannina, 45110 Ioannina, Greece

D. Mietton, Unit of Carcinogen Identification and Evaluation

G. Nordberg[1], Unit of Carcinogen Identification and Evaluation

C. Partensky, Unit of Carcinogen Identification and Evaluation

I. Peterschmitt, Unit of Carcinogen Identification and Evaluation, Geneva, Switzerland

D. Shuker, Unit of Environmental Carcinogenesis and Host Factors

L. Shuker, Unit of Carcinogen Identification and Evaluation

L. Tomatis, Director

H. Vainio, Unit of Carcinogen Identification and Evaluation

J. Wilbourn, Unit of Carcinogen Identification and Evaluation

H. Yamasaki, Unit of Mechanisms of Carcinogenesis

Secretarial assistance

J. Cazeaux

M. Lézère

M. Mainaud

S. Reynaud

---

[1]Present address: Department of Environmental Medicine, University of Umeå, S-901 87 Umeå, Sweden

PREAMBLE

IARC MONOGRAPHS PROGRAMME ON THE EVALUATION OF CARCINOGENIC RISKS TO HUMANS[1]

PREAMBLE

1. BACKGROUND

In 1969, the International Agency for Research on Cancer (IARC) initiated a programme to evaluate the carcinogenic risk of chemicals to humans and to produce monographs on individual chemicals. The *Monographs* programme has since been expanded to include consideration of exposures to complex mixtures of chemicals (which occur, for example, in some occupations and as a result of human habits) and of exposures to other agents, such as radiation and viruses. With Supplement 6(1), the title of the series was modified from *IARC Monographs on the Evaluation of the Carcinogenic Risk of Chemicals to Humans* to *IARC Monographs on the Evaluation of Carcinogenic Risks to Humans*, in order to reflect the widened scope of the programme.

The criteria established in 1971 to evaluate carcinogenic risk to humans were adopted by the working groups whose deliberations resulted in the first 16 volumes of the *IARC Monographs* series. Those criteria were subsequently re-evaluated by working groups which met in 1977(2), 1978(3), 1979(4), 1982(5) and 1983(6). The present preamble was prepared by two working groups which met in September 1986 and January 1987, prior to the preparation of Supplement 7(7) to the *Monographs* and was modified by a working group which met in November 1988(8).

2. OBJECTIVE AND SCOPE

The objective of the programme is to prepare, with the help of international working groups of experts, and to publish in the form of monographs, critical

---

[1]This project is supported by PHS Grant No. 5 UO1 CA33193-08 awarded by the US National Cancer Institute, Department of Health and Human Services, and with a subcontract to Tracor Technology Resources, Inc. Since 1986, this programme has also been supported by the Commission of the European Communities.

reviews and evaluations of evidence on the carcinogenicity of a wide range of human exposures. The *Monographs* may also indicate where additional research efforts are needed.

The *Monographs* represent the first step in carcinogenic risk assessment, which involves examination of all relevant information in order to assess the strength of the available evidence that certain exposures could alter the incidence of cancer in humans. The second step is quantitative risk estimation, which is not usually attempted in the *Monographs*. Detailed, quantitative evaluations of epidemiological data may be made in the *Monographs*, but without extrapolation beyond the range of the data available. Quantitative extrapolation from experimental data to the human situation is not undertaken.

These monographs may assist national and international authorities in making risk assessments and in formulating decisions concerning any necessary preventive measures. The evaluations of IARC working groups are scientific, qualitative judgements about the degree of evidence for carcinogenicity provided by the available data on an agent. These evaluations represent only one part of the body of information on which regulatory measures may be based. Other components of regulatory decisions may vary from one situation to another and from country to country, responding to different socioeconomic and national priorities. *Therefore, no recommendation is given with regard to regulation or legislation, which are the responsibility of individual governments and/or other international organizations.*

The *IARC Monographs* are recognized as an authoritative source of information on the carcinogenicity of chemicals and complex exposures. A users' survey, made in 1988, indicated that the *Monographs* are consulted by various agencies in 57 countries. Each volume is generally printed in 4000 copies for distribution to governments, regulatory bodies and interested scientists. The *Monographs* are also available *via* the Distribution and Sales Service of the World Health Organization.

3. SELECTION OF TOPICS FOR MONOGRAPHS

Topics are selected on the basis of two main criteria: (a) that they concern agents and complex exposures for which there is evidence of human exposure, and (b) that there is some evidence or suspicion of carcinogenicity. The term agent is used to include individual chemical compounds, groups of chemical compounds, physical agents (such as radiation) and biological factors (such as viruses) and mixtures of agents such as occur in occupational exposures and as a result of personal and cultural habits (like smoking and dietary practices). Chemical analogues and compounds with biological or physical characteristics similar to those of suspected carcinogens may also be considered, even in the absence of data on carcinogenicity.

The scientific literature is surveyed for published data relevant to an assessment of carcinogenicity; the IARC surveys of chemicals being tested for carcinogenicity(9) and directories of on-going research in cancer epidemiology(10) often indicate those exposures that may be scheduled for future meetings. Ad-hoc working groups convened by IARC in 1984 and 1989 gave recommendations as to which chemicals and exposures to complex mixtures should be evaluated in the *IARC Monographs* series(11,12).

As significant new data on subjects on which monographs have already been prepared become available, re-evaluations are made at subsequent meetings, and revised monographs are published.

4. DATA FOR MONOGRAPHS

The *Monographs* do not necessarily cite all the literature concerning the subject of an evaluation. Only those data considered by the Working Group to be relevant to making the evaluation are included.

With regard to biological and epidemiological data, only reports that have been published or accepted for publication in the openly available scientific literature are reviewed by the working groups. In certain instances, government agency reports that have undergone peer review and are widely available are considered. Exceptions may be made on an ad-hoc basis to include unpublished reports that are in their final form and publicly available, if their inclusion is considered pertinent to making a final evaluation (see pp. 29 *et seq*.). In the sections on chemical and physical properties and on production, use, occurrence and analysis, unpublished sources of information may be used.

5. THE WORKING GROUP

Reviews and evaluations are formulated by a working group of experts. The tasks of this group are five-fold: (i) to ascertain that all appropriate data have been collected; (ii) to select the data relevant for the evaluation on the basis of scientific merit; (iii) to prepare accurate summaries of the data to enable the reader to follow the reasoning of the Working Group; (iv) to evaluate the results of experimental and epidemiological studies; and (v) to make an overall evaluation of the carcinogenicity of the exposure to humans.

Working Group participants who contributed to the considerations and evaluations within a particular volume are listed, with their addresses, at the beginning of each publication. Each participant who is a member of a working group serves as an individual scientist and not as a representative of any organization, government or industry. In addition, representatives from national and international agencies and industrial associations are invited as observers.

6. WORKING PROCEDURES

Approximately one year in advance of a meeting of a working group, the topics of the monographs are announced and participants are selected by IARC staff in consultation with other experts. Subsequently, relevant biological and epidemiological data are collected by IARC from recognized sources of information on carcinogenesis, including data storage and retrieval systems such as CHEMICAL ABSTRACTS, MEDLINE and TOXLINE—including EMIC and ETIC for data on genetic and related effects and teratogenicity, respectively.

The major collection of data and the preparation of first drafts of the sections on chemical and physical properties, on production and use, on occurrence, and on analysis are carried out under a separate contract funded by the US National Cancer Institute. Efforts are made to supplement this information with data from other national and international sources. Representatives from industrial associations may assist in the preparation of sections on production and use.

Production and trade data are obtained from governmental and trade publications and, in some cases, by direct contact with industries. Separate production data on some agents may not be available because their publication could disclose confidential information. Information on uses is usually obtained from published sources but is often complemented by direct contact with manufacturers.

Six months before the meeting, reference material is sent to experts, or is used by IARC staff, to prepare sections for the first drafts of monographs. The complete first drafts are compiled by IARC staff and sent, prior to the meeting, to all participants of the Working Group for review.

The Working Group meets in Lyon for seven to eight days to discuss and finalize the texts of the monographs and to formulate the evaluations. After the meeting, the master copy of each monograph is verified by consulting the original literature, edited and prepared for publication. The aim is to publish monographs within nine months of the Working Group meeting.

7. EXPOSURE DATA

Sections that indicate the extent of past and present human exposure, the sources of exposure, the persons most likely to be exposed and the factors that contribute to exposure to the agent, mixture or exposure circumstance are included at the beginning of each monograph.

Most monographs on individual chemicals or complex mixtures include sections on chemical and physical data, and production, use, occurrence and analysis. In other monographs, for example on physical agents, biological factors, occupational exposures and cultural habits, other sections may be included, such as: historical perspectives, description of an industry or habit, exposures in the work place or chemistry of the complex mixture.

The Chemical Abstracts Services Registry Number, the latest Chemical Abstracts Primary Name and the IUPAC Systematic Name are recorded. Other synonyms are given, but the list is not necessarily comprehensive.

Information on chemical and physical properties and, in particular, data relevant to identification, occurrence and biological activity are included. A separate description of technical products gives relevant specifications and includes available information on composition and impurities. Trade names are given; some of the trade names may be those of mixtures in which the agent being evaluated is only one of the ingredients.

The dates of first synthesis and of first commercial production of an agent or mixture are provided; for agents which do not occur naturally, this information may allow a reasonable estimate to be made of the date before which no human exposure to the agent could have occurred. The dates of first reported occurrence of an exposure are also provided. In addition, methods of synthesis used in past and present commercial production and different methods of production which may give rise to different impurities are described.

Data on production, foreign trade and uses are obtained for representative regions, which usually include Europe, Japan and the USA. It should not, however, be inferred that those areas or nations are necessarily the sole or major sources or users of the agent being evaluated.

Some identified uses may not be current or major applications, and the coverage is not necessarily comprehensive. In the case of drugs, mention of their therapeutic uses does not necessarily represent current practice nor does it imply judgement as to their clinical efficacy.

Information on the occurrence of an agent or mixture in the environment is obtained from data derived from the monitoring and surveillance of levels in occupational environments, air, water, soil, foods and animal and human tissues. When available, data on the generation, persistence and bioaccumulation are also included. In the case of mixtures, industries, occupations or processes, information is given about all agents present. For processes, industries and occupations, a historical description is also given, noting variations in chemical composition, physical properties or levels of occupational exposure with time.

Statements concerning regulations and guidelines (e.g., pesticide registrations, maximal levels permitted in foods, occupational exposure limits) are included for some countries as indications of potential exposures, but they may not reflect the most recent situation, since such limits are continuously reviewed and modified.

The absence of information on regulatory status for a country should not be taken to imply that that country does not have regulations with regard to the exposure.

The purpose of the section on analysis is to give the reader an overview of current methods cited in the literature, with emphasis on those widely used for regulatory purposes. No critical evaluation or recommendation of any of the methods is meant or implied. Methods for monitoring human exposure are also given, when available. The IARC publishes a series of volumes, *Environmental Carcinogens: Methods of Analysis and Exposure Measurement(13)*, that describe validated methods for analysing a wide variety of agents and mixtures.

8. BIOLOGICAL DATA RELEVANT TO THE EVALUATION OF CARCINOGENICITY TO HUMANS

The term 'carcinogen' is used in these monographs to denote an agent or mixture that is capable of increasing the incidence of malignant neoplasms; the induction of benign neoplasms may in some circumstances (see p. 22) contribute to the judgement that the exposure is carcinogenic. The terms 'neoplasm' and 'tumour' are used interchangeably.

Some epidemiological and experimental studies indicate that different agents may act at different stages in the carcinogenic process, probably by fundamentally different mechanisms. In the present state of knowledge, the aim of the *Monographs* is to evaluate evidence of carcinogenicity at any stage in the carcinogenic process independently of the underlying mechanism involved. There is as yet insufficient information to implement classification according to mechanisms of action(6).

Definitive evidence of carcinogenicity in humans can be provided only by epidemiological studies. Evidence relevant to human carcinogenicity may also be provided by experimental studies of carcinogenicity in animals and by other biological data, particularly those relating to humans.

The available studies are summarized by the Working Group, with particular regard to the qualitative aspects discussed below. In general, numerical findings are indicated as they appear in the original report; units are converted when necessary for easier comparison. The Working Group may conduct additional analyses of the published data and use them in their assessment of the evidence and may include them in their summary of a study; the results of such supplementary analyses are given in square brackets. Any comments are also made in square brackets; however, these are kept to a minimum, being restricted to those instances in which it is felt that an important aspect of a study, directly impinging on its interpretation, should be brought to the attention of the reader.

For experimental studies with mixtures, consideration is given to the possibility of changes in the physicochemical properties of the test substance during collection, storage, extraction, concentration and delivery. Either chemical or toxicological interactions of the components of mixtures may result in nonlinear dose-response relationships.

An assessment is made as to the relevance to human exposure of samples tested in experimental systems, which may involve consideration of: (i) physical and chemical characteristics, (ii) constituent substances that indicate the presence of a class of substances, (iii) tests for genetic and related effects, including genetic activity profiles, (iv) DNA adduct profiles, (v) oncogene expression and mutation; suppressor gene inactivation.

9. EVIDENCE FOR CARCINOGENICITY IN EXPERIMENTAL ANIMALS

For several agents (e.g., 4-aminobiphenyl, bis(chloromethyl)ether, diethylstilboestrol, melphalan, 8-methoxypsoralen (methoxsalen) plus ultra-violet radiation, mustard gas and vinyl chloride), evidence of carcinogenicity in experimental animals preceded evidence obtained from epidemiological studies or case reports. Information compiled from the first 41 volumes of the *IARC Monographs*(14) shows that, of the 44 agents and mixtures for which there is *sufficient* or *limited evidence* of carcinogenicity to humans (see p. 30), all 37 that have been tested adequately experimentally produce cancer in at least one animal species. Although this association cannot establish that all agents and mixtures that cause cancer in experimental animals also cause cancer in humans, nevertheless, *in the absence of adequate data on humans, it is biologically plausible and prudent to regard agents and mixtures for which there is sufficient evidence (see p. 31) of carcinogenicity in experimental animals as if they presented a carcinogenic risk to humans*.

The monographs are not intended to summarize all published studies. Those that are inadequate (e.g., too short a duration, too few animals, poor survival; see below) or are judged irrelevant to the evaluation are generally omitted. They may be mentioned briefly, particularly when the information is considered to be a useful supplement to that of other reports or when they provide the only data available. Their inclusion does not, however, imply acceptance of the adequacy of the experimental design or of the analysis and interpretation of their results. Guidelines for adequate long-term carcinogenicity experiments have been outlined (e.g., 15).

The nature and extent of impurities or contaminants present in the agent or mixture being evaluated are given when available. Mention is made of all routes of exposure that have been adequately studied and of all species in which relevant experiments have been performed. Animal strain, sex, numbers per group, age at start of treatment and survival are reported.

Experiments in which the agent or mixture was administered in conjunction with known carcinogens or factors that modify carcinogenic effects are also reported. Experiments on the carcinogenicity of known metabolites and derivatives may be included.

(a) Qualitative aspects

An assessment of carcinogenicity involves several considerations of qualitative importance, including (i) the experimental conditions under which the test was performed, including route and schedule of exposure, species, strain, sex, age, duration of follow-up; (ii) the consistency of the results, for example, across species and target organ(s); (iii) the spectrum of neoplastic response, from benign tumours to malignant neoplasms; and (iv) the possible role of modifying factors.

Considerations of importance to the Working Group in the interpretation and evaluation of a particular study include: (i) how clearly the agent was defined and, in the case of mixtures, how adequately the sample characterization was reported; (ii) whether the dose was adequately monitored, particularly in inhalation experiments; (iii) whether the doses used were appropriate and whether the survival of treated animals was similar to that of controls; (iv) whether there were adequate numbers of animals per group; (v) whether animals of both sexes were used; (vi) whether animals were allocated randomly to groups; (vii) whether the duration of observation was adequate; and (viii) whether the data were adequately reported. If available, recent data on the incidence of specific tumours in historical controls, as well as in concurrent controls, should be taken into account in the evaluation of tumour response.

When benign tumours occur together with and originate from the same cell type in an organ or tissue as malignant tumours in a particular study and appear to represent a stage in the progression to malignancy, it may be valid to combine them in assessing tumour incidence. The occurrence of lesions presumed to be pre-neoplastic may in certain instances aid in assessing the biological plausibility of any neoplastic response observed.

Of the many agents and mixtures that have been studied extensively, few induced only benign neoplasms. Benign tumours in experimental animals frequently represent a stage in the evolution of a malignant neoplasm, but they may be 'endpoints' that do not readily undergo transition to malignancy. However, if an agent or mixture is found to induce only benign neoplasms, it should be suspected of being a carcinogen and it requires further investigation.

(b) Quantitative aspects

The probability that tumours will occur may depend on the species and strain, the dose of the carcinogen and the route and period of exposure. Evidence of an increased incidence of neoplasms with increased level of exposure strengthens the inference of a causal association between the exposure and the development of neoplasms.

The form of the dose-response relationship can vary widely, depending on the particular agent under study and the target organ. Since many chemicals require metabolic activation before being converted into their reactive intermediates, both metabolic and pharmacokinetic aspects are important in determining the dose-response pattern. Saturation of steps such as absorption, activation, inactivation and elimination of the carcinogen may produce nonlinearity in the dose-response relationship, as could saturation of processes such as DNA repair(16,17).

(c) *Statistical analysis of long-term experiments in animals*

Factors considered by the Working Group include the adequacy of the information given for each treatment group: (i) the number of animals studied and the number examined histologically, (ii) the number of animals with a given tumour type and (iii) length of survival. The statistical methods used should be clearly stated and should be the generally accepted techniques refined for this purpose(17,18). When there is no difference in survival between control and treatment groups, the Working Group usually compares the proportions of animals developing each tumour type in each of the groups. Otherwise, consideration is given as to whether or not appropriate adjustments have been made for differences in survival. These adjustments can include: comparisons of the proportions of tumour-bearing animals among the 'effective number' of animals alive at the time the first tumour is discovered, in the case where most differences in survival occur before tumours appear; life-table methods, when tumours are visible or when they may be considered 'fatal' because mortality rapidly follows tumour development; and the Mantel-Haenszel test or logistic regression, when occult tumours do not affect the animals' risk of dying but are 'incidental' findings at autopsy.

In practice, classifying tumours as fatal or incidental may be difficult. Several survival-adjusted methods have been developed that do not require this distinction(17), although they have not been fully evaluated.

10. OTHER RELEVANT DATA IN EXPERIMENTAL SYSTEMS AND HUMANS

(a) *Structure-activity considerations*

This section describes structure-activity correlations that are relevant to an evaluation of the carcinogenicity of an agent.

(b) *Absorption, distribution, excretion and metabolism*

Concise information is given on absorption, distribution (including placental transfer) and excretion. Kinetic factors that may affect the dose-reponse relationship, such as saturation of uptake, protein binding, metabolic activation, detoxification and DNA repair processes, are mentioned. Studies that indicate the metabolic fate of the agent in experimental animals and humans are summarized briefly, and comparisons of data from animals and humans are made when possible. Comparative information on the relationship between exposure and the dose that reaches the target site may be of particular importance for extrapolation between species.

(c) *Toxicity*

Data are given on acute and chronic toxic effects (other than cancer), such as organ toxicity, immunotoxicity, endocrine effects and preneoplastic lesions. Effects on reproduction, teratogenicity, feto- and embryotoxicity are also summarized briefly.

(d) *Genetic and related effects*

Tests of genetic and related effects may indicate possible carcinogenic activity. They can also be used in detecting active metabolites of known carcinogens in human or animal body fluids, in detecting active components in complex mixtures and in the elucidation of possible mechanisms of carcinogenesis.

The adequacy of the reporting of sample characterization is considered and, where necessary, commented upon. The available data are interpreted critically by phylogenetic group according to the endpoints detected, which may include DNA damage, gene mutation, sister chromatid exchange, micronuclei, chromosomal aberrations, aneuploidy and cell transformation. The concentrations (doses) employed are given and mention is made of whether an exogenous metabolic system was required. When appropriate, these data may be represented by bar graphs (activity profiles), with corresponding summary tables and listings of test systems, data and references. Detailed information on the preparation of these profiles is given in an appendix to those volumes in which they are used.

Positive results in tests using prokaryotes, lower eukaryotes, plants, insects and cultured mammalian cells suggest that genetic and related effects (and therefore possibly carcinogenic effects) could occur in mammals. Results from such tests may also give information about the types of genetic effect produced and about the involvement of metabolic activation. Some endpoints described are clearly genetic in nature (e.g., gene mutations and chromosomal aberrations), others are to a greater or lesser degree associated with genetic effects (e.g., unscheduled DNA synthesis). In-vitro tests for tumour-promoting activity and for cell transformation may detect changes that are not necessarily the result of genetic alterations but that may have specific relevance to the process of carcinogenesis. A critical appraisal of these tests has been published(15).

Genetic or other activity detected in the systems mentioned above is not always manifest in whole mammals. Positive indications of genetic effects in experimental mammals and in humans are regarded as being of greater relevance than those in other organisms. The demonstration that an agent or mixture can induce gene and chromosomal mutations in whole mammals indicates that it may have the potential for carcinogenic activity, although this activity may not be detectably expressed in any or all species tested. Relative potency in tests for mutagenicity and related effects is not a reliable indicator of carcinogenic potency. Negative results in tests for mutagenicity in selected tissues from animals treated *in vivo* provide less weight, partly because they do not exclude the possibility of an effect in tissues other than those examined. Moreover, negative results in short-term tests with genetic endpoints cannot be considered to provide evidence to rule out carcinogenicity of agents or mixtures that act through other mechanisms. Factors may arise in many tests that could give misleading results; these have been discussed in detail elsewhere(15).

The adequacy of epidemiological studies of reproductive outcomes and genetic and related effects in humans is evaluated by the same criteria as are applied to epidemiological studies of cancer.

11. EVIDENCE FOR CARCINOGENICITY IN HUMANS

(a) Types of studies considered

Three types of epidemiological studies of cancer contribute data to the assessment of carcinogenicity in humans—cohort studies, case-control studies and correlation studies. Rarely, results from randomized trials may be available. Case reports of cancer in humans are also reviewed.

Cohort and case-control studies relate individual exposures under study to the occurrence of cancer in individuals and provide an estimate of relative risk (ratio of incidence in those exposed to incidence in those not exposed) as the main measure of association.

In correlation studies, the units of investigation are usually whole populations (e.g., in particular geographical areas or at particular times), and cancer frequency is related to a summary measure of the exposure of the population to the agent, mixture or exposure circumstance under study. Because individual exposure is not documented, however, a causal relationship is less easy to infer from correlation studies than from cohort and case-control studies.

Case reports generally arise from a suspicion, based on clinical experience, that the concurrence of two events—that is, a particular exposure and occurrence of a cancer—has happened rather more frequently than would be expected by chance. Case reports usually lack complete ascertainment of cases in any population, definition or enumeration of the population at risk and estimation of the expected number of cases in the absence of exposure.

The uncertainties surrounding interpretation of case reports and correlation studies make them inadequate, except in rare instances, to form the sole basis for inferring a causal relationship. When taken together with case-control and cohort studies, however, relevant case reports or correlation studies may add materially to the judgement that a causal relationship is present.

Epidemiological studies of benign neoplasms and presumed preneoplastic lesions are also reviewed by working groups. They may, in some instances, strengthen inferences drawn from studies of cancer itself.

(b) Quality of studies considered

It is necessary to take into account the possible roles of bias, confounding and chance in the interpretation of epidemiological studies. By 'bias' is meant the operation of factors in study design or execution that lead erroneously to a stronger or weaker association than in fact exists between disease and an agent, mixture or exposure circumstance. By 'confounding' is meant a situation in which the relationship with disease is made to appear stronger or to appear weaker than it truly is as a result of an association between the apparent causal factor and another factor that is associated with either an increase or decrease in the incidence of the disease. In evaluating the extent to which these factors have been minimized in an individual study, working groups consider a number of aspects of design and analysis as described in the report of the study. Most of these considerations apply equally to case-control, cohort and correlation studies. Lack of clarity of any of these aspects in the reporting of a study can decrease its credibility and its consequent weighting in the final evaluation of the exposure.

Firstly, the study population, disease (or diseases) and exposure should have been well defined by the authors. Cases in the study population should have been identified in a way that was independent of the exposure of interest, and exposure should have been assessed in a way that was not related to disease status.

Secondly, the authors should have taken account in the study design and analysis of other variables that can influence the risk of disease and may have been related to the exposure of interest. Potential confounding by such variables should have been dealt with either in the design of the study, such as by matching, or in the analysis, by statistical adjustment. In cohort studies, comparisons with local rates of disease may be more appropriate than those with national rates. Internal comparisons of disease frequency among individuals at different levels of exposure should also have been made in the study.

Thirdly, the authors should have reported the basic data on which the conclusions are founded, even if sophisticated statistical analyses were employed. At the very least, they should have given the numbers of exposed and unexposed cases and controls in a case-control study and the numbers of cases observed and expected in a cohort study. Further tabulations by time since exposure began and other temporal factors are also important. In a cohort study, data on all cancer sites and all causes of death should have been given, to avoid the possibility of reporting bias. In a case-control study, the effects of investigated factors other than the exposure of interest should have been reported.

Finally, the statistical methods used to obtain estimates of relative risk, absolute cancer rates, confidence intervals and significance tests, and to adjust for confounding should have been clearly stated by the authors. The methods used should preferably have been the generally accepted techniques that have been refined since the mid-1970s. These methods have been reviewed for case-control studies(19) and for cohort studies(20).

(c) *Quantitative considerations*

Detailed analyses of both relative and absolute risks in relation to age at first exposure and to temporal variables, such as time since first exposure, duration of exposure and time since exposure ceased, are reviewed and summarized when available. The analysis of temporal relationships can provide a useful guide in formulating models of carcinogenesis. In particular, such analyses may suggest whether a carcinogen acts early or late in the process of carcinogenesis(6), although such speculative inferences cannot be used to draw firm conclusions concerning the mechanism of action and hence the shape (linear or otherwise) of the dose-response relationship below the range of observation.

(d) *Criteria for causality*

After the quality of individual epidemiological studies has been summarized and assessed, a judgement is made concerning the strength of evidence that the agent, mixture or exposure circumstance in question is carcinogenic for humans. In making their judgement, the Working Group considers several criteria for causality. A strong association (i.e., a large relative risk) is more likely to indicate causality than a weak association, although it is recognized that relative risks of small magnitude do not imply lack of causality and may be important if the disease is common. Associations that are replicated in several studies of the same design or using different epidemiological approaches or under different circumstances of exposure are more likely to represent a causal relationship than isolated observations from single studies. If there are inconsistent results among investigations, possible reasons are sought (such as differences in amount of exposure), and results of studies judged to be of high quality are given more weight than those from studies judged to be methodologically less sound. When suspicion of carcinogenicity arises largely from a single study, these data are not combined with those from later studies in any subsequent reassessment of the strength of the evidence.

If the risk of the disease in question increases with the amount of exposure, this is considered to be a strong indication of causality, although absence of a graded response is not necessarily evidence against a causal relationship. Demonstration of a decline in risk after cessation of or reduction in exposure in individuals or in whole populations also supports a causal interpretation of the findings.

Although a carcinogen may act upon more than one target, the specificity of an association (i.e., an increased occurrence of cancer at one anatomical site or of one morphological type) adds plausibility to a causal relationship, particularly when excess cancer occurrence is limited to one morphological type within the same organ.

Although rarely available, results from randomized trials showing different rates among exposed and unexposed individuals provide particularly strong evidence for causality.

When several epidemiological studies show little or no indication of an association between an exposure and cancer, the judgement may be made that, in the aggregate, they show evidence of lack of carcinogenicity. Such a judgement requires first of all that the studies giving rise to it meet, to a sufficient degree, the standards of design and analysis described above. Specifically, the possibility that bias, confounding or misclassification of exposure or outcome could explain the observed results should be considered and excluded with reasonable certainty. In addition, all studies that are judged to be methodologically sound should be consistent with a relative risk of unity for any observed level of exposure and, when considered together, should provide a pooled estimate of relative risk which is at or near unity and has a narrow confidence interval, due to sufficient population size. Moreover, no individual study nor the pooled results of all the studies should show any consistent tendency for relative risk of cancer to increase with increasing level of exposure. It is important to note that evidence of lack of carcinogenicity obtained in this way from several epidemiological studies can apply only to the type(s) of cancer studied and to dose levels and intervals between first exposure and observation of disease that are the same as or less than those observed in all the studies. Experience with human cancer indicates that, in some cases, the period from first exposure to the development of clinical cancer is seldom less than 20 years; latent periods substantially shorter than 30 years cannot provide evidence for lack of carcinogenicity.

12. SUMMARY OF DATA REPORTED

In this section, the relevant experimental and epidemiological data are summarized. Only reports, other than in abstract form, that meet the criteria outlined on p. 17 are considered for evaluating carcinogenicity. Inadequate studies are generally not summarized: such studies are usually identified by a square-bracketed comment in the text.

(a) *Exposures*

Human exposure is summarized on the basis of elements such as production, use, occurrence in the environment and determinations in human tissues and body fluids. Quantitative data are given when available.

(b) *Experimental carcinogenicity data*

Data relevant to the evaluation of carcinogenicity in animals are summarized. For each animal species and route of administration, it is stated whether an increased incidence of neoplasms was observed, and the tumour sites are indicated. If the agent or mixture produced tumours after prenatal exposure or in single-dose experiments, this is also indicated. Dose-response and other quantitative data may be given when available. Negative findings are also summarized.

(c) *Human carcinogenicity data*

Results of epidemiological studies that are considered to be pertinent to an assessment of human carcinogenicity are summarized. When relevant, case reports and correlation studies are also considered.

(d) *Other relevant data*

Structure-activity correlations are mentioned when relevant.

Toxicological information and data on kinetics and metabolism in experimental animals are given when considered relevant. The results of tests for genetic and related effects are summarized for whole mammals, cultured mammalian cells and nonmammalian systems.

Data on other biological effects in humans of particular relevance are summarized. These may include kinetic and metabolic considerations and evidence of DNA binding, persistence of DNA lesions or genetic damage in exposed humans.

When available, comparisons of such data for humans and for animals, and particularly animals that have developed cancer, are described.

13. EVALUATION

Evaluations of the strength of the evidence for carcinogenicity arising from human and experimental animal data are made, using standard terms.

It is recognized that the criteria for these evaluations, described below, cannot encompass all of the factors that may be relevant to an evaluation of carcinogenicity. In considering all of the relevant data, the Working Group may assign the agent, mixture or exposure circumstance to a higher or lower category than a strict interpretation of these criteria would indicate.

(a) Degrees of evidence for carcinogenicity in humans and in experimental animals and supporting evidence

It should be noted that these categories refer only to the strength of the evidence that an exposure is carcinogenic and not to the extent of its carcinogenic activity (potency) nor to the mechanism involved. A classification may change as new information becomes available.

An evaluation of degree of evidence, whether for a single substance or a mixture, is limited to the materials tested, and these are chemically and physically defined. When the materials evaluated are considered by the Working Group to be sufficiently closely related, they may be grouped for the purpose of a single evaluation of degree of evidence.

(i) Human carcinogenicity data

The applicability of an evaluation of the carcinogenicity of a mixture, process, occupation or industry on the basis of evidence from epidemiological studies depends on the variability over time and place of the mixtures, processes, occupations and industries. The Working Group seeks to identify the specific exposure, process or activity which is considered most likely to be responsible for any excess risk. The evaluation is focused as narrowly as the available data on exposure and other aspects permit.

The evidence relevant to carcinogenicity from studies in humans is classified into one of the following categories:

Sufficient evidence of carcinogenicity: The Working Group considers that a causal relationship has been established between exposure to the agent, mixture or exposure circumstance and human cancer. That is, a positive relationship has been observed between the exposure and cancer in studies in which chance, bias and confounding could be ruled out with reasonable confidence.

Limited evidence of carcinogenicity: A positive association has been observed between exposure to the agent, mixture or exposure circumstance and cancer for which a causal interpretation is considered by the Working Group to be credible, but chance, bias or confounding could not be ruled out with reasonable confidence.

Inadequate evidence of carcinogenicity: The available studies are of insufficient quality, consistency or statistical power to permit a conclusion regarding the presence or absence of a causal association.

Evidence suggesting lack of carcinogenicity: There are several adequate studies covering the full range of levels of exposure that human beings are known to encounter, which are mutually consistent in not showing a positive association between exposure to the agent, mixture or exposure circumstance and any studied cancer at any observed level of exposure. A conclusion of 'evidence suggesting lack of carcinogenicity' is inevitably limited to the cancer sites, conditions and levels of exposure and length of observation covered by the available studies. In addition, the possibility of a very small risk at the levels of exposure studied can never be excluded.

In some instances, the above categories may be used to classify the degree of evidence for carcinogenicity for specific organs or tissues.

(ii) *Experimental carcinogenicity data*

The evidence relevant to carcinogenicity in experimental animals is classified into one of the following categories:

Sufficient evidence of carcinogenicity: The Working Group considers that a causal relationship has been established between the agent or mixture and an increased incidence of malignant neoplasms or of an appropriate combination of benign and malignant neoplasms (as described on p. 22) in (a) two or more species of animals or (b) in two or more independent studies in one species carried out at different times or in different laboratories or under different protocols.

Exceptionally, a single study in one species might be considered to provide sufficient evidence of carcinogenicity when malignant neoplasms occur to an unusual degree with regard to incidence, site, type of tumour or age at onset.

In the absence of adequate data on humans, it is biologically plausible and prudent to regard agents and mixtures for which there is *sufficient evidence* of carcinogenicity in experimental animals as if they presented a carcinogenic risk to humans.

Limited evidence of carcinogenicity: The data suggest a carcinogenic effect but are limited for making a definitive evaluation because, e.g., (a) the evidence of carcinogenicity is restricted to a single experiment; or (b) there are unresolved questions regarding the adequacy of the design, conduct or interpretation of the study; or (c) the agent or mixture increases the incidence only of benign neoplasms or lesions of uncertain neoplastic potential, or of certain neoplasms which may occur spontaneously in high incidences in certain strains.

Inadequate evidence of carcinogenicity: The studies cannot be interpreted as showing either the presence or absence of a carcinogenic effect because of major qualitative or quantitative limitations.

Evidence suggesting lack of carcinogenicity: Adequate studies involving at least two species are available which show that, within the limits of the tests used, the agent or mixture is not carcinogenic. A conclusion of evidence suggesting lack of carcinogenicity is inevitably limited to the species, tumour sites and levels of exposure studied.

(iii) *Supporting evidence of carcinogenicity*

Other evidence judged to be relevant to an evaluation of carcinogenicity and of sufficient importance to affect the overall evaluation is then described. This may include data on tumour pathology, genetic and related effects, structure-activity relationships, metabolism and pharmacokinetics, physicochemical parameters, chemical composition and possible mechanisms of action. For complex exposures, including occupational and industrial exposures, the potential contribution of carcinogens known to be present as well as the relevance of materials tested are considered by the Working Group in its overall evaluation of human carcinogenicity. The Working Group also determines to what extent the materials tested in experimental systems are relevant to those to which humans are exposed. The available experimental evidence may help to specify more precisely the causal factor(s).

(b) *Overall evaluation*

Finally, the body of evidence is considered as a whole, in order to reach an overall evaluation of the carcinogenicity to humans of an agent, mixture or circumstance of exposure.

An evaluation may be made for a group of chemical compounds that have been evaluated by the Working Group. In addition, when supporting data indicate that other, related compounds for which there is no direct evidence of capacity to induce cancer in animals or in humans may also be carcinogenic, a statement describing the rationale for this conclusion is added to the evaluation narrative; an additional evaluation may be made for this broader group of compounds if the strength of the evidence warrants it.

The agent, mixture or exposure circumstance is described according to the wording of one of the following categories, and the designated group is given. The categorization of an agent, mixture or exposure circumstance is a matter of scientific judgement, reflecting the strength of the evidence derived from studies in humans and in experimental animals and from other relevant data.

Group 1—The agent (mixture) is carcinogenic to humans.
The exposure circumstance entails exposures that are carcinogenic to humans.

This category is used only when there is *sufficient evidence* of carcinogenicity in humans.

Group 2

This category includes agents, mixtures and exposure circumstances for which, at one extreme, the degree of evidence of carcinogenicity in humans is almost sufficient, as well as those for which, at the other extreme, there are no human data but for which there is experimental evidence of carcinogenicity. Agents, mixtures and exposure circumstances are assigned to either 2A (probably carcinogenic) or 2B (possibly carcinogenic) on the basis of epidemiological, experimental and other relevant data.

Group 2A—The agent (mixture) is probably carcinogenic to humans.
The exposure circumstance entails exposures that are probably carcinogenic to humans.

This category is used when there is *limited evidence* of carcinogenicity in humans and *sufficient evidence* of carcinogenicity in experimental animals. Exceptionally, an agent, mixture or exposure circumstance may be classified into this category solely on the basis of *limited evidence* of carcinogenicity in humans or of *sufficient evidence* of carcinogenicity in experimental animals strengthened by supporting evidence from other relevant data.

Group 2B—The agent (mixture) is possibly carcinogenic to humans.
The exposure circumstance entails exposures that are possibly carcinogenic to humans.

This category is generally used for agents, mixtures and exposure circumstances for which there is *limited evidence* of carcinogenicity in humans in the absence of *sufficient evidence* of carcinogenicity in experimental animals. It may also be used when there is *inadequate evidence* of carcinogenicity in humans or when human data are nonexistent but there is *sufficient evidence* of carcinogenicity in experimental animals. In some instances, an agent, mixture or exposure circumstance for which there is *inadequate evidence* of or no data on carcinogenicity in humans but *limited evidence* of carcinogenicity in experimental animals together with supporting evidence from other relevant data may be placed in this group.

Group 3—The agent (mixture, exposure circumstance) is not classifiable as to its carcinogenicity to humans.

Agents, mixtures and exposure circumstances are placed in this category when they do not fall into any other group.

Group 4—The agent (mixture, exposure circumstance) is probably not carcinogenic to humans.

This category is used for agents, mixtures and exposure circumstances for which there is *evidence suggesting lack of carcinogenicity* in humans together with *evidence suggesting lack of carcinogenicity* in experimental animals. In some instances, agents, mixtures or exposure circumstances for which there is *inadequate evidence* of or no data on carcinogenicity in humans but *evidence suggesting lack of carcinogenicity* in experimental animals, consistently and strongly supported by a broad range of other relevant data, may be classified in this group.

References

1. IARC (1987) *IARC Monographs on the Evaluation of Carcinogenic Risks to Humans*, Supplement 6, *Genetic and Related Effects: An Updating of Selected* IARC Monographs *from Volumes 1 to 42*, Lyon
2. IARC (1977) *IARC Monographs Programme on the Evaluation of the Carcinogenic Risk of Chemicals to Humans. Preamble* (IARC intern. tech. Rep. No. 77/002), Lyon

3. IARC (1978) *Chemicals with Sufficient Evidence of Carcinogenicity in Experimental Animals*—IARC Monographs *Volumes 1-17* (IARC intern. tech. Rep. No. 78/003), Lyon
4. IARC (1979) *Criteria to Select Chemicals for* IARC Monographs (IARC intern. tech. Rep. No. 79/003), Lyon
5. IARC (1982) *IARC Monographs on the Evaluation of the Carcinogenic Risk of Chemicals to Humans*, Supplement 4, *Chemicals, Industrial Processes and Industries Associated with Cancer in Humans (IARC Monographs, Volumes 1 to 29)*, Lyon
6. IARC (1983) *Approaches to Classifying Chemical Carcinogens According to Mechanism of Action* (IARC intern. tech. Rep. No. 83/001), Lyon
7. IARC (1987) *IARC Monographs on the Evaluation of Carcinogenic Risks to Humans*, Supplement 7, *Overall Evaluations of Carcinogenicity: An Updating of* IARC Monographs *Volumes 1 to 42*, Lyon
8. IARC (1988) *Report of an IARC Working Group to Review the Approaches and Processes Used to Evaluate the Carcinogenicity of Mixtures and Groups of Chemicals* (IARC intern. tech. Rep. No. 88/002), Lyon
9. IARC (1973-1990) *Information Bulletin on the Survey of Chemicals Being Tested for Carcinogenicity*, Numbers 1-14, Lyon

    Number 1 (1973)   52 pages
    Number 2 (1973)   77 pages
    Number 3 (1974)   67 pages
    Number 4 (1974)   97 pages
    Number 5 (1975)   88 pages
    Number 6 (1976)   360 pages
    Number 7 (1978)   460 pages
    Number 8 (1979)   604 pages
    Number 9 (1981)   294 pages
    Number 10 (1983) 326 pages
    Number 11 (1984) 370 pages
    Number 12 (1986) 385 pages
    Number 13 (1988) 404 pages
    Number 14 (1990) 369 pages

10. Coleman, M. & Wahrendorf, J., eds (1991) *Directory of On-going Studies in Cancer Epidemiology 1991* (IARC Scientific Publications No. 110), Lyon, IARC [and previous annual volumes]
11. IARC (1984) *Chemicals and Exposures to Complex Mixtures Recommended for Evaluation in* IARC Monographs *and Chemicals and Complex Mixtures Recommended for Long-term Carcinogenicity Testing* (IARC intern. tech. Rep. No. 84/002), Lyon
12. IARC (1989) *Chemicals, Groups of Chemicals, Mixtures and Exposure Circumstances to be Evaluated in Future IARC Monographs, Report of an ad hoc Working Group* (IARC intern. tech. Rep. No. 89/004), Lyon

13. *Environmental Carcinogens. Methods of Analysis and Exposure Measurement*:
    Vol. 1. *Analysis of Volatile Nitrosamines in Food* (IARC Scientific Publications No. 18). Edited by R. Preussmann, M. Castegnaro, E.A. Walker & A.E. Wasserman (1978)
    Vol. 2. *Methods for the Measurement of Vinyl Chloride in Poly(vinyl chloride), Air, Water and Foodstuffs* (IARC Scientific Publications No. 22). Edited by D.C.M. Squirrell & W. Thain (1978)
    Vol. 3. *Analysis of Polycyclic Aromatic Hydrocarbons in Environmental Samples* (IARC Scientific Publications No. 29). Edited by M. Castegnaro, P. Bogovski, H. Kunte & E.A. Walker (1979)
    Vol. 4. *Some Aromatic Amines and Azo Dyes in the General and Industrial Environment* (IARC Scientific Publications No. 40). Edited by L. Fishbein, M. Castegnaro, I.K. O'Neill & H. Bartsch (1981)
    Vol. 5. *Some Mycotoxins* (IARC Scientific Publications No. 44). Edited by L. Stoloff, M. Castegnaro, P. Scott, I.K. O'Neill & H. Bartsch (1983)
    Vol. 6. *N-Nitroso Compounds* (IARC Scientific Publications No. 45). Edited by R. Preussmann, I.K. O'Neill, G. Eisenbrand, B. Spiegelhalder & H. Bartsch (1983)
    Vol. 7. *Some Volatile Halogenated Hydrocarbons* (IARC Scientific Publications No. 68). Edited by L. Fishbein & I.K. O'Neill (1985)
    Vol. 8. *Some Metals: As, Be, Cd, Cr, Ni, Pb, Se, Zn* (IARC Scientific Publications No. 71). Edited by I.K. O'Neill, P. Schuller & L. Fishbein (1986)
    Vol. 9. *Passive Smoking* (IARC Scientific Publications No. 81). Edited by I.K. O'Neill, K.D. Brunnemann, B. Dodet & D. Hoffmann (1987)
    Vol. 10. *Benzene and Alkylated Benzenes* (IARC Scientific Publications No. 85). Edited by L. Fishbein & I.K. O'Neill (1988)
14. Wilbourn, J., Haroun, L., Heseltine, E., Kaldor, J., Partensky, C. & Vainio, H. (1986) Response of experimental animals to human carcinogens: an analysis based upon the IARC Monographs Programme. *Carcinogenesis*, 7, 1853-1863
15. Montesano, R., Bartsch, H., Vainio, H., Wilbourn, J. & Yamasaki, H., eds (1986) *Long-term and Short-term Assays for Carcinogenesis—A Critical Appraisal* (IARC Scientific Publications No. 83), Lyon, IARC
16. Hoel, D.G., Kaplan, N.L. & Anderson, M.W. (1983) Implication of nonlinear kinetics on risk estimation in carcinogenesis. *Science*, 219, 1032-1037
17. Gart, J.J., Krewski, D., Lee, P.N., Tarone, R.E. & Wahrendorf, J. (1986) *Statistical Methods in Cancer Research*, Vol.3, *The Design and Analysis of Long-term Animal Experiments* (IARC Scientific Publications No. 79), Lyon, IARC
18. Peto, R., Pike, M.C., Day, N.E., Gray, R.G., Lee, P.N., Parish, S., Peto, J., Richards, S. & Wahrendorf, J. (1980) Guidelines for simple, sensitive significance tests for carcinogenic effects in long-term animal experiments. In: *IARC Monographs on the Evaluation of the Carcinogenic Risk of Chemicals to Humans*, Supplement 2, *Long-term and Short-term Screening Assays for Carcinogens: A Critical Appraisal*, Lyon, pp. 311-426
19. Breslow, N.E. & Day, N.E. (1980) *Statistical Methods in Cancer Research*, Vol. 1, *The Analysis of Case-control Studies* (IARC Scientific Publications No. 32), Lyon, IARC

20. Breslow, N.E. & Day, N.E. (1987) *Statistical Methods in Cancer Research*, Vol. 2, *The Design and Analysis of Cohort Studies* (IARC Scientific Publications No. 82), Lyon, IARC

GENERAL REMARKS

In this fifty-second volume of *IARC Monographs*, evaluations are made of the evidence in relation to the carcinogenicity of chlorinated water supplies, of chemical by-products of the chlorination of drinking-water, of some other chemicals found in drinking-water and of cobalt and cobalt compounds. By-products of the chlorination process found in chlorinated drinking-water that have been evaluated previously in this series are listed in Table 1.

Table 1. Organic compounds formed during chlorination of drinking-water supplies that have been evaluated in the *IARC Monographs*

Compound	Year	Degree of evidence for carcinogenicity[a]		Overall evaluation of carcinogenicity to humans[a]
		Human	Animal	
Acetaldehyde	1987	I	S	2B
Chloroform	1987	I	S	2B
2,4-Dichlorophenol (see Chlorophenols)	1986	L	–	2B
Formaldehyde	1987	L	S	2A
2,4,6-Trichlorophenol	1987	L	S	2B
Pentachlorophenol	1990	I	S	2B

[a]I, inadequate evidence; S, sufficient evidence; L, limited evidence; ND, no data; see pp. 30-33 of the Preamble for definitions of these classifications.

Chlorinated drinking-water

The maintenance of an adequate supply of unpolluted water is a requirement for both human health and good environmental quality. Our demands upon the planet's water are great, and in some regions dangerously so. Water is taken for drinking, irrigation and industry and is returned as industrial discharge, agricultural run-off and microbiologically contaminated, treated or untreated sewage. Water quality varies according to these discharges, the seasons and the geology of an area. The most critical characteristic of water for human health is its microbiology.

The microbiological quality of water can be controlled effectively by disinfection methods, which normally involve the introduction of chemical oxidants to the water supply. Chemicals used on a substantial scale as disinfectants are chlorine, hypochlorite, chloramine, chlorine dioxide and ozone. Chlorination is almost universally accepted as the method of choice for purifying water supplies. It was first used on a continuous basis for this purpose at the beginning of the twentieth century. It is also used for sewage treatment in a few countries. Since some water suppliers have difficulty in maintaining acceptable water quality, particularly with regard to taste and odour, chlorine may be used in combination with ozone, chlorine dioxide, ammonia and activated charcoal. These treatments are sometimes followed by dechlorination, for example with sulfur dioxide.

There are substantial and irrefutable benefits of disinfection of water supplies by chlorination. Any major change to these programmes would need to be evaluated fully as to its costs and benefits with regard not only to the need to maintain microbiological safety but also to the possible long-term adverse effects of alternatives to chlorination. Nonetheless, it is now known that the interaction of chlorine with naturally occurring humic and fulvic acids in water supplies results in the formation of by-products such as trihalomethanes, halogenated acids and aldehydes, some of which are either known or suspected carcinogens. Other compounds that occur at much lower concentrations include substances that are mutagenic primarily in bacterial and in-vitro systems but have not been demonstrated to produce cancer in man or experimental animals.

The investigation of possible risks of cancer from consuming chlorinated drinking-water in human populations is difficult. There are a number of methodological obstacles. Chlorination may produce quite different profiles of chemical by-products in different areas. Characterization of a person's water consumption is complicated by the fact that in some parts of the world people change residence during life, and the nature of their domestic water source changes in consequence. Furthermore, people may consume water not only at home but also at work and in other places, and may drink unchlorinated water, bottled water, boiled water and other liquids, which will greatly influence their exposure to chlorination by-products. Exposure to constituents of water other than by ingestion—by inhalation or skin absorption—may also occur. Even if associations can be demonstrated between human cancer risk and exposure to residential chlorinated water supplies, they may be due to other constituents of the water that is chlorinated or to particular characteristics of the populations who live in areas served by chlorinated water supplies.

Evaluation of chlorinated water for carcinogenicity in experimental animals is similarly challenging. Indeed, few studies have yet been conducted in which the effects, if any, of constituents of chlorinated water have been compared with those of constituents of water from the same supply collected before chlorination.

This volume includes monographs related to the chlorination rather than other methods of disinfection of water, because, firstly, chlorination is the most commonly used disinfection process, and, secondly, potentially carcinogenic by-products have been measured in chlorinated water (although not uniquely so, as they have also been identified in water disinfected by other methods).

Cobalt and cobalt compounds

Some explanation is necessary with regard to the inclusion of a monograph on cobalt and cobalt compounds in this volume. It was originally prepared for inclusion in Volume 49 of the *Monographs* (*Chromium, Nickel and Welding*); however, during the preparation of that meeting, it became clear that there would be insufficient time to consider this topic adequately. Consequently, it was decided to postpone the evaluation of cobalt and cobalt compounds until the earliest convenient *Monographs* meeting, which was the present one.

Cobalt has widespread and important uses, particularly in alloys with chromium, nickel and several other metals, and it is therefore a component of some industrial environments. Alloys that contain cobalt in combination with chromium, molybdenum, nickel, tungsten and other metals are commonly used in orthopaedic prostheses and in other implanted medical or dental devices. Since, on a worldwide basis, hundreds of thousands of patients receive such implants each year, concern has arisen about the potential carcinogenic hazards of wear particles and of cobalt and other metals that are gradually solubilized from the surface of implants.

Cobalt is a component of an essential micronutrient, vitamin B_{12} (cyanocobalamin), which is synthesized by certain microorganisms and ingested in the diet of humans and animals. Since the kinetics and metabolism of vitamin B_{12} are distinct from those of other cobalt compounds, however, it was decided not to consider vitamin B_{12} itself. Similarly, except for cobalt naphthenate, which is widely used industrially, other organic cobalt compounds have not been considered.

THE MONOGRAPHS

CHLORINATED DRINKING-WATER

CHLORINATED DRINKING-WATER

1. Description of the Process

1.1 History of chlorination of drinking-water

Chlorine in one form or another is by far the most commonly used chemical for the disinfection of water supplies. It is also active for other purposes associated with water treatment and supply, such as prevention of algal, bacterial and general slime growths in treatment plants and pipeworks, control of tastes and odours, and removal of iron, manganese and colour (White, 1986).

The history and use of chlorine in the treatment of water has been reviewed in detail (White, 1986), and the following summary is based largely on that work.

Chlorine was discovered in 1774 by Karl W. Scheele and identified as an element in 1810 by Humphrey Davy. Javel water (a solution of potassium hypochlorite) was introduced in 1785 by Berthollet, and the commercially important development of a cheap, stable bleaching powder, calcium hypochlorite, was achieved by Tennant in 1798.

One of the first reported uses of chlorination for the disinfection of water supplies was in 1897, when bleach solution was used to disinfect a water main in Maidstone, Kent, UK, following an outbreak of typhoid. Regular use in water treatment began around the beginning of the twentieth century. Probably, the first continuous application was in 1902 at Middelkerke, Belgium, where ferric chloride, used for 'coagulation' (see p. 47) was mixed with calcium hypochlorite, producing hypochlorous acid; in 1903, at Ostende, Belgium, chlorine was generated from potassium chlorate and oxalic acid. In the UK, the first known regular use (of sodium hypochlorite) was in 1905 in Lincoln after a typhoid epidemic. In 1908 in Chicago, IL, USA, George A. Johnson instituted chlorination by adding 'chloride of lime' to contaminated river water. Chlorination of a river water supply to Jersey City, USA, at the turn of the century was significant in that, in the litigation that developed, objections regarding the ineffectiveness, potential hazards and general undesirability of the addition of chlorine to water supplies were overcome. These developments were quickly followed by similar examples in most industrialized countries. As a result, most large-scale public water supplies are now disinfected chemically by chlorine (White, 1986), although there are many small local supplies (small wells, private springs) that are not disinfected by any means.

Prior to the successful widespread introduction of chlorination, water treatment techniques existed that included filtration, followed by chemical precipitation and sedimentation techniques. These methods alone, however, could not guarantee a bacteriologically safe water supply.

The main diseases can be controlled (to varying extents) by good physical/chemical water treatment and chemical disinfection include typhoid fever, cholera, amœbic dysentry, bacterial gastroenteritis, shigellosis, salmonellosis, *Campylobacter* enteritis, *Yersina* enteritis, *Pseudomonas* infections, schistosomiasis, giardiasis and various viral diseases, such as hepatitis A (National Research Council, 1980; Hoff & Akin, 1986; White, 1986).

The early use of chlorine to disinfect drinking-water involved hypochlorite solutions. In 1910-20, it became possible to store and transport liquid chlorine, and the development of suitable chlorinator installations led to increased use of chlorine itself for this purpose, providing easier control and monitoring and better disinfection than the various hypochlorite solutions. Notable in these and subsequent developments in the field of water treatment chlorinators were Wallace and Tiernan, who patented a variety of control and safety devices (White, 1986). The introduction of chlorine-resistant plastics in the 1950s and increased understanding of the chemistry of chlorination hastened the process. Further major developments were the use of ammonia-chlorine reactions and the breakpoint phenomenon (see p. 51) to minimize the taste and odour of chlorine, precise control of chlorine residues by dechlorination with sulfur dioxide and, more recently, concern over organic chemical by-products and the possible need for their control. These developments are discussed in the following sections.

1.2 Overview of the addition of chlorine during drinking-water treatment

Before discussing the addition of chlorine during water treatment, it is useful to review the important stages of water treatment and the chemistry of chlorination.

(a) Drinking-water treatment

The fundamental purpose of water treatment is to protect the consumer from impurities that may be offensive or injurious to human health. A secondary purpose is to deal with impurities which, although not directly harmful to health, may cause problems such as corrosion and discoloration. These purposes are achieved by setting up barriers such as coagulation and filtration, which remove impurities by precipitation and particle capture. The final barrier is disinfection.

The main purpose of treatment prior to disinfection is to prepare the water for effective and reliable disinfection, for example by removing suspended solids which can impair disinfection efficiency.

Surface water sources, i.e., those exposed to air on the surface of the Earth, comprise waters of widely varying quality, from high quality waters containing little known contamination (such as treated or untreated wastewater) to lowland rivers that contain appreciable contamination from a variety of sources. Deep groundwaters, i.e., the water that is naturally contained in and saturates the subsoil, are normally of high quality. However, some groundwaters, particularly those that are shallow and those in highly permeable strata, are vulnerable to specific localized contamination by a variety of substances—especially volatile chlorinated hydrocarbons such as trichloroethylene (see IARC, 1987). Springs constitute a water source in which the groundwater meets an impermeable rock stratum and is 'forced' out of the ground; they are usually of high purity.

Surface waters are more prone to contamination than groundwaters and so more often need pretreatment. Some pretreatment may be afforded by storing the water in a reservoir, which can result in sedimentation of suspended solids and a significant reduction in the numbers of any pathogenic organisms present. Various additional pretreatment methods are used, generally to remove suspended solids and naturally occurring coloured impurities. The principles involved in these processes are discussed below. Apart from disinfection, high quality groundwaters need no or minimal physical or chemical treatment.

(i) *Coagulation, sedimentation and filtration*

Coagulation: Some impurities in natural waters cannot be removed by settling alone, either because they are dissolved or because they occur in a very finely divided ('colloidal') state. The addition of a chemical coagulant is needed to create large particles that can settle, called 'flocs'. The coagulants most commonly used are aluminium and ferric salts. When these chemicals are added, a precipitate of the metal hydroxide forms which removes suspended solids, algae and colour by a number of mechanisms, including adsorption and trapping. Mechanical or hydraulic mixing causes the hydroxide precipitate, together with impurities, to agglomerate into flocs a few millimetres in diameter. Other chemicals, called polyelectrolytes, can be used in addition to, or in place of, aluminium or iron coagulants to produce stronger or larger flocs. Once formed, the flocs are removed from the water by filtration, generally preceded by sedimentation.

Sedimentation and flotation: Sedimentation is used to remove the bulk of the flocs, so as to reduce the load on downstream filters. Sedimentation may take place in rectangular or circular, horizontal basins in which discrete settling of flocs occurs or, commonly in some European designs, in 'floc blanket' clarifiers, in which the water flows upwards through a fluidized bed of flocs and treated water is taken from the top of the clarifier. Flocs have a density only marginally greater than water, so treatment rates must be low. Typical tank loadings are < 1-5 m/h.

An alternative process to sedimentation is dissolved air flotation. In this process, water saturated with air under pressure is released into the water containing flocs, and tiny air bubbles become attached to the flocs and float them to the surface of the water. This is a faster process than sedimentation; typical loadings being 5-12 m/h. Dissolved air flotation may be particularly suitable for the treatment of coloured, low-turbidity waters and algal-laden waters.

Filtration: Deep-bed filtration through sand is employed to remove the remaining particulate matter. Water is passed through a bed of sand, typically composed of grains 0.5-1.0 mm in diameter, one-metre deep. Particles are trapped within the bed by a variety of mechanisms including straining, sedimentation, interception and electrostatic adhesion. Filtration rates are typically 4-10 m/h. As particles are trapped within the bed, the resistance to flow increases, necessitating a greater head of water (pressure) to maintain a constant rate of flow. Once a limiting head loss is reached, or solids start to be released from the filter, the filter is cleaned by backflushing with clean water.

In the treatment of turbid waters, filtration is almost always preceded by sedimentation, and filters are of the open 'gravity filter' type. With some low turbidity waters, including coloured moorland waters, the sedimentation stage may be omitted and direct filtration employed. With direct filtration, pressure filtration can be used to conserve a hydrostatic head.

(ii) *Slow sand filtration*

Slow sand filtration, which is a well-established process, is an alternative to the coagulation process for waters with little colour and a moderately low concentration of suspended solids. A slow sand filter consists of a 0.5-1.5-m-deep bed of fine (0.15-0.35 mm) sand, supported on a layer of gravel by a system of underdrains. At the low flow rates used (0.1-0.3 m/h), solids settle onto the surface of the sand. The layer formed, known as the 'Schmutzdecke', contains mud, organic waste, bacterial matter and algae and is biologically active. The mechanisms involved in slow sand filtration are: removal of colloidal material by straining, adsorption and bacterial action; destruction of pathogenic organisms by bacterial action; and purification of the water above the filter by bacterial action, flocculation and pathogen death.

As filtration progresses, the head loss through the bed increases to the point at which the required flow rate cannot be maintained. The filter is then taken out of action and the top layer is skimmed off manually or mechanically. The sand is washed for re-use. Eventually the depth of sand in the filter becomes insufficient for effective filtration, and more sand is added.

(iii) *Other processes*

A number of other processes may be employed prior to disinfection of water; these processes are applicable to groundwaters as well as surface waters.

Aeration may be employed for a variety of reasons, including removal of volatile taste- and odour-producing compounds, precipitation of iron and manganese and removal of carbon dioxide.

Oxidation may be used for purposes other than disinfection; these include precipitation of iron and manganese, taste and odour control, colour removal and oxidation of trace organic compounds. The principal oxidizing agents employed in water treatment are chlorine, chloramine, ozone and chlorine dioxide (White, 1986).

pH Adjustment, usually to more alkaline levels, is used to achieve optimal values for other processes, including coagulation and disinfection, as well as to reduce the corrosiveness of the water supply. pH can be increased by adding chemicals such as lime, caustic soda or soda ash or by placing the water in contact with a bed of sparingly soluble material, such as marble.

The pH of drinking-water is typically in the range 6.5-8.5, but levels up to 9.5 can occur.

Softening: Hardness in water results from the presence of calcium and magnesium compounds. When hardness is excessive, it can be reduced by precipitation softening or ion exchange. In precipitation softening, lime (and sometimes soda ash) is added to precipitate calcium as calcium carbonate, which is removed in a sedimentation tank. Ion-exchange softening is used only for groundwaters; the water is passed through a bed of cationic resin which exchanges sodium for calcium and magnesium. When the resin is fully loaded with calcium and magnesium, it is regenerated using a strong brine solution.

Activated carbon may be employed to remove natural and synthetic organic chemicals. It is produced by the controlled combustion of wood, coal and other material to produce a porous material with a large surface area and a high affinity for organic compounds. A slurry of powder can be added to the water and then removed by subsequent treatment processes, such as coagulation. Alternatively, granular-activated carbon can be employed in purpose-built adsorbers, or as a replacement for some of the sand in a rapid gravity filter.

(b) *General chemistry of the addition of chlorine*

The basic chemistry of water chlorination has been studied by a large number of workers and has been reviewed (National Academy of Sciences, 1979; National Research Council, 1980; White, 1986). The main features are as follows.

Chlorine dissolves rapidly in water to establish an equilibrium with hypochlorous acid (HOCl) and hydrochloric acid (HCl):

$$Cl_2 + H_2O \leftrightarrows HOCl + H^+ + Cl^- \quad (pK_a = 9.5).$$

In dilute solutions and at pH levels above 4.0, the equilibrium is displaced to the right and very little molecular chlorine exists in solution. Between pH 6.0 and 8.5, hypochlorous acid dissociates almost completely to form the hypochlorite ion (OCl^-):

$$HOCl \leftrightarrows OCl^- + H^+.$$

At pH levels above 9.0, hypochlorite ions are the dominant species.

Alternative sources of hypochlorite ions are calcium hypochlorite and sodium hypochlorite. Essentially the same active species and equilibria are established whether the source of chlorine is liquid or gaseous or a hypochlorite compound.

The total concentration of molecular chlorine, hypochlorous acid and hypochlorite ion is defined as 'free available chlorine'. Total available chlorine may be defined as the mass equivalent of chlorine contained in all chemical species that contain chlorine in an oxidized state. Combined available chlorine can be defined as the difference between total available chlorine and free available chlorine and represents the amount of chlorine that is in chemical association with various compounds (usually amino- or ammoniacal nitrogen) but that is also capable of disinfecting. Free chlorine species are generally more effective disinfectants than combined chlorine species.

Raw (untreated) water may contain a large number of compounds that can react with chlorine species, including inorganic reducing agents (H_2S, SO_3^{2-}, NO_2^-, Fe^{2+} and Mn^{2+}, which are oxidized to, for example, SO_4^{2-}, NO_3^-, Fe^{3+} and MnO_2); ammonia and amino-nitrogen groups; and organic substances.

The principal effects of these side-reactions are the formation of by-products and a loss of disinfection efficiency as active chlorine species are reduced to less active combined species, particularly the non-bactericidal chloride. The most significant side-reactions, in terms of chlorine demand, are those involving ammonia or amino-nitrogen groups.

The reaction between hypochlorous acid and ammonia in dilute aqueous solution yields, successively, monochloramine (NH_2Cl), dichloramine ($NHCl_2$) and trichloramine (more commonly known as nitrogen trichloride, NCl_3):

$$NH_3 + HOCl \leftrightarrows NH_2Cl + H_2O$$
$$NH_2Cl + HOCl \leftrightarrows NHCl_2 + H_2O$$
$$NH_2Cl + HOCl \leftrightarrows NCl_3 + H_2O.$$

Hypochlorous acid and ammonia may also react to yield nitrogen:

$$2NH_3 + 3HOCl \leftrightarrows N_2 + 3HCl + 3H_2O.$$

These reactions are dependent on pH, temperature and the initial ratio of chlorine to ammoniacal nitrogen.

An important reaction that often occurs in the chlorination of water is the formation of hypobromous acid from bromide:

$$HOCl + Br^- \rightarrow HOBr + Cl^-.$$

Even at low bromide concentrations, this reaction leads to readily detectable levels of brominated organic by-products, such as brominated trihalomethanes, due to the reactivity of hypobromous acid. Bromide concentrations in untreated water vary widely: for example, in nine rivers in various regions of the USA, bromide levels ranged from 10 to 245 µg/l (Amy *et al.*, 1985). The occasional detection of iodinated halomethanes is probably due to a similar mechanism involving iodides.

Organic chloramines are formed when chlorine reacts with amines, amino acids, proteinaceous material and other forms of organic nitrogen involving amino groups or linkages. Organic chloramines are usually formed at slower rates than inorganic chloramines and are not considered to be effective disinfectants. While some organic chloramines are stable, others are not and degrade to many other by-products.

Addition of chlorine to waters containing dissolved organic compounds can result in three possible reactions, which are classified as:

(i) addition,
(ii) ionic substitution and
(iii) oxidation.

While all of these reactions result in an increase in the oxidation state of the substrate, (iii) results only in unchlorinated products (Pierce, 1978). The amount of organic matter in untreated water varies considerably. Typically, high quality groundwater contains up to 1 mg/l (as organic carbon), river water contains 1-10 mg/l (as organic carbon), while upland water may contain up to 20 mg/l (as organic carbon) which is almost entirely of natural origin (in humic substances). The total organic matter present would be roughly double these concentrations.

The use of ammonia with chlorine in water treatment, often called the 'chloramination' or 'chloramine' process, is designed to convert fully or partially the free chlorine to chloramine. The chloramine produced has a disinfectant action. Although it is less effective than chlorine, it has a lesser tendency to react with organic matter to form by-products: it generates less chlorophenolic taste from phenol and, of more recent interest, fewer by-products such as trihalomethanes. Chloramines are also more persistent in the drinking-water distribution system. The development of the chloramination process has been reviewed (White, 1986).

Chloramination was popular until the discovery and understanding of the 'breakpoint phenomenon'. In breakpoint chlorination, the aim is to maintain an optimal free residue of chlorine; to achieve this, any ammonia in the water is destroyed by addition of sufficient chlorine. As described above, chlorine reacts rapidly with ammonia in water to form monochloramine, dichloramine and trichloramine, depending on the ratio of chlorine to ammonia and other factors, such as pH. In practice, as the molar ratio of chlorine to ammonia increases towards 1:1, the combined chlorine residue in the water increases steadily. Beyond this ratio, i.e., with more added chlorine, the combined residue decreases quickly to a point beyond which further addition of chlorine produces a steady increase in free chlorine residue. This point (theoretically at around 1.5 mol chlorine to 1.0 mol ammonia) is the so-called breakpoint. For many waters, addition of sufficient chlorine to exceed the breakpoint, thus achieving a combined residue (free chlorine plus chloramines) containing about 85% free chlorine, produces the most satisfactory palatability. It was found recently that these levels of free chlorine often enhance levels of organic chemical by-products such as trihalomethanes; consequently, breakpoint chlorination has been replaced at some treatment works by other processes (White, 1986).

The concentration of chlorine entering the distribution system is often reduced slightly, to conform to operational requirements, by the addition of a small quantity of a reducing agent; typically, sulfur dioxide is used.

(c) *Addition of chlorine during water treatment*

Current drinking-water treatments reflect other objectives of chlorination, in addition to killing pathogenic organisms. These objectives include the destruction of substances and organisms that confer taste and odour on the supply and foul equipment, such as filters and pipelines, and the oxidation of undesirable chemical substances such as Fe^{2+} and Mn^{2+} in raw water.

Additions of chlorine during the treatment and distribution of drinking-water can be summarized as follows:
—prechlorination of raw water (i.e., prior to any treatment),
—addition at various points in the treatment process,
—addition after treatment but before distribution (i.e., final works disinfection),
—addition during distribution, and
—miscellaneous use during maintenance activities.

Prechlorination has been used extensively for the treatment of lower quality surface water. The amount of chlorine added is usually in the range of 1-10 mg/l—typically around 5 mg/l, although much higher levels have been used. Such additions of relatively large amounts of chlorine directly to raw water can produce high levels of by-products such as trihalomethanes; consequently, efforts have been made to reduce the level of prechlorination or to abandon it completely.

Chlorine (typically less than 5 mg/l) may also be added after coagulation/before sedimentation or after sedimentation/before filtration, generally to maintain improved flow by preventing build-up of slimes and bacterial growth. At some works, chlorine is added (at 2-5 mg/l) to oxidize ferrous sulfate to ferric sulfate, which is then used as a coagulant.

The quantity of chlorine added for disinfection after treatment depends on the actual treatment process, but generally sufficient chlorine is added to provide the desired chlorine residue (free chlorine and chloramine), usually in the range of 0.5-1 mg/l. Higher levels have been used (e.g., up to 5 mg/l; White, 1986) when difficulties in maintaining a residue in distribution are experienced, for example, with long pipelines.

Within large distribution systems, further chlorine may be added to maintain a desired residue at consumer taps. The quantity of chlorine added, usually at a covered water storage reservoir, varies but is typically in the range of 0.5-2 mg/l.

High doses of chlorine (about 50 mg/l) are used for disinfecting new or repaired equipment such as distribution pipes; however, such highly chlorinated water is usually flushed to waste.

In Europe, the USA and in other industrialized countries, where most water supplies are disinfected, usually with chlorine, high-quality groundwater sources usually receive minimal treatment and relatively low doses of chlorine (up to around 1 mg/l) for disinfection. Surface waters generally receive more chlorine, depending on the quality of the source water, as discussed above.

1.3 Impurities in chlorine gas and liquid

Various processes have been used for the commercial production of chlorine gas and liquid; the relative popularity of each has often been governed by economic aspects—particularly the cost and availability of starting chemicals from other industrial processes. Most of the current production of chlorine is accomplished electrolytically from brine using diaphragm, mercury or membrane cells. To a lesser extent, hydrochloric acid is used instead of brine. Some chlorine is also produced by the catalytic oxidation of hydrochloric acid and the action of nitric acid on sodium chloride, known as the salt process.

The main impurities in chlorine that are of possible relevance to the quality of drinking-water are carbon tetrachloride (see IARC, 1987) and bromine. Generally, the level of carbon tetrachloride is such that the residual concentrations in drinking-water, if any, are very low. A detectable level (1 mg/l) that was reported appeared to be due to unsuitable chlorine manufacture (carbon tetrachloride was used in this particular process). Consequent to this incident, the American Water Works Association set a maximum level for carbon tetrachloride in chlorine at 150 mg/l (White, 1986).

Bromine in chlorine gas or liquid could result in brominated by-products. The levels of bromine in commercial chlorine available in the UK for drinking-water treatment are, however, low (maxima, 850 and 2500 ppm (w/w) in two sources) (ICI Chemicals and Polymers Ltd, 1988), and typical levels in the USA are 50-125 ppm (maximum, 200 ppm) (The Chlorine Institute, USA, 1990).

1.4 Alternative disinfectants for drinking-water

Although chlorine is by far the most commonly used disinfectant (and oxidant) in drinking-water treatment, other chemicals, particularly ozone and chlorine dioxide, have been used for many years. Concern over possible risks to health due to the by-products of chlorination has led to a wider interest in alternatives.

Ozone is a powerful oxidant and an excellent disinfectant. It is used for treating drinking-water at many waterworks throughout the world, particularly in certain countries, for example France. It must be generated on site, and consequently it is less suited than chlorine to application at small treatment works. It does not leave a residue in the distribution system, since it decays quickly in water; therefore, if a residue is required, ozone must be used in conjunction with a disinfectant that gives such a residue (White, 1986).

Ozone produces a range of by-products, particularly aldehydes and organic acids (White, 1986), and it can generate low levels of bromoform (see monograph, p. 213) (Jacangelo *et al.*, 1989) by oxidation of bromide to hypobromous acid (Amy *et al.*, 1985). Evidence concerning the bacterial mutagenicity of ozonation by-products is conflicting; in general, ozone generates less mutagenicity than chlorine, but different mutagens are likely to be produced (National Academy of Sciences, 1979; National Research Council, 1980; Fielding & Horth, 1988).

Chlorine dioxide is used at a number of waterworks, particularly for water sources in which chlorophenolic tastes result from the use of chlorine (due to chlorination of phenol). It does not form trihalomethanes and it persists in drinking-water, which means that it provides a residue in the distributed supply. In use, however, it produces chlorite and chlorate, which must be carefully controlled, as they are relatively toxic species. The by-products of chlorine dioxide are not well characterized. In general, chlorine dioxide produces low levels of bacterial mutagenicity, but, as in the case of ozone, the mutagens involved are probably different from those produced by chlorine (National Academy of Sciences, 1979; National Research Council, 1980; Fielding & Horth, 1988).

Monochloramine is a less powerful disinfectant than chlorine, ozone or chlorine dioxide, but it is more persistent in drinking-water and has been used to maintain a low residual level in a distribution system over many years. Recently, interest in its use on a more substantial scale has been raised because it does not lead to high levels of trihalomethanes (Jacangelo *et al.*, 1989). Little information is available on the by-products of chloramine; it generates bacterial mutagenicity, but less consistently and at a lower level than does chlorine.

2. Occurrence and Analysis of Compounds Formed by the Chlorination of Drinking-water

2.1 Occurrence

The composition of chlorinated drinking-water to which the consumer is exposed varies according to location. The variables of established importance in the production of potentially toxic compounds are total organic carbon concentrations, pH, ammonium and bromide ion concentrations and the qualitative composition of the organic matter. Minor constituents are other inorganic ions such as nitrate, additives to drinking-water and other treatment processes.

The pK_a of HOCl (one of the forms that chlorine assumes in aqueous solution) is 9.5 (see equation on p. 50). At pH < pK_a, chlorination reactions are more prominent (White, 1986). Many by-products produced at low pH (2-7) are unstable at neutral to alkaline pH. This is particularly true of mutagenic chemicals formed on chlorination (Meier *et al.*, 1983, 1985). The concentrations of other by-products (e.g., dichloroacetic acid) appear to be more or less independent of the pH (Krasner *et al.*, 1989), while others decrease markedly at high pH (e.g., trichloroacetic acid and chloral). Conversely, the amount of trihalomethanes increases markedly as the pH becomes more alkaline. The pH of drinking-water is sometimes altered during the course of treatment (e.g., lime softening).

The relationship between chlorine dose and the amount of organic carbon that is present greatly affects the by-products formed. This becomes a critical issue in assessing whether the chlorine residue commonly maintained in chlorinated waters or the by-products of chlorination that are formed are responsible for any effects observed epidemiologically. The chlorine:carbon molar ratios normally found during the chlorination reaction in drinking-water treatment are very different from those found in ingested water in the gastrointestinal tract. In drinking-water, the ratio is typically in the range of 1.0-1.5, and that in the gastrointestinal tract is much lower. As a consequence, data gathered in the USA, where fairly high residual levels of chlorine remain in treated water as it is consumed at the tap (0.5-2 mg/l), may not be applicable to practice in other parts of the world where residues are deliberately maintained at low levels (< 0.1 mg/l). Finally, it is important to recognize that the actual practice in many locations is to maintain residues as 'combined residuals' (e.g., by adding ammonia to form chloramine) after using chlorine or other chemicals for primary disinfection.

As a consequence, chlorinated water in different locations cannot be considered to be the same entity. This fact has added a complex dimension to evaluation of the carcinogenic hazard for humans of chlorinated water that is not ordinarily encountered in these *Monographs*. Nevertheless, it was the view of the Working Group that this issue was of great importance to public health. Consequently, it endeavoured to make as objective an evaluation as is possible, given the vagaries of the data. The Working Group considered it important that the appropriate public health and regulatory authorities recognize the need to clarify the *broad* issue of drinking-water disinfection with appropriate research efforts in the near future. This issue must be resolved in a way that first protects against the waterborne infectious diseases observed in past centuries and then provides for minimizing or even eliminating any carcinogenic hazards that are secondary to this primary goal.

The addition of chlorine to waters containing dissolved organic compounds results in complex reactions that lead to chlorination by-products. The nature and extent of reaction of organic substrates in natural waters with chlorine is controlled by several factors, particularly pH and the chlorine:substrate ratio. An additional factor of importance is the presence of bromide in the untreated water (see p. 54), which can lead to brominated compounds.

Improvements in analytical techniques over recent years have revealed a complex range of organic substances in water supplies (Commission of the European Communities, 1989), and it has become apparent that many of these are generated during water chlorination. The probable organic precursors of these substances occur commonly and are of natural origin; they include humic substances and organic nitrogen compounds, such as amino acids (White, 1986).

The following sections summarize the available information on groups of halogenated by-products, selected mainly on the basis of the frequency of their occurrence in chlorinated water.

(a) Trihalomethanes

The production of chloroform (see IARC, 1987) during chlorination of natural waters was first observed by Bellar *et al.* (1974) and Rook (1974); this finding initiated many investigations into the identity, source and significance of chlorination by-products. Subsequently, a variety of additional trihalomethanes was detected (for example, Fawell *et al.*, 1986; Fielding & Horth, 1986), which include bromodichloromethane (see monograph, p. 179), chlorodibromomethane (see monograph, p. 243), tribromomethane (bromoform) (see monograph, p. 213), iodadichloromethane, iododibromomethane, bromochloroiodomethane and chlorodiiodomethane. Total trihalomethane levels in treated drinking-water were reported in one survey in the UK (Water Research Centre, 1980): Chlorinated water derived from a lowland river contained a mean level of 89.2 µg/l (SD, 0.9-3.9), and that from an upland reservoir, 18.7 µg/l (SD, 0.2-1.3). The study also showed that chlorinated groundwater was contaminated by trihalomethanes to a significantly lesser extent than chlorinated surface waters. Chloroform was the predominant trihalomethane.

The occurrence of chloroform in drinking-water was reviewed in an earlier monograph (see IARC, 1979a), which indicated that unchlorinated waters contain low concentrations (typically < 1 µg/l), but chlorinated waters in several countries invariably contain chloroform at levels up to 311 µg/l (Symons et al., 1975). Similar findings were reported in later surveys (for example, Brass et al., 1977; Water Research Centre, 1980).

Chloroform was measured in a range of surface, reservoir, lake and groundwaters in the USA (Krasner et al., 1989). The median values (according to season) ranged from 9.6 to 15 µg/l for chloroform, 4.1-10 µg/l for bromodichloromethane, 2.6-4.5 µg/l for chlorodibromomethane and 0.33-0.88 µg/l for bromoform. Concentrations of chloroform in 100 US surface waters were 0.1-1 µg/l (39%), 1-10 µg/l (49%), 10-100 µg/l (12%) and 100-1000 µg/l (< 1%) (Perwak et al., 1980). Quaghebeur and De Wulf (1980) found mean total concentrations of trihalomethanes in Belgium of 7.7 µg/l in groundwater and 78 µg/l in surface water; chloroform was the predominant trihalomethane in treated surface waters. In the USA, three of 13 groundwater supplies had levels of < 0.2, 2.6 and 83 µg/l chloroform, while in the other 10 surface water supplies the levels ranged from 1.3 to 130 µg/l (Reding et al., 1989).

Nicholson et al. (1984) reported chloroform concentrations in drinking-water from 17 countries at levels ranging from not detected to 823 µg/l; levels of bromodichloromethane ranged from not detected to 228 µg/l; those of chlorodibromomethane ranged from not detected to 288 µg/l; and those of bromoform from not detected to 289 µg/l.

Bromodichloromethane levels have been reported in many studies. In treated drinking-water, concentrations typically range from 1 to 50 µg/l, with higher or lower values at some locations compared with those in untreated water samples, which are typically less than 1 µg/l (see monograph, p. 179). Surface and groundwater samples showed a similar range; however, in certain groundwaters, the concentrations were higher than those in surface waters. An analysis of 19 550 water samples in the USA revealed a mean bromodichloromethane concentration of 11 µg/l (range, 0-10 133 µg/l) (US Environmental Protection Agency, 1985). [The Working Group noted that the very high concentrations seen may be misleading, since no information was available on possible contamination by wastewater or on measures of quality control.] Concentrations of bromodichloromethane in 118 surface waters in the USA ranged from 0.1 to 1 µg/l in 66% of the samples, 1-10 µg/l in 31% and 10-100 µg/l in 3% (Perwak et al., 1980).

Chlorodibromomethane levels have also been reported in many studies. In treated drinking-water, concentrations typically ranged from 1 to 20 µg/l, with higher or lower values at some locations compared with those in untreated waters, which are typically less than 1 µg/l. Treated groundwater samples showed, in general, higher chlorodibromomethane concentrations than treated surface water (see monograph, p. 243). An analysis of 18 616 water samples in the USA revealed a mean chlorodibromomethane concentration of 10 µg/l (range, 0-10 133 µg/l) (US Environmental Protection Agency, 1985). [The Working Group noted that the very high concentrations may be misleading, since no information was available on possible contamination by wastewater or on measures of analytical quality control.] Concentrations of chlorodibromomethane in 115 surface waters in the USA ranged from 0.1 to 1 µg/l in 80% of samples and from 1 to 10 µg/l in 20% (Perwak et al., 1980).

Bromoform has been determined in many chlorinated drinking-water samples (see monograph, p. 213). It was not usually found (< 1 µg/l) in untreated water sources in the USA (Symons et al., 1975). Concentrations in surface water in the USA typically ranged from 1 to 10 µg/l, with a median of about 4 µg/l (Brass et al., 1977; Perwak et al., 1980). Maximal levels in chlorinated groundwaters tend to be higher (up to 240 µg/l) (Glaze & Rawley, 1979; Page, 1981). Levels of bromoform in chlorinated groundwater vary widely, probably because of variations in the natural bromide content; at high bromide levels, the median value for bromoform was 72 µg/l (Krasner et al., 1989).

Heating or boiling drinking-water containing trihalomethanes causes the concentrations to decrease significantly (Table 1) (Lahl et al., 1982).

Table 1. Effect of heating and boiling water on trihalomethane content[a]

Compound	Level (µg/l)				
	Original tap water	80°C: 1 min	100°C: 0 min	Boiling: 1 min	Boiling: 5 min
Chloroform	45.6	23.2	12.3	9.4	4.1
Bromodichloromethane	44.6	24.1	13.5	10.8	4.6
Chlorodibromomethane	42.3	24.1	14.4	12.3	5.5
Bromoform	35.9	21.3	13.9	13.5	6.8

[a]From Lahl et al. (1982)

Since the late 1970s, many countries have endeavoured to control the levels of trihalomethanes in water supplies to meet national standards, which range from 25 to 250 µg/l (World Health Organization, 1988). The World Health Organization (1984) set a guideline value for chloroform in drinking-water of 30 µg/l.

(b) Halogenated acetic acids

Halogenated acetic acids, although not investigated as thoroughly as trihalomethanes, are probably major chlorination by-products in drinking-water. Table 2 summarizes the levels reported.

Table 2. Halogenated acetic acids in chlorinated drinking-water

Water type (location)	Compound	Concentration range (µg/l)	Reference
Two chlorinated surface waters (USA)	Monochloroacetic acid	1 and 4	Jacangelo et al. (1989)
	Dichloroacetic acid	9.4 and 23	
	Trichloroacetic acid	7.4 and 22	
	Monobromoacetic acid	< 0.5 and 3.8	
	Dibromoacetic acid	0.7 and 11	
Range of surface, reservoir, lake, and groundwaters (USA)	Monochloroacetic acid	< 1–1.2[a]	Krasner et al. (1989)
	Dichloroacetic acid	5.0–7.3[a]	
	Trichloroacetic acid	4.0–6.0[a]	
	Monobromoacetic acid	< 0.5–1.6[b]	
	Dibromoacetic acid	0.9–19[b]	
Tap water (reservoir) (USA)	Dichloroacetic acid	63.1–133	Uden & Miller (1983)
	Trichloroacetic acid	33.6–161	
Tap water (Germany)	Trichloroacetic acid	Not detected–3	Lahl et al. (1984)
Surface waters (USA)	Trichloroacetic acid	4.23–53.8	Norwood et al. (1986)
Treated water (Australia)	Trichloroacetic acid	200 max	Nicholson et al. (1984)
	Dichloroacetic acid	(similar max)	

[a]Median value
[b]High bromide level

(c) Halogenated acetonitriles

A variety of halogenated acetonitriles (see monograph, p. 269) have been detected in chlorinated drinking-water samples, formed by the action of chlorine on natural organic matter in water (Oliver, 1983; Jacangelo et al., 1989; Krasner et al., 1989; Peters et al., 1989). The levels found vary; the highest total concentration found was 42 µg/l in a survey in Florida (Trehy & Bieber, 1981). Halogenated acetonitriles were not detected in raw water (Oliver, 1983).

The most abundant of the acetonitriles is dichloroacetonitrile. In surveys in the USA, this compound was found in most chlorinated water supplies at concentrations of up to 24 µg/l, with a median of 1.2 µg/l. Bromochloroacetonitrile was found at concentrations up to 10 µg/l, with a median of 0.5 µg/l. Dibromoacetonitrile was found in some water supplies at maximum concentrations of 11 µg/l, with a median of 0.5 µg/l (Krasner *et al.*, 1989; Reding *et al.*, 1989).

(d) *Chlorinated ketones*

A range of chlorinated ketones are produced during chlorination (Table 3). Other chlorinated ketones that have been detected in drinking-water but have not been quantified, include 1,1,3,3-tetrachloropropanone, 3,3-dichloro-2-butanone, 1,1-dichloro-2-butanone, 1,1,1-trichloro-2-butanone and 2,2-dichloro-3-pentanone (Coleman *et al.*, 1984).

Table 3. Chlorinated ketones in chlorinated drinking-water

Water type (location)	Compound	Concentration range (µg/l)	Reference
Range of surface, reservoir, lake and groundwaters (USA)	1,1-Dichloropropanone	0.46–0.55[a] 2.2 (max)	Krasner *et al.* (1989)
	1,1,1-Trichloropropanone	0.35–0.80[a] 2.4 (max)	
Two chlorinated surface waters (USA)	1,1-Dichloropropanone 1,1,1-Trichloropropanone	0.16–0.24 1.1–1.8	Jacangelo *et al.* (1989)
Drinking-water (Australia)	1,1,1-Trichloropropanone	20 (max)	Nicholson *et al.* (1984)

[a]Median

(e) *Halogenated phenols*

Chloro-, chlorobromo- and bromophenols can be formed from phenol during chlorination. They add objectionable tastes or odours to drinking-water when present at levels over a few micrograms per litre. Although high concentrations may occur during phenol pollution of untreated water, typical levels in drinking-water are kept low to avoid consumer complaints. A recent investigation of drinking-water gave the following levels (µg/l): 2-chlorophenol, < 0.004-0.065; 4-chlorophenol, < 0.004-0.127; 2,4-dichlorophenol (see IARC, 1986), < 0.002-0.072; 2,6-dichlorophenol, < 0.002-0.033; 2,4,6-trichlorophenol (see IARC, 1987), < 0.008-0.719; pentachlorophenol (see IARC, 1987), < 0.004-0.034; bromodichlorophenol, < 0.002-0.78; chlorodibromophenol, < 0.004-0.022; 2,4-dibromophenol, < 0.002-0.084; and 2,4,6-tribromophenol, < 0.004-0.022 (Sithole & Williams, 1986).

(f) *Other halogenated hydrocarbons*

Other halogenated hydrocarbons have been detected in chlorinated drinking-water; although accurate quantitative data are not available, levels are typically < 1 µg/l (McKinney *et al.*, 1976; Suffet *et al.*, 1980; Anon., 1983; Coleman *et al.*, 1984; Kopfler *et al.*, 1985; Fielding & Horth, 1986; Fawell *et al.*, 1987; Horth *et al.*, 1989). These compounds include bromoethane (see monograph, p. 299), bromobutane, bromochloromethane, bromochloropropanes, bromopentachloroethane, bromopropane, bromopentane, bromotrichloroethylene, chlorobutane, chloroethane (see monograph, p. 315), dibromomethane, dichloromethane (see IARC, 1987), 1,1-dichloroethane, 1,2-dichloroethane (see IARC, 1979b), dichloropropene (see IARC, 1987), hexachloroethane (see IARC, 1979c), hexachlorocyclopentadiene, iodoethane, tetrachloromethane (carbon tetrachloride) (see IARC, 1987) and pentachloropropene. It is not clear, however, to what extent, if any, these compounds result from chlorination of water.

(g) *Chlorinated furanones and related compounds*

Studies on the possible identity of chemical mutagens formed during chlorination (see p. 71) have led to the detection in drinking-water (Kronberg & Vartiainen, 1988; Horth *et al.*, 1989; Fawell & Horth, 1990) of 3-chloro-4-(dichloromethyl)-5-hydroxy-2(5*H*)-furanone, referred to as MX, and E-2-chloro-3-(dichloromethyl)-4-oxobutenoic acid, referred to as E-MX (see also the section on genetic and related effects, p. 66). Levels of MX and E-MX that have been detected are given in Table 4.

Table 4. Concentrations of MX and E–MX in chlorinated drinking-water

Water type (location)	Compound	Concentration range (µg/l)	Reference
Surface treated and chlorinated waters (Finland)	MX E-MX	< 0.004–0.067 0.002–0.059	Kronberg & Vertiainen (1988)
Treated and chlorinated lowland rivers (UK)	MX	Not detected–0.006	Fawell & Horth (1990)
Treated and chlorinated upland waters (UK)	MX	Not detected–0.041	Fawell & Horth (1990)

MX and E-MX are though to be related in the following manner:

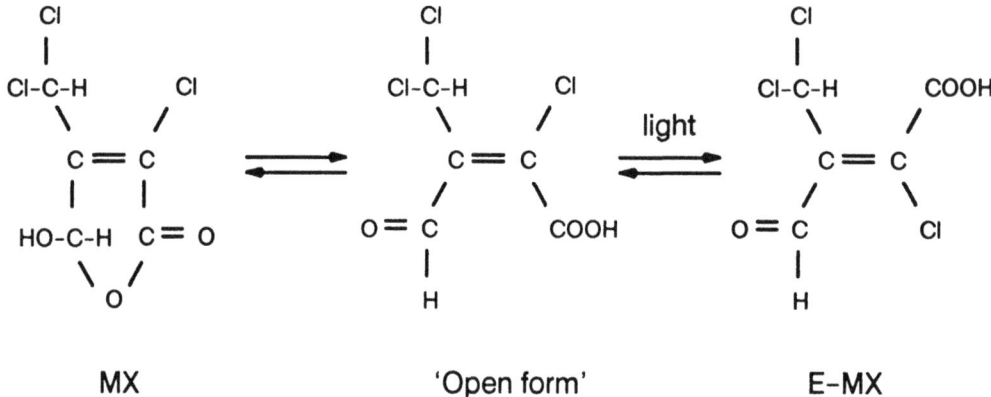

(h) Miscellaneous compounds found in chlorinated water

Other compounds that have been reported to be present in chlorinated drinking-water are listed in Table 5.

Table 5. Concentrations of miscellaneous chlorination products in chlorinated drinking-water

Water type (location)	Compound	Concentration range (µg/l)	Reference
Eight treated waters (UK)	5-Chlorouracil 5-Chlorouridine 4-Chlororesorcinol 5-Chlorosalicylic acid	0.1–14.1 0.7–26.7 1.6–4.7 2.3–12.5	Crathorne *et al.* (1979)
Six treated waters (USA)	Chloral (hydrate)	7.2–18.2	Uden & Miller (1983)
Two utilities (USA)	Chloral (hydrate)	6.3–19	Jacangelo *et al.* (1989)
Range of surface, reservoirs, lake and groundwaters (USA)	Chloral (hydrate)	1.7–3	Krasner *et al.* (1989)
Range of surface, reservoir, lake and groundwaters (France, UK, USA)	Chloropicrin	0.07–1	Duguet *et al.* (1985); Fawell *et al.* (1986, 1987); Jacangelo *et al.*, 1989)
		0.10–0.16	Krasner *et al.* (1989)

Table 5 (contd)

Water type (location)	Compound	Concentration range (µg/l)	Reference
Range of surface and groundwaters (UK)	Bromodichloronitro-methane Bromochloronitromethane	Not quantified Not quantified	Fawell et al. (1986)
Not stated (USA)	Trichloropropenenitrile	Not quantified	Coleman et al. (1984)
Range of surface and groundwaters (UK)	Benzyl cyanide Chlorohydroxybenzyl cyanide	Not quantified Not quantified	Fawell et al. (1986, 1987)
Range of surface, reservoir, lake and groundwaters (USA)	Formaldehyde Acetaldehyde	2.1–17[a] 2.1–7.1[a]	Krasner et al. (1989)

[a]Due to presence in untreated water and increase during chlorination

Formaldehyde and acetaldehyde were found in untreated water and were found at higher levels in water treated with various disinfectants, including chlorine. Ozone produced the highest levels (Krasner et al., 1989).

(i) *Adsorbable organic halide*

The total halogenated matter generated by chlorination has been estimated by measuring adsorbable organic halide (halogenated organic compounds that can be adsorbed onto activated carbon; see p. 49). A recent survey of drinking-water (Krasner et al., 1989) found median levels in the range of 150-250 µg/l. The relationships among the individual chlorination by-products covered by this measurement and between individual products and halogenated organic compounds vary substantially.

(j) *Sources of chlorination by-products*

At present, it is not possible to analyse all of the by-products of chlorination or other disinfectants/oxidants. In order to understand the production of by-products and to identify unknown by-products, many workers have studied substances occurring in raw water that could react with chlorine. Such studies have revealed by-products that have been found in water supplies and others that could be present but have not, as yet, been detected.

Rook (1977) suggested that humic substances are involved as precursors. These naturally occurring substances are an ill-defined mixture of chemically and microbiologically degraded plant residues, bound together by chemical and physical processes, and are characterized as refractory, yellow-to-black materials. They are complex, high-molecular-weight, ubiquitous constituents of natural waters, where they consist mainly of humic and fulvic acids, the latter normally predominating. They vary in character to some extent from site to site and according to season; the organic matter in upland, coloured, natural water is mostly humic substances. They are extracted from water in several ways but usually by adsorption onto resins. Humic acids are defined operationally as becoming insoluble at pH < 2. Fulvic acids, however, are soluble in water at all pHs.

Several research groups have studied the chlorination of humic substances extracted from soil and water and confirmed that chloroform and dichloro- and trichloroacetic acids are produced as major reaction products. A variety of other products and intermediates have also been characterized. Christman *et al.* (1983) studied the chlorination of humic and fulvic substances extracted from water and found a wide variety of chlorinated saturated and unsaturated aliphatic acids. In a recent review, Christman *et al.* (1989) gave the significant products detected as: chloroform ($CHCl_3$), bromodichloromethane ($CHBrCl_2$), chloral (CCl_3–CHO), chloroacetic acid (H_2CCl–COOH), dichloroacetic acid (H_2CCl–COOH) and trichloroacetic acid (CCl_3–COOH), which are found in chlorinated drinking-water. Others produced in the laboratory are 2,2-dichloropropanoic acid (CH_3–CCl_2–COOH), 3,3-dichloropropenoic acid (CCl_2=CH–COOH), 2,3,3-trichloropropenoic acid (CCl_2=CCl–COOH), dichloropropanedioic acid (HOOC–CCl_2–COOH), butanedioic acid (HOOC–$(CH_2)_2$–COOH), chlorobutanedioic acid (HOOC–CH_2–CHCl–COOH), 2,2-dichlorobutanedioic acid (HOOC–CCl_2–CH_2–COOH), *cis*-chlorobutenedioic acid (HOOC–CH=CCl–COOH), *cis*-dichlorobutenedioic acid (HOOC–CCl=CCl–COOH) and *trans*-dichlorobutenedioic acid (HOOC–CCl=CCl–COOH).

Nonchlorinated products—for example, benzene carboxylic acids, carboxyphenylglyoxylic acids and mono- and dibasic alkanoic acids—were also reported. Depending on reaction conditions, chloroform, dichloro- and trichloroacetic acids accounted for over 50% of the adsorbable organic halides (see p. 63) produced during chlorination of humic substances. de Leer (1987) identified over 100 products of the chlorination of humic acids. These were mainly those found by previous workers, but, in addition, he described various other chlorinated carboxylic acids, cyano-alkanoic acids and trichloromethyl precursors of chloroform. Examples of the many precursors detected are: 3,3,3-trichloro-2-hydroxypropanoic acid (Cl_3C–CH(OH)–COOH), 4,4,4-trichloro-3-hydroxybutanoic acid (Cl_3C–CH(OH)–CH_2–COOH) and 2-chloro-3-(trichloroacetyl)butenedioic acid (COOH–(CCl_3–CO)C=CCl–COOH). These by-products have not been detected in drinking-water but are probably reaction intermediates. The chloroform precursors may form chloroform in the following manner:

$$Cl_3C-CO-CCl=CCl-CCl_2-COOH \xrightarrow{H_2O} CHCl_3 + HOOC-CCl=CCl-CCl_2-COOH.$$

The cyanoalkanoic acids (which are presumably derived from nitrogen-containing elements of humic substances) were examined further by de Leer (1987). Cyanopropanoic acid and cyanoacetic acid (the latter was not detected as a chlorination by-product but its presence was postulated) reacted readily with chlorine. The following chlorination products were identified after reaction of cyanoacetic acid at pH 10: dichloroacetic acid ($CHCl_2-COOH$) and trichloroacetic acid (CCl_3-COOH), which are found in chlorinated drinking-water; and 2,2-dichloroacetamide ($CHCl_2-CONH_2$), 2,2-dichloro-N-hydroxyethaneimidoyl chloride ($CHCl_2-CCl=NOH$), 2,2-dichloro-2-carboxyacetamide ($HOOC-CCl_2-CONH_2$), 2,2-dichloropropanedioic acid ($HOOC-CCl_2-COOH$), 2,2,2-trichloro-N-hydroxyethaneimidoyl chloride ($CCl_3-CCl=NOH$) and 2,2-dichloro-2-carboxy-N-hydroxyethaneimidoyl chloride ($HOOC-CCl_2-CCl=NOH$). At lower pH, conversion to dichloroacetonitrile, dichloroacetic acid and dichlorosuccinic acid was favoured.

Several workers have concluded that most chlorination products are formed by a reaction involving 1,3-dihydroxybenzene (resorcinol) structures within the humic structure (Rook, 1980; Boyce & Hornig, 1983; de Leer, 1987; Christman *et al.*, 1989).

Unsaturated organic compounds, alkenes and unsaturated fatty acids such as oleic and linoleic acids, can react with chlorine in the laboratory under conditions similar to those of water treatment to form chlorohydrins (Gibson *et al.*, 1986); however, their presence in chlorinated drinking-water has not been investigated.

Amino acids are common constituents of raw water. Although they normally occur at low concentrations (typically up to 100 μg/l), bound amino acids, such as proteins and peptides, may predominate (Le Cloirec & Martin, 1985; Thurman, 1985).

The general reaction of amino acids with chlorine in aqueous solution has been known for many years, and reviews have been published (for example, Glaze *et al.*, 1982). It is now known that most, if not all, amino acids of the type $R-CH_2-CH(COOH)NH_2$ react readily with chlorine and initially form monochloramines ($R-CH_2-CH(COOH)NHCl$) and, depending on the conditions, dichloramines ($R-CH_2-CH(COOH)NCl_2$). Further reaction leads to nitriles ($R-CH_2CN$) and/or aldehydes ($R-CH_2CHO$). Le Cloirec and Martin (1985) postulated the mechanism involved. Amino acids have been shown to generate a wide range of other by-products during chlorination (Horth, 1989).

(k) Mutagenic by-products

Mutagenicity assays have been used in many countries to study the mutagenic potential of drinking-water samples (see p. 70). A number of the substances found in chlorinated drinking-water have been shown to be bacterial mutagens (Table 6). Only the chlorinated furanones and related compounds (see p. 61) are sufficiently potent and occur in sufficiently high concentrations to account for a significant proportion of the mutagenicity measured in some extracts of chlorinated drinking-water (Kronberg & Vartiainen, 1988; Horth, 1989). Many mutagens are generated during laboratory chlorination of humic substances and amino acids and during chlorination of wood pulp in experiments designed to indicate those substances that may be formed in the chlorination of natural waters; however, not all of these have been detected in drinking-water.

Table 6. Studies in which bacterial (*Salmonella typhimurium* TA100 without an exogenous metabolic system) mutagens were identified in chlorinated drinking-water, chlorinated solutions of humic substances and amino acids and in chlorinated wood pulp effluent

Mutagen	Reference[a]			
	Drinking-water	Humic substances	Amino acids	Wood pulp
Halo-alkanes				
Bromoform	1	ND	ND	ND
Bromochloromethane	1	ND	ND	ND
Bromodichloromethane	1	1,7	ND	2
Dibromomethane	1	ND	ND	2
Chlorodibromomethane	1	ND	ND	2
Dichloromethane	15	ND	ND	3
Bromoethane	1	ND	ND	ND
1-Bromopropane	1	ND	ND	ND
1-Bromobutane	1	ND	ND	ND
1,2-Dichloroethane	1	ND	ND	3
1,1,1-Trichloroethane	ND	ND	ND	3
1,1,2,2-Tetrachloroethane	ND	ND	ND	3
Iodoethane	1	ND	ND	ND
Chloro-alkenes				
Trichloroethylene	ND	ND	ND	2
Tetrachloroethylene	ND	ND	ND	2
Dichloropropene	1	ND	ND	ND
Tetrachloropropene	ND	ND	ND	2
Pentachloropropene	ND	4	ND	2

Table 6 (contd)

Mutagen	Reference[a]			
	Drinking-water	Humic substances	Amino acids	Wood pulp
Chloro–ketones				
1,1-Dichloropropanone	5	4,5	ND	ND
1,3-Dichloropropanone	ND	4,5	ND	2
1,1,1-Trichloropropanone	5	4,5	ND	ND
1,1,3-Trichloropropanone	ND	4,5	ND	ND
3,5,5-Trichloropent-4-ene-2-one	ND	ND	ND	2
1,1,3,3-Tetrachloropropanone	5	4,5	ND	2,3
Pentachloropropanone	ND	4,5	ND	3
Hexachloropropanone	ND	6	ND	2,3
Chloro-aldehydes/furanones				
Chloral (trichloroethanal)	1	16	7	ND
Chloroacetaldehyde	ND	ND	ND	2
2-Chloropropenal	ND	4	ND	2
Dichloropropanal	ND	4	ND	ND
2,3-Dichloropropenal	ND	4	ND	ND
3,3-Dichloropropenal	ND	4,5	ND	ND
Trichloropropanal	ND	4	ND	ND
2,3,3-Trichloropropenal	ND	4,5	ND	ND
2-Phenyl-2,2-dichloroethanal	ND	ND	7	ND
E-2-Chloro-3-[dichloromethyl]-4-oxo-butenoic acid (E-MX)	8	9	7	10
3-Chloro-4-[dichloromethyl]-5-hydroxy-2(5H)-furanone (MX)	8,11	9,10	7	2,12
3,4-Dichloro-5-[dichloromethyl]-5-hydroxy-2-furanone	ND	ND	ND	13
3-Chloro-4-[bromochloromethyl]-5-hydroxy-2(5H)-furanone (BMX-1)	ND	ND	14	ND
3-Chloro-4-[dibromomethyl]-5-hydroxy-2(5H)-furanone (BMX-2)	ND	ND	14	ND
3-Bromo-4-[dibromomethyl]-5-hydroxy-2(5H)-furanone (BMX-3)	ND	ND	14	ND
Halo–nitriles				
Bromochloroacetonitrile	1	ND	ND	ND
Dichloroacetonitrile	1	5,7	7	ND

Table 6 (contd)

Mutagen	Reference[a]			
	Drinking-water	Humic substances	Amino acids	Wood pulp
Miscellaneous				
Chloropicrin[b]	1	ND	ND	ND
Trichloro-1,2,3-trihydroxybenzene	ND	ND	ND	2
Benzyl chloride[c]	ND	ND	7	3
Benzoyl chloride[c]	ND	ND	7	ND
Bromo-*para*-cymene	ND	ND	ND	2
Dichloro-*para*-cymene	ND	ND	ND	2

[a]References: 1, Fielding & Horth (1986); 2, Rapson *et al.* (1985); 3, McKague *et al.* (1981); 4, Kopfler *et al.* (1985); 5, Meier *et al.* (1985); 6, de Leer (1987); 7, Horth (1989); 8, Kronberg & Vartiainen (1988); 9, Kronberg *et al.* (1990); 10, Holmbom *et al.* (1990); 11, Hemming *et al.* (1986); 12, Holmbom *et al.* (1984); 13, Strömberg *et al.* (1987); 14, Fawell & Horth (1990); 15, Anon. (1983); 16, Coleman *et al.* (1984)
[b]With S9 activation
[c]Tentative identification
ND, not detected

2.2 Analytical methods

Methods for the analysis of chlorinated compounds produced during the chlorination of drinking-water can be divided into three basic types: techniques for identifying unknown or suspected substances—not necessarily specific for halogenated compounds; specific techniques for the analysis of known or suspected halogenated compounds; and techniques designed for a gross estimate of halogenated organic matter in chlorinated water.

(a) Analysis of unknown chlorination by-products

In the 1970s, concern over the presence of organic micropollutants in drinking-water together with the emergence of powerful, sensitive analytical techniques for separating and identifying these substances, such as capillary column gas chromatography-mass spectrometry (GC-MS) led to the identification of a large number of organic compounds in drinking-water at low concentrations (Commission of the European Communities, 1989). Techniques involving GC-MS have been used extensively to analyse drinking-water for unknown and known chlorination by-products in addition to contaminants in general. With the exception of grossly contaminated drinking-water, concentrations of organic chemicals are such that direct application of identification techniques is usually

impossible and, consequently, some form of isolation/concentration process is required. The mixture of organic chemicals isolated is so complex that considerable separation (invariably by some form of chromatography) is also needed prior to application of instrumental techniques capable of providing structural information. Thus, the overall analytical technique deployed usually consists of:
  (i)   isolation/concentration (not necessarily as one step),
  (ii) separation (of the components in the complex mixtures isolated) and
  (iii) detection and structural analysis.

Various methods for isolating and concentrating organic chemicals, such as chlorination by-products, from drinking-water exist, and a number of validated methods have emerged that are based upon solvent extraction, adsorption (usually by XAD resin), followed by solvent elution of the adsorbent, headspace analysis and related methods. Application of these techniques is virtually routine, and examples abound in the literature (Keith, 1976; Coleman *et al.*, 1980; Keith, 1981; Fawell *et al.*, 1986).

(b) Analysis of known or suspected chlorination by-products

Analytical methods have been developed for a range of identified chlorination by-products. The following is a summary of those used for the substances discussed above.

  (i) *Trihalomethanes*

Trihalomethanes were shown to be present in drinking-water as a result of chlorination (Bellar *et al.*, 1974; Rook, 1974) using purge and trap and solvent extraction methods of concentration, followed by GC with electron capture detection (ECD; Croll *et al.*, 1986). Subsequently, a number of analytical methods for the determination of trihalomethanes in drinking-water have been published (for review, see Croll *et al.*, 1986); they include direct aqueous injection (Nicholson *et al.*, 1977; Peters, 1980; Grob & Habich, 1983), liquid-liquid extraction (Dressman *et al.*, 1979; US Environmental Protection Agency, 1979a; Standing Committee of Analysts, 1980), purge and trap (Bellar *et al.*, 1974; Dressman *et al.*, 1979; US Environmental Protection Agency, 1979b) and headspace analysis (Rook, 1974; Otson *et al.*, 1979; Croll *et al.*, 1986) with separation and detection by GC-ECD. More information is given in the respective monographs about the analysis of bromodichloromethane, chlorodibromomethane and bromoform.

  (ii) *Halogenated acetic acids*

Halogenated acetic acids are common by-products of water chlorination. Most of the analytical methods involve extraction into a solvent at low pH (0.5-2) with addition of sodium chloride to salt out the substances, derivatization and then detection by GC-ECD (Krasner *et al.*, 1989; Uden & Miller, 1983; Lahl *et al.*, 1984), GC with microwave plasma detection (Miller *et al.*, 1982) or isotope dilution MS (Norwood *et al.*, 1986).

(iii) *Halogenated acetonitriles*

Dichloroacetonitrile and other halogenated analogues have been determined in drinking-water by solvent extraction with salting out using sodium chloride or sodium sulfate followed by GC-ECD (Oliver, 1983; Italia & Uden, 1988; Krasner *et al.*, 1989). More information on analytical methods for halogenated acetonitriles is given in the monograph.

(iv) *Chlorophenols*

Chlorophenols are well-known chlorination by-products, since they can confer objectionable tastes and odours in the supply. A variety of techniques have been developed for their analysis, which usually involve derivatization to methyl, acetyl or pentafluorobenzoyl derivatives, followed by GC-ECD (Renberg, 1981; Abrahamsson & Xie, 1983; Standing Committee of Analysts, 1985, 1988) or, in some cases, GC-MS with specific-ion monitoring (Sithole & Williams, 1986).

(v) *Chlorouracil, chlorouridine, chlororesorcinol and chlorosalicylic acid*

These unchlorinated substances occur in natural waters and can become chlorinated during water treatment. They have been determined after freeze-drying or vacuum evaporation, extraction with methanol and examination by high-performance liquid chromatography with confirmation by GC-MS (Crathorne *et al.*, 1979).

(vi) *Organic chloramines*

Chlorine can react extensively with organic amines, amino acids and related substances in water to produce chloramines. Specific analysis of organic chloramines in drinking-water is difficult, and, consequently, there is little detailed information on their presence and concentrations. In recent years, some specific methods have appeared which are based on derivatization followed by high-performance liquid chromatography with fluorescence detection (Scully *et al.*, 1984) or ultraviolet/electrochemical detection (Lukasewycz *et al.*, 1989).

(c) *Mutagens and mutagenicity in chlorinated drinking-water*

The presence of mutagenic chemicals in concentrated extracts of drinking-water is inferred from the positive results obtained in bacterial mutagenicity assays such as the *Salmonella*/microsome mutagenicity assay (Ames & Yanofsky, 1971; Ames *et al.*, 1975).

Organic compounds present at low concentrations in drinking-water must be extracted and concentrated prior to assays for mutagenicity. No single technique is capable of extracting all organic material from water, and therefore several methods have been used in combination with bacterial mutagenicity assays, including reverse osmosis or freeze drying, followed by extraction of the solids with organic solvent, or adsorption on resins followed by elution with solvents (for review, see Wilcox et al., 1986). The most widely used technique involves adsorption on XAD macroreticular resin. Although a small proportion of the organic matter in drinking-water is recovered, the level of mutagenic activity of the extracts is high (Fielding & Horth, 1986; Ringhand et al., 1987). Different compounds may be recovered by altering the pH of the water prior to XAD adsorption. Some groups have reported considerably higher levels of mutagenic activity in low pH/XAD extracts than in extracts obtained at sample pH (near neutral) (Kronberg et al., 1985a; Ringhand et al., 1987; Horth et al., 1989). With all the methods, it is essential to check that mutagens are not generated as artefacts by the process itself, as impurities in solvents and other materials used or even their reaction products with free chlorine or chloramine in the water samples being processed.

Studies of the compounds responsible for the mutagenicity detected have led to the identification of strong bacterial mutagens, especially MX. The levels of this chlorination by-product have been determined in drinking-water by a method based on adsorption on resin at low pH followed by desorption with solvent, methylation and GC-MS with selected-ion monitoring (Hemming et al., 1986; Horth et al., 1989).

(d) Measurement of total halogenated organic matter in drinking-water

Methods have been used to estimate total (as near as possible) organically bound halogen in chlorinated drinking-water (for review, see Oake & Anderson, 1984). The basis of the most commonly used technique, which involves measuring adsorbable organic halogen, includes extraction of organic chlorine (or bromine) compounds from water by adsorption onto activated carbon, removal of inorganic halide by washing the carbon with a nitrate solution, conversion of organically bound halogen to inorganic halide (usually by combustion, although other approaches exist) and, finally, measurement of the halide (usually by microcoulometry). The terms 'total organic halide' and 'adsorbable organic halide' tend to be used in practice (Krasner et al., 1989); however, the latter is preferable, since very polar and very volatile halogenated compounds would not be recovered quantitatively by the usual methods.

3. Biological Data Relevant to the Evaluation of Carcinogenic Risk to Humans

3.1 Carcinogenicity studies in animals

Most of the studies reported here were designed to investigate the effects of organic extracts of drinking-water. These studies did not address the potential effects of by-products of disinfection, since that variable was not controlled for, i.e., generally, no control group of animals treated with extracts of raw water was included. Furthermore, the methods used to extract organic material from water were somewhat selective and would not result in equal concentration of all components (see p. 69); in particular, volatile substances may be lost. The extracts studied, therefore, may not be completely representative of the substances found in chlorinated water. Finally, the potential for introducing impurities into organic extracts by the interaction of free chlorine and chloramine in drinking-water with a solvent or resin may also be a confounding factor. Notwithstanding the difficulties in designing studies that control for this variable, it must be considered in their interpretation. Although of limited relevance to evaluating the carcinogenicity of chlorinated drinking-water, the studies of organic extracts are included for completeness.

(a) Oral administration

Mouse: Groups of 25 male and 25 female CFLP Han mice [age unspecified] were administered a chloroform (triple distillate) extract of disinfected river water from France (treatment procedure: flocculation, filtration, prechlorination, ozonization and postchlorination), dissolved in agar at a weight ratio of 1:20, prepared every two weeks and added to the diet for 104 weeks. The river water was collected over a two-year period. The treatment doses of organic material [1.2 and 2.4 mg/kg bw per day, respectively] corresponded to 100 and 200 times the calculated human dose, based on an assumed human consumption of 3 l/60 kg bw per day. The average yield of organic extract was 0.24 mg/l (mean of 10 samples). A group of 50 males and 50 females served as controls [control diet not specified]. No control receiving unchlorinated water was included. Increased mortality was observed in animals of each sex in the treated group [details not given]. The frequency of malignant tumours in males (predominantly thyroid gland tumours and lymphosarcomas) was: control, 4.9%; low-dose, 11.1%; and high-dose groups, 11.1%. The frequency of malignant tumours among females (predominantly mammary gland and ovarian adenocarcinomas and lymphosarcomas) was: control, 14.3%; low-dose, 43.8%; and high-dose, 45.0% (Truhaut *et al.*, 1979). [The Working Group noted the lack of an adequate control group to test for the extraction procedure, and that the incidence of individual tumour types and the incidence of benign tumours were not given.]

Groups of 50 male and 50 female B6C3F$_1$ mice, six to eight weeks of age, received solutions of either chlorinated humic acids (carbon:chlorine ratio, 1:1 or 1:0.3), produced by the addition of sodium hypochlorite to a commercial preparation of humic acid, or unchlorinated humic acids in the drinking-water, prepared freshly once a week, for two years. The average daily intake of total organic carbon was 2.8-2.9 mg/mouse for males and 2.1-2.2 mg/mouse for females. Similar numbers of male and female mice received sodium chloride (daily intake, 26.4 mg/male mouse and 22 mg/female mouse) in the drinking-water. As a positive control, equal numbers of mice of each sex were given dibromoethane at doses of 1.4 mg/male and 1.2 mg/female. A group of 100 male and 100 female mice received no treatment. Surviving animals were killed at 24 months, with the exception of the dibromoethane-treated groups, which were killed at 18 months. At two years, more than 78% of treated and control animals were still alive, except among males given dibromoethane. There was no difference in the percentage or the number of tumour-bearing animals in the treated groups. Several types of tumours occurred at higher incidence in the groups treated with 1:1 chlorinated humic acids, 1:0.3 chlorinated humic acids or unchlorinated humic acids, when compared to the untreated group, but the incidences were not increased when compared to the sodium chloride-treated control group (Van Duuren *et al.*, 1986).

Rat: Groups of 25 male and 25 female Sprague-Dawley rats [age unspecified] were administered the same extract of disinfected river water described for CFLP Han mice, above, at the same treatment doses. A group of 25 males and 25 females served as controls [control diet not specified]. No control receiving unchlorinated water was included. A dose-dependent increase in mortaliity was observed in animals of each sex [details not given]. The frequency of malignant tumours in males (thyroid gland tumours and lymphosarcomas) was significantly increased: 0 in controls, 33.3% with the low dose and 50% with the high dose. The frequency of malignant tumours in females (mammary gland and ovarian adenocarcinomas and lymphosarcomas) was 4.5% in controls, 40% in low-dose animals and 57.1% in high-dose groups (Truhaut *et al.*, 1979). [The Working Group noted the lack of a control group to test for the extraction procedure, that exact incidences of individual tumours types were not given, that tumours of different origins were combined for analysis, and that the incidences of benign tumours were not given.]

Groups of 50 male and 50 female Wistar rats (RIV:Tox(M)), weighing 165 g and 130 g, respectively, were administered organic extracts of surface tap water from the Netherlands [disinfection procedure unspecified] in nonmutagenic drinking-water for 106 weeks. Water consumption was measured weekly. Extraction and concentration were carried out on XAD-4/8 resin, and elution with dimethylsulfoxide, such that a 0.11-ml sample of concentrate contained 115 µg organic material, which corresponded to 1 l tap water. Daily dose levels were calculated as multiples of the expected human exposure based upon a daily consumption of 2 l water per 70 kg bw: 0, 4.5 times (11 mg/kg bw organic matter), 14 times (34 mg/kg bw), 40 times (97 mg/kg bw) for males and 0, 7 times (17 mg/kg bw), 22 times (53 mg/kg bw) and 68 times (165 mg/kg bw) for females. A slight increase in mortality was observed in the exposed groups. The numbers of animals with tumours (benign and malignant combined) were: males—control, 29/50; low-dose, 23/47; mid-dose, 27/50; and high-dose, 34/50; and females—control, 36/49; low-dose, 30/47; mid-dose, 33/47; and high-dose, 35/50. The frequency and types of tumours were similar in the treated and control groups (Kool et al., 1985a). [The Working Group noted that no control group to test for the extraction procedure was used and that several contaminants are unstable in dimethylsulfoxide (see p. 82; Meier et al., 1987; Kronberg & Vartiainen, 1988; Fielding & Horth, 1988).]

(b) *Skin application*

Mouse: Two groups of 40 male C57Bl mice, eight to ten weeks old, received skin applications of one drop of a tap water (collected over a period of one year) extract (preparation: US river water was treated by breakpoint chlorination, coagulation, filtration, concentration of organic compounds by passing through activated carbon, extraction of adsorbed organic matter with diethyl ether, removal of the ether by evaporation), either undiluted or diluted with methyl ethyl ketone (1:1) (Braus et al., 1951). The original tap water contained 0.1-1 mg/l organic material. The extracts were painted on a 1-cm^2 area of shaved shoulder twice a week for five months. The two groups of mice were then combined, and the animals received one drop of undiluted sample twice a week for a further a 12-13 months, when survivors were killed. A vehicle control group of 25 male mice received one drop of methyl ethyl ketone on shaved skin twice a week for four months. One skin papilloma developed among the treated mice, whereas no skin tumour was observed among vehicle controls. Amyloidosis of the spleen, liver and kidneys was observed in several animals (Hueper & Ruchhoft, 1954). [The Working Group noted the lack of information on the quantity of tap water used for extraction, the quantity of organic material in the extracts and the lack of an adequate control group].

As part of a larger experiment, groups of 36 male and 36 female C57Bl mice, two months of age, received skin applications on the shaved neck region of either one drop of undiluted chloroform extract condensate prepared by passing chlorinated US tap water through activated carbon and elution with chloroform (yield, 1 g/620 gallons [2347 litres] water) once every two weeks (total of 28 applications), or one drop of undiluted ethanol-extract condensate prepared by passing chlorinated tap water through activated carbon (yield, 1 g/890 gallons [3369 litres] water) once every two weeks (total of 20 applications). Forty male and female mice were untreated. The 12 surviving animals in the chloroform-extract group and the three in the ethanol-extract group were killed at 18 months. No tumour developed among the treated mice either locally or in distant organs (Hueper & Payne, 1963). [The Working Group noted the infrequent application of the material and the lack of an adequate control group.]

(c) *Subcutaneous administration*

Mouse: Groups of 36 male and 36 female C57Bl mice, two months of age, were injected subcutaneously in the neck region at two-week intervals with 4 mg condensate prepared by passing chlorinated tap-water from the USA through activated carbon and extraction with chloroform (yield, 1 g/620 gallons [2347 litres] water) in 0.05 ml tricaprylin (total of 28 injections [total dose, 112 mg/mouse]). Additional groups of 36 male and 36 female C57Bl mice were similarly treated with 4 mg of tap water condensate prepared by passing chlorinated drinking-water through activated carbon and extraction with ethanol (yield, 1 g/890 gallons [3369 litres] water) in 0.5 ml ethanol (total of 20 injections [total dose, 80 mg/mouse]). Three animals in the chloroform-extract group survived to 18 months, and ten animals in the ethanol-extract group survived to 15 months. One skin papilloma at the site of injection and one leukaemia/lymphoma developed in the chloroform extract-treated animals [sex unspecified]. One leukaemia/lymphoma was observed among the ethanol extract-treated mice (Hueper & Payne, 1963). [The Working Group noted the infrequent application of the test material and the lack of adequate control groups.]

Six groups of 50-72 non-inbred albino mice received a subcutaneous injection of an extract of drinking-water collected every two weeks from three water-treatment plants in the USA, based on surface water sources (5000 gallons [18 927 litres]; treatment procedure: coagulation, sedimentation, filtration and chlorination with free chlorine). The water was passed through an activated carbon filter, and the adsorbate was eluted either with chloroform or with ethanol; eluates were pooled to obtain a one-year representative sample for each type of eluate. Median organic yields from the three sources were 45-78 µg/l for chlorofom and 98-122 µg/l for ethanol extracts. The mice received three injections of one of the extracts in 0.025 ml diluted propylene glycol (1:1 with isotonic saline) on the following dosing schedule: shortly after birth (4-18 h), 0.5 mg; at 10 days of age, 1.0 mg; and at 20 days of age, 3.5 mg/mouse (total dose, 5 mg). Control mice received either diluted propylene glycol or saline by a similar injection schedule. High mortality was observed in the neonatal period in each of the six treated and two control groups; however, more animals died in the chloroform extract-treated groups than in the ethanol extract-treated and vehicle control groups. The total number of surviving animals at week 4 ranged between 43 and 53. Survivors were observed for 78 weeks and were killed at 1.5 years of age. No tumour had developed at the injection sites. The types and numbers of other tumours were similar in the experimental and control groups (Dunham *et al.*, 1967). [The Working Group noted the infrequent injection schedule, the high early mortality, the short duration of the experiment and the lack of an adequate control group.]

(d) Administration with known carcinogens

Groups of 50 Sencar mice [sex unspecified], aged six to nine weeks, were given six subcutaneous injections over two weeks (to give a total dose of 1.5 ml) of water from a US river, disinfected, after settling, coagulation and filtration, by either chlorine (2.0-2.5 mg/l), chloramine (2.0-3.0 mg/l), chlorine dioxide (2.0-3.0 mg/l) or ozone (1.0-3.0 mg/l), then concentrated 100-180 fold using reverse osmosis. Treated but not disinfected water, similarly concentrated [organic material not quantified] served as control water. Equal numbers of mice received isotonic saline by the same treatment schedule. As a positive control, 7.5 µg 7,12-dimethylbenz[*a*]anthracene (DMBA) in 10% Emulphor were administered subcutaneously to 50 mice. Two weeks after the last initiating dose, 25 mice in each group received topical applications of 2.5 µg 12-*O*-tetradecanoylphorbol 13-acetate (TPA) in 0.2 ml acetone three times a week for 18 weeks, and the remaining 25 received 0.2 ml acetone without TPA; all mice were then observed for an additional 28 weeks. The numbers of animals injected with concentrate followed by topical application of TPA that had macroscopic skin tumours at one year were: non-disinfected water condensate, 0/25; chlorine disinfected water condensate, 4/25; chloramine disinfected water condensate, 5/25; chlorine dioxide disinfected water concentrate, 0/25; ozone disinfected water concentrate, 7/25; saline control, 1/25; DMBA positive control, 16/25. Histologically verified skin tumours (papillomas and carcinomas) were observed at the end of the study in 1/25, 4/25, 3/25, 0/25, 4/25, 1/25 and 9/25 in these groups, respectively (Bull *et al.*, 1982). [The Working Group noted that no information was provided on skin tumour frequency in the acetone-treated control groups.]

In two experiments, groups of 60 Sencar mice [sex unspecified], aged six to nine weeks, were given six subcutaneous injections over two weeks (to give a total dose of 1.5 ml) of the same samples described above but which were concentrated 400 fold using reverse osmosis followed by freeze-drying. Equal numbers of animals received isotonic saline by the same treatment schedule. As a positive control, groups of mice [numbers unspecified] received 7.5 or 25 µg/mouse DMBA or 9 mg/mouse urethane. Two weeks after the last initiating dose, 40 mice in each group received topical applications of 1 µg/mouse TPA in 0.2 ml acetone three times a week for 20 weeks; the remaining 20 animals in each group received applications of 0.2 ml acetone without TPA and were observed for an additional 28 weeks. The incidence of skin papillomas observed macroscopically at one year was similar in the treated and saline control group in both experiments (Bull *et al.*, 1982). [The Working Group noted that data were not provided on tumour incidence in the positive control groups or in the acetone-treated groups.]

Groups of 60 male Sencar mice, 8-10 weeks of age, received six subcutaneous injections over two weeks of two types of drinking-water concentrates obtained from five water-treatment plants with different water sources [method of disinfection unspecified]. One sample (ROE) was obtained by reverse osmosis followed by extraction with pentane and dichloromethane. The other sample (XAD) was obtained by passing the aqueous residue of the reverse osmosis extraction through XAD-2 resin and eluting with ethanol. The extracts were administered in 0.1 ml Emulphor to give a total dose of 150 mg/kg bw. A vehicle control group received 0.1 ml Emulphor alone. As a positive control, a total dose of 25 µg/mouse DMBA in 0.1 ml Emulphor was injected in six subcutaneous injections over a two-week period. Two weeks after the last initiating dose, 40 mice from each group received topical applications of 0.1 µg/mouse TPA in 0.1 ml acetone three times a week for 20 weeks; the remaining 20 animals in each group received 0.1 ml acetone only. Surviving animals were sacrified one year after completion of promotion. Skin tumours that persisted for three weeks or more were included in a cumulative count. There was a statistically significant increase in the number of skin papillomas per mouse in one group treated with the ROE sample from one source plus TPA, and in one group treated with the XAD sample from another source plus TPA, as compared with the vehicle control and the TPA control (Robinson *et al.*, 1981).

3.2 Other relevant data

As chlorine dissolves in water to produce hypochlorous acid and hypochlorite, the Working Group summarized experiments that utilized high concentrations of chlorine, hypochlorous acid and hypochlorite in the monograph on hypochlorite. Only studies that were directed specifically to by-products isolated from chlorinated drinking-water (in comparison to non-chlorinated water from the same source) or sought to model processes that are known to occur in chlorinated drinking-water are discussed here.

(*a*) *Experimental systems*

(i) *Toxic effects*

Organic material recovered from chlorinated water by reverse osmosis (reduced in volume by 100 and 400 times) and given to 10 male and 10 female CD-1 mice in each experimental group as drinking-water was compared in a 30-day study with non-disinfected water and water treated with other disinfectants. A significant increase in liver weights was reported in female but not male mice given the high dose of chlorinated water concentrate in comparison with the concentrate from non-chlorinated control water. Male mice had reduced lung weights at both doses and decreased testicular weights at the high dose. No histological examination was performed (Miller *et al.*, 1986).

Organic chemicals from the same waters, recovered on XAD resin, were administered as 0.3 ml of a 1000- or 4000-times concentrate by gavage three times per week for four weeks to 10 CD-1 mice of each sex per group. Treatment had no effect on organ weights in animals of either sex; however, water that had been chlorinated, filtered through granular activated carbon and then rechlorinated increased liver weight in male mice at both doses and decreased lung weight at the high dose. Both doses reduced ovary weight in female mice. No histological finding was reported (Miller *et al.*, 1986).

A combined acid and neutral fraction of organic chemicals recovered on XAD resin, dissolved in dimethyl sulfoxide (corresponding to 100 l of chlorinated drinking-water), was administered intraperitoneally on two consecutive days to 10-day-old Wistar rats and once to 20-day-old rats. This treatment resulted in 50% mortality at 48 h in 10-day-old rats and in 30% mortality in 20-day-old rats; it also induced various alterations in drug metabolizing enzyme activities in liver fractions obtained from surviving animals. The most consistent effect was an increase in the level of hepatic 7-ethoxyresorufin-*O*-deethylase compared to solvent-treated controls (Liimatainen *et al.*, 1988). [The Working Group noted that no non-chlorinated water control was available.]

In a 90-day study in groups of 15 rats, humic acids dissolved at concentrations of 0.1, 0.5 and 1 g/l in distilled water and chlorinated with a 1:1 ratio of chlorine equivalents to organic carbon were given as drinking-water (pH 3). Renal weights were increased relative to body weight at 0.5 and 1 g/l, and there was a small increase in blood urea nitrogen. Haematuria was seen with the 1.0 g/l dose, which appeared to be related to the deposition of crystals [composition unspecified] in the renal pelvis (Condie *et al.*, 1985).

Chlorine reacts very rapidly with purified DNA and RNA (Hayatsu *et al.*, 1971). It chlorinates uracil to produce 5-chlorouracil at low chlorine:carbon ratios and dichlorouracil and ring cleavage at higher ratios (Gould *et al.*, 1984). Relatively stable organic chloramines are formed with cytosine. Purines (modelled by caffeine) are chlorinated to a very small extent, with ring cleavage to a complex array of products (Gould & Hay, 1982). A peroxide of adenosine 5'-monophosphate (AMP) has been shown to form at physiological pH, and this reaction is dependent on the NaOCl:AMP ratio, reaching a plateau when this reatio is less than 1 (Bernofsky *et al.*, 1987).

(ii) *Effects on reproduction and prenatal toxicity*

McKinney *et al.* (1976) reported an increased incidence of dead implantations and of litters with malformed fetuses among Swiss CD-1 mice given chlorinated tap water from Durham, NC, USA. The effect was reported to be seasonal [data not shown]. The control group in this study was given the same water purified by filtration (to reduce organic material and remove microparticulates), demineralization and distillation.

These observations stimulated a series of teratogenicity studies in which Durham city tap water was compared with water purified by the same method as described above. Using much larger group sizes than McKinney *et al.* (1976), Staples *et al.* (1979) found no significant overall difference in the reproductive status of pregnant mice given tap water or purified water. Month-by-month comparisons over a nine-month period (including the critical winter months suspected of being important by McKinney *et al.*) indicated occasionally improved reproductive performance only in the tap water group. Chernoff *et al.* (1979) also found no significant effect on any fetal parameter in CD-1 mice, except for an increased incidence of supernumerary ribs, which the authors considered to be spurious, in the groups given Durham tap water. They considered the possibility that drinking-water quality had changed during the intervening years since the study by McKinney *et al.*

Kavlock *et al.* (1979) evaluated the effects of concentrates of organic materials from the drinking-water of five US cities representative of major sources of raw water. Because the reverse osmosis method used for concentrating these materials does not retain organohalides with a molecular weight of less than 200, an artificially constituted organohalide mixture was also prepared and evaluated. Groups of Swiss mice were given 300, 1000 or 3000 times the anticipated human dose of these materials by gavage on gestation days 7-14. No adverse effect on embryonal or fetal development was observed.

[The Working Group noted that these studies were designed to study the effects of the drinking-water of individual cities but not to investigate the developmental toxicity of chlorinated drinking-water, nor did they include a non-chlorinated water control.]

(iv) *Genetic and related effects*

The results obtained in a variety of short-term tests for samples of chlorinated water have been reviewed (Kraybill, 1980; Loper, 1980; Alink, 1982; Kool *et al.*, 1982a; Nestmann, 1983; Bull, 1985; Degraeve, 1986; Fielding & Horth, 1986; Meier, 1988). Many of the studies were concerned with the mutagenicity of drinking-water and not with the influence of chlorination. As the source of mutagenicity in drinking-water may also be polluted raw water, the role of water chlorination cannot be evaluated unless a comparison is made with an unchlorinated sample. Papers lacking this aspect and those in which no reference is made to the disinfection agent used are not summarized here. In many papers, data were available to allow comparison of unchlorinated and chlorinated waters, and when the authors did not do this, the Working Group drew their own conclusions.

The Working Group also limited themselves to studies of water samples disinfected with chlorine or hypochlorite; studies on water samples disinfected with chlorine dioxide, monochloramine or ozone alone were not considered.

By far the majority of studies were with *Salmonella typhimurium* strains TA98 and TA100.

(1) *Chlorinated water* (Table 7)

Chlorination did not increase the mutagenicity of drinking-water prepared from surface or spring water, as studied in fluctuation tests with *S. typhimurium* strains TA98 and TA100. Mutations were not induced in *S. typhimurium* TA100 when samples of chlorinated water were used to prepare bottom agar for the test plates. [The Working Group noted that volatile substances would be lost if the water were autoclaved.] Chlorination of a tap water sample derived from surface water did not increase the number of micronuclei in *Tradescantia* pollen mother cells. Chromosomal aberrations were induced in *Allium cepa*, however, by a river water sample chlorinated in the laboratory. Cell transformation was not induced by chlorinated tap water in cultured Syrian hamster embryo cells or by finished drinking-water from a surface water source in mouse embryo cells.

(2) *Concentrates of chlorinated water* (Tables 8-10)

The most widely used method for isolating organic material from water samples is adsorption to macroreticular resin (various types of Amberlite XAD) followed by elution with an organic solvent. Liquid-liquid extraction with an organic solvent is also commonly used. The different concentration methods used for mutagenicity studies have been discussed (Forster & Wilson, 1981; Harrington *et al.*, 1983; Maruoka & Yamanaka, 1983; Monarca *et al.*, 1985a,b; Wigilius *et al.*, 1985; Vartiainen *et al.*, 1987a). The methods are more or less selective and do not concentrate all organic materials, e.g., XAD adsorption and liquid-liquid extraction techniques may result in the loss of highly polar compounds. Concentration of an extract invariably means that volatile substances are removed with the solvent. The extent of loss depends upon the solvent used: use of low-boiling-point solvents, e.g., ether and acetone, leads to smaller losses (compounds with boiling-points of about 120°C should be retained unless evaporation is to dryness), while use of high-boiling-point solvents, e.g., ethyl acetate and dimethyl sulfoxide, leads to greater losses.

Chlorination of surface water usually resulted in increased mutagenicity of concentrated samples towards *S. typhimurium*, particularly strains TA100 and TA98 (Tables 8 and 10). In the few studies in which these strains were not used, negative responses were obtained, perhaps because the most sensitive organism

Table 7. Summary of the influence of chlorine disinfection on the genetic and related effects of unconcentrated drinking-water samples in comparison with unchlorinated water

Source of water; disinfection method[a]	Test system	Result		Dose or dose range	Reference
		Without exogenous metabolic system	With exogenous metabolic system		
PROKARYOTES					
Italy; lake; NaOCl, flocculation, RSF, NaOCl	Mutation, *S. typhimurium* TA100 fluctuation test	−	0	5–100%	Monarca *et al.* (1985b)
Italy; river; NaOCl, flocculation, SSF, NaOCl	Mutation, *S. typhimurium* TA100 fluctuation test	−	0	5–100%	Monarca *et al.* (1985b)
Italy; spring water; NaOCl	Mutation, *S. typhimurium* TA100 fluctuation test	−	0	5–100%	Monarca *et al.* (1985b)
USA; chlorinated drinking-water from two supply systems	Mutation, *S. typhimurium* TA100	−	−	1–20 ml/plate	Schwartz *et al.* (1979)
PLANTS					
Macomb (IL, USA); chlorinated tap water from city reservoir	Micronuclei, *Tradescantia* clone 03, pollen mother cells	−	0	Cuttings placed in sample	Ma *et al.* (1985)
Sava river (Zagreb, Yugoslavia); NaOCl 1–10 mg Cl/l in laboratory	Chromosomal aberations, *Allium cepa*	(+)	0	Roots suspended in sample	Al-Sabti & Kurelec (1985)
MAMMALIAN CELLS *IN VITRO*					
Pretoria (South Africa); reclaimed tap water; activated sludge, clarification, Cl$_2$, clarification, alum, sand filtration, Cl$_2$, active carbon, Cl$_2$	Transformation, golden hamster embryo cells, colony morphology	−	0	Media made up from sample	Kfir & Prozesky (1982)
Mississippi river (USA); 115 finished water samples	Transformation, mouse embryo R846–DP-6 cells, growth pattern	−	0	72% of sample in medium	Pelon *et al.* (1980)

[a]RSF, rapid sand filtration; SSF, slow sand filtration

was not used. Inclusion of a metabolic activation system usually resulted in a reduced response or totally abolished it. Mutagenic effects were consistently found in samples of surface water that had a high content of natural organic compounds at the time of the chlorination. Water samples in which the organic content had been reduced before chlorination by water treatment procedures tended to show reduced or no mutagenicity.

Chlorinated ground- and spring water samples (Table 9) were less frequently mutagenic than chlorinated surface water samples (Table 10).

Much of the bacterial mutagenicity of concentrated chlorinated surface water samples is probably due to chlorination of natural constituents, such as humic and fulvic acids. Chlorination of aqueous solutions of fulvic and humic acids resulted in the formation of mutagenic compounds (Meier *et al.*, 1983; Kowbel *et al.*, 1984; Kopfler *et al.*, 1985; Kronberg *et al.*, 1985a,b; Meier *et al.*, 1985; Kowbel *et al.*, 1986; Maruoka, 1986; Meier *et al.*, 1986; Van Duuren *et al.*, 1986; Agarwal & Neton, 1989; Horth, 1989; Pommery *et al.*, 1989). The mutagenicity of chlorinated water samples is not due to the volatile trihalomethanes known to be formed at chlorination; much of the mutagenicity is due to nonvolatile acidic and polar substances. Such compounds require acidic conditions for efficient extraction by non-polar solvents. In several studies, the greatest mutagenic activity was seen when concentration was performed at low pH (e.g., pH 2) (Kool *et al.*, 1981; Van Der Gaag *et al.*, 1982; Kronberg *et al.*, 1985a,b; Vartiainen & Liimatainen, 1986; Ringhand *et al.*, 1987; Fawell & Horth, 1990). A single organic compound, MX, has been shown to be responsible for a significant portion of the bacterial mutagenicity of some concentrated chlorinated surface water samples. This compound is unstable at high pH and in dimethyl sulfoxide (Meier *et al.*, 1987; Kronberg & Vartiainen, 1988; Fielding & Horth, 1988).

Some concentrates of chlorinated tap water prepared from surface waters, groundwater or their mixture induced more sister chromatid exchange in Chinese hamster ovary cells than concentrates of the respective raw waters. In the only study of its kind, concentrates of chlorinated river water that had undergone extensive water treatment procedures did not increase the incidence of *hprt* locus mutations in Chinese hamster V79 cells. Chlorination was associated with an increase in the frequency of micronuclei in Chinese hamster ovary cells exposed to some samples of concentrated chlorinated tap water prepared from surface water and mixed ground- and surface water but not in those exposed to concentrated chlorinated groundwater. Concentrates prepared from chlorinated water from a river and a reservoir induced chromosomal aberrations in Chinese hamster ovary cells.

No studies in mammals *in vivo* were available.

Table 8. Summary of the influence of chlorine disinfection on the genetic and related effects of surface water concentrates in comparison with concentrates of unchlorinated water

Source of water; disinfection method[a]	Concentration and extraction method (concentration factor)[b]	Test system	Result Without exogenous metabolic system	Result With exogenous metabolic system	Dose or dose range[c]	Reference
		PROKARYOTES				
Belgium; rechlorination (0.5 mg/l) of contact water dechlorinated totally with sulfur dioxide	Freeze-drying, methanol	Mutation, *S. typhimurium*, fluctuation test TA100 TA98	+ –	0 0	0.02–0.1 l/ml 0.02–0.1 l/ml	Wilcox & Denny (1985)
UK; chlorinated water water from lowland rivers[d]	Freeze-drying	Mutation, *S. typhimurium*, fluctuation test TA100 TA98	+ 4/5 + 2/5	(+)* +* 2/5	0.00	Fielding & Horth (1988)
UK; chlorinated water water from upland reservoirs[d]	Freeze-drying	Mutation, *S. typhimurium*, fluctuation test TA100 TA98	+ 2/3	(–) (+)	0.00	Fielding & Horth (1988)
Savojärovi (Finland); humic lake water; Cl_2 21 mg/l	XAD 4/8, ethyl acetate	Mutation, *S. typhimurium* TA100 TA98 TA97	+ + +	0 0 0	10–200 ml/pl 10–200 ml/pl 10–200 ml/pl	Backlund et al. (1985)
Mississippi River (USA); lime and alum, CO_2, activated carbon powder, Cl_2 4–8 ppm, alum[d]	XAD–4, acetone, dichloromethane	Mutation, *S. typhimurium* TA100	+	0	0.1–0.6 l/pl	Cheh et al. (1980)
Oise River (France); O_3, storage, coagulation, flocculation, decantation, filtration, O_3, GAC, O_3, Cl_2 0.9 mg/l	XAD–4 and XAD–8, DMSO (700–1000x)	Mutation, *S. typhimurium*[e] TA100 TA98	– –	– –	0.00 0.00	Bourbigot et al. (1983)

Table 8 (contd)

Source of water; disinfection method[a]	Concentration and extraction method (concentration factor)[b]	Test system	Result - Without exogenous metabolic system	Result - With exogenous metabolic system	Dose or dose range[c]	Reference
Mississipi River (USA); lime and alum, CO_2, activated carbon powder, Cl_2 4–8 ppm, Na_2SO_3, alum[d]	XAD-4, acetone, dichloromethane	Mutation, *S. typhimurium* TA100	+	0	0.1–0.6 l/pl	Cheh *et al.* (1980)
Seine River (France); pulsation, RSF, GAC, Cl_2 residual 0.2 mg/l	XAD-2 and XAD-8, CH_2Cl_2 or CH_3OH	Mutation, *S. typhimurium*[e] TA98	−	−	1 ml/pl	Cognet *et al.* (1986, 1987)
Seine River (France); pulsation, RSF, O_3, GAC, Cl_2 residual 0.2 mg/l	XAD-2 and XAD-8, CH_2Cl_2 or CH_3OH	Mutation, *S. typhimurium*[e] TA98	−	−	1 ml/p	Cognet *et al.* (1986, 1987)
Houlle River (France); Cl_2 5 ppm, coagulation, flotation, GAC	XAD-2 and XAD-8, CH_2Cl_2 or CH_3OH	Mutation, *S. typhimurium* TA98	+	0	1–5 l/pl	Cognet *et al.* (1986)
Arno River (Florence, Italy); NaOCl 2.5–7.5 g Cl/m^3, activated carbon, coagulation, flocculation	XAD-2, CH_2Cl_2 and $CHCl_3$	Mutation, *S. typhimurium* TA100 TA1538	+ +	+* +	0.375–10 l/pl 10 l/pl	Dolara *et al.* (1981)
Ottawa (Canada); chlorinated tap water from Ottawa River	XAD-2, hexane:acetone (200 000× stock)	Mutation, *S. typhimurium* TA100 TA98	+} Toxicity +} observed	+ +	0.3–2 mg/pl 0.3–2 mg/pl	Nestmann *et al.* (1979)
Ontario (Canada); chlorinated tap water from a river	XAD-2, hexane:acetone	Mutation, *S. typhimurium* TA100	+	−	92–756 µg/pl DD: (2.3 l eq/ml)	Douglas *et al.* (1986)
Ontario (Canada); chlorinated tap water from a river	XAD-2, hexane:acetone	Mutation, *S. typhimurium* TA98	+	0	DD: (1.8 l/ml)	Douglas *et al.* (1986)

Table 8 (contd)

Source of water, disinfection method[a]	Concentration and extraction method (concentration factor)[b]	Test system	Result Without exogenous metabolic system	Result With exogenous metabolic system	Dose or dose range[c]	Reference
Ontario (Canada); chlorinated tap water from mixed ground- and surface water	XAD-2, hexane:acetone	Mutation, S. typhimurium TA98	+	0	DD: (1.5 l/ml)	Douglas et al. (1986)
Ontario (Canada); chlorinated tap water from two lakes	XAD-2, hexane:acetone	Mutation, S. typhimurium TA100	+	0	DD: (2.2–4.1 l/ml)	Douglas et al. (1986)
Calumet River (Indiana, USA); Cl_2 10 mg/l 2 h, 0.2–1 mg/l residual Cl_2	XAD-2, diethyl ether, CH_3OH	Mutation, S. typhimurium TA1538	+	•	0.06–0.5 l/pl	Flanagan & Allen (1981)
Calumet River (Indiana, USA); Cl_2 10 mg/l 2 h, alum and polymer addition, flocculation, sedimentation	XAD-2, diethyl ether, CH_3OH	Mutation, S. typhimurium TA1538	+	+	0.06–0.5 l/pl	Flanagan & Allen (1981)
Calumet River (Indiana, USA); alum and polymer addition, flocculation, sedimentation, Cl_2 10 mg/l 2 h	XAD-2, diethyl ether, CH_3OH	Mutation, S. typhimurium TA1538	+	•	0.06–0.5 l/pl	Flanagan & Allen (1981)
Fox River (Illinois, USA); Cl_2 10 mg/l 2 h, residual Cl 0.2–1 mg/l	XAD-2, diethyl ether, CH_3OH	Mutation, S. typhimurium TA1538	–	0	0.00	Flanagan & Allen (1981)
Fox River (Illinois, USA); Cl_2 10 mg/l 2 h, alum and polymer addition, flocculation, sedimentation	XAD-2, diethyl ether, CH_3OH	Mutation, S. typhimurium TA1538	–	0	0.00	Flanagan & Allen (1981)
Fox River (Illinois, USA); alum and polymer addition, flocculation, sedimentation (?), Cl_2 10 mg/l 2 h	XAD-2, diethyl ether, CH_3OH	Mutation, S. typhimurium TA1538	–	0	0.00	Flanagan & Allen (1981)

Table 8 (contd)

Source of water; disinfection method[a]	Concentration and extraction method (concentration factor)[b]	Test system	Result		Dose or dose range[c]	Reference
			Without exogenous metabolic system	With exogenous metabolic system		
Parys (South Africa); water from Vaal River; alum flocculation, sedimentation, sand filtration, chlorination, residual Cl_2 0.4–0.5 mg/l	Liquid-liquid extraction, Cl_2CH_2 (10 000x), neutral	Mutation, *S. typhimurium* TA100 TA98 TA1535	+ + –	– (+)* –	112–190 µg/pl 112–190 µg/pl 112–190 µg/pl (doses equal to 1 l/pl used in comparison)	Grabow et al. (1981)
Parys (South Africa); water from Vaal River; alum flocculation, sedimentation, sand filtration, chlorination, residual Cl_2 0.4–0.5 mg/l	Liquid-liquid extraction, Cl_2CH_2 (10 000x), acidic	Mutation, *S. typhimurium* TA100 TA98 TA1535	– (+) –	– – –	36–125 µg/pl 36–125 µg/pl 36–125 µg/pl (doses equal to 1 l/pl used in comparison)	Grabow et al. (1981)
Parys (South Africa); water from Vaal River; alum flocculation, sedimentation, sand filtration, chlorination, residual Cl_2 0.4–0.5 mg/l	Liquid-liquid extraction, Cl_2CH_2 (10 000x), basic	Mutation, *S. typhimurium* TA100 TA98 TA1535	– – –	– – –	<1–25 µg/pl <1–25 µg/pl <1–25 µg/pl (doses equal to 1 l/pl used in comparison)	Grabow et al. (1981)
Des Moines (Iowa, USA); chlorinated and fluoridated water from a river and an infiltration gallery	XAD–4, diethyl ether (>200 000x)	Mutation, *S. typhimurium* TA100 TA98	+ +	+* (+)	0.00 0.00	Grimm-Kibalo et al. (1981)
Des Moines (Iowa, USA); chlorinated and fluoridated water from a river and an infiltration gallery	XAD–4, ethanol after diethyl ether (>200 000x)	Mutation, *S. typhimurium* TA100 TA98	+ (+)	+* (+)	0.00 0.00	Grimm-Kibalo et al. (1981)

Table 8 (contd)

Source of water; disinfection method[a]	Concentration and extraction method (concentration factor)[b]	Test system	Result Without exogenous metabolic system	Result With exogenous metabolic system	Dose or dose range[c]	Reference
Como (Italy) outlet; mixed from ground water and Lake Como; conventional treatment and NaOCl	XAD-2 and XAD-7, acetone	Mutation, *S. typhimurium* TA100 TA98	+ +	+ +	0.5–5.0 l/pl 0.5–5.0 l/pl	Galassi *et al.* (1989)
Netherlands; Rhine River; filtration, ferric chloride, filtration, pH adjustment, Cl$_2$ (NaOCl) 5–15.7 mg/l, pH 6.2	XAD-4/8, DMSO (1500x)	Mutation, *S. typhimurium* TA100 TA98	(+) (+)	0 0	0.25–0.5 ml/pl 0.25–0.5 ml/pl	de Greef *et al.* (1980)
Netherlands; dune infiltrated river water after transport chlorination	XAD-4/8, DMSO (8000x)	Mutation, *S. typhimurium* TA98	+	–	1.5 l/pl	Kool *et al.* (1981)
Netherlands; river water; transport chlorination; end of transport system	XAD-4/8, DMSO (8000x)	Mutation, *S. typhimurium* TA98	+	+	1.5 l/pl	Kool *et al.* (1981)
Netherlands; river water; transport chlorination, RSF, filtration	XAD-4/8, DMSO (8000x)	Mutation, *S. typhimurium* TA98	+	+	1.5 l/pl	Kool *et al.* (1981)
Netherlands; dune infiltrated river water; transport chlorination, RSF	XAD-4/8, DMSO (8000x)	Mutation, *S. typhimurium* TA98	–	–	1.5 l/pl	Kool *et al.* (1981)
Netherlands; Rhine River; 5–15 mg Cl$_2$/l	XAD-4/8, DMSO (1800x)	Mutation, *S. typhimurium* TA100 TA98	– +	0 0	0.25–0.5 ml/pl 0.25–0.5 ml/pl	Kool *et al.* (1981)
Netherlands; Meuse River; breakpoint chlorination	XAD-4/8, DMSO (7000x)	Mutation, *S. typhimurium* TA100 TA98	+ + + +	– +	1.5 l/pl 1.5 l/pl	Kool *et al.* (1982b)

Table 8 (contd)

Source of water; disinfection method[a]	Concentration and extraction method (concentration factor)[b]	Test system	Result without exogenous metabolic system	Result with exogenous metabolic system	Dose or dose range[c]	Reference
Netherlands; Meuse River; breakpoint chlorination and activated carbon	XAD-4/8, DMSO (7000x)	Mutation, S. typhimurium[e] TA100 TA98	+ –	– –	1.5 l/pl 1.5 l/pl	Kool et al. (1982b)
Netherlands; Meuse River; post-chlorination	XAD-4/8, DMSO (7000x)	Mutation, S. typhimurium[e] TA100 TA98	+ + + +	– –	1.5 l/pl 1.5 l/pl	Kool et al. (1982b)
Netherlands; Meuse River; transport chlorination	XAD-4/8, DMSO (8000x)	Mutation, S. typhimurium[e] TA98	+	+	2 l/pl	Kool et al. (1982b)
Netherlands; Meuse River; transport chlorination, dune infiltration or activated carbon with RSF and SSF	XAD-4/8, DMSO (8000x)	Mutation, S. typhimurium[e] TA98	–	–	2 l/pl	Kool et al. (1982b)
Netherlands; Rhine River; chlorination	XAD-4/8, acetone, XAD-4/8 (45 000x), TLC fraction	Mutation, S. typhimurium TA98	+	–	0.00	Kool et al. (1982b)
Netherlands; Meuse River; breakpoint chlorination, O_3, activated carbon, post-chlorination	XAD-4/8, DMSO (7000x)	Mutation, S. typhimurium[e] TA100 TA98	+ –	– –	3 l/p 3 l/p	Kool et al. (1982b)
Netherlands; Meuse or Rhine River; prechlorination 5 mg Cl_2/l	XAD-4/8, DMSO (2000–4000x)	Mutation, S. typhimurium TA98	+	+*	0.25–0.5 ml/pl	Zoeteman et al. (1982)
Nieuwegein (Netherlands); Rhine River; dune recharge, Cl_2 0.2 mg/l after 20 min	XAD-4, ethanol, cyclohexane/ethanol, pH 7 (4000x)	Mutation, S. typhimurium TA100	+	+*	1–3 l/pl	Van Der Gaag et al. (1982)

Table 8 (contd)

Source of water; disinfection method[a]	Concentration and extraction method (concentration factor)[b]	Test system	Result Without exogenous metabolic system	Result With exogenous metabolic system	Dose or dose range[c]	Reference
Nieuwegein (Netherlands); Rhine River; dune recharge, Cl_2 0.2 mg/l after 20 min	XAD-4, ethanol, cyclohexane/ethanol, pH 2 (4000x)	Mutation, *S. typhimurium* TA100	+	+*	1–3 l/pl	Van Der Gaag et al. (1982)
Nieuwegein (Netherlands); Rhine River; active carbon filtration, Cl_2 0.2 mg/l after 20 min	XAD-4, ethanol, cyclohexane/ethanol, pH 7 (4000x)	Mutation, *S. typhimurium* TA100	–	–	1–4 l/pl	Van Der Gaag et al. (1982)
Nieuwegein (Netherlands); Rhine River; active carbon filtration, Cl_2 0.2 mg/l after 20 min	XAD-4, ethanol, cyclohexane/ethanol, pH 2 (4000x)	Mutation, *S. typhimurium* TA100	+	–	1.4 l/pl	Van Der Gaag et al. (1982)
Netherlands; Meuse River, Cl_2 1.5 mg/l	XAD-4/8, DMSO or acetone (7000x)	Mutation, *S. typhimurium* TA100 TA98 TA100NR– TA98NR–	+ + – +	(–) + – –	0.1–0.2 ml 0.1–0.2 ml 0.1–0.2 ml 0.1–0.2 ml	Kool et al. (1985b)
Netherlands; Meuse River, Cl_2 5–15 mg/l	XAD-4/8, DMSO (4000x)	Mutation, *S. typhimurium* TA100 TA98	– +	– +*	2 l/pl 2 l/pl	Kool et al. (1985c)
Netherlands; Meuse River, transport chlorination Cl_2 1–2 mg/l	XAD-4/8, DMSO	Mutation, *S. typhimurium* TA98	+	–	1.5 l/pl	Kool & Hrubec (1986); Kool et al. (1985b)
Netherlands; Meuse River, prechlorination Cl_2 1.8 mg/l	XAD-4/8, DMSO (7000x)	Mutation, *S. typhimurium*[e] TA100 TA98	+ +	+* +*	3.5 l/pl 3.5 l/pl	Kool & Hrubec (1986); Kool & van Kreijl (1984); Kool et al. (1985b)

Table 8 (contd)

Source of water; disinfection method[a]	Concentration and extraction method (concentration factor)[b]	Test system	Result without exogenous metabolic system	Result with exogenous metabolic system	Dose or dose range[c]	Reference
Netherlands; Meuse or Rhine River; postchlorination 0.15 mg Cl_2 after 20 min	XAD–4/8, DMSO (7000x)	Mutation, S. typhimurium[e] TA100 TA98	+ +	+ +*	3.5 l/pl 3.5 l/pl	Kool & Hrubec (1986); Kool & van Kreijl (1984); Kool et al. (1985b)
Netherlands; Meuse River; Cl_2 1–5 mg/l	XAD–4/8, DMSO (3500x)	Mutation, S. typhimurium TA100 TA98	+ +	0 +*	1.7 l/pl 1.7 l/pl	Kool & Hrubec (1986)
Netherlands; Rhine River; Cl_2 1–5 mg/l	XAD–4/8, DMSO (3500x)	Mutation, S. typhimurium TA100 TA98	– (+)	0 (+)	0.9 l/pl 0.9 l/pl	Kool & Hrubec (1986)
Netherlands; surface water from one city; Cl_2 1 mg/l	XAD–4/8, DMSO (7000x), neutral fraction	Mutation, S. typhimurium TA100 TA98	– +	– +	3.5 l/pl 3.5 l/pl	Kool & Hrubec (1986)
Netherlands; surface water from one city; Cl_2 1 mg/l	XAD–4/8, DMSO (7000x), acidic fraction	Mutation, S. typhimurium TA100 TA98	– –	– +	3.5 l/pl 3.5 l/pl	Kool & Hrubec (1986)
Netherlands; surface water from one city; Cl_2 1 mg/l	XAD–4/8, DMSO (7000x), neutral fraction	Mutation, S. typhimurium TA100 TA98	+ +	– +	3.5 l/pl 3.5 l/pl	Kool et al. (1985c)
Netherlands; surface water from one city; Cl_2 1 mg/l	XAD–4/8, DMSO (7000x), acidic fraction	Mutation, S. typhimurium TA100 TA98	– +	– –	3.5 l/pl 3.5 l/pl	Kool et al. (1985c)
Cincinnati (Ohio, USA); Ohio River; presettling with aluminium sulfate, Cl_2, lime, F, ferric sulfate, coagulation, flocculation, sedimentation, RSF	XAD–2, hexane-acetone (10 000x)	Mutation, S. typhimurium TA100 TA98	+ +	+* +*	0.25–1.5 l/pl 0.25–1.5 l/pl	Loper et al. (1985)

Table 8 (contd)

Source of water, disinfection method[a]	Concentration and extraction method (concentration factor)[b]	Test system	Result		Dose or dose range[c]	Reference
			Without exogenous metabolic system	With exogenous metabolic system		
Cincinnati (Ohio, USA); tap water; presettling with aluminium sulfate, Cl_2, lime, F, ferric sulfate, coagulation, flocculation, sedimentation, RSF	XAD-2, hexane-acetone (10 000x)	Mutation, *S. typhimurium* TA100 TA98	+ +	(+)* (+)*	0.25–1.5 l/pl 0.25–1.5 l/pl	Loper *et al.* (1985)
Cincinnati (Ohio, USA); Ohio River; presettling with aluminium sulfate, Cl_2, lime, F, ferric sulfate, coagulation, flocculation, sedimentation, RSF, GAC	XAD-2, hexane-acetone (10 000x)	Mutation, *S. typhimurium* TA100 TA98	– –	0 0	0.25–1.5 l/pl 0.25–1.5 l/pl	Loper *et al.* (1985)
Cincinnati (Ohio, USA); Ohio River; presettling with aluminium sulfate, Cl_2, lime, F, ferric sulfate, coagulation, flocculation, sedimentation, RSF, GAC, Cl_2 2.6 mg/l	XAD-2, hexane-acetone (1000x)	Mutation, *S. typhimurium* TA100 TA98	– –	– –	0.25–3 l/pl 0.25–2 l/pl	Loper *et al.* (1985)
Jefferson Parish (Louisiana, USA); pilot plant; Cl_2	XAD-2 and XAD-8, acetone (4000x), pH 2	Mutation, *S. typhimurium* TA100 TA98	+ +	0 0	0.00 0.00	Miller *et al.* (1986)
Jefferson Parish (Louisiana, USA); pilot plant; Cl_2, fresh GAC	XAD-2 and XAD-8, acetone (4000x)	Mutation, *S. typhimurium* TA100 TA98	– –	0 0	0.00 0.00	Miller *et al.* (1986)
Jefferson Parish (Louisiana, USA); pilot plant, Cl_2, GAC after 14 months	XAD-2 and XAD-8, acetone (4000x)	Mutation, *S. typhimurium* TA100 TA98	– +	0 0	0.00 0.00	Miller *et al.* (1986)
Jefferson Parish (Louisiana, USA); pilot plant; Cl_2, fresh GAC, Cl_2	XAD-2 and XAD-8, acetone (4000x)	Mutation, *S. typhimurium* TA100 TA98	– –	0 0	0.00 0.00	Miller *et al.* (1986)

Table 8 (contd)

Source of water; disinfection method[a]	Concentration and extraction method (concentration factor)[b]	Test system	Result Without exogenous metabolic system	Result With exogenous metabolic system	Dose or dose range[c]	Reference
Jefferson Parish (Louisiana, USA); pilot plant; Cl_2, GAC after 6 months, Cl_2	XAD–2 and XAD–8, acetone (4000x)	Mutation, S. typhimurium TA100 TA98	+ +	0 0	0.00 0.00	Miller et al. (1986)
Jefferson Parish (Lopuisiana, USA); pilot plant; Cl_2	Reverse osmosis (400x)	Mutation, S. typhimurium TA100 TA98	– –	– –	0.025–1 ml/pl 0.025–1 ml/pl	Miller et al. (1986)
Jefferson Parish (Louisiana, USA); pilot plant, Mississippi River; clarification, settling, F, sand filtration, Cl_2 0.2–7.5 ppm	XAD–2 and XAD–4, acetone pH 2	Mutation, S. typhimurium TA100	+	0	0.1–1.6 l/pl	Ringhand et al. (1987)
Jefferson Parish (Louisiana, USA); pilot plant; Mississipi River; clarification, settling, F, sand filtration, Cl_2 0.2–7.5 ppm	XAD–2 and XAD–4, acetone pH 8	Mutation, S. typhimurium TA100	(+)	0	0.1–1.6 l/pl	Ringhand et al. (1987)
Cincinnati (Ohio, USA); pilot plant, Ohio River; clarification, coagulation, flocculation, sedimentation, RSF, Cl_2 0.2–7.5 ppm	XAD–2 and XAD–4, acetone pH 2	Mutation, S. typhimurium TA100	+	0	0.1–1.6 l/pl	Ringhand et al. (1987)
Cincinnati (Ohio, USA); pilot plant, Ohio River; clarification, sedimentation, coagulation, flocculation, sedimentation, RSF, Cl_2 0.2–7.5 ppm	XAD–2 and XAD–4, acetone pH 8	Mutation, S. typhimurium TA100	(+)	0	0.1–1.6 l/pl	Ringhand et al. (1987)
USA; chlorinated drinking-water	Polyurethane foam column, acetone, benzene (30 000x)	Mutation, S. typhimurium TA98	–	–	0.1– 1 l/pl	Schwartz et al. (1979)

Table 8 (contd)

Source of water; disinfection method[a]	Concentration and extraction method (concentration factor)[b]	Test system	Result		Dose or dose range[c]	Reference
			Without exogenous metabolic system	With exogenous metabolic system		
Italy; lake; Cl_2, flocculation, RSF, Cl_2	Liquid-liquid extraction, CH_2Cl_2, neutral	Mutation, *S. typhimurium* TA100 TA98	– –	– –	0.00 0.00	Monarca *et al.* (1985a)
Italy; lake; Cl_2, flocculation, RSF, Cl_2	Liquid-liquid extraction, CH_2Cl_2, acidic	Mutation, *S. typhimurium* TA100 TA98	+ –	– –	0.00 0.00	Monarca *et al.* (1985a)
Italy; lake; Cl_2, flocculation, RSF, Cl_2	Liquid-liquid extraction, CH_2Cl_2, basic	Mutation, *S. typhimurium* TA100 TA98	+ –	– –	0.00 0.00	Monarca *et al.* (1985a)
Italy; river, Cl_2, flocculation, SSF, Cl_2	Liquid-liquid extraction, CH_2Cl_2, neutral	Mutation, *S. typhimurium* TA100 TA98	– +	– –	0.00 0.00	Monarca *et al.* (1985a)
Italy; river, Cl_2, flocculation, SSF, Cl_2	Liquid-liquid extraction, CH_2Cl_2, acidic	Mutation, *S. typhimurium* TA100 TA98	+ +	– –	0.00 0.00	Monarca *et al.* (1985a)
Italy; river, Cl_2, flocculation, SSF, Cl_2	Liquid-liquid extraction, CH_2Cl_2, basic	Mutation, *S. typhimurium* TA100 TA98	– –	– –	0.00 0.00	Monarca *et al.* (1985a)
Italy; lake; Cl_2, flocculation, RSF, Cl_2	XAD-2, acetone	Mutation, *S. typhimurium* TA100 TA98	– –	– –	0.00 0.00	Monarca *et al.* (1985a)
Italy; river, Cl_2, flocculation, SSF, Cl_2	XAD-2, acetone	Mutation, *S. typhimurium* TA100 TA98	– –	– –	0.00 0.00	Monarca *et al.* (1985a)
Italy; lake; NaOCl, flocculation, RSF, NaOCl	XAD-2, acetone	Mutation, *S. typhimurium*, fluctuation test TA100	–	0	0.1–1 l/test	Monarca *et al.* (1985b)

Table 8 (contd)

Source of water; disinfection method[a]	Concentration and extraction method (concentration factor)[b]	Test system	Result		Dose or dose range[c]	Reference
			Without exogenous metabolic system	With exogenous metabolic system		
Italy; river; NaOCl, flocculation, SSF, NaOCl	XAD-2, acetone	Mutation, *S. typhimurium*[e] fluctuation test TA100	–	0	0.1–1 l/test	Monarca *et al.* (1985b)
Italy; lake; NaOCl, flocculation, RSF, NaOCl	Liquid-liquid extraction, CH_2Cl_2, neutral	Mutation, *S. typhimurium* fluctuation test TA100	–	0	0.1–1 l/test	Monarca *et al.* (1985b)
Italy; lake; NaOCl, flocculation, RSF, NaOCl	Liquid-liquid extraction, CH_2Cl_2, acidic	Mutation, *S. typhimurium* fluctuation test TA100	+	0	0.1–1 l/test	Monarca *et al.* (1985b)
Italy; lake; NaOCl, flocculation, RSF, NaOCl	Liquid-liquid extraction, CH_2Cl_2, basic	Mutation, *S. typhimurium* fluctuation test TA100	–	0	0.1–1 l/test	Monarca *et al.* (1985b)
Italy; river; NaOCl, flocculation, SSF, NaOCl	Liquid-liquid extraction, CH_2Cl_2, neutral	Mutation, *S. typhimurium* fluctuation test TA100	–	0	0.1–1 l/test	Monarca *et al.* (1985b)
Italy; river; NaOCl, flocculation, SSF, NaOCl	Liquid-liquid extraction, CH_2Cl_2, acidic	Mutation, *S. typhimurium* fluctuation test TA100	+	0	0.1–1 l/test	Monarca *et al.* (1985b)
Italy; river; NaOCl, flocculation, SSF, NaOCl	Liquid-liquid extraction, CH_2Cl_2, basic	Mutation, *S. typhimurium* fluctuation test TA100	–	0	0.1–1 l/test	Monarca *et al.* (1985b)
Italy; surface water; NaOCl, flocculation, sand filtration, NaOCl	Sep-Pak, methanol	Mutation, *S. typhimurium* fluctuation test TA100	+	0	from 0.1 l/test	Monarca *et al.* (1985b)

Table 8 (contd)

Source of water; disinfection method[a]	Concentration and extraction method (concentration factor)[b]	Test system	Result Without exogenous metabolic system	Result With exogenous metabolic system	Dose or dose range[c]	Reference
Kuopio, Finland; Kallavesi Lake; Ca(OH)$_2$, Al$_2$(SO$_4$)$_3$, Cl$_2$ 1 mg/l, CO$_2$, flotation, sand filtration, Ca(OH)$_2$, Cl$_2$ 1 mg/l, F	Liquid-liquid extraction, diethyl ether	Mutation, *S. typhimurium* TA100 TA98	+ +	+* +*	3–33 ml/pl 4–35 ml/pl	Vartiainen & Liimatainen (1986)
Kuopio, Finland; Kallavesi Lake; Ca(OH)$_2$, Al$_2$(SO$_4$)$_3$, Cl$_2$ 1 mg/l, CO$_2$, flotation, sand filtration, Ca(OH)$_2$, Cl$_2$ 1 mg/l, F	Liquid-liquid extraction, CH$_2$Cl$_2$	Mutation, *S. typhimurium* TA100 TA98	+ +	0 0	100–300 ml/pl 100–300 ml/pl	Vartiainen & Liimatainen (1986)
Kuopio, Finland; Kallavesi Lake; Ca(OH)$_2$, Al$_2$(SO$_4$)$_3$, Cl$_2$ 1 mg/l, CO$_2$, flotation, sand filtration, Ca(OH)$_2$, Cl$_2$ 1 mg/l, F	XAD 8, ethyl acetate	Mutation, *S. typhimurium* TA100 TA98	+ +	0 0	4–40 ml/pl 4–40 ml/pl	Vartiainen & Liimatainen (1986)
Varkaus, Finland; lake; chlorinated drinking-water	Liquid-liquid extraction, CH$_2$Cl$_2$	Mutation, *S. typhimurium* TA100 TA98	+ +	0 0	0.00 0.00	Vartiainen & Liimatainen (1986)
Kuopio, Finland; Kallavesi Lake; Ca(OH)$_2$, Cl$_2$ 0.7 g/m^3, Al$_2$(SO$_4$)$_3$, CO$_2$, mixing, flotation or sedimentation, sand filtration, Cl$_2$ 1.2 g/m^3, F, Ca(OH)$_2$	XAD 8, ethyl acetate, acidic/neutral	Mutation, *S. typhimurium* TA100 TA98 TA97	+ + +	0 0 0	0.00 0.00 0.00	Vartiainen et al. (1987b)
Kuopio, Finland; Kallavesi Lake; Ca(OH)$_2$, Cl$_2$ 0.7 g/m^3, Al$_2$(SO$_4$)$_3$, CO$_2$, mixing, flotation or sedimentation, sand filtration, Cl$_2$ 1.2 g/m^3, F, Ca(OH)$_2$	XAD 8, ethyl acetate, basic	Mutation, *S. typhimurium* TA100 TA98 TA97	– – –	0 0 0	0.00 0.00 0.00	Vartiainen et al. (1987b)

Table 8 (contd)

Source of water; disinfection method[a]	Concentration and extraction method (concentration factor)[b]	Test system	Result		Dose or dose range[c]	Reference
			Without exogenous metabolic system	With exogenous metabolic system		
Kuopio, Finland; artificially recharged water from Kallavesi Lake; aeration, Ca(OH)$_2$, ClO$^-$ 1 mg/l, mixing, Al$_2$(SO$_4$)$_3$, flocculation, sedimentation, sand filtration, postchlorination 0.5 mg/l	XAD 8, ethyl acetate, acidic/neutral	Mutation, *S. typhimurium* TA100 TA98 TA97	+ + +	0 0 0	0.00 0.00 0.00	Vartiainen *et al.* (1987b)
Kuopio, Finland; artificially recharged water from Kallavesi Lake; aeration, Ca(OH)$_2$, ClO$^-$ 1 mg/l, mixing, Al$_2$(SO$_4$)$_3$, flocculation, sedimentation, sand filtration, postchlorination 0.5 mg/l	XAD 8, ethyl acetate, basic	Mutation, *S. typhimurium* TA100 TA98 TA97	− + +	0 0 0	0.00 0.00 0.00	Vartiainen *et al.* (1987b)
Kuopio, Finland; artificially recharged water from Kallavesi Lake; aeration, Ca(OH)$_2$, ClO$^-$ 1 mg/l, mixing, Al$_2$(SO$_4$)$_3$, flocculation, sedimentation, sand filtration	XAD 8, ethyl acetate, acidic/neutral	Mutation, *S. typhimurium* TA100 TA98 TA97	+ + +	0 0 0	0.00 0.00 0.00	Vartiainen *et al.* (1987b)
Kuopio, Finland; artificially recharged water from Kallavesi Lake; aeration, Ca(OH)$_2$, ClO$^-$ 1 mg/l, mixing, Al$_2$(SO$_4$)$_3$, flocculation, sedimentation, sand filtration	XAD 8, ethyl acetate, basic	Mutation, *S. typhimurium* TA100 TA98 TA97	− + +	0 0 0	0.00 0.00 0.00	Vartiainen *et al.* (1987b)
Kuopio, Finland; artificially recharged water from Kallavesi Lake; aeration, Ca(OH)$_2$, ClO$^-$ 1 mg/l, KMnO$_4$, mixing, Al$_2$(SO$_4$)$_3$, flocculation, sedimentation, sand filtration	XAD 8, ethyl acetate, acidic/neutral	Mutation, *S. typhimurium* TA100 TA98 TA97	+ + + + −	0 0 0	0.00 0.00 0.00	Vartiainen *et al.* (1987b)

Table 8 (contd)

Source of water; disinfection method[a]	Concentration and extraction method (concentration factor)[b]	Test system	Result Without exogenous metabolic system	Result With exogenous metabolic system	Dose or dose range[c]	Reference
Kuopio, Finland; artificially recharged water from Kallavesi Lake; aeration, Ca(OH)$_2$, ClO$^-$ 1 mg/l, KMnO$_4$, mixing, Al$_2$(SO$_4$)$_3$, flocculation, sedimentation, sand filtration	XAD 8, ethyl acetate, basic	Mutation, *S. typhimurium* TA100 TA98 TA97	− − −	0 0 0	0.00 0.00 0.00	Vartiainen *et al.* (1987b)
Kuopio, Finland; artificially recharged water from Kallavesi Lake; aeration, ClO$^-$ 2 mg/l	XAD-8, ethyl acetate, acidic/neutral	Mutation, *S. typhimurium* TA100 TA98 TA97	+ + +	0 0 0	0.00 0.00 0.00	Vartiainen *et al.* (1987b)
Kuopio, Finland; artificially recharged water from Kallavesi Lake; aeration, ClO$^-$ 2 mg/l	XAD 8, ethyl acetate, basic	Mutation, *S. typhimurium* TA100 TA98 TA97	− + −	0 0 0	0.00 0.00 0.00	Vartiainen *et al.* (1987b)
Finland; nine artificially recharged waters; alum coagulation, clarification, sand filtration, pH adjustment, Cl$_2$ 1 ± 0.9 mg/l	XAD 8 at pH 2, ethyl acetate	Mutation, *S. typhimurium*[e] TA100 TA98 TA97	+ (+) −	0 0 0	0.00 0.00 0.00	Vartiainen *et al.* (1988)
Finland; 14 surface waters; filtration, Cl$_2$ 0.6 ± 0.6 mg/l	XAD 8 at pH 2, ethyl acetate	Mutation, *S. typhimurium*[e] TA100 TA98 TA97	+ + +	0 0 0	0.00 0.00 0.00	Vartiainen *et al.* (1988)
Finland; 22 surface waters; alum coagulation with or without Fe$_2$(SO$_4$)$_3$, clarification, sand filtration, pH adjustment, Cl$_2$ 1.3 ± 0.9 mg/l	XAD 8 at pH 2, ethyl acetate	Mutation, *S. typhimurium*[e] TA100 TA98 TA97	+ + +	0 0 0	0.00 0.00 0.00	Vartiainen *et al.* (1988)

Table 8 (contd)

Source of water; disinfection method[a]	Concentration and extraction method (concentration factor)[b]	Test system	Result Without exogenous metabolic system	Result With exogenous metabolic system	Dose or dose range[c]	Reference
Finland; 22 surface waters; Cl_2 1.7 ± 1.2 mg/l, alum coagulation with or without $Fe_2(SO_4)_3$, clarification, sand filtration, pH adjustment, Cl_2 1 ± 0.5 mg/l	XAD 8 at pH 2, ethyl acetate	Mutation, *S. typhimurium*[e] TA100 TA98 TA97	+ + −	0 0 0	0.00 0.00 0.00	Vartiainen et al. (1988)
Finland; three surface waters; $KMnO_4$, alum coagulation, clarification, sand filtration, pH adjustment, Cl_2 1.4 ± 0.3 mg/l	XAD 8 at pH 2, ethyl acetate	Mutation, *S. typhimurium*[e] TA100 TA98 TA97	+ (+) +	0 0 0	0.00 0.00 0.00	Vartiainen et al. (1988)
Taipei (Taiwan); chlorinated river water; total Cl_2 before XAD 1.2–13.4 ppm	XAD-2, acetone, pH 7, 6.9 or 6	Mutation, *S. typhimurium* TA100 TA98	+ −	− −	0.00 0.00	Wei et al. (1984)
Taipei (Taiwan); chlorinated river water; total Cl_2 before XAD 36 ppm	XAD-2, acetone, pH 5.2	Mutation, *S. typhimurium* TA100 TA98	+ −	+* −	0.25–1 l/pl 0.00	Wei et al. (1984)
Taipei (Taiwan); chlorinated river water; total Cl_2 0.1–13.3 ppm, boiling before or after Cl_2	XAD-2, acetone, pH 6.5–8.5	Mutation, *S. typhimurium* TA100 TA98	− −	− −	0.00 0.00	Wei et al. (1984)
Como (Italy) outlet; mixed from groundwater and Lake Como; conventional treatment, NaOCl	XAD-2 and XAD-7, acetone	Gene conversion, *Saccharomyces cerevisiae* 6117 cyh2 locus	+	0	0.2–5.0 l/ml	Galassi et al. (1989)

Table 8 (contd)

Source of water; disinfection method[a]	Concentration and extraction method (concentration factor)[b]	Test system	Result Without exogenous metabolic system	Result With exogenous metabolic system	Dose or dose range[c]	Reference
		MAMMALIAN CELLS IN VITRO				
Oise River (France), O_3, storage, coagulation, flocculation, decantation, filtration, O_3, GAC, O_3, Cl_2 0.9 mg/l	XAD-4 and XAD-8, DMSO	Mutation, Chinese hamster V79 cells[f], hprt resistance including 'initiator and promoter activity'	–	0	0.00	Bourbigot et al. (1983)
Ontario (Canada); river; chlorinated tap water	XAD-2, hexane:acetone	Sister chromatid exchange, Chinese hamster CHO cells	+	(–)	DD: 1.2 l/ml	Douglas et al. (1986)
Ontario (Canada); river; chlorinated tap water	XAD-2, hexane:acetone	Sister chromatid exchange, Chinese hamster CHO cells	–	0	DD: 0.8 l/ml	Douglas et al. (1986)
Ontario (Canada); lake; chlorinated tap water	XAD-2, hexane:acetone	Sister chromatid exchange, Chinese hamster CHO cells	–	0	DD: 2.0 l/ml	Douglas et al. (1986)
Ontario (Canada); lake; chlorinated tap water	XAD-2, hexane:acetone	Sister chromatid exchange, Chinese hamster CHO cells	(+)	0	DD: 0.9 l/ml	Douglas et al. (1986)
Ontario (Canada); mixed surface and groundwater; chlorinated tap water	XAD-2, hexane:acetone	Sister chromatid exchange, Chinese hamster CHO cells	+	0	DD: 1 l/ml	Douglas et al. (1986)
UK; lowland river; chlorinated	XAD-2, acetone (10 000x)	Chromosomal aberrations, Chinese hamster CHO cells	+	0	0.5–2 l/ml	Wilcox & Williamson (1986)
UK; upland reservoir; chlorinated	XAD-2, acetone (10 000x)	Chromosomal aberrations, Chinese hamster CHO cells	+	0	1–4 l/ml	Wilcox & Williamson (1986)

Table 8 (contd)

Source of water; disinfection method[a]	Concentration and extraction method (concentration factor)[b]	Test system	Result without exogenous metabolic system	Result with exogenous metabolic system	Dose or dose range[c]	Reference
Ontario (Canada); river; chlorinated tap water	XAD-2 hexane:acetone	Micronuclei, Chinese hamster CHO cells	(+)	–	4–27 µg/ml	Douglas et al. (1986)
Ontario (Canada); river; chlorinated tap water	XAD-2 hexane:acetone	Micronuclei, Chinese hamster CHO cells	–	0	0.00	Douglas et al. (1986)
Ontario (Canada); lake; chlorinated tap water	XAD-2 hexane:acetone	Micronuclei, Chinese hamster CHO cells	–	0	0.00	Douglas et al. (1986)
Ontario (Canada); lake; chlorinated tap water	XAD-2 hexane:acetone	Micronuclei, Chinese hamster CHO cells	+	0	DD: 2.8 l/ml	Douglas et al. (1986)
Ontario (Canada); mixed ground- and surface waters; chlorinated tap water	XAD-2 hexane:acetone	Micronuclei, Chinese hamster CHO cells	+	0	DD: 0.9 l/ml	Douglas et al. (1986)

[a]GAC, granular activated carbon; RSF, rapid sand filtration; SSF, slow sand filtration
[b]DMSO, dimethyl sulfoxide; TLC, thin-layer chromatography
[c]pl, plate; DD, doubling dose. Doses given in litres per unit are litre equivalents of the original water sample per that unit. Other units refer to the amount of concentrate added per plate, millilitre, etc.
[d]Treatment performed in laboratory instead of water treatment plant
[e]Mean of net revertants/litre compared to mean of net revertants/litre in raw waters
[f]Comparison to previous stage in the treatment process
*Lower effect than without metabolic activation
**Data not given but lower effect than without metabolic activation

Table 9. Summary of the influence of chlorination upon the genetic and related effects of groundwater and spring water concentrates in comparison with concentrates of unchlorinated water

Source of water; disinfection method[a]	Concentration and extraction method (concentration factor)[b]	Test system	Result without exogenous metabolic system	Result with exogenous metabolic system	Dose or dose range[c]	Reference
Netherlands; groundwater from five cities; Cl$_2$ 1 mg/l	XAD-4/8, DMSO (7000x), neutral	Mutation, S. typhimurium TA100 TA98	(+) (1/5)[d] + (3/5)	− + (2/5)	3.5 l/pl 3.5 l/pl	Kool & Hrubec (1986)
Netherlands; groundwater from five cities; Cl$_2$ 1 mg/l	XAD-4/8, DMSO (7000x), acidic	Mutation, S. typhimurium TA100 TA98	+ (2/5) + (2/5)	− (+) (1/5)	3.5 l/pl 3.5 l/pl	Kool & Hrubec (1986)
Netherlands; groundwater from five cities; Cl$_2$ 1 mg/l	XAD-4/8, DMSO (7000x), neutral	Mutation, S. typhimurium TA100 TA98	+ (2/5) + (2/5)	− (+) (1/5)	3.5 l/pl 3.5 l/pl	Kool et al. (1985c)
Netherlands; groundwater from five cities; Cl$_2$ 1 mg/l	XAD-4/8, DMSO (7000x), acidic	Mutation, S. typhimurium TA100 TA98	(+) (1/5) + (3/5)	− (+) (3/5)	3.5 l/pl 3.5 l/pl	Kool et al. (1985c)
Italy; spring water; Cl$_2$	Liquid-liquid extraction, CH$_2$Cl$_2$, neutral, acidic, basic	Mutation, S. typhimurium TA100 TA98	− −	− −	0.00 0.00	Monarca et al. (1985a)
Italy; spring water; Cl$_2$	XAD-2, acetone	Mutation, S. typhimurium TA100 TA98	− −	− −	0.00 0.00	Monarca et al. (1985a)
Italy; spring water; NaOCl	XAD-2, acetone	Mutation, S. typhimurium, fluctuation test TA100	−	0	0.1–1 l/test	Monarca et al. (1985b)

Table 9 (contd)

Source of water; disinfection method[a]	Concentration and extraction method (concentration factor)[b]	Test system	Result Without exogenous metabolic system	Result With exogenous metabolic system	Dose or dose range[c]	Reference
Italy; spring water; NaOCl	Liquid-liquid extraction, CH_2Cl_2, neutral, acidic, basic	Mutation, *S. typhimurium*, fluctuation test TA100	–	0	0.1–1 l/test	Monarca *et al.* (1985b)
Italy; spring water; NaOCl	Sep-Pak, methanol	Mutation, *S. typhimurium*, fluctuation test TA100	–	0.00	0	Monarca *et al.* (1985b)
Ontario (Canada); groundwater; chlorinated tap water	XAD-2, hexane:acetone	Mutation, *S. typhimurium* TA100	+	0	DD: (15.5 l/ml)	Douglas *et al.* (1986)
Siilinjärvi, Finland; groundwater; chlorinated drinking-water	Liquid-liquid extraction, CH_2Cl_2	Mutation, *S. typhimurium* TA100 TA98	– –	0 0	0.00 0.00	Vartiainen & Liimatainen (1986)
Siilinjärvi, Finland; groundwater; chlorinated Cl_2, 2 or 20 mg/l[g]	Liquid-liquid extraction, CH_2Cl_2	Mutation, *S. typhimurium*[e] TA100 TA98	+[f] –	0 0	0.00 0.00	Vartiainen & Liimatainen (1986)
UK; groundwater; chlorinated	Freeze-drying	Mutation, *S. typhimurium*, fluctuation test TA100 TA98	– +	– +		Fielding & Horth (1988)

Table 9 (contd)

Source of water; disinfection method[a]	Concentration and extraction method (concentration factor)[b]	Test system	Result		Dose or dose range[c]	Reference
			Without exogenous metabolic system	With exogenous metabolic system		
Ontario (Canada); ground-water; chlorinated tap water	XAD-2, hexane:acetone	Sister chromatic exchange, Chinese hamster CHO cells	(+)	0	DD: 1.9 l/ml	Douglas et al. (1986)
Ontario (Canada); ground-water; chlorinated tap water	XAD-2, hexane:acetone	Micronuclei, Chinese hamster CHO cells	–	0	0.00	Douglas et al. (1986)

[a]GAC, granular activated carbon; RSF, rapid sand filtration; SSF, slow sand filtration.
[b]DMSO, dimethyl sulfoxide
[c]pl, plate; DD, doubling dose. Doses given in l per unit are litre equivalents of the original water sample per that unit. Other units refer to the amount of concentrate added per plate, millilitre, etc.
[d]In parentheses, number of cities
[e]Comparison to previous stage in the treatment process
[f]Only 20 mg Cl$_2$/l positive
[g]Treatment performed in laboratory and not in water treatment plant

Table 10. Summary of the influence of chlorination in combination with either chlorine dioxide or ozone treatment upon the genetic activity of surface water concentrates in comparison with concentrates of unchlorinated water

Source of water; disinfection method	Concentration and extraction method (concentration factor)	Test system	Result without exogenous metabolic system	Result with exogenous metabolic system	Dose or dose range[a]	Reference
Savojärvi, Finland; humic lake water; O_3 5.7–33.2 mg/l, Cl_2 21 mg/l	XAD 4/8, ethyl acetate	Mutation, *S. typhimurium* TA100 / TA98 / TA97	+ / + / +	0 / 0 / 0	10–50 ml/pl / 10–50 ml/pl / 10–50 ml/pl	Backlund *et al.* (1985)
Savojärvi, Finland; humic lake water; Cl_2 10.5 mg/l, ClO_2 10.5 mg/l	XAD 4/8, ethyl acetate	Mutation, *S. typhimurium* TA100 / TA98 / TA97	+ / + / –	0 / 0 / 0	10–50 ml/pl / 10–50 ml/pl / 10–50 ml/pl	Backlund *et al.* (1985)
Savojärvi, Finland; humic lake water; alum flocculation, Cl_2 6.5 mg/l or O_3 2.9 mg/l, Cl_2 6.5 mg/l	XAD 4/8, ethyl acetate	Mutation, *S. typhimurium* TA100 / TA98 / TA97	+ / + / +	0 / 0 / 0	10–200 ml/pl / 10–200 ml/pl / 10–200 ml/pl	Backlund *et al.* (1985)
Savojärvi, Finland; humic lake water; alum flocculation, Cl_2 3.25 mg/l, ClO_2 3.25 mg/l	XAD 4/8, ethyl acetate	Mutation, *S. typhimurium* TA100 / TA98 / TA97	(+) / – / –	0 / 0 / 0	10–200 ml/pl / 10–200 ml/pl / 10–200 ml/pl	Backlund *et al.* (1985)
Arno River (Florence, Italy); NaOCl 2.5–7.5 g Cl/m^3, activated carbon, coagulation, flocculation, decantation, O_3	XAD-2, CH_2Cl_2, $CHCl_3$	Mutation, *S. typhimurium* TA100 / TA1538	(+) / +	+ / +	0.375–10 l/pl / 10 l/pl	Dolara *et al.* (1981)
Finland, seven surface waters; ClO_2 1 ± 0.5 mg Cl_2/l, alum coagulation, clarification, sand filtration, pH adjustment, Cl_2 0.7 ± 0.2 mg/l	XAD 8, pH 2, ethyl acetate	Mutation, *S. typhimurium* TA100 / TA98 / TA97	+ / – / +	0 / 0 / 0	0.00 / 0.00 / 0.00	Vartiainen *et al.* (1988)

Table 10 (contd)

Source of water; disinfection method[a]	Concentration and extraction method (concentration factor)[b]	Test system	Result		Dose or dose range[a]	Reference
			Without exogenous metabolic system	With exogenous metabolic system		
Finland, seven surface waters; O$_3$, alum coagulation, clarification, sand filtration, pH adjustment, Cl$_2$ 0.7 ± 0.3 mg/l	XAD 8, pH 2, ethyl acetate	Mutation, *S. typhimurium* TA100 TA98 TA97	+ (+) +	0 0 0	0.00 0.00 0.00	Vartiainen *et al.* (1988)

[a]pl, plate

(b) Humans

(i) *Toxic effects*

The possible effect of drinking-water on serum lipids was examined in a cross-sectional study of about 1500 healthy persons who had resided for at least 10 years in 46 different communities in Wisconsin, USA. Alcohol consumption, smoking habits, dietary fat intake, dietary calcium intake and body mass were considered in the analyses. The public water supply varied in magnesium and calcium content (water hardness), whether it had been chlorinated or not. The prevalence of women whose serum cholesterol level exceeded 270 mg/dl was greater in communities served by chlorinated drinking-water (odds ratio, 2.0); the mean serum cholesterol concentrations in women on chlorinated and on nonchlorinated drinking-water supplies were 247.9 mg/dl and 239.8 mg/dl, respectively (a significant difference). A smaller difference in men was not significant. Water hardness did not influence the serum cholesterol levels in either women or men (Zeighami *et al.*, 1990).

(ii) *Effects on reproduction and prenatal toxicity*

Rausch (1980) evaluated pregnancy outcomes during 1968-77 in several villages in New York State, USA, served by nonchlorinated groundwater, chlorinated groundwater or chlorinated surface water supplies. A significantly greater incidence of late fetal deaths and neonatal deaths was observed in villages on nonchlorinated groundwater; and a significantly higher prevalence of anencephaly was seen in villages using surface water as compared to villages using groundwater. There was, however, no difference in the prevalence of anencephaly between villages using chlorinated groundwater and those using nonchlorinated groundwater. The confounders analysed were season and year of birth, sex and cause of death of the fetus or newborn, maternal age, education, previous reproductive history and prenatal care and hospital where delivery occurred.

In a retrospective study described in detail in the monograph on sodium chlorite, Tuthill *et al.* (1982) compared neonatal morbidity and mortality in two similar communities in the USA, one of which used chlorination and the other of which used chlorine dioxide for disinfecting potable water. The number of infants that were judged by the attending physician to be premature or to have greater weight loss after birth was significantly greater in the community with chlorine dioxide-treated water. [The Working Group noted the difficulties associated with establishing prematurity and poor weight gain after birth, especially in a retrospective study, and that confounding factors were not controlled for.]

In a case-control study of spontaneous abortions in relation to tap-water consumption in northern California, USA, Hertz-Picciotto *et al.* (1989) observed a crude odds ratio of 1.7 (95% confidence interval (CI), 1.2-2.3) for drinkers of tap water as compared with drinkers of bottled water. After controlling for a large number of confounders, including demographic, reproductive and life-style variables, the results were still significant.

(iv) *Genetic and related effects*

No data were available to the Working Group.

3.3 Epidemiological studies of carcinogenicity in humans

The epidemiological investigation of the relation between exposure to chlorinated drinking-water and cancer occurrence is problematic because any increase in relative risk over that in people drinking unchlorinated water is likely to be small and therefore difficult to detect in epidemiological studies. It is particularly important to obtain valid assessment of disease status, of confounding factors (see also Preamble, p. 26) and, most relevantly, of the level of exposure to chlorinated water.

Relevant exposure to chlorinated water is particularly difficult to measure. A number of surrogates, such as use of surface water, depth of wells and residence in a community with a chlorinated water supply, have been used. To the extent that they do not reflect exposure to chlorinated water during the possibly relevant time periods for the etiology of the cancers in question, they will result in misclassification of subjects by exposure and will introduce bias. In some studies, concentrations of particular chlorination by-products have been modelled retrospectively; the assumptions underlying such models are, however, unproven.

Correlation studies are generally of uncertain validity, because exposure variables assessed for whole communities do not necessarily reflect the exposure of individuals. Such studies have been used extensively in relation to chlorinated drinking-water, however, as exposure may vary less within geographical units (such as towns) than between them.

Case-control studies are generally considered to provide greater opportunity for valid inference than correlation studies, because in these studies exposure and outcome are correlated at the individual level. Some of the case-control studies available for review were, however, based on exposure measured at the community level, because of the difficulties in assessing exposure of individuals.

Moreover, in many of these correlation and case-control studies, information on the nature of the water source and its chlorination status was obtained subsequently to or contemporaneously with the period over which cancer occurrence was measured. Because there are usually long latent periods between exposure and disease, cancer rates should be correlated with the characteristics of water supplies that were current before the cancers occurred. Most of these studies also did not address the problem of migration in and out of communities over time: the degree of exposure misclassification consequent on population mobility can vary between geographical areas and thus lead to unpredictable bias.

In a small number of case-control studies of cancer incidence, detailed information was collected about the residential histories of subjects and their exposure to chlorinated water over long periods, estimated by reference to historical data on water supplies. The accuracy of such exposure measurements depends on the accuracy of recall by study subjects and the availability of relevant water supply records. Moreover, water consumed outside the home and the daily quantity of water consumed have rarely been taken into account. Thus, even in the best studies, errors in exposure measurement may still be a problem.

An additional problem encountered in assessing the effects of chlorinated water is that the profile of chemical exposures resulting from chlorination depends on local conditions and may vary from place to place and from time to time. It is possible therefore that one criterion for assessing causality—consistency of findings among epidemiological studies—may not be entirely appropriate.

Comparisons of populations living in communities served by chlorinated water supplies and populations living in towns served by unchlorinated sources could be confounded by many factors, including the constituents of water supplies other than chlorination by-products; socioeconomic, industrial and cultural (e.g., smoking, diet, use of medications) characteristics of the populations; and the medical facilities available for diagnosing cancer. In most of the studies, very few, if any, of these factors were mentioned. Virtually all of the studies reviewed are therefore susceptible to bias from confounding, to some degree.

(a) *Correlation studies*

(i) *Surface* versus *groundwater*

In the studies described below, the surrogate measure of exposure to chlorinated water was exposure to surface water, although the status of neither the surface nor the groundwater was known.

Page *et al.* (1976) studied 64 parishes in Louisiana, USA, in which 32% of the population were supplied by the Mississippi River, 56% by groundwater and 12% by other surface supplies. Age-adjusted 20-year mortality rates (Mason & McKay, 1974) for cancers of the gastrointestinal tract, urinary tract (these two groupings were necessary owing to small numbers of deaths from cancers at related individual sites in some counties), breast and prostate and for cancers at all sites were analysed by multiple regression, including as independent variables the percentage of the parish population drinking Mississippi River water, rurality, income, and occupation in the petroleum and coal, chemical and mining industries, in four subgroups: white men, white women, non-white men and non-white women. The proportion of the parish population using Mississippi River water was positively ($p < 0.05$) associated with cancer of the gastrointestinal tract in all four groups, for urinary tract cancer in white men and non-white women and for cancers at all sites in white men, non-white men and non-white women. [The Working Group noted that the parishes using Mississippi River water were all located in the southern part of Louisiana, and the possible effects of water supply type and social and cultural differences cannot be separated.]

A further analysis of these and other data (DeRouen & Diem, 1977) also took account of region, i.e., northern or southern Louisiana, to allow for cultural and occupational differences. In a multiple regression analysis, percentage use of Mississippi River water was significantly associated with mortality from cancers at all sites for white men, non-white men and non-white women, with gastrointestinal tract cancer mortality for non-white men and non-white women, with urinary tract cancer for non-white women, and with cancer of the lung for non-white men. Southern Louisiana parishes in which part of the population was supplied with Mississippi River water showed significantly higher mortality from all cancers for non-white women, cancer of the stomach for non-white women, cancer of the colon for both white and non-white women, cancer of the rectum for white men, cancer of the urinary bladder for white men and cancer of the lung for non-white men. In these parishes, there was significantly lower mortality from cancer of the lung in white women and from cancer of the liver in white men.

Kuzma *et al.* (1977) classified 88 Ohio (USA) counties by ground- or surface water source on the basis of a survey of water supplies conducted in 1960. A substantial proportion of the population was served by sources not included in the survey: in only 39 counties was more than 50% of the population on a water source that was covered. Average annual age-adjusted cancer mortality rates among whites from 1950 to 1969 (Mason & McKay, 1974) were obtained for cancers of the stomach, large intestine, rectum, biliary passages and liver, pancreas and urinary bladder; for all cancers at all sites; for lung cancer in men; and for breast cancer in women. Analysis of covariance included the water classification variable as a factor and percentage urbanization, median income, population size and percentage of the male population in manufacturing activity and agriculture-forestry-fishery activity as covariables. Adjusted mean mortality in counties classified as supplied by surface water significantly ($p < 0.05$) exceeded that in those supplied by groundwater for cancers at all sites combined in men, for stomach cancer in both men and women and for bladder cancer in men. When the 39 counties in which more than 50% of the population was on a water source covered by the survey were analysed, similar results were obtained. [The Working Group noted that in the absence of information on county population size and water source, it was impossible to confirm the adequacy of covariance analysis to control for population size.]

Bean et al. (1982) compared cancer incidence rates in Iowa (USA) municipalities served by water from surface sources with those from groundwater sources. They omitted municipalities with populations of fewer than 1000 and those in which less than 90% of the water used was from the classified source. Only municipalities receiving water from a single source type in 1965-79 were included. Cancer incidence data were obtained for 1969-71 and 1973-78, from the Third National Cancer Survey and from the Surveillance, Epidemiology and End Results Program, respectively. Age-standardized, sex-specific incidence rates were calculated for cancers of the urinary bladder, breast, colon, lung, rectum, prostate and stomach. Details of socioeconomic status and occupation were obtained from the 1970 census and from the Directory of Iowa Manufacturers. Using a previously conducted population-based case-control study of urinary bladder cancer which included subjects in Iowa (Hoover & Strasser, 1980), the authors derived information on several variables, including education, income, manufacturing, labour force, change in population between 1960 and 1970 and smoking habits, for residents of towns on groundwater and on surface water. The case-control study had shown that 63% of the controls over 55 years of age had been on the same water supply for at least 20 years before onset of their cancer, and 77% had been on the same supply for at least 10 years before onset. Analyses were based on log-linear models. After adjustment for population size, the incidences of lung cancer and rectal cancer were significantly greater [details not given] for men and women served by surface water than for those drinking groundwater. Trends of risk over three categories of well depth were not significant.

Kool et al. (1981) studied 19 cities in the Netherlands, representing approximately one-third of the population of that country. Directly standardized, sex-specific mortality rates for cancers of the bladder, lung, oesophagus, stomach, colon, rectum and liver were calculated for 1964-76. Organic constituents of tap water were determined in 1976. Correlation coefficients between source type (surface/groundwater) were calculated, and a transformation of the rates showed that mortality from liver and urinary bladder cancer in men and lung cancer in both men and women was significantly greater ($p < 0.05$) in cities supplied with surface water. [The Working Group noted that no potential confounding factor was taken into account, and the statistical methods were not adequately described.]

(ii) *Chlorination and chlorination by-products*

Cantor et al. (1978) calculated directly age-standardized, sex-specific, cancer mortality rates by site for whites for 1968-71 in 923 US counties with more than 50% urbanization in 1970. Chloroform and total trihalomethane (THM) levels were obtained from two drinking-water surveys carried out in 1975 by the US Environmental Protection Agency, and levels of bromine-containing trihalomethanes (BTHM) were calculated by subtraction. The proportion of each county served in 1960 by the sampled municipal water supplies was estimated. The correlation between chloroform and BTHM levels was 0.54. Weighted linear regression using all 923 counties was used to predict sex- and site-specific cancer rates in 1970 by including the following variables in the model: urbanization (%; 1970), education (1970), population size (1970), ratio of 1970:1950 population, workforce in manufacturing (%; 1970), population in each of 10 ethnic groups (%) and region. The differences between the observed and predicted values (residuals) were correlated with log-THM in the 76 counties where 50% or more of the population was served by a water supply included in either of the two surveys and in the 25 counties where 85% or more of the population was so served. All cancer sites for which the sex-specific mortality rate was more than $1.5/10^5$ per year were studied. In the analysis of the 76 counties, the only significant correlation found was between the residual mortality rates for lung cancer in females and level of total THM (correlation coefficient, $r = 0.22; p = 0.05$). In the analysis of the subset of 25 counties, there were significant correlations between kidney cancer in men and chloroform level ($r = 0.42, p = 0.04$) and between urinary bladder cancer in women and BTHM level ($r = 0.45, p = 0.02$); whereas the correlations for kidney cancer in women and lung cancer in men were very low or negative, the correlation coefficient for male urinary bladder cancer and BTHM level was 0.38 ($p = 0.06$). Partial correlations controlling for high-risk occupation were calculated for cancers of the urinary bladder and lung. After allowance for lung cancer mortality [presumably a proxy for cigarette smoking], the partial correlations of urinary bladder cancer with log-BTHM level in counties in which 85-100% of the population was served by sampled supplies were 0.33 and 0.42 for men and women, respectively. Adjustment for occupational exposures left the correlations unchanged.

Hogan *et al.* (1979) used cancer mortality rates in US counties for 1950-69 (Mason & McKay, 1974) for white men and women in a multiple regression analysis. The cancers considered were of the tongue, oesophagus, stomach, large intestine, rectum, biliary passages and liver, pancreas, breast, ovary, kidney, urinary bladder and other urinary, thyroid and bone and cancers at all sites. Data on exposure to chloroform were taken from the two surveys carried out by the US Environmental Protection Agency in 1975 and referred to by Cantor *et al.* (1978). Weighted and unweighted analyses were carried out, which included the following independent variables: population density, percentage of urbanization, percentage of non-white, percentage of foreign born, median income, education, percentage in manufacturing industry, population size (all in 1960), region and chloroform in finished water. There were substantial differences in the results of three analyses based on different methods of weighting the units of observation. Consistent positive associations were found with chloroform exposure level in men and women in all analyses (with at least one significant result, $p < 0.05$) for cancers of the urinary bladder, breast, rectum and large intestine using data for the counties covered in the first survey and for cancers of the liver and tongue using data from the second survey. A significant negative association was obtained for pancreatic cancer using data from the second survey. Finally, in an analysis restricted to counties in which 50% or more of the population was on a sampled supply, significant associations were found for cancer of the large intestine in both men and women and for urinary bladder cancer in women. [The Working Group noted that the geographical areas and exposure measures were similar to those used by Cantor *et al.* (1978), and these results, therefore, do not provide independent evidence.]

Carlo and Mettlin (1980) obtained age-adjusted cancer incidence rates from the New York State Tumor Registry for 218 census tracts in Erie County, NY, USA, between 1973 and 1976. Nine census tracts with rates greater than three standard deviations from the mean or large institutions were excluded; cases with incomplete residence data were excluded. The cancer sites studied were oesophagus, stomach, colon, rectum, urinary bladder and pancreas; socioeconomic factors, mobility, percentage of non-white, urbanicity and occupation (only for bladder cancer) were controlled for. Total THM, derived from State records, and type of water source were included in the analysis. Use of surface water was significantly ($p < 0.05$) associated with the incidence of oesophageal and pancreatic cancer; total THM was not significantly related to any cancer site studied. [The Working Group noted that the quality of data on THM levels could not be assessed; it is probable that a number of neighbouring census tracts shared the same water supply; and the statistical procedure used was unclear.]

Tuthill and Moore (1980) studied communities in Massachusetts (USA) served by surface water in 1949, excluding those with a population of fewer than 10 000 persons in 1970 or with a growth rate exceeding 25% between 1950 and 1970. They calculated sex-specific standardized mortality ratios (SMRs) for 1969-76 for nine digestive and urinary tract cancers and ten other cancer sites thought unlikely to be related to water quality. Correlations were made between SMRs and three measures of water quality: average past (1949-51) chlorine dose, recent chlorine dose and recent total THM level. Data analysis included correlation and stepwise multiple regression. Potential confounding variables included were ethnic group, income, education, percentage of foreign-born, occupation in the textile, printing and chemical industries and population growth between 1950 and 1970. There was no significant correlation between sex-specific SMRs and average chlorine dose in 1949-51. For recent chlorine dose and recent total THM level, there was a significant ($p < 0.05$) positive association with stomach cancer for women and for rectal cancer for men. For recent total THM level, there was a significant negative association with stomach cancer for men. After multiple regression analysis allowing for sociodemographic factors, the significant associations disappeared. [The Working Group noted that the number of communities studied was not explicitly stated and the methods of analysis were not fully presented.]

In the study by Kool *et al.* (1981) (described on p. 110), no significant relationship was found between levels of THM and cancer mortality in 19 cities in the Netherlands.

Isacson *et al.* (1983) extended the analysis of their earlier study (Bean *et al.*, 1982; see p. 110) to include the chlorination status of the water supply (chlorinated prior to 1966 or never chlorinated). Directly standardized incidence rates for rectal cancer in men were significantly lower in municipalities with chlorinated water in both categories of well depth (< 150 feet and ≥150 feet [< or ≥45.7 m]), but this difference was no longer significant after adjustment for potential confounding by other methods of water treatment (aeration, filtration, coagulation and sedimentation). When water sources were classified by chloroform content (0-96, 100-230 and 260-900 μg/l), nonsignificant increases were observed across these levels for cancers of the colon, rectum and urinary bladder in men and cancers of colon and rectum in women. [The Working Group noted that no test for trend with increasing chloroform level was presented.]

Zierler *et al.* (1986) compared mortality rates from cancers of the stomach, colon, rectum, urinary bladder, breast, lung, pancreas, kidney and lymphatic system in 23 Massachusetts (USA) communities provided with chlorinated water with mortality from these cancers in Massachusetts as a whole. There were higher mortality rates in communities served by a chlorinated water source for cancer of the stomach among both males and females and for cancer of the lung among males. [The Working Group noted that the exposed group provided a substantial proportion of the reference population.]

(iii) *Time-trend study*

Cech *et al.* (1987) used the introduction of a new water supply in Houston, TX, USA, as a natural experiment. Lake Houston was constructed in 1954, and much of the population of Houston, previously supplied with water from lightly chlorinated underground sources, thereafter received heavily chlorinated surface water. Fifty-six census tracts were studied: group A (138 697 residents) had used groundwater over the whole period of the study; group B (46 394) changed from ground- to surface water in 1954; and groups C (84 159) and D (163 466) changed from ground- to surface water after 1954. THM levels were measured in 1978-79, and average concentrations in source areas A, B, C and D were 4, 111, 129 and 50 μg/l, respectively. The outcome measure was age-adjusted five-year average mortality from urinary tract cancer in 1940-74; control causes of death were respiratory cancer, bronchitis-emphysema and homicide. Trends in death rates over time showed little variation that could be related to the change in water supply. The slopes of the regression lines for urinary cancer rates for 1940-59 and for 1960-74 in area B showed a significant ($p < 0.05$) decrease for white men and a significant increase for white women; no such difference was found in other areas. Adjustment of mortality rates for education, population density, percentage population employed in high-risk industries, percentage population foreign-born, presence of metal or petroleum industries, and presence of hospitals with oncological units had no effect on the results. A cohort analysis was also carried out: there was some evidence of a birth cohort effect for urinary cancer in white women in area B. The authors concluded that there was little evidence of an effect of chlorination.

(b) *Case-control studies*

(i) *Community exposure data*

Alavanja *et al.* (1978) carried out a death certificate case-control study in seven counties of New York State (USA), chosen because the water supplies were diverse (including chlorinated surface and chlorinated and nonchlorinated groundwater); individual supplies had been stable for at least 15 years before the date of the study, and immigration had been low during the same period. In all, 3446 deaths occurring in 1968-70 from cancers of the gastrointestinal tract (oesophagus, stomach, small intestine, large intestine, rectum, liver, intrahepatic bile ducts, gall-bladder and bile ducts, pancreas and peritoneum and retroperitoneal tissue) and urinary tract (bladder, kidney, renal pelvis, ureter and other unspecified urinary organs) and 1416 lung cancer deaths were individually matched to noncancer deaths [not further defined] by year of death, race, sex, birthplace and county of residence. The 'usual place of residence' on the death certificate was taken as the place of residence. Water distribution maps [from an unspecified period] were used to locate the water supply for each case and control individually. The odds ratios associated with chlorination for gastrointestinal and urinary tract cancers combined were 1.44 for women and 2.09 for men (both $p < 0.005$). For lung cancer in all urban and rural areas combined, the odds ratios were 1.55 (not significant) for women and 1.83 ($p < 0.005$) for men. For individual cancer sites among men, all odds ratios were significantly greater than 1; only the odds ratio for stomach cancer was significantly raised for women. Random samples of cases and controls were taken in order to compare possible occupational exposures; male cases were more likely to have had occupational exposure to carcinogens than male controls (odds ratio, 1.25; not statistically significant). [The Working Group noted that no allowance for potential confounders (such as occupational exposure and smoking) was made in the analysis, the statistical analysis of the data is inadequately described, and it is likely that the matching was not dealt with appropriately.]

Brenniman *et al.* (1980) carried out a death certificate case-control study incorporating 3208 deaths from gastrointestinal and urinary tract cancer occurring in Illinois, USA, between 1973 and 1976 and 43 666 non-cancer deaths as controls, excluding deaths from complications of pregnancy, congenital anomalies, perinatal disorders, mental disorders, senility and infectious diseases. The study was restricted to whites and to communities served by groundwater—272 chlorinated and 270 nonchlorinated. Data on water supply were obtained from an inventory of municipal water facilities published in 1963 and verified, where possible, by a questionnaire sent to the water supply source. Allowance was made in the analysis for age, sex, urbanicity and residence in a standard metropolitan statistical area. Cancers of the oesophagus, stomach, large intestine, rectum, liver, gall-bladder and bile ducts, pancreas, bladder and other urinary organs were studied for men and women separately. No significantly elevated odds ratio was found for any individual site.

Results from a study based on southern Louisiana (USA) parishes were presented in three articles (Gottlieb *et al.*, 1981, 1982; Gottlieb & Carr, 1982). Gottlieb *et al.* (1982) carried out a case-control study of 10 205 cancer deaths in 13 parishes in southern Louisiana. Deaths from the following cancers formed the case series: urinary bladder, colon, kidney, liver, non-Hodgkin's lymphoma, rectum, stomach, breast, brain, oesophagus, pancreas, Hodgkin's disease, leukaemia, lung, malignant melanoma, multiple myeloma and prostate. For each case, a control matched on sex, age, race and year of death was selected from among deaths from causes other than cancer, excluding causes related to each cancer. Analyses of surface *versus* groundwater were carried out (i) according to water source at death and (ii) restricted to subjects on the same source type at birth and at death (lifetime exposure). The former analysis revealed only three significant odds ratios: 1.79 for rectal cancer, 1.21 for breast cancer and 0.70 for multiple myeloma. The analysis of lifetime water use gave significant odds ratios of 2.50 for rectal cancer and 1.30 for breast cancer. [Confidence intervals were not given.] A dose-related response was seen for each of these cancers in the categories lifetime surface water, some surface water (only birth or death in a parish served by surface water) and lifetime groundwater use. Odds ratios were elevated for rectal cancer among men according to water use at death (2.21; 95% CI, 1.57-3.12) and for men and women according to lifetime water use (3.18; 1.96-5.19 in men and 1.73; 0.97-3.10 in women). There were also elevated risks for lung cancer among men on surface water at death (1.30; 1.05-1.62) and for breast cancer among women both on surface water at death (1.21; 1.00-1.46) and with lifetime surface water use (1.30; 1.00-1.69). An additional analysis based on 11 349 case-control pairs from 20 parishes (Gottlieb & Carr, 1982) did not provide different results. [The Working Group noted some inconsistencies in the number of parishes studied; the analysis is inappropriate since matching was broken, and the results are not presented in an understandable format.] A further analysis of these data (Gottlieb *et al.*, 1981) was restricted to a sample of 692 deaths from rectal cancer and 1167 from colon cancer; 1859 controls were selected from strata based on age at death (± five years), race, sex, year of death and parish. Four categories of estimated lifetime surface water use were derived as in the earlier work: mostly surface (birth and death in a surface water parish), some surface (either birth or death in a surface water parish), possible surface (death in a groundwater parish, birth outside the study area), least surface (birth and death in a groundwater parish). Of the total population, 99.2% could be classified into one of the four surface water exposure levels. For rectal cancer, relative risks of 1.61 (95% CI, 0.91-2.85) and 2.11 (1.17-3.84) were found for a residence served by surface water for 10-19 and > 30 years, respectively, as compared with residence served by groundwater. A significantly increasing trend with a relative risk of 2.07 (95% CI, 1.49-2.88) was also found for lifetime consumption of 'mostly surface water' relative to the 'least surface water' category. No elevation in odds ratios for any of the variables of interest was found for colon cancer. [The Working Group noted some internal inconsistencies in these papers in the numbers of cases and controls, and the method of selection of controls and whether they were individually matched is unclear.]

A study of cancer mortality in Wisconsin, USA, in relation to water chlorination was reported by Young *et al.* (1981), Kanarek and Young (1982) and Young and Kanarek (1983). Young *et al.* (1981) carried out a matched case-control study based on the death certificates of white females who had died of cancer in Wisconsin during 1972-77, restricted to the 28 counties in which the population was relatively stable and in which there were both chlorinated and unchlorinated water supplies. A total of 8029 deaths due to cancers of the gastrointestinal and urinary tracts, lung, breast and brain were matched to white female noncancer deaths on county of residence, year of death and birth date. The water supply serving the usual place of residence as recorded on each subject's death certificate was obtained from a 1970 survey of the 202 water sources serving the study areas, gathered from a postal questionnaire to the water utilities. These data included type of water source (surface or ground), presence or absence of environmental factors that might influence organic content (e.g., rural run-off) and mean daily chlorination doses over the previous 20 years in four levels: none, low (< 1.00 ppm), medium (1.00-1.70 ppm) and high (1.71-7.00 ppm). In an unmatched analysis, adjusted for marital status, urban residence and high-risk occupations specific for certain cancer sites, significantly high odds ratios were found for colon cancer: 1.53 (95% CI, 1.11-2.11), 1.53 (1.08-2.00) and 1.51 (1.06-2.14) for the low, medium and high categories, respectively. No significant increase in risk was found for other cancers. In areas with a rural run-off into the water supply, the odds ratios for colon cancer were higher, and these increased slightly after adjustment for depth of groundwater source and purification. In an additional analysis using matched data, similar results were obtained (Young & Kanarek, 1983). In a further analysis of these data (Kanarek & Young, 1982), in which organic contamination, source depth and purification were taken into account, the odds ratio for colon cancer among persons using chlorinated in relation to that for people using unchlorinated water sources (1.43; $p < 0.02$) increased to 1.81 ($p = 0.03$) for chlorinated sources contaminated by organic compounds and to 2.81 ($p = 0.01$) for chlorinated surface water.

In a case-control study of multiple cancer sites based on death certificates, Zierler *et al.* (1986) (see p. 113) compared communities in Massachusetts (USA) in which surface water was disinfected by chloramine treatment (see p. 51) (20 communities) or chlorine treatment (23 communities). More than 50 000 deaths from cancers of the urinary bladder, colon, kidney, pancreas, rectum, stomach, lung and breast occurring between 1969 and 1983 in persons aged 45 years or more were identified. Over 200 000 deaths from lymphatic cancer, cardiovascular disease, cerebrovascular disease, pulmonary disease and pneumonia/influenza were used as controls. Exposure was defined as residence in a community supplied with chlorinated water at the time of death; nonexposure, in a community supplied with chloraminated water. No elevated risk for cancer at any site was observed.

(ii) *Individual exposure data*

Lawrence *et al.* (1984) carried out a case-control study based on death certificates of white women who had been members of the New York State (USA) Teachers' Retirement System and had died from cancers of the colon and rectum. After geographical restrictions and other exclusions, 395 deaths occurring between 1962 and 1978 were included. Controls (395) were selected randomly from deaths due to any cause except malignant tumours and matched to the cases by age and year of death (within two years). Information was obtained on residence and employment 20 years prior to death, and water records were abstracted for both home and work addresses over the 20-year period. A model-based estimate of exposure to THM was derived from a study of New York State surface water systems. Potential confounding factors included in the matched and unmatched logistic regression analyses were population density, marital status, age and year of death. Only analyses for grouped colon and rectal cancers were reported. Results from the matched and unmatched analyses were identical. There was no significant finding in relation either to source type (odds ratio, 1.07; 90% CI, 0.79-1.43), to 20 years' cumulative chloroform dose or to five other measures of exposure to chlorine or THM.

Cantor et al. (1985, 1987) conducted a population-based case-control interview study of urinary bladder cancer in 10 areas of the USA comprising 2982 cases aged 21-84 who had been newly diagnosed in 1978 (73% of the eligible pool). A total of 5782 controls were selected by random-digit dialling for those age 21-64 and by Health Care Financing Agency listings for those 65 and older. Interviews in the homes of the subjects gathered information on residential history, fluid consumption and potential confounders (smoking, occupation, lower urinary tract infection, artificial sweetener use, use of hair dyes). Data on water source and treatment since 1900 was obtained from an independent survey of water utilities. Year-by-year profiles of water source (surface and ground) and water treatment (chlorinated and not) were derived for the lifetime of each respondent by merging individual residential and water utility information; 76% of all person-years could be related to a known water source. Reported consumption of drinking-water was added to the intake of other home beverages containing tap water to estimate total daily ingestion of tap water. Since one goal of the study was to estimate the risk associated with consumption of chlorinated surface water in comparison with nonchlorinated groundwater, the primary analyses were restricted to a subset of respondents who had lived at least 50% of their lifetime prior to interview at residences served by one or both of these two types of water source [59% of all cases and controls]. Analysis was by logistic regression. In initial analyses (Cantor et al., 1985) that did not consider tap-water consumption levels, an association was found between duration of residence with a chlorinated surface source and risk of urinary bladder cancer. Only among nonsmokers was there a significant odds ratio for those exposed for more than 60 years (odds ratio, 2.3; 95% CI, 1.3-4.2); there was a nonsignificant inverse trend for current smokers. For all groups combined (controlling for smoking), odds ratios for the duration measure were close to one. In subsequent analyses (Cantor et al., 1987), current total fluid and tap-water consumption were considered in conjunction with duration of exposure to chlorinated surface water. Total fluid consumption was related to urinary bladder cancer risk, and tap water was the main risk factor (test for trend: males, $\chi^2 = 22.6$, $p < 0.0001$; females, $\chi^2 = 3.15$, $p = 0.08$). These findings were not modified by extent of disease. When respondents were grouped by duration of chlorinated surface water use, significant trends with tap-water intake were restricted to persons who had consumed chlorinated water for 40-59 years and ≥ 60 years. The odds ratios for the highest (≥ 1.96 l/day) versus the lowest (≤ 0.80 l/day) quintiles of intake in these two duration strata were 1.7 and 2.0, respectively, with significant trends ($p = 0.006$ and $p = 0.014$, respectively). The trends in odds ratios with tap-water intake were nonsignificant for up to 39 years' duration. There was a significant trend with duration of residence with a chlorinated surface water supply, but only among women whose tap-water consumption was above the median ($p = 0.02$). The overall increase in the odds ratio with duration seen among nonsmokers in the previous analysis (Cantor et al., 1985) was more exaggerated among respondents whose tap-water consumption was above the median ($p = 0.01$ for trend) than in those whose consumption was below the median ($p = 0.40$ for trend).

Lynch et al. (1989) conducted an analysis of the Iowa respondents in the study of Cantor et al. (1987), comprising 354 cases of urinary bladder cancer and 752 controls. Chlorination was quantified in four ways, with increasing levels of specificity: (i) assuming that the respondent's lifetime was spent consuming the type of water provided by the community of his or her most recent place of residence; (ii) assuming that the person's most recently used water supply (whether or not his or her community's supply) was used for life; (iii) applying the most recent water supply to the number of years of actual residence at this place; and (iv) using the entire lifetime residential/water supply history. For methods (ii), (iii) and (iv), there were significant trends with exposure to chlorination, the highest odds ratio being found for method (iv) (test for trend: $\chi^2 = 10.90, p = 0.001$). The odds ratios for 1-25, 26-50 and > 50 years of exposure to chlorination relative to no exposure using method (iv) were 1.42, 1.70 ($p < 0.01$) and 2.14 ($p < 0.01$). After adjustment for age and smoking, the odds ratio for history of any exposure to chlorinated water was 1.47. The highest unadjusted odds ratio [no adjusted odds ratio reported] was found for exposure only to prechlorinated or prefiltered surface or shallow groundwater (odds ratio, 2.95; 95% CI, 1.52-5.75). In this study subset of Iowa respondents, cigarette smokers (more than 25 pack-years) who had had exposure to chlorinated drinking-water had a higher odds ratio (4.48; 95% CI, 2.47-8.13) than smokers never so exposed (2.89; 1.41-5.89), relative to nonsmokers never exposed to chlorinated drinking-water. This result contrasts with the findings from the overall study in which smokers who had used chlorinated surface water were not at excess risk (Cantor et al., 1985).

Cragle et al. (1985) identified 200 cases of colon cancer newly diagnosed between 1978 and 1980 at seven hospitals in North Carolina (USA) who had had at least 10 years' residence in the state. At least two hospital controls were matched to each case by age, race, sex, vital status, date of diagnosis and hospital. Information on residential history and a variety of potential confounding factors was collected from the respondents by either a personal interview or by mail questionnaire. Each subject's residence history for 1953-78 was related to data from the water company to derive estimates of the duration of residence on chlorinated and nonchlorinated supplies. A logistic regression analysis was carried out which included a chlorination variable and several potential confounders. The authors concluded that there was an association between chlorination and colon cancer in people over the age of 60. [The Working Group noted that a number of details of the study design are not adequately described: it is not stated how many deceased cases and controls were selected and what procedure was used for obtaining data on these subjects; it is not clear how the chlorination variable was treated in the analysis; and, in spite of the matched nature of the design, an unmatched analysis was apparently carried out.]

A study in Wisconsin (USA), reported by Young *et al.* (1987) was designed to estimate the risk for colon cancer associated with chronic ingestion of THMs occurring as by-products of water chlorination. White men and women aged 50-90 were included. Cases were identified from a state-wide hospital tumour registry over a two-year period; 347 cases (45% of those sampled) were included in the analysis. Two sets of controls were used: 639 cancer controls identified from the same source as the cases, and 611 population controls identified from driver's license records, representing 48% of controls sampled. Self-completed questionnaires, supplemented with medical records, were used to obtain lifetime histories of residence, water use and medical and occupational histories. Water company records and contemporary measurements of THM were used to estimate the THM content of all types of water source in the past and then to construct estimates of lifetime ingestion of THM for each subject. Odds ratios for colon cancer relative to population controls, adjusted for age, sex and population size were 1.10 (95% CI, 0.68-1.78) for estimated cumulative exposure to 100-300 mg THM and 0.73 (0.44-1.21) for 300 mg or more, relative to the baseline group (less than 100 mg lifetime ingestion of THM). Analyses comparing surface with groundwater sources and chlorinated with nonchlorinated sources also showed no association with colon cancer risk. [The Working Group considered that the response rate in this study was too low to permit reliable inferences to be made.]

Zierler *et al.* (1988) carried out a case-control study of urinary bladder cancer based on death certificates of residents of 43 Massachusetts (USA) communities served by surface water disinfected by chlorine or chloramine. A total of 1057 deaths from urinary bladder cancer in people aged 45 or more occurring between 1978 and 1984 were identified. Controls were obtained from an age-stratified sample of deaths from the following causes: lung cancer, lymphoma, cardiovascular disease, cerebrovascular disease and chronic obstructive pulmonary disease (total, 2144). A large number of the cases and controls included in this study were also included in a previous case-control study carried out by the same authors (Zierler *et al.*, 1986; see p. 117). Informants were found for 614 (58%) of the cases and 1074 (50%) of controls and were interviewed about the decedents' residential and smoking history. Each subject's residential history was linked to historical data on water source obtained from the US Environmental Protection Agency and State water authorities. Four categories of lifetime exposure to chlorinated water were defined, and each individual was placed into one of these. Information on socioeconomic status and high-risk occupations was obtained indirectly at the level of the community. Odds ratios for usual and lifetime exposure to chlorinated water with respect to lifetime exposure to chloramine were 1.4 (95% CI, 1.1-1.8) and 1.6 (1.2-2.1). When analysis was restricted to 30 communities each supplied by a single authority, the odds ratio for lifetime exposure with respect to lifetime nonexposure was 1.6 (1.1.-2.4). [The Working Group noted that the response rate was very low. It was unclear whether information on water supplies was obtained when individuals resided outside the 43 communities. The choice of controls may not have been appropriate. Confounding by city size was not addressed. Differences between the results of this study and those of Zierler *et al.* (1986) may be due to the fact that the exposure information in this study was more precise or to selection biases due to low response rates.]

(c) Cohort study

Wilkins and Comstock (1981) conducted a cohort study in Washington County, Maryland (USA) on a population of 14 553 white men and 16 227 white women over 25 years of age, who were resident in 1963. Follow-up over a 12-year period to mid-1975 was through death certificate records, the cancer registry and medical records at Washington County Hospital. [No information was given on completeness of follow-up.] Data on personal and socioeconomic variables in 1963 (age, marital status, education, smoking history, length of residence, frequency of church attendance, adequacy of housing and persons per room, source of drinking-water) were available. Sex- and site-specific incidence rates were calculated for cancers of the biliary passages and liver, kidney and urinary bladder. Mortality rates were calculated for the same sites and also for cancers of the oesophagus, stomach, colon, rectum, pancreas, lung, breast, cervix, ovary, prostate and brain, and leukaemia, and non-cancer causes of death (cirrhosis of the liver, bronchitis and emphysema, pneumonia, aortic aneurysm, road accident, fall, suicide, arteriosclerotic heart disease, hypertension, stroke and all causes.) Water sources were classified into three groups according to the subjects' residence in 1963: high exposure (23 727 urban residents served by chlorinated surface water systems; average chloroform concentration, 107 μg/l), low exposure (2231 users of unchlorinated, deep wells), and an intermediate group of 4842 residents of four small towns served by combined chlorinated surface and groundwater. In the incidence study, the only consistent results for men and women were adjusted relative risks (high *versus* low exposure) of 1.80 and 1.60 for urinary bladder cancer based on five and two cases in the low-exposure category (both $p > 0.05$). Only for urinary bladder cancer in men was there a relationship with duration of exposure (relative risk, 6.46; 95% CI, 1.00- > 100 for 12 or more years at one address). In the mortality study, a significant result was obtained only for breast cancer (2.27; 1.16-4.89); however, when the relative risks were ranked, three of the four highest were for sites for which there was an a-priori suspicion of an association with organic contamination of drinking-water (liver: 2.98, 0.92-14.84; kidney: 2.76, 0.67-23.06; and urinary bladder: 2.20, 0.71-9.39). Relative risks were 0.89 (0.57-1.43) for cancer of the colon and 1.42 (0.70-3.16) for cancer of the rectum. [The Working Group noted that the large number of liver cancer deaths may indicate the inclusion of secondary liver cancers.]

Studies relevant to the evalution

Table 11 gives a summary of the results from those studies on which the final evaluation was based. Some studies were excluded because of the methodological limitations described on pp. 107-108 and in the square brackets following the descriptions of some studies; some were excluded because they largely overlapped with other studies included in the Table. For correlation studies, only an indication of the direction of the results is given; odds ratios or relative risks (with 95% CI when available) are given for case-control studies and for the cohort study.

4. Summary of Data Reported and Evaluation

4.1 Exposure data

Water supplies were first chlorinated at the turn of the century, and over the following two decades chlorination was introduced for disinfection of drinking-water in most industrialized countries. In the chlorination process, chlorine reacts mainly with natural water constituents to produce a complex mixture of by-products, including a wide variety of halogenated compounds, the actual levels of which depend on the amount of chlorine added and the type of water source. In general, groundwaters produce lower levels, while surface waters often tend to produce higher levels of chlorination by-products; however, there is some evidence that groundwaters can give higher levels of brominated substances, probably due to higher levels of bromide in the untreated water. Estimates of the total halogenated organic matter generated during chlorination suggest typical levels in the range < 10-250 µg/l as chlorine. The main chlorination by-products are trihalomethanes and chlorinated acetic acids, which usually occur in the range 1-100 µg/l (although higher levels have been reported). Many products occur in the range 1-10 µg/l, while a large number can be detected at levels of < 1 µg/l. The

Table 11. Summary of results of selected epidemiological studies[a]

Author, year	Exposure variable	Bladder M	Bladder F	Colon M	Colon F	Rectum M	Rectum F	Stomach M	Stomach F	Lung M	Lung F
Correlation studies											
DeRouen & Diem (1977)	Surface vs groundwater										
	Whites	(+)[b]	(+)[b]	(+)[c]	(+)[c]						
	Nonwhites	(+)[b]	+[b]	+[c]	+[c]						
	River vs non-river										
	Whites	+	(-)	(+)	+	+	(+)	(-)	(-)	(+)	-
	Nonwhites	(+)	(+)	(+)	+	(+)	(+)	(+)	+	+	(-)
Kuzma et al (1977)	Surface vs groundwater	+	(+)	(+)	+/-	(+)	+/-	+	+	(+)	
Bean et al (1982)	Surface vs groundwater	+/-	(-)	+/-	+/-	+	+	+/-	+/-	+	+
Cantor et al (1978)	Chlorinated surface vs unchlorinated groundwater	(+)	+	+/-	+/-	+/-	+/-	+/-	(-)	(+)	(+)
Tuthill & Moore (1980)	Trihalomethanes in 1978	+/-	(+)	+/-	+/-	+	(+)	-	+	+/-	+/-
	Chlorine dose in 1950	+/-	+/-	+/-	+/-	(-)	+/-	(-)	(-)	+/-	(-)
Isacson et al (1983)	Chlorinated vs unchlorinated	(+)	(-)	(+)	(-)	-	(+)	(-)	(-)	(+)	(+)
Time–trend studies											
Cech et al (1987)	Chlorinated vs unchlorinated	(-)[b]	(+)[b]							(-)	(+)
Case–control studies, community-based exposure definition											
Brenniman et al (1980)	Chlorinated vs unchlorinated	0.99	0.95	1.04	1.17	1.14	1.35	0.91	1.07		
Gottlieb et al (1982)	Lifetime use of surface vs groundwater.	1.32 (0.88–1.97)	1.02 (0.57–1.82)	0.90 (0.60–1.37)	1.05 (0.73–1.51)	3.18[*d] (1.96–5.19)	1.73[*d] (0.97–3.10)	1.25 (0.85–1.84)	1.01 (0.61–1.66)	1.05 (0.77–1.43)	1.39 (0.63–3.11)
Young & Kanarek (1983)	Chlorinated vs unchlorinated[e]		1.08		1.41*		1.19		0.72		0.86
Zierler et al (1986)	Chlorinated vs chloraminated	1.04 (0.94–1.16)	1.05 (0.92–1.21)	0.85 (0.80–0.90)	0.92 (0.87–0.97)	0.98 (0.88–1.09)	0.94 (0.84–1.05)	0.95 (0.87–1.03)	1.01 (0.92–1.10)	0.91 (0.86–0.96)	0.95 (0.91–0.98)

Table 11 (contd)

Author, year	Exposure variable	Bladder		Colon		Rectum		Stomach		Lung	
		M	F	M	F	M	F	M	F	M	F
Case–control studies, individual exposure definition											
Lawrence et al. (1984)	Surface vs groundwater[a]					1.07 (0.79–1.43)[f]					
Cantor et al. (1987)	60 years or more on chlorinated water, water consumption > median	1.2 (0.7–2.1)	3.2*[s] (1.2–8.7)								
Zierler et al. (1988)	Chlorinated vs chloraminated, lifetime exposure		1.6* (1.2–2.1)								
Cohort study											
Wilkins & Comstock (1981)	Chlorinated vs unchlorinated	1.80 (0.80–4.75)	1.60 (0.54–6.32)								

[a](+), positive association; +, positive association, $p < 0.05$; (–) negative association; –, negative association, $p < 0.05$; +/–, no association
[b]Urinary tract
[c]Gastrointestinal tract
[d]Significant trend (both sexes combined) across two levels of exposure (source at death, lifetime source)
[e]Women only
[f]Colorectal cancer; 90%confidence interval
[w]Women only
[s]Significant trend across five levels of duration of residence with a chlorinated surface drinking-water source (0, 1–19, 20–39, 40–59 and ≥ 60 years)
*$p < 0.05$

by-products responsible for most of the bacterial mutagenicity found in chlorinated drinking-water, 3-chloro-4-(dichloromethyl)-5-hydroxy-2[5H]-furanone (MX) and associated substances, are present at very low concentrations (< 0.1 µg/l).

4.2 Experimental carcinogenicity data

Two series of studies were considered to provide evidence that could support an evaluation of the potential carcinogenicity of chlorinated drinking-water.

Samples of material concentrated from treated and undisinfected or treated and chlorinated water samples were tested in mice in three initiation-promotion experiments (by subcutaneous injection followed by topical application of 12-O-tetradecanoylphorbol 13-acetate). None of the concentrates derived from the chlorinated water induced a significantly increased incidence of skin tumours when compared with concentrates derived from undisinfected water samples or with saline.

In one experiment in mice, oral administration of chlorinated humic acids in the drinking-water did not increase the incidence of tumours over that in animals receiving unchlorinated humic acids or in saline-treated controls.

4.3 Human carcinogenicity data

Seven case-control studies conducted in the USA were considered to provide evidence that could support an evaluation. Four of these had community exposure data, and three had individually derived exposure data. The four studies with community exposure data each included several cancer sites. One study showed a significant increase in risk for colon cancer only; another showed a significant increase only for rectal cancer; the other two studies showed no excess risk for cancer.

Of the three case-control studies with individual exposure data, one was a population-based study of urinary bladder cancer carried out by interview in 10 areas of the USA. Many potential confounding factors, including smoking, were taken into account in the analyses. An early analysis of the study showed a significant association between long-term use at home of a chlorinated surface water source (as compared to an unchlorinated groundwater source) and urinary bladder cancer in nonsmokers only. In a subsequent analysis, tap-water intake was considered in addition to home water source, and consumption level of tap water was significantly associated with urinary bladder cancer; this effect was substantially confined to those who had lived for 40 years or more in a house with a chlorinated surface water source. There were significant and increasing trends in urinary bladder cancer risk with duration of residence in a house with a chlorinated surface water source for both women and nonsmokers whose tap-water consumption was above the median. In a further report based only on Iowa participants in this study, risk for urinary bladder cancer was associated with duration of use of a chlorinated water source, and the association became stronger with increasing accuracy of the exposure measure.

In the second of these case-control studies, carried out in Massachusetts (USA), the authors reported an excess risk for mortality from urinary bladder cancer among people who had lived in areas with chlorinated water supplies as compared with chloraminated supplies. Some confounding factors, including smoking, were taken into account; however, the proportion of eligible subjects for whom exposure could be ascertained was low.

In a third case-control study, based on deaths among members of the New York State Teachers' Retirement System, no association was found between deaths from cancers of the colon and rectum combined and estimated use of surface water or intake of chloroform from domestic and workplace water supplies over the 20 years prior to death. Few confounding variables were taken into acount.

A cohort of the general population in a county in Maryland (USA) was enrolled and surveyed in 1963 and followed up to 12 years. Urinary bladder cancer incidence was found to be higher in both men and women residents supplied mainly by a chlorinated surface water source compared with county residents who obtained their drinking-water from unchlorinated deep wells; but the effects of chlorinated drinking-water could not be distinguished from factors related to urbanicity, and the numbers were too small to rule out a chance effect.

Six correlation studies and one time-trend study were considered by the Working Group to provide some useful data. These studies showed moderately consistent patterns of a positive correlation between use of surface water or of chlorinated water and cancers of the stomach, colon, rectum, urinary bladder and lung, with the most consistent patterns for cancers of the urinary bladder and rectum.

The studies that were considered informative, and therefore included in this summary, were nevertheless difficult to interpret in an evaluation of the carcinogenicity of chlorinated drinking-water. The water variables studied— whether surface or groundwater and others—were generally imperfect surrogates for the subject of this monograph. There is cause for some scepticism about the estimates of exposure to chlorinated drinking-water in all of these studies. Furthermore, very few attempted to document exposure over long periods of the subjects' lives. Chlorination by-products differ according to local conditions and practices of chlorination, and the health effects found in one place may not be found elsewhere. Many variables, such as smoking habits, dietary practices and environmental conditions, influence the risks for cancer, and they may differ between populations served by chlorinated and unchlorinated water supplies. Such factors should ideally be taken into account in an epidemiological study; however, in most of the studies evaluated, there was little if any information available about them. When the data are examined on the basis of individual cancer sites, the evidence of elevated risk is strongest for cancer of the urinary bladder. The strongest study of cancer at this site supports the hypothesis of an elevated risk due to drinking chlorinated surface water compared with unchlorinated groundwater. However, the sum of the evidence from other studies, although showing some degree of consistency, is severely compromised by the weaknesses outlined above.

4.4 Other relevant data

Elevated serum cholesterol levels were reported in women but not in men living in communities served by chlorinated *versus* nonchlorinated water supplies in one study. No difference in the prevalence of ancephaly was observed between villages served by chlorinated and nonchlorinated groundwater in another study.

In regard to studies of genetic and related effects, only those reports were included in which the role of chlorination could be evaluated. Samples of unconcentrated chlorinated drinking-water were not genotoxic in bacteria or in a micronucleus assay in plants and did not induce morphological transformation in cultured mammalian cells. Samples of organic material concentrated from chlorinated surface waters were usually genotoxic in bacteria and induced sister chromatid exchange, micronuclei and chromosomal aberrations in single studies with cultured mammalian cells. In a single study, no activity was observed in a mammalian cell assay for mutation.

Samples of organic material concentrated from chlorinated groundwaters were less frequently mutagenic in bacteria than those from chlorinated surface waters; in a single study, they induced sister chromatid exchange but not micronuclei in cultured mammalian cells.

Samples of organic material concentrated from surface water treated with either chlorine dioxide or ozone followed by chlorination induced mutation in bacteria in some studies.

4.5 Evaluation[1]

There is *inadequate evidence* for the carcinogenicity of chlorinated drinking-water in humans.

---

[1]For definition of the italicized terms, see Preamble, pp. 30-33.

There is *inadequate evidence* for the carcinogenicity of chlorinated drinking-water in experimental animals.

Overall evaluation

Chlorinated drinking-water *is not classifiable as to its carcinogenicity to humans (Group 3)*.

5. References

Abrahamsson, K. & Xie, T.M. (1983) Direct determination of trace amounts of chlorophenols in fresh water, waste water and sea water. *J. Chromatogr.*, 279, 199-208

Agarwal, S.C. & Neton, J. (1989) Mutagenicity and alkylating activity of the aqueous chlorination products of humic acid and their molecular weight fractions. *Sci. total Environ.*, 79, 69-83

Alavanja, M., Goldstein, I. & Susser, M. (1978) A case control study of gastrointestinal and urinary tract cancer mortality and drinking water chlorination. In: Jolley, R.J., Gorchev, H. & Hamilton, D.H., Jr, eds, *Water Chlorination: Environmental Impact and Health Effects*, Vol. 2, Ann Arbor, MI, Ann Arbor Science, pp. 395-409

Alink, G.M. (1982) Genotoxins in waters. In: Sorsa, M. & Vainio, H., eds, *Mutagens in Our Environment*, New York, Alan R. Liss, pp. 261-276

Al-Sabti, K. & Kurelec, B. (1985) Chromosomal aberrations in onion (*Allium cepa*) induced by water chlorination by-products. *Bull. environ. Contam. Toxicol.*, 34, 80-88

Ames, B.N. & Yanofsky, C. (1971) The detection of chemical mutagens with enteric bacteria. In: Hollaender, A., ed., *Chemical Mutagens. Principles and Methods for Their Detection*, Vol. 1, New York, Plenum Press, pp. 267-282

Ames, B.N., McCann, J. & Yamasaki, E. (1975) Methods for detecting carcinogens and mutagens with the *Salmonella*/mammalian-microsome mutagenicity test. *Mutat. Res.*, 31, 347-364

Amy, G.L., Chadik, P.A., Chowdhury, Z.K., King, P.H. & Cooper, W.J. (1985) Factors affecting incorporation of bromide into brominated trihalomethanes during chlorination. In: Jolley, R.L., Bull, R.J., Davis, W.P., Katz, S., Roberts, M.H., Jr & Jacobs, V.A., eds, *Water Chlorination: Chemistry, Environmental Impact and Health Effects*, Vol. 5, Chelsea, MI, Lewis Publishers, pp. 907-922

Anon. (1983) Water substances spur pleas for regulation. *J. Commerce, 30 September*, 22b1

Backlund, P., Kronberg, L., Pensar, G. & Tikkanen, L. (1985) Mutagenic activity in humic water and alum flocculated humic water treated with alternative disinfectants. *Sci. total Environ.*, 47, 257-264

Bean, J.A., Isacson, P., Hausler, W.J., Jr & Kohler, J. (1982) Drinking water and cancer incidence in Iowa. I. Trends and incidence by source of drinking water and size of municipality. *Am. J. Epidemiol.*, 116, 912-923

Bellar, T.A., Lichtenberg, J.J. & Kroner, R.C. (1974) The occurrence of organohalides in chlorinated drinking waters. *J. Am. Water Works Assoc.*, *66*, 703-706

Bernofsky, C., Strauss, S.L. & Hinojosa, O. (1987) Binding of hypochlorite-modified adenosine 5'-monophosphate (AMP) to protein and nucleic acid and its possible role in cytotoxicity. *Biochem. Arch.*, *3*, 95-101

Bourbigot, M.M., Paquin, J.L., Pottenger, L.H., Blech, M.F. & Hartemann, P. (1983) Study of mutagenic activity of water in a progressive ozonation unit (Fr.). *Aqua*, *3*, 99-102

Boyce, S.D. & Hornig, J.F. (1983) Reaction pathways of trihalomethane formation from the halogenation of dihydroxyaromatic model compounds for humic acid. *Environ. Sci. Technol.*, *17*, 202-211

Brass, H.J., Feige, M.A., Halloran, T., Mello, J.W., Munch, D. & Thomas, R.F. (1977) The national organic monitoring survey: samplings and analyses for purgeable organic compounds. In: Pojasek, R.B., ed., *Drinking Water Quality Enhancement through Source Protection*, Ann Arbor, MI, Ann Arbor Science, pp. 393-416

Braus, H., Middleton, F.M. & Walton, G. (1951) Organic chemical compounds in raw and filtered surface waters. *Anal. Chem.*, *23*, 1160-1164

Brenniman, G.R., Vasilomanolakis-Lagos, J., Amsel, J., Namekata, T. & Wolf, A.H. (1980) Case-control study of cancer deaths in Illinois communities served by chlorinated or nonchlorinated water. In: Jolley, R.L., Brungs, W.A., Cumming, R.B. & Jacobs, V.A., eds, *Water Chlorination: Environmental Impact and Health Effects*, Vol. 3, Ann Arbor, MI, Ann Arbor Science, pp. 1043-1057

Bull, R.J. (1985) Carcinogenic and mutagenic properties of chemicals in drinking water. *Sci. total Environ.*, *47*, 385-413

Bull, R.J., Robinson, M., Meier, J.R. & Stober, J. (1982) Use of biological assay systems to assess the relative carcinogenic hazards of disinfection by-products. *Environ. Health Perspect.*, *46*, 215-227

Cantor, K.P., Hoover, R., Mason, T.J. & McCabe, L.J. (1978) Associations of cancer mortality with halomethanes in drinking water. *J. natl Cancer Inst.*, *61*, 979-985

Cantor, K.P., Hoover, R., Hartge, P., Mason, T.J., Silverman, D.T. & Levin, L.I. (1985) Drinking water source and risk of bladder cancer: a case-control study. In: Jolley, R.L., Bull, R.J., Davis, W.P., Katz, S., Roberts, M.H., Jr & Jacobs, V.A., eds, *Water Chlorination: Chemistry, Environmental Impact and Health Effects*, Vol. 5, Chelsea, MI, Lewis Publishers, pp. 145-152

Cantor, K.P., Hoover, R., Hartge, P., Mason, T.J., Silverman, D.T., Altman, R., Austin, D.F., Child, M.A., Key, C.R., Marrett, L.D., Myers, M.H., Narayana, A.S., Levin, L.I., Sullivan, J.W., Swanson, G.M., Thomas, D.B. & West, D.W. (1987) Bladder cancer, drinking water source, and tap water consumption: a case-control study. *J. natl Cancer Inst.*, *79*, 1269-1279

Carlo, G.L. & Mettlin, C.J. (1980) Cancer incidence and trihalomethane concentrations in a public drinking water system. *Am. J. public Health*, *70*, 523-525

Cech, I., Holguin, A.H., Littell, A.S., Henry, J.P. & O'Connell, J. (1987) Health significance of chlorination byproducts in drinking water: the Houston experience. *Int. J. Epidemiol.*, *16*, 198-207

Cheh, A.M., Skochdopole, J., Koski, P. & Cole, L. (1980) Nonvolatile mutagens in drinking water: production by chlorination and destruction by sulfite. *Science*, *207*, 90-92

Chernoff, N., Rogers, E., Carver, B., Kavlock, R. & Gray, E. (1979) The fetotoxic potential of municipal drinking water in the mouse. *Teratology*, *19*, 165-170

The Chlorine Institute, USA (1990) *Bromine Levels in Chlorine Used for Disinfection Purposes*, Washington DC

Christman, R.F., Norwood, D.L., Millington, D.S., Johnson, J.D. & Stevens, A.A. (1983) Identity and yields of major halogenated products of aquatic fulvic acid chlorination. *Environ. Sci. Technol.*, *17*, 625-628

Christman, R.F., Norwood, D.L., Seo, Y. & Frimmel, F.H. (1989) Oxidative degradation of humic substances from freshwater environments. In: Hayes, H.B., MacCarthy, P., Malcolm, R.L. & Swift, R.S., eds, *Humic Substances II: In Search of Structure*, Chichester, John Wiley, pp. 33-67

Cognet, L., Courtois, Y. & Mallevialle, J. (1986) Mutagenic activity of disinfection by-products. *Environ. Health Perspect.*, *69*, 165-175

Cognet, L., Duguet, J.P., Courtois, Y., Bordet, J.P. & Mallevialle, J. (1987) Mutagenic activity of various drinking water treatment lines. *Adv. Chem. Ser.*, *214*, 627-640

Coleman, W.E., Melton, R.G., Kopfler, F.C., Barone, K.A., Aurand, T.A. & Jellison, M.G. (1980) Identification of organic compounds in a mutagenic extract of a surface drinking water by a computerized gas chromatography/mass spectrometry system (GC/MS/COM). *Environ. Sci. Technol.*, *14*, 576-588

Coleman, W.E., Munch, J.W., Kaylor, W.H., Streicher, R.P., Ringhand, H.P. & Meier, J.R. (1984) Gas chromatography/mass spectroscopy analysis of mutagenic extracts of aqueous chlorinated humic acid. A comparison of the byproducts to drinking water contaminants. *Environ. Sci. Technol.*, *18*, 674-681

Commission of the European Communities (1989) COST Project 641: Organic Micropollutants in the Aquatic Environment. *Newsletter*, *4*, 4

Condie, L.W., Laurie, R.D. & Bercz, J.P. (1985) Subchronic toxicology of humic acid following chlorination in the rat. *J. Toxicol. environ. Health*, *15*, 305-314

Cragle, D.L., Shy, C.M., Struba, R.J. & Siff, E.J. (1985) A case-control study of colon cancer and water chlorination in North Carolina. In: Jolley R.J., Bull, R.J., Davis, W.P., Katz, S., Roberts, M.H., Jr & Jacobs, V.A., eds, *Water Chlorination: Chemistry, Environmental Impact and Health Effects*, Vol. 5, Chelsea, MI, Lewis Publishers, pp. 153-159

Crathorne, B., Watts, C.D. & Fielding, M. (1979) Analysis of non-volatile organic compounds in water by high-performance liquid chromatography. *J. Chromatogr.*, *185*, 671-690

Croll, B.T., Summer, M.E. & Leathard, D.A. (1986) Determination of trihalomethanes in water using gas syringe injection of headspace vapours and electron-capture gas chromatography. *Analyst*, *111*, 73-76

Degraeve, N. (1986) Genotoxic effects related to chlorination of drinking-water (Fr.). *Aqua*, *6*, 333-335

DeRouen, T.A. & Diem, J.E. (1977) Relationships between cancer mortality in Louisiana drinking-water source and other possible causative agents. In: Hiatt, H.H., Watson, J.D. & Winsten, J.A., eds, *Origins of Human Cancer*, Book A, *Incidence of Cancer in Humans*, Cold Spring Harbor, NY, CSH Press, pp. 331-345

Dolara, P., Ricci, V., Burrini, D. & Griffini, O. (1981) Effect of ozonation and chlorination on the mutagenic potential of drinking water. *Bull. environ. Contam. Toxicol.*, 27, 1-6

Douglas, G.R., Nestmann, E.R. & Lebel, G. (1986) Contribution of chlorination to the mutagenic activity of drinking water extracts in *Salmonella* and Chinese hamster ovary cells. *Environ. Health Perspect.*, 69, 81-87

Dressman, R.C., Stevens, A.A., Fair, J. & Smith, B. (1979) Comparison of methods for determination of trihalomethanes in drinking water. *J. Am. Water Works Assoc.*, 71, 392-396

Duguet, J.P., Tsutsumi, Y., Bruchet, A. & Mallevialle, J. (1985) Chloropicrin in potable water: conditions of formation and production during treatment processes. In: Jolley, R.L., Bull, R.J., Davis, W.P., Katz, S., Roberts, M.H., Jr & Jacobs, V.A., eds, *Water Chlorination: Chemistry, Environmental Impact and Health Effects*, Vol. 5, Chelsea, MI, Lewis Publishers, pp. 1201-1213

Dunham, L.J., O'Gara, R.W. & Taylor, F.B. (1967) Studies on pollutants from processed water: collection from three stations and biologic testing for toxicity and carcinogenesis. *Am. J. public Health*, 57, 2178-2185

Fawell, J.K. & Horth, H. (1990) Assessment and identification of genotoxic compounds in water. In: Waters, M.D., Daniel, F.B., Lewtas, J., Moore, M. & Nesnow, S., eds, *Genetic Toxicology of Complex Mixtures* (Environmental Science Research Series Vol. 39), New York, Plenum Press, pp. 197-214

Fawell, J.K., Fielding, M., Horth, H., James, H., Lacey, R.F., Ridgway, J.W., Wilcox, P. & Wilson, I. (1986) *Health Aspects of Organics in Drinking Water* (Publication No. TR 231), Medmenham, Water Research Centre

Fawell, J.K., Fielding, M. & Ridgway, J.W. (1987) Health risks of chlorination. Is there a problem? *J. Inst. Water environ. Manage.*, 1, 61-66

Fielding, M. & Horth, H. (1986) Formation of mutagens and chemicals during water treatment chlorination. *Water Supply*, 4, 103-126

Fielding, M. & Horth, H. (1988) The formation and removal of chemical mutagens during drinking water treatment. In: Angeletti, G. & Bjørseth, A., eds, *Organic Micropollutants in the Aquatic Environment*, Dordrecht, Kluwer Academic Publishers, pp. 284-298

Flanagan, E.P. & Allen, H.E. (1981) Effect of water treatment on mutagenic potential. *Bull. environ. Contam. Toxicol.*, 27, 765-772

Forster, R. & Wilson, I. (1981) The application of mutagenicity testing to drinking water. *J. Inst. Water Eng. Sci.*, 35, 259-274

Galassi, S., Guzzella, L. & Sora, S. (1989) Mutagenic potential of drinking waters from surface supplies in northern Italy. *Environ. Toxicol. Chem.*, 8, 109-116

Gibson, T.M., Haley, J., Righton, M. & Watts, C.D. (1986) Chlorination of fatty acids during water treatment disinfection: reactivity and product identification. *Environ. Technol. Lett.*, 7, 365-372

Glaze, W.H. & Rawley, R.A. (1979) A preliminary survey of trihalomethane levels in selected East Texas water supplies. *J. Am. Water Works Assoc.*, 71, 509-515

Glaze, W.H., Burleson, J.L., Henderson, J.E., IV, Jones, P.C., Kinstley, W., Peyton, G.R., Rawley, R., Saleh, F.Y. & Smith, G. (1982) *Analysis of Chlorinated Organic Compounds Formed During Chlorination of Wastewater Products* (EPA-600/4-82-072), Athens, GA, US Environmental Protection Agency

Gottlieb, M.S. & Carr, J.K. (1982) Case-control cancer mortality study and chlorination of drinking water in Louisiana. *Environ. Health Perspect.*, 46, 169-177

Gottlieb, M.S., Carr, J.K. & Morris, D.T. (1981) Cancer and drinking water in Louisiana: colon and rectum. *Int. J. Epidemiol.*, 10, 117-125

Gottlieb, M.S., Carr, J.K. & Clarkson, J.R. (1982) Drinking water and cancer in Louisiana. A retrospective mortality study. *Am. J. Epidemiol.*, 116, 652-667

Gould, J.P. & Hay, T.R. (1982) The nature of the reactions between chlorine and purine and pyrimidine bases: products and kinetics. *Water Sci. Technol.*, 14, 629-640

Gould, J.P., Richards, J.T. & Miles, M.G. (1984) The kinetics and primary products of uracil chlorination. *Water Res.*, 18, 205-212

Grabow, W.O.K., Van Rossum, P.G., Grabow, N.A. & Denkhaus, R. (1981) Relationship of the raw water quality to mutagens detectable by the Ames *Salmonella*/microsome assay in a drinking-water supply. *Water Res.*, 15, 1037-1043

de Greef, E., Morris, J.C., van Kreijl, C.F. & Morra, C.F.H. (1980) Health effects in the chemical oxidation of polluted waters. In: Jolley, R.L., Brungs, W.A., Cumming, R.B. & Jacobs, V.A., eds, *Water Chlorination: Environmental Impact and Health Effects*, Vol. 3, Ann Arbor, MI, Ann Arbor Science, pp. 913-924

Grimm-Kibalo, S.M., Glatz, B.A. & Fritz, J.S. (1981) Seasonal variation of mutagenic activity in drinking water. *Bull. environ. Contam. Toxicol.*, 26, 188-195

Grob, K. & Habich, A. (1983) Trace analysis of halocarbons in water; direct aqueous injection with electron capture detection. *J. high Resolut. Chromatogr.*, 6, 11-15

Harrington, T.R., Nestmann, E.R. & Kowbel, D.J. (1983) Suitability of the modified fluctuation assay for evaluating the mutagenicity of unconcentrated drinking water. *Mutat. Res.*, 120, 97-103

Hayatsu, H., Pan, S.-K. & Ukita, T. (1971) Reaction of sodium hypochlorite with nucleic acids and their constituents. *Chem. pharm. Bull.*, 19, 2189-2192

Hemming, J., Holmbom, B., Reunanen, M. & Kronberg, L. (1986) Determination of the strong mutagen 3-chloro-4-(dichloromethyl)-5-hydroxy-2(5H)-furanone in chlorinated drinking and humic waters. *Chemosphere*, 15, 549-556

Hertz-Picciotto, I., Swan, S.H., Neutra, R.R. & Samuels, S.J. (1989) Spontaneous abortions in relation to consumption of tap-water: an application of methods from survival analysis to a pregnancy follow-up study. *Am. J. Epidemiol.*, 130, 79-93

Hoff, J.C. & Akin, E.W. (1986) Microbial resistance to disinfectants: mechanisms and significance. *Environ. Health Perspect.*, 69, 7-13

Hogan, M.D., Chi, P.-Y., Hoel, D.G. & Mitchell, T.J. (1979) Association between chloroform levels in finished drinking water supplies and various site-specific cancer mortality rates. *J. environ. Pathol. Toxicol.*, 2, 873-887

Holmbom, B., Voss, R.H., Mortimer, R.D. & Wong, A. (1984) Fractionation, isolation, and characterisation of Ames mutagenic compounds in kraft chlorination effluents. *Environ. Sci. Technol.*, 18, 333-337

Holmbom, B., Kronberg, L., Backlund, P., Längvik, V.-A., Hemming, J., Reunanen, M., Smeds, A. & Tikkanen, L. (1990) Formation and properties of 3-chloro-4-(dichloromethyl)-5-hydroxy-2(5H)-furanone, a potent mutagen, in chlorinated waters. In: Jolley, R.L., Condie, L.W., Johnson, J.D., Katz, S., Minear, R.A., Mattice, J.S. & Jacobs, V.A., eds, *Water Chlorination: Chemistry, Environmental Impact and Health Effects*, Vol. 6, Chelsea, MI, Lewis Publishers, pp. 125-135

Hoover, R.N. & Strasser, P.H. (1980) Artificial sweeteners and human bladder cancer. Preliminary results. *Lancet*, i, 837-840

Horth, H. (1989) Identification of mutagens in drinking water. *Aqua*, 38, 80-100

Horth, H., Fielding, M., Gibson, T., James, H.A. & Ross, H. (1989) *Identification of Mutagens in Drinking Water* (Publication No. PRD 2038-M), Medmenham, Water Research Centre

Hueper, W.C. & Payne, W.W. (1963) Carcinogenic effects of adsorbates of raw and finished water supplies. *Am. J. clin. Pathol.*, 39, 475-481

Hueper, W.C. & Ruchhoft, C.C. (1954) Carcinogenic studies on adsorbates of industrially polluted raw and finished water supplies. *Arch. ind. Hyg.*, 9, 488-495

IARC (1979a) *IARC Monographs on the Evaluation of the Carcinogenic Risk of Chemicals to Humans*, Vol. 20, *Some Halogenated Hydrocarbons*, Lyon, pp. 401-427

IARC (1979b) *IARC Monographs on the Evaluation of the Carcinogenic Risk of Chemicals to Humans*, Vol. 20, *Some Halogenated Hydrocarbons*, Lyon, pp. 429-448

IARC (1979c) *IARC Monographs on the Evaluation of the Carcinogenic Risk of Chemicals to Humans*, Vol. 20, *Some Halogenated Hydrocarbons*, Lyon, pp. 467-476

IARC (1986) *IARC Monographs on the Evaluation of the Carcinogenic Risk of Chemicals to Humans*, Vol. 41, *Some Halogenated Hydrocarbons and Pesticide Exposures*, Lyon, p. 320

IARC (1987) *IARC Monographs on the Evaluation of Carcinogenic Risks to Humans*, Suppl. 7, *Overall Evaluations of Carcinogenicity: An Updating of IARC Monographs Volumes 1 to 42*, Lyon

ICI Chemicals and Polymers Ltd (1988) *Chlor-Chemicals*, Runcorn

Isacson, P., Bean, J.A. & Lynch, C. (1983) Relationship of cancer incidence rates in Iowa municipalities to chlorination status of drinking water. In: Jolley R.J., Brungs, W.A., Cotruvo, J.A., Cumming, R.B., Mattice, J.S. & Jacobs, V.A., eds, *Water Chlorination: Environmental Impact and Health Effects*, Vol. 4, Book 2, *Environment, Health, and Risk*, Ann Arbor, MI, Ann Arbor Science, pp. 1353-1364

Italia, M.P. & Uden, P.C. (1988) Comparison of volatile halogenated compounds formed in the chloramination and chlorination of humic acid by gas chromatography-electron-capture detection. *J. Chromatogr.*, 449, 326-330

Jacangelo, J.G., Patania, N.L., Reagan, K.M., Aieta, E.M., Krasner, S.W. & McGuire, M.J. (1989) Ozonation: assessing its role in the formation and control of disinfection by-products. *J. Am. Water Works Assoc.*, *81*, 74-84

Kanarek, M.S. & Young, T.B. (1982) Drinking water treatment and risk of cancer death in Wisconsin. *Environ. Health Perspect.*, *46*, 179-186

Kavlock, R., Chernoff, N., Carver, B. & Kopfler, F. (1979) Teratology studies in mice exposed to municipal drinking-water concentrates during organogenesis. *Food Cosmet. Toxicol.*, *17*, 343-347

Keith, L.H., ed. (1976) *Identification and Analysis of Organic Pollutants in Water*, Ann Arbor, MI, Ann Arbor Science

Keith, L.H. (1981) *Advances in the Identification and Analysis of Organic Pollutants in Water*, Vols 1 and 2, Ann Arbor, MI, Ann Arbor Science

Kfir, R. & Prozesky, O.W. (1982) Detection of potential carcinogens and toxicants in tap and reclaimed water by the golden hamster cell transformation assay. *Water Res.*, *16*, 1561-1568

Kool, H.J. & Hrubec, J. (1986) The influence of an ozone, chlorine and chlorine dioxide treatment on mutagenic activity in (drinking) water. *Ozone Sci. Eng.*, *8*, 217-234

Kool, H.J. & van Kreijl, C.F. (1984) Formation and removal of mutagenic activity during drinking water preparation. *Water Res.*, *18*, 1011-1016

Kool, H.J., van Kreijl, C.F., van Kranen, H.J. & de Greef, E. (1981) Toxicity assessment of organic compounds in drinking water in the Netherlands. *Sci. total Environ.*, *18*, 135-153

Kool, H.J., van Kreijl, C.F. & Zoeteman, B.C.J. (1982a) Toxicology assessment of organic compounds in drinking water. *Crit. Rev. environ. Control*, *12*, 307-357

Kool, H.J., van Kreijl, C.F., de Greef, E. & van Kranen, H.J. (1982b) Presence, introduction and removal of mutagenic activity during the preparation of drinking water in the Netherlands. *Environ. Health Perspect.*, *46*, 207-214

Kool, H.J., Kuper, F., van Haeringen, H. & Koeman, J.H. (1985a) A carcinogenicity study with mutagenic organic concentrates of drinking-water in the Netherlands. *Food chem. Toxicol.*, *23*, 79-85

Kool, H.J., Hrubec, J., van Kreijl, C.F. & Piet, G.J. (1985b) Evaluation of different treatment processes with respect to mutagenic activity in drinking water. *Sci. total Environ.*, *47*, 229-256

Kool, H.J., van Kreijl, C.F. & Hrubec, J. (1985c) Mutagenic and carcinogenic properties of drinking water. In: Jolley, R.L., Bull, R.J., Davis, W.P., Katz, S., Roberts, M.H., Jr & Jacobs, V.A., eds, *Water Chlorination: Chemistry, Environmental Impact and Health Effects*, Vol. 5, Chelsea, MI, Lewis Publishers, pp. 187-205

Kopfler, F.C., Ringhand, H.P., Coleman, W.E. & Meier, J.R. (1985) Reactions of chlorine in drinking water, with humic acids and *in vivo*. In: Jolley, R.L., Bull, R.J., Davis, W.P., Katz, S., Roberts, M.H., Jr & Jacobs, V.A., eds, *Water Chlorination: Chemistry, Environmental Impact and Health Effects*, Vol. 5, Chelsea, MI, Lewis Publishers, pp. 161-173

Kowbel, D.J., Malaiyandi, M., Paramasigamani, V. & Nestmann, E.R. (1984) Chlorination of ozonated soil fulvic acid: mutagenicity studies in *Salmonella*. *Sci. total Environ.*, 37, 171-176

Kowbel, D.J., Ramaswamy, S., Malaiyandi, M. & Nestmann, E.R. (1986) Mutagenicity studies in *Salmonella*: residues of ozonated and/or chlorinated water fulvic acids. *Environ. Mutagenesis*, 8, 253-262

Krasner, S.W., McGuire, M.J., Jacangelo, J.G., Patania, N.L., Reagan, K.M. & Aieta, E.M. (1989) The occurrence of disinfection by-products in US drinking water. *J. Am. Water Works Assoc.*, 81, 41-53

Kraybill, H.F. (1980) Evaluation of public health aspects of carcinogenic/mutagenic biorefractories in drinking water. *Prev. Med.*, 9, 212-218

Kronberg, L. & Vartiainen, T. (1988) Ames mutagenicity and concentration of the strong mutagen 3-chloro-4-(dichloromethyl)-5-hydroxy-2(5H)-furanone and of its geometric isomer E-2-chloro-3-(dichloromethyl)-4-oxo-butenoic acid in chlorine-treated tap waters. *Mutat. Res.*, 206, 177-182

Kronberg, L., Holmbom, B. & Tikkanen, L. (1985a) Mutagenic activity in drinking water and humic water after chlorine treatment. *Vatten*, 41, 106-109

Kronberg, L., Holmbom, B. & Tikkanen, L. (1985b) Fractionation of mutagenic compounds formed during chlorination of humic water. *Sci. total Environ.*, 47, 343-347

Kronberg, L., Holmbom, B. & Tikkanen, L. (1990) Identification of the strong mutagen, 3-chloro-4-(dichloromethyl)-5-hydroxy-2(5H)-furanone and of its geometric isomer E-2-chloro-3-(dichloromethyl)-4-oxobutenoic acid, in mutagenic fractions of chlorine-treated humic water and drinking waters. In: Jolley, R.L., Condie, L.W., Johnson, J.D., Katz, S., Mincar, R.A., Mattice, J.S. & Jacobs, V.A., eds, *Water Chlorination: Chemistry, Environmental Impact and Health Effects*, Vol. 6, Chelsea, MI, Lewis Publishers, pp. 137-146

Kuzma, R.J., Kuzma, C.M. & Buncher, C.R. (1977) Ohio drinking water source and cancer rates. *Am. J. public Health*, 67, 725-729

Lahl, U., Cetinkaya, M., Duszeln, J.V., Gabel, B., Stachel, B. & Thiemann, W. (1982) Health risks for infants caused by trihalomethane generation during chemical disinfection of feeding utensils. *Ecol. Food Nutr.*, 12, 7-17

Lahl, U., Stachel, B., Schroer, W. & Zeschman, B. (1984) Determination of halogenated organic acids in water samples. *Z. Wasser Abwasser Forsch.*, 17, 45-49

Lawrence, C.E., Taylor, P.R., Trock, B.J. & Reilly, A.A. (1984) Trihalomethanes in drinking water and human colorectal cancer. *J. natl Cancer Inst.*, 72, 563-568

Le Cloirec, C. & Martin, G. (1985) Evolution of amino acids in water treatment plants and the effect of chlorination on amino acids. In: Jolley, R.L., Bull, R.J., Davis, W.P., Katz, S., Roberts, M.H., Jr & Jacobs, V.A., eds, *Water Chlorination: Chemistry, Environmental Impact and Health Effects*, Vol. 5, Chelsea, MI, Lewis Publishers, pp. 821-834

de Leer, E.W.B. (1987) *Aqueous Chlorination Products: The Origin of Organochlorine Compounds in Drinking and Surface Waters*, Delft, University Press

Liimatainen, A., Müller, D., Vartiainen, T., Jahn, F., Kleeberg, U., Klinger, W. & Hänninen, O. (1988) Chlorinated drinking water is mutagenic and causes 3-methylcholanthrene type induction of hepatic monooxygenase. *Toxicology, 51*, 281-289

Loper, J.C. (1980) Mutagenic effects of organic compounds in drinking water. *Mutat. Res., 76*, 241-268

Loper, J.C., Tabor, M.W., Rosenblum, L. & DeMarco, J. (1985) Continuous removal of both mutagens and mutagen-forming potential by an experimental full-scale granular activated carbon treatment system. *Environ. Sci. Technol., 19*, 333-339

Lukasewycz, M.T., Bieringer, C.M., Liukkonen, R.J., Fitzsimmons, M.E., Corcoran, H.F., Lin, S. & Carlson, R.M. (1989) Analysis of inorganic and organic chloramines: derivatization with 2-mercaptobenzothiazole. *Environ. Sci. Technol., 23*, 196-199

Lynch, C.F., Woolson, R.F., O'Gorman, T. & Cantor, K.P. (1989) Chlorinated drinking water and bladder cancer: effect of misclassification on risk estimates. *Arch. environ. Health, 44*, 252-259

Ma, T.-H., Anderson, V.A., Harris, M.M., Neas, R.E. & Lee, T.-S. (1985) Mutagenicity of drinking water detected by the *Tradescantia* micronucleus test. *Can. J. Genet. Cytol., 27*, 143-150

Maruoka, S. (1986) Analysis of mutagenic by-products produced by chlorination of humic substances by thin layer chromatography and high-performance liquid chromatography. *Sci. total Environ., 54*, 195-205

Maruoka, S. & Yamanaka, S. (1983) Mutagenic potential of laboratory chlorinated river water. *Sci. total Environ., 29*, 143-154

Mason, T.-J. & McKay, F.W. (1974) *US Cancer Mortality by County: 1950-1969*, Washington DC, US Government Printing Office

McKague, A.B., Lee, E.G.-H. & Douglas, G.R. (1981) Chloroacetones: mutagenic constituents of bleached kraft chlorination effluent. *Mutat. Res., 91*, 301-306

McKinney, J.D., Maurer, R.R., Hass, J.R. & Thomas, R.O. (1976) Possible factors in the drinking water of laboratory animals causing reproductive failure. In: Keith, L.H., ed., *Advances in Identification and Analysis of Organic Pollutants in Water*, Ann Arbor, MI, Ann Arbor Science, pp. 417-432

Meier, J.R. (1988) Genotoxic activity of organic chemicals in drinking water. *Mutat. Res., 196*, 211-245

Meier, J.R., Lingg, R.D. & Bull, R.J. (1983) Formation of mutagens following chlorination of humic acid: a model for mutagen formation during drinking water treatment. *Mutat. Res., 118*, 25-41

Meier, J.R., Ringhand, H.P., Coleman, W.E., Munch, J.W., Streicher, R.P., Kaylor, W.H. & Schenk, K.M. (1985) Identification of mutagenic compounds formed during chlorination of humic acid. *Mutat. Res., 157*, 111-122

Meier, J.R., Ringhand, H.P., Coleman, W.E., Schenck, K.M., Munch, J.W., Streicher, R.P., Kaylor, W.H. & Kopfler, F.C. (1986) Mutagenic by-products from chlorination of humic acid. *Environ. Health Perspect., 69*, 101-107

Meier, J.R., Knohl, R.B., Coleman, W.E., Ringhand, H.P., Munch, J.W., Kaylor, W.H., Streicher, R.P. & Kopfler, F.C. (1987) Studies on the potent bacterial mutagen, 3-chloro-4-(dichloromethyl)-5-hydroxy-2(5H)-furanone: aqueous stability, XAD recovery and analytical determination in drinking water and in chlorinated humic acid solutions. *Mutat. Res.*, *189*, 363-373

Miller, J.W., Uden, P.C. & Barnes, R.M. (1982) Determination of trichloroacetic acid at the part-per-billion level in water by precolumn trap enrichment gas chromatography with microwave plasma emission detection. *Anal. Chem.*, *54*, 485-488

Miller, R.G., Kopfler, F.C., Condie, L.W., Pereira, M.A., Meier, J.R., Ringhand, H.P., Robinson, M. & Casto, B.C. (1986) Results of toxicological testing of Jefferson Parish pilot plant samples. *Environ. Health Perspect.*, *69*, 129-139

Monarca, S., Pasquini, R. & Scasselati Sforzolini, G. (1985a) Mutagenicity assessment of different drinking water supplies before and after treatments. *Bull. environ. Contam. Toxicol.*, *34*, 815-823

Monarca, S., Pasquini, R. & Arcaleni, P. (1985b) Detection of mutagens in unconcentrated and concentrated drinking water supplies before and after treatment using a microscale fluctuation test. *Chemosphere*, *14*, 1069-1080

National Academy of Sciences (1979) *The Chemistry of Disinfectants in Water: Reactions and Products* (NTIS PB-292 776), Washington DC, US Environmental Protection Agency

National Research Council (1980) *Drinking Water and Health*, Vol. 2, Washington DC, National Academy Press

Nestmann, E. (1983) Mutagenic activity of drinking water. In: Stich, H.F., ed., *Carcinogens and Mutagens in the Environment*, Vol. 3, *Naturally Occurring Compounds: Epidemiology and Distribution*, Boca Raton, FL, CRC Press, pp. 137-147

Nestmann, E.R., LeBel, G.L., Williams, D.T. & Kowbel, D.J. (1979) Mutagenicity of organic extracts from Canadian drinking water in the *Salmonella*/mammalian-microsome assay. *Environ. Mutagenesis*, *1*, 337-345

Nicholson, A.A., Meresz, O. & Lemyk, B. (1977) Determination of free and total potential haloforms in drinking water. *Anal. Chem.*, *49*, 814-819

Nicholson, B.C., Hayes, K.P. & Bursill, D.B. (1984) By-products of chlorination. *Water, September*, 11-15

Norwood, D.L., Christman, R.F., Johnson, J.D. & Hass, J.R. (1986) Using isotope dilution mass spectrometry to determine aqueous trichloroacetic acid. *J. Am. Water Works Assoc.*, *78*, 175-180

Oake, R.J. & Anderson, I.M. (1984) *The Determination of Carbon Adsorbable Organo-halide in Waters* (Publ. No. TR 217), Medmenham, Water Research Centre

Oliver, B.G. (1983) Dihaloacetonitriles in drinking water: algae and fulvic acid as precursors. *Environ. Sci. Technol.*, *17*, 80-83

Otson, R., Williams, D.T. & Bothwell, P.D. (1979) A comparison of dynamic and static head head space and solvent extraction techniques for the determination of trihalomethanes in water. *Environ. Sci. Technol.*, *13*, 936-939

Page, G.W. (1981) Comparison of groundwater and surface water for patterns and levels of contamination by toxic substances. *Environ. Sci. Technol.*, *15*, 1475-1481

Page, T., Harris, R.H. & Epstein, S.S. (1976) Drinking water and cancer mortality in Louisiana. *Science*, *193*, 55-57

Pelon, W., Beasley, T.W. & Lesley, D.E. (1980) Transformation of the mouse clonal cell line R846-DP8 by Mississippi river, raw, and finished water samples from southeastern Louisiana. *Environ. Sci. Technol.*, *14*, 723-726

Perwak, J., Goyer, M., Harris, J., Schimke, G., Scow, K., Wallace, D. & Slimak, M. (1980) *An Exposure and Risk Assessment for Trihalomethanes: Chloroform, Bromoform, Bromodichloromethane, Dibromochloromethane* (EPA-440/4-81-018; NTIS PB 85-211977), Washington DC, Office of Water Regulations and Standards, US Environmental Protection Agency

Peters, C.J. (1980) *Trihalomethane Formation Arising from the Chlorination of Potable Waters*, PhD Thesis, London, Imperial College of Science and Technology

Peters, R.J.B., Versteegh, J.F.M. & de Leer, E.W.B. (1989) Dihaloacetonitriles in drinking water in the Netherlands (Dutch). H_2O, *22*, 800-804

Pierce, R.C. (1978) *The Aqueous Chlorination of Organic Compounds: Chemical Reactivity and Effects on Environmental Quality* (NRCC Publ. No. 16450), Ottawa, National Research Council of Canada

Pommery, J., Imbenotte, M., Urien, A.F., Marzin, D. & Erb, F. (1989) SOS chromotest study concerning some appreciation criteria of humic substances' genotoxic potency. *Mutat. Res.*, *223*, 183-189

Quaghebeur, D. & De Wulf, E. (1980) Volatile halogenated hydrocarbons in Belgian drinking waters. *Sci. total Environ.*, *14*, 43-52

Rapson, W.H., Isacovics, B. & Johnson, C.I. (1985) Mutagenicity produced by aqueous chlorination of tyrosine. In: Jolley, R.L., Bull, R.J., Davis, W.P., Katz, S., Roberts, M.H., Jr & Jacobs, V.A., eds, *Water Chlorination: Chemistry, Environmental Impact and Health Effects*, Vol. 5, Chelsea, MI, Lewis Publishers, pp. 237-249

Rausch, L.L. (1980) *Chlorination of Drinking Water and Pregnancy Outcomes in New York Villages, 1968-1977*, PhD Thesis, Columbia University, New York City, NY

Reding, R., Fair, P.S., Shipp, C.J. & Brass, H.J. (1989) Measurement of dihaloacetonitriles and chloropicrin in US drinking water. In: *Disinfection By-products: Current Perspectives*, Denver, CO, American Water Works Association, pp. 11-22

Renberg, L. (1981) Gas chromatographic determination of phenolic compounds in water as their pentafluorobenzoyl derivatives. *Chemosphere*, *10*, 767-773

Ringhand, H.P., Meier, J.R., Kopfler, F.C., Schenck, K.M., Kaylor, W.H. & Mitchell, D.E. (1987) Importance of sample pH on recovery of mutagenicity from drinking water by XAD resins. *Environ. Sci. Technol.*, *21*, 382-387

Robinson, M., Glass, J.W., Cmehil, D., Bull, R.J. & Orthoefor, J.G. (1981) The initiation and promoting activity of chemicals isolated from drinking waters in the SENCAR mouse: a five-city survey. In: Waters, M.D., Sandhu, S.S., Huisingh, J.L., Claxton, L. & Nesnow, S., eds, *Short-term Bioassays in the Analysis of Complex Environmental Mixtures. II* (Environmental Science Research Series, Vol. 22), New York, Plenum Press, pp. 177-188

Rook, J.J. (1974) Formation of haloforms during chlorination of natural waters. *Water Treat. Exam.*, *23*, 234-243

Rook, J.J. (1977) Chlorination reactions of fulvic acids in natural waters. *Environ. Sci. Technol.*, *11*, 478-482

Rook, J.J. (1980) Possible pathways for the formation of chlorinated degradation products during chlorination of humic acids and resorcinol. In: Jolley, R.L., Brungs, W.A., Cumming, R.B. & Jacobs, V.A., eds, *Water Chlorination: Chemistry, Environmental Impact and Health Effects*, Vol. 3, Ann Arbor, MI, Ann Arbor Science, pp. 85-98

Schwartz, D.J., Saxena, J. & Kopfler, F.C. (1979) Water distribution system, a new source of mutagens in drinking waters. *Environ. Sci. Technol.*, *13*, 1138-1141

Scully, F.E., Jr, Yang, J.P., Mazina, K. & Daniel, F.B. (1984) Derivatization of organic and inorganic *N*-chloramines for high performance liquid chromatographic analysis of chlorinated water. *Environ. Sci. Technol.*, *18*, 787-792

Sithole, B.B. & Williams, D.T. (1986) Halogenated phenols in water at forty Canadian potable water treatment facilities. *J. Assoc. off. anal. Chem.*, *69*, 807-810

Standing Committee of Analysts (1980) *Chloro and Bromo Trihalogenated Methanes in Water* (Methods for the Examination of Waters and Associated Materials), London, Her Majesty's Stationery Office

Standing Committee of Analysts (1985) *Chlorophenoxy Acidic Herbicides, Trichloroacetic Acid, Chlorophenols, Triazines and Glyphosate in Water 1985* (Methods for the Examination of Waters and Associated Materials), London, Her Majesty's Stationery Office

Standing Committee of Analysts (1988) *The Determination of Microgram and Submicrogram Amounts of Individual Phenols in River and Potable Waters* (Methods for the Examination of Waters and Associated Materials), London, Her Majesty's Stationery Office

Staples, R.E., Worthy, W.C. & Marks, T.A. (1979) Influence of drinking water—tap versus purified—on embryo and fetal development in mice. *Teratology*, *19*, 237-244

Strömberg, L.M., de Sousa, F., Ljungquist, P., McKague, B. & Kringstad, K.P. (1987) An abundant chlorinated furanone in the spent chlorination liquor from pulp bleaching. *Environ. Sci. Technol.*, *21*, 754-756

Suffet, I.H., Brenner, L. & Cairo, P.R. (1980) GC/MS identification of trace organics in Philadelphia drinking waters during a 2-year period. *Water Res.*, *14*, 853-867

Symons, J.M., Bellar, T.A., Carswell, J.K., DeMarco, J., Kropp, K.L., Robeck, G.G., Seeger, D.R., Slocum, C.J., Smith, B.L. & Stevens, A.A. (1975) National organics reconnaissance survey for halogenated organics. *J. Am. Water Works Assoc.*, *67*, 634-647

Thurman, E.M. (1985) *Organic Geochemistry of Natural Waters*, Dordrecht, Martinus Nijhoff/Dr W. Junk Publishers

Trehy, M.L. & Bieber, T.I. (1981) Detection, identification and quantitative analysis of dihaloacetonitriles in chlorinated natural waters. In: Keith, L.H., ed., *Advances in the Identification and Analysis of Organic Pollutants in Water*, Vol. 2, Ann Arbor, MI, Ann Arbor Science, pp. 941-975

Truhaut, R., Gak, J.C. & Graillot, C. (1979) Studies on the risks that may result from chemical pollution of drinking-water—I. Study of the long-term toxicity in rats and mice of organic micropollutants extracted from drinking water with chloroform (Fr.). *Water Res.*, 13, 689-697

Tuthill, R.W. & Moore, G.S. (1980) Drinking water chlorination: a practice unrelated to cancer mortality. *J. Am. Water Works Assoc.*, 72, 570-573

Tuthill, R.W., Giusti, R.A., Moore, G.S. & Calabrese, E.J. (1982) Health effects among newborns after prenatal exposure to ClO_2-disinfected drinking water. *Environ. Health Perspect.*, 46, 39-45

Uden, P.C. & Miller, J.W. (1983) Chlorinated acids and chloral in drinking water. *J. Am. Water Works Assoc.*, 75, 524-427

US Environmental Protection Agency (1979a) *EPA 501-2*, Cincinnati, OH, Environmental Monitoring and Support Laboratory

US Environmental Protection Agency (1979b) *EPA 501-1*, Cincinnati, OH, Environmental Monitoring and Support Laboratory

US Environmental Protection Agency (1985) *Health and Environmental Effects Profile for Bromochloromethanes* (EPA-600/X-85-397; NTIS PB 88-174610), Cincinnati, OH, Environmental Criteria Assessment Office

Van Der Gaag, M.A., Noordsij, A. & Oranje, J.P. (1982) Presence of mutagens in Dutch surface water and effects of water treatment processes for drinking water preparation. In: Sorsa, M. & Vainio, H., eds, *Mutagens in Our Environment*, New York, Alan R. Liss, pp. 277-286

Van Duuren, B.L., Melchionne, S., Seidman, I. & Pereira, M.A. (1986) Chronic bioassays of chlorinated humic acids in $B6C3F_1$ mice. *Environ. Health Perspect.*, 69, 109-117

Vartiainen, T. & Liimatainen, A. (1986) High levels of mutagenic activity in chlorinated drinking water in Finland. *Mutat. Res.*, 169, 29-34

Vartiainen, T., Liimatainen, A., Jääskeläinen, S. & Kauranen, P. (1987a) Comparison of solvent extractions and resin adsorption for isolation of mutagenic compounds from chlorinated drinking water with high humus content. *Water Res.*, 21, 773-779

Vartiainen, T., Liimatainen, A., Keränen, P., Ala-Peijari, T. & Kalliokoski, P. (1987b) Effect of permanganate oxidation and chlorination on the mutagenic activity and other quality parameters of artificially recharged ground water processed from humus-rich surface water. *Chemosphere*, 16, 1489-1499

Vartiainen, T., Liimatainen, A., Kauranen, P. & Hiisvirta, L. (1988) Relations between drinking water mutagenicity and water quality parameters. *Chemosphere*, 17, 189-202

Water Research Centre (1980) *Trihalomethanes in Water: Seminar Held at the Lorch Foundation, Lane End, Buckinghamshire, UK, 16-17 January 1980*, Medmenham, UK

Wei, R.-D., Chang, S.-C. & Jeng, M.-H. (1984) Mutagenicity of organic extracts from drinking water in the Taipei area. *Natl Sci. Council Monthly ROC*, 11, 1565-1572

White, G.C., ed. (1986) *The Handbook of Chlorination*, 2nd ed., New York, Van Nostrand Reinhold

Wigilius, B., Borén, H., Carlberg, G.E., Grimvall, A. & Möller, M. (1985) A comparison of methods for concentrating mutagens in drinking water—recovery aspects and their implications for the chemical character of major unidentified mutagens. *Sci. total Environ.*, *47*, 265-272

Wilcox, P. & Denny, S. (1985) Effect of dechlorinating agents on the mutagenic activity of chlorinated water samples. In: Jolley, R.L., Bull, R.J., Davis, W.P., Katz, S., Roberts, M.H., Jr & Jacobs, V.A. eds, *Water Chlorination: Chemistry, Environmental Impact and Health Effects*, Vol. 5, Chelsea, MI, Lewis Publishers, pp. 1341-1353

Wilcox, P. & Williamson, S. (1986) Mutagenic activity of concentrated drinking water samples. *Environ. Health Perspect.*, *69*, 141-149

Wilcox, P., van Hoof, F. & Van Der Gaag, M. (1986) Isolation and characterization of mutagens from drinking water (Abstract). In: Léonard, A. & Kirsch-Volders, M., eds, *Proceedings of the XVIth Annual Meeting of the European Environmental Mutagen Society, 23-30 August 1986, Brussels*, Brussels, Free University, pp. 92-103

Wilkins, J.R., III & Comstock, G.W. (1981) Source of drinking water at home and site-specific cancer incidence in Washington County, Maryland. *Am. J. Epidemiol.*, *114*, 178-190

World Health Organization (1984) *Guidelines for Drinking Water Quality*, Vols 1 and 2, Geneva

World Health Organization (1988) *Introduction to National Seminars on Drinking-water Quality*, Geneva, p. V-7

Young, T.B. & Kanarek, M.S. (1983) Matched pair case control study of drinking water chlorination and cancer mortality. In: Jolley R.J., Brungs, W.A., Cotruvo, J.A., Cumming, R.B., Mattice, J.S. & Jacobs, V.A., eds, *Water Chlorination: Environmental Impact and Health Effects*, Vol. 4, Book 2, *Environmental Health and Risk*, Ann Arbor, MI, Ann Arbor Science, pp. 1365-1380

Young, T.B., Kanarek, M.S. & Tsiatis, A.A. (1981) Epidemiologic study of drinking water chlorination and Wisconsin female cancer mortality. *J. natl Cancer Inst.*, *67*, 1191-1198

Young, T.B., Wolf, D.A. & Kanarek, M.S. (1987) Case-control study of colon cancer and drinking water trihalomethanes in Wisconsin. *Int. J. Epidemiol.*, *16*, 190-197

Zeighami, E.A., Watson, A.P. & Craun, G.F. (1990) Chlorination, water hardness and serum cholesterol in forty-six Wisconsin communities. *Int. J. Epidemiol.*, *19*, 49-58

Zierler, S., Danley, R.A. & Feingold, L. (1986) Type of disinfectant in drinking water and patterns of mortality in Massachusetts. *Environ. Health Perspect.*, *69*, 275-279

Zierler, S., Feingold, L., Danley, R.A. & Craun, G. (1988) Bladder cancer in Massachusetts related to chlorinated and chloraminated drinking water: a case-control study. *Arch. environ. Health*, *43*, 195-200

Zoeteman, B.C.J., Hrubec, J., de Greef, E. & Kool, H.J. (1982) Mutagenic activity associated with by-products of drinking water disinfection by chlorine, chlorine dioxide, ozone and UV-irradiation. *Environ. Health Perspect.*, *46*, 197-205

SOME CHEMICALS USED IN THE CHLORINATION OF DRINKING-WATER

SODIUM CHLORITE

1. Chemical and Physical Data

1.1 Synonyms and molecular formulae and weights

Sodium chlorite

Chem. Abstr. Services Reg. No: 7758-19-2
Synonym: Chlorous acid, sodium salt
Molecular formula: $NaClO_2$
Molecular weight: 90.44

Sodium chlorite trihydrate

Chem. Abstr. Services Reg. No.: 49658-21-1
Synonym: Chlorous acid, sodium salt, trihydrate
Molecular formula: $NaClO_2 \cdot 3H_2O$
Molecular weight: 144.49

1.2 Chemical and physical properties of the pure substance

From Weast (1989) unless otherwise specified

Sodium chlorite

(a) *Description*: White, hygroscopic crystals
(b) *Melting-point*: Decomposes at 180-200°C
(c) *Solubility*: Soluble in water (g/l): 390 at 17°C, 550 at 60°C
(d) *Stability*: Stable in the absence of oxidizable organic matter (Budavari, 1989)
(e) *Reactivity*: Oxidizes many organic and inorganic substances; can explode in response to physical shock or heat when mixed with combustibles (Canadian Centre for Occupational Health and Safety, 1989). In aqueous

alkaline solutions, chlorite ion is very stable, even at 100°C; in acid solutions, chlorite forms chlorous acid ($HClO_2$), which rapidly disproportionates to chlorine dioxide, chlorate and chloride (Aieta & Roberts, 1985).

Sodium chlorite, trihydrate

(a) *Description*: Triclinic leaflets

(b) *Stability*: Becomes anhydrous at 38°C or in a dessicator over potassium hydroxide at room temperature (Budavari, 1989)

1.3 Technical products and impurities

Trade names: Neo Silox D; Textone

Sodium chlorite is the only chlorite salt marketed and used in commercially significant quantities. The technical-grade usually contains approximately 80% sodium chlorite (Noack & Doerr, 1979). Products are available in dry form or as aqueous solutions containing 25-80% sodium chlorite (Olin Corporation, 1989). A small amount of sodium hydroxide is usually retained to stabilize the product, and sodium chloride is usually added to reduce the sodium chlorite content to a maximum of about 80%; this is necessary to ensure safety and minimize reactivity (Noack & Doerr, 1979).

2. Production, Use, Occurrence and Analysis

2.1 Production and use

(a) *Production*

Sodium chlorite was described in 1843 by N. Milan; it was first produced in the USA by the Mathieson Company in 1937. In 1948, Degussa AG became the first German producer (Eul, 1989).

The commercial manufacture of sodium chlorite depends entirely on chlorine dioxide made from sodium chlorate. Generally, chlorine dioxide is absorbed in caustic soda containing hydrogen peroxide as a reducing agent to produce sodium chlorite. An excess of hydrogen peroxide is important to prevent disproportionation, which yields sodium chlorate. Sodium chloride may be added to decrease the sodium chlorite content to 80% of dry product weight (Noack & Doerr, 1979).

Commercially significant quantities of sodium chlorite are produced mainly in Japan, the European Economic Community and the USA. Accurate production statistics are not available. The Japanese Ministry of International Trade and Industry reported production of 6200 tonnes in 1973 and 5400 tonnes in 1974, although it is unclear whether these statistics refer strictly to sodium chlorite (Noack & Doerr, 1979). Japanese production of sodium chlorite (100%) was 4000 tonnes in 1984, 3815 tonnes in 1985, 3634 tonnes in 1986, 3811 tonnes in 1987 and 3503 tonnes in 1988 (Anon., 1985; Ministry of International Trade and Industry, 1989).

Sodium chlorite is produced by one company each in Belgium, Brazil, France, Italy, Spain and the UK, by two companies in Germany and the USA, and by five companies in Japan (Chemical Information Services Ltd, 1988).

(b) Use

Most of the sodium chlorite used in the USA is in the production of aqueous chlorine dioxide solutions at the site of use. This conversion can be carried out by the disproportionation of chlorous acid formed from chlorite in aqueous hydrochloric acid solution, but is more commonly achieved by the oxidation of chlorite by chlorine or hypochlorous acid (Aieta & Roberts, 1985; White, 1986). Chlorine dioxide is generated to bleach and strip textiles, to bleach wood pulp in paper processing, to eliminate tastes and odours in potable water, to reduce loads of adsorbable organic halogenated compound in industrial effluents, to control microbiological growth in paper mills, oil wells and petroleum systems and food processing plant flume water, to bleach (upgrade) fats and oils, to disinfect sewage, to treat factory wastes, where it converts simple phenolic compounds, simple cyanides and sulfides, to bleach natural foliage and to control algae in industrial cooling towers. Sodium chlorite is also used in the electronics industry for etching printed circuits because it oxidizes copper metal to copper[II] directly (Olin Corporation, 1984, 1989). The compound is used for the same processes in Europe and Asia, but more sodium chlorite is used to bleach textiles (Noack & Doerr, 1979).

2.2 Occurrence

(a) Natural occurrence

Sodium chlorite is not known to occur naturally.

(b) Occupational exposure

Industrial processes in which workers may be exposed to sodium chlorite include generation of chlorine dioxide in aqueous solutions at pulp and textile mills and the treatment of drinking-water; however, no data on occupational exposures were available to the Working Group. No regulatory standards or guidelines have been published for exposures to sodium chlorite.

2.3 Analysis

A method for the quantitative analysis of chlorite in water, which distinguishes chlorite from other oxychlorine species, has been reported (White, 1986).

3. Biological Data Relevant to the Evaluation of Carcinogenic Risk to Humans

3.1 Carcinogenicity studies in animals (Table 1)

(a) Oral administration

Mouse: Groups of 50 male and 50 female B6C3F$_1$ mice, six weeks old, were given 0.0, 0.025 or 0.05% sodium chlorite (82-87% pure [impurities unspecified]) in the drinking-water for 80 weeks. Survival at 85 weeks was: males—control, 35/50; low-dose, 47/50; high-dose, 43/50; females—control, 47/50; low-dose, 50/50; high-dose, 50/50. Hyperplastic nodules in the liver occurred in 6/35 control, 14/47 low-dose and 11/43 high-dose males, and hepatocellular carcinomas occurred in 4/35 control, 8/47 low-dose and 6/43 high-dose males; these differences are not significant. The incidences of liver tumours in control and treated females were similar. Adenomas of the lung occurred in 0/35 control, 2/47 low-dose and 5/43 high-dose male mice ($p < 0.05$, chi-square test), and adenocarcinomas occurred in 0/35 control, 1/47 low-dose and 2/43 high-dose males; the incidences of lung tumours were similar in control and treated females. The proportion of high-dose female mice with malignant lymphoma/leukaemia was significantly smaller than that in controls (control, 7/47; low-dose, 5/50; high-dose, 1/50; $p < 0.05$, chi-square test) (Yokose *et al.*, 1987). [The Working Group noted that survival-adjusted statistical analyses were not reported.]

Rat: Groups of 50 male and 50 female Fischer 344 rats, six weeks old, were given 0.0, 0.03 or 0.06% sodium chlorite [purity unspecified] in the drinking-water for 85 weeks. The study was terminated early because of pneumonia associated with Sendai virus infection. Survival at 85 weeks was: males—control, 34/50; low-dose, 30/50; high-dose, 43/50; females—control, 47/50; low-dose, 44/50; high-dose, 50/50. The proportions of low- and high-dose rats with tumours were not significantly greater than that of controls. Hyperplastic foci of the liver occurred in 0/34 control, 0/30 low-dose and 4/43 high-dose male rats (Shimoyama *et al.*, 1985).

(b) Skin application

Mouse: A group of 20 female Sencar mice, six weeks old, was given 0.2-ml topical applications of sodium chlorite [purity unspecified] at 20 mg/ml in acetone

(4 mg per mouse) twice a week for 51 weeks. A group of 15 female mice given topical applications of acetone was used as controls. All mice survived to the end of the study; no skin tumour was observed in the treated or control groups (Kurokawa et al., 1984). [The Working Group noted the small number of animals used.]

In an initiation-promotion study, a group of 20 female Sencar mice, six weeks old, was given a single topical application of 20 nmol (5.1 µg) 7,12-dimethylbenz[a]anthracene in acetone followed by 0.2-ml applications of sodium chlorite [purity unspecified] at 20 mg/ml in acetone twice a week for 51 weeks. A group of 15 female mice given a single application of 7,12-dimethylbenz[a]anthracene followed by applications of acetone was used as controls. The effective number of mice was 20 [number of survivors not given]. Squamous-cell carcinomas of the skin were seen in 5/20 treated mice and no control (Kurokawa et al., 1984).

3.2 Other relevant data

(a) Experimental systems

(i) Absorption, distribution, metabolism and excretion

Radiolabel (^{36}Cl) derived from sodium chlorite given to rats by gavage (0.15 mg/kg bw) was absorbed and appeared in serum at a rate constant of 0.2/h relative to an absorption constant of 0.04/h for sodium chloride also given by gavage. The radiolabel was eliminated from blood with a half-time of 35 h, compared to 52 h for sodium chloride. The authors reported that 32% of the original dose of sodium chlorite was eliminated as chloride in the urine, while 6% was eliminated as chlorite within 72 h. Of the recovered dose, 83% was found in urine and 13% in faeces. After periods ranging from 72 to 120 h, the highest concentration of radiolabel derived from sodium chlorite appeared in blood (Abdel-Rahman et al., 1982; Suh & Abdel-Rahman, 1983; Abdel-Rahman et al., 1984; Abdel-Rahman, 1985).

Haem catalysed the chlorination of monochlorodimedone (a model substrate for myeloperoxidase) by sodium chlorite *in vitro* (Wilson et al., 1983).

(ii) Toxic effects

The toxicological effects of chlorite have been reviewed (National Academy of Sciences, 1980, 1987).

Sodium chlorite induced haemolysis, decreased haemoglobin concentrations and loss of packed cell volume in the blood of male Sprague-Dawley rats treated for 30 and 60 days with 100-500 mg/l chlorite in drinking-water but did not induce methaemoglobinaemia at concentrations up to 500 mg/l chlorite. After 90 days of treatment, red blood cell counts and haemoglobin concentrations returned to normal; however, red blood cell glutathione concentrations remained significantly

Table 1. Summary of carcinogenicity studies of sodium chlorite in experimental animals

Reference	Species/strain	Sex	Dose schedule	Experimental parameter/observation	Group				Significance	Comments
					0	1	2	3		
Yokose et al. (1987)	Mouse B6C3F$_1$	M	Ad-lib. drinking-water, 80 weeks	Dose (%)	0	0.025	0.05	–		
				Survival (85 weeks)	35/50	47/50	43/50			
				Liver						
				Hyperplastic nodule	6/35	14/47	11/43			
				Carcinoma	4/35	8/47	6/43			
				Lung						
				Adenoma	0/35	2/47	5/43		$p < 0.05$	Increase
				Adenocarcinoma	0/35	1/47	2/43			
		F		Dose (%)	0	0.025	0.05	–		
				Survival (85 weeks)	47/50	50/50	50/50			
				Lymphoma/leukaemia	7/47	5/50	1/50		$p < 0.05$	Decrease
Kurokawa et al. (1984)	Mouse Sencar	F	2 d/week skin appl. acetone, 51 weeks	Dose (mg/mouse)	0	4	–	–		No skin tumour
				Survival (51 weeks)	15/15	20/20				
Shimoyama et al. (1985)	Rat	M	Ad-lib. drinking-water, 85 weeks	Dose (%)	0	0.03	0.06	–		
				Survival (85 weeks)	34/50	30/50	43/50			
				Hepatic hyperplasia	0/34	0/30	4/43			
		F		Dose (%)	0	0.03	0.06			
				Survival (85 weeks)	47/50	44/50	50/50			

depressed and 2,3-diphosphoglycerate levels elevated in animals treated with concentrations of chlorite as low as 50 mg/l (Heffernan *et al.*, 1979). Similar haematological effects have been observed in cats, rats, mice and monkeys exposed to similar concentrations (Heffernan *et al.*, 1979a; Abdel-Rahman *et al.*, 1980; Moore & Calabrese, 1980; Bercz *et al.*, 1982). Bercz *et al.* (1982) saw no effect on thyroid function in monkeys exposed to sodium chlorite for 30-60 days in drinking-water at doses up to 60 mg/kg per day.

Intraperitoneal injection of sodium chlorite at doses of 10 mg/kg and above produced methaemoglobinaemia in rats (Musil *et al.*, 1964; Heffernan *et al.*, 1979).

(iii) *Effects on reproduction and prenatal toxicity*

Reproductive effects: Female A/J mice were given sodium chlorite (100 mg/l) in distilled drinking-water throughout gestation and through 28 days of lactation (Moore *et al.*, 1980). Control mice received distilled drinking-water without sodium chlorite. The proportion of mice that became pregnant was smaller among treated than control mice. There was no change in litter size, weight at birth or neonatal survival through lactation, but pups in the sodium chlorite group had a significant reduction in weight gain.

Male and female Long-Evans rats were given 0, 1, 10 or 100 mg/l sodium chlorite in deionized drinking-water (Carlton *et al.*, 1987). Males were treated for 56 days before mating and during 10 days of mating; females were treated for 14 days before mating, throughout the mating period and gestation and through to day 21 of lactation. Additional males were given 0, 10, 100 or 500 mg/l sodium chlorite in drinking-water for 72-76 days to confirm the observed changes in sperm count, morphology and movement. There was no significant effect on the body weight of adults or offspring, but water consumption was decreased at the 500 mg/l dose level. There was no effect on fertility, litter size or survival of neonates or on the weight of the testis, epididymis or cauda epididymis. On histological examination of organs of the male and female reproductive tract, no chemically induced toxicity was seen. With 100 and 500 mg/l sodium chlorite, there was a subtle but reproducible increase in the number of abnormal sperm and a decrease in progressive sperm motility, as observed by videomicrography. Adverse effects in neonates were limited to a significant decrease in the levels of the thyroid hormones triiodothyronine and thyroxine in the serum on postnatal days 21 and 40.

Developmental toxicity: Groups of 4-10 Sprague-Dawley rats were treated with sodium chlorite as 0.1, 0.5 or 2% in drinking-water or by daily intraperitoneal injections of 10, 20 or 50 mg/kg bw, or by gavage at 200 mg/kg bw, on gestation days 8-15. Toxicity, including high mortality in some groups, was observed in groups receiving sodium chlorite by intraperitoneal injection at 20 or 50 mg/kg bw or by gavage, but not in the groups given sodium chlorite in the drinking-water or at 10 mg/kg bw intraperitoneally. The group receiving 0.1% in the drinking-water had a significant increase in body weight compared to controls, whereas significant decreases were observed in groups that received higher doses in the drinking-water and in the group receiving 20 mg/kg sodium chlorite intraperitoneally. There was no evidence of a teratogenic effect, and there was no effect on litter size, postnatal growth or survival (Couri *et al.*, 1982). [The Working Group noted the small group sizes.]

Female Sprague-Dawley rats were given sodium chlorite in the drinking-water at 0, 1 or 10 mg/l for 2.5 months and were then mated with untreated males. Treatment of the mated females continued through to gestation day 20. There were six to nine pregnant females per group. No apparent maternal toxicity and no change in litter size or weight was seen, but a dose-related increase in variants of the sternum and an increase in crown-rump length of fetuses occurred at the high-dose level (Suh *et al.*, 1983). [The Working Group noted the small group sizes.]

(iv) *Genetic and related effects* (Table 2)

Positive responses were reported with sodium chlorite in a *Salmonella typhimurium* reverse mutation assay, in a chromosomal aberration test with the Chinese hamster CHL cell line and in a micronucleus test in mice after a single intraperitoneal treatment. In mouse bone-marrow micronucleus tests, a weak response was observed following a single oral treatment, while no effect was observed after multiple oral or intraperitoneal administration. In mice, multiple oral treatments did not induce chromosomal aberration or aneuploidy in bone marrow or abnormal sperm morphology. [The Working Group noted that the conflicting findings obtained in tests for micronucleus induction in mice reflect differences in dose.]

(b) *Humans*

(i) *Absorption, distribution, excretion and metabolism*

No data were available to the Working Group.

(ii) *Toxic effects*

Michael *et al.* (1981) took blood samples from the population of a village in which the water disinfectant system was changed to chlorine dioxide as a disinfectant during the summer months to avoid taste and odour problems produced by chlorine. A population of 197 individuals was screened and compared with 112 unexposed individuals. Chlorine dioxide concentrations ranged from 0.33 to 1.11 mg/l, chlorite levels from 3.19 to 6.96 mg/l and chlorate levels from 0.34 to 1.82 mg/l; total available chlorine (chlorine was used to generate the chlorine dioxide from chlorite) ranged from 8.79 to 20.83 mg/l. No significant difference was

Table 2. Genetic and related effects of sodium chlorite

Test system	Result		Dose LED/HID	Reference
	Without exogenous metabolic system	With exogenous metabolic system		
SA0, *Salmonella typhimurium* TA100, reverse mutation	+	+	100.000	Ishidate et al. (1984)
CIC, Chromosomal aberrations, Chinese hamster CHL fibroblasts	+	0	20.000	Ishidate (1987)
MVM, Micronucleus test, CD-1 mice *in vivo*	–	0	30.000	Meier et al. (1985)
MVM, Micronucleus test, ddY mice	+	0	15.000	Hayashi et al. (1988)
MVM, Micronucleus test, ddY mice	–	0	15.000	Hayashi et al. (1988)
MVM, Micronucleus test, ddY mice	(+)	0	150.000	Hayashi et al. (1988)
CBA, Chromosomal aberrations, bone–marrow cells of CD-1 mice *in vivo*	–	0	30.000	Meier et al. (1985)
AVA, Aneuploidy, bone–marrow cells of CD-1 mice *in vivo*	–	0	30.000	Meier et al. (1985)
SPM, Sperm morphology, B6C3F$_1$, mice *in vivo*	–	0	30.000	Meier et al. (1985)

observed in haematocrit, haemoglobin, red blood cell count, white cell count, reticulocyte count, mean corpuscular volume, methaemoglobin, blood urea nitrogen, serum creatinine or total bilirubin 70 days after chlorine dioxide treatment was initiated. A single glucose-6-phosphate dehydrogenase-deficient male identified in the population displayed a reduction of haemoglobin concentrations from 14.7 to 12.9 g/dl, of haematocrit from 46 to 40% and of red blood cell count from 4.78 to 4.11 × 10^6 ml. [The Working Group noted that the last observed effect was mild and that a causal association with chlorite could not be established.]

A clinical study of the effects of chlorite, chlorate and chlorine dioxide was conducted in three parts; tolerance to a rising dose by normal male volunteers (Lubbers & Bianchine, 1984), a 12-week study (Lubbers *et al.*, 1984a) and a study in three glucose-6-phosphate dehydrogenase-deficient male volunteers (Lubbers *et al.*, 1984b). In the study of rising doses, the maximal concentrations of chlorite and chlorate were 2.4 mg/l and that of chlorine dioxide, 24 mg/l; each treatment involved drinking two 500-ml portions. The authors concluded that although there were some significant trends, no physiological importance could be attached to the changes (Lubbers & Bianchine, 1984). The 12-week study required subjects to drink 500 ml of water containing 5 mg/l chlorine dioxide, chlorite or chlorate within a 15-min period each day. Again, there were significant, but small trends in certain parameters. No physiological significance could be attributed to any of these changes (Lubbers *et al.*, 1984a). The study of the three glucose-6-phosphate dehydrogenase-deficient individuals was conducted under circumstances similar to those outlined above, and they were followed for an additional eight weeks after treatment. No clinically significant treatment-related change was identified (Lubbers *et al.*, 1984b).

(iii) *Effects on reproduction and prenatal toxicity*

In a retrospective epidemiological study, Tuthill *et al.* (1982) compared neonatal morbidity and mortality in two communities in the USA, one of which employed chloride dioxide for water disinfection in 1945 and the other, chlorination (see also the monograph on chlorination of drinking-water, p. 106). Exposure was estimated from the amount of chlorite added to the water; the monthly average was 0.32 mg/l (high, 0.56 mg/l), which gave rise to an average concentration of residual chlorine species in excess of 0.3 mg/l. In subsequent years, the mean level of sodium chlorite added to the water declined to 0.16 mg/l. The number of infants that were judged by the attending physician to be premature or to have greater weight loss after birth was significantly greater in the community utilizing chlorine dioxide compared with the community utilizing chlorination. The rates of jaundice, birth defects and neonatal mortality did not differ between the communities. [The

SODIUM CHLORITE

Working Group noted the difficulties associated with establishing prematurity and poor weight gain after birth, especially in a retrospective study, and that confounding factors were not controlled for.]

(iv) *Genetic and related effects*

No data were available to the Working Group.

3.3 Case reports and epidemiological studies of carcinogenicity to humans

No data were available to the Working Group.

4. Summary of Data Reported and Evaluation

4.1 Exposure data

Sodium chlorite is the only chlorite salt produced commercially in significant quantities. It is used mainly for the generation of chlorine dioxide *in situ* for bleaching textiles, in pulp and paper processing, and for disinfection. Sodium chlorite is used in a small number of water treatment plants to generate chlorine dioxide; this may result in low residual concentrations of chlorite in drinking-water.

No information was available on occupational exposures to sodium chlorite.

4.2 Experimental carcinogenicity data

Sodium chlorite was tested for carcinogenicity in male and female $B6C3F_1$ mice and Fischer 344 rats by oral administration in the drinking-water and in a limited study by skin application in female Sencar mice. It was further tested for promoting effects in female Sencar mice by skin application following a single application of 7,12-dimethylbenz[*a*]anthracene. Oral administration of sodium chlorite to mice was associated with a marginal increase in the incidence of lung tumours in treated males. In the study in rats, no significant increase in tumour incidence at any site was seen in treated animals. Skin tumours did not occur in Sencar mice following skin application of sodium chlorite. In the initiation/promotion study, sodium chlorite had a marginal promoting effect.

4.3 Human carcinogenicity data

No data were available to the Working Group.

4.4 Other relevant data

Sodium chlorite has been shown to produce haemolytic anaemia in several animal species at concentrations in drinking-water of 100 mg/l or higher. No sign of such effects was seen in humans at much lower doses.

Minimal adverse reproductive effects were observed in rats and mice given sodium chlorite in the drinking-water at concentrations of 100 mg/l or higher.

Single studies indicated that sodium chlorite induced mutations in bacteria and chromosomal aberrations in cultured mammalian cells. In mice treated *in vivo*, conflicting results were obtained with regard to the induction of micronuclei, while a single study showed no induction of aneuploidy, chromosomal aberrations or abnormal sperm morphology.

4.5 Evaluation[1]

There is *inadequate evidence* for the carcinogenicity of sodium chlorite in experimental animals.

No data were available from studies in humans on the carcinogenicity of sodium chlorite.

Overall evaluation

Sodium chlorite *is not classifiable as to its carcinogenicity to humans (Group 3)*.

5. References

Abdel-Rahman, M.S. (1985) Pharmacokinetics of chlorine obtained from chlorine dioxide, chlorine, chloramine, and chloride. In: Jolley, R.L., Bull, R.J., Davis, W.P., Katz, S., Roberts, M.H., Jr & Jacobs, V.A., eds, *Water Chlorination: Chemistry, Environmental Impact and Health Effects*, Vol. 5, Ann Arbor, MI, Lewis Publishers, pp. 281-293

Abdel-Rahman, M.S., Couri, D. & Bull, R.J. (1980) Kinetics of ClO_2 and effects of ClO_2, ClO_2^-, and ClO_3^- in drinking water on blood glutathione and hemolysis in rat and chicken. *J. environ. Pathol. Toxicol.*, 3, 431-449

Abdel-Rahman, M.S., Couri, D. & Bull, R.J. (1982) Metabolism and pharmacokinetics of alternative drinking water disinfectants. *Environ. Health Perspect.*, 46, 19-23

Abdel-Rahman, M.S., Couri, D. & Bull, R.J. (1984) The kinetics of chlorite and chlorate in the rat. *J. Am. Coll. Toxicol.*, 3, 261-267

Aieta, E.M. & Roberts, P.V. (1985) The chemistry of oxo-chlorine compounds relevant to chlorine dioxide generation. In: Jolley, R.L., Bull, R.J., Davis, W.P., Katz, S., Roberts, M.H., Jr & Jacobs, V.A., eds, *Water Chlorination: Chemistry, Environmental Impact and Health Effects*, Vol. 5, Ann Arbor, MI, Lewis Publishers, pp. 783-794

---

[1]For definition of the italicized terms, see Preamble, pp. 30-33.

Anon. (1985) Chemical industry trends in 1984. In: Japan Chemical Week, ed., *Japan Chemical Annual 1985*, Tokyo, The Chemical Daily Co., pp. 14-18

Bercz, J.P., Jones, L., Garner, L., Murray, D., Ludwig, D.A. & Boston, J. (1982) Subchronic toxicity of chlorine dioxide and related compounds in drinking water in the nonhuman primate. *Environ. Health Perspect.*, 46, 47-55

Budavari, S., ed. (1989) *The Merck Index*, 11th ed., Rahway, NJ, Merck & Co., pp. 1359-1360

Canadian Centre for Occupational Health and Safety (1989) *Cheminfo Database Summary, Sodium Chlorite*, Hamilton, Ontario, pp. 1-15

Carlton, B.D., Habash, D.L., Basaran, A.H., George, E.L. & Smith, M.K. (1987) Sodium chlorite administration in Long-Evans rats: reproductive and endocrine effects. *Environ. Res.*, 42, 238-245

Chemical Information Services Ltd (1988) *Directory of World Chemical Producers—1989/1990 Edition*, Oceanside, NY, p. 519

Couri, D., Miller, C.H., Jr, Bull, R.J., Delphia, J.M. & Ammar, E.M. (1982) Assessment of maternal toxicity, embryotoxicity and teratogenic potential of sodium chlorite in Sprague-Dawley rats. *Environ. Health Perspect.*, 46, 25-29

Eul, W.L. (1989) Applications of chlorine dioxide produced from sodium chlorite in Europe. In: *Chlorine Dioxide: Scientific, Regulatory and Application Issues, Denver, 1-2 November 1989*, Denver, CO, American Water Works Association, pp. 45-58

Hayashi, M., Kishi, M., Sofuni, T. & Ishidate, M., Jr (1988) Micronucleus tests in mice on 39 food additives and eight miscellaneous chemicals. *Food chem. Toxicol.*, 26, 487-500

Heffernan, W.P., Guion, C. & Bull, R.J. (1979) Oxidative damage to the erythrocyte induced by sodium chlorite, *in vivo*. *J. environ. Pathol. Toxicol.*, 2, 1487-1499

Ishidate, M., Jr (1987) *Data Book of Chromosomal Aberration Test In Vitro*, rev. ed., Tokyo, Life Science Information Center, p. 376

Ishidate, M., Jr, Sofuni, T., Yoshikawa, K., Hayashi, M., Nohmi, T., Sawada, M. & Matsuoka, A. (1984) Primary mutagenicity screening of food additives currently used in Japan. *Food chem. Toxicol.*, 22, 623-636

Kurokawa, Y., Takamura, N., Matsushima, Y., Imazawa, T. & Hayashi, Y. (1984) Studies on the promoting and complete carcinogenic activities of some oxidizing chemicals in skin carcinogenesis. *Cancer Lett.*, 24, 299-304

Lubbers, J.R. & Bianchine, J.R. (1984) Effects of the acute rising dose administration of chlorine dioxide, chlorate and chlorite to normal healthy adult male volunteers. *J. environ. Pathol. Toxicol. Oncol.*, 5, 215-228

Lubbers, J.R., Chauhan, S., Miller, J.K. & Bianchine, J.R. (1984a) The effects of chronic administration of chlorine dioxide, chlorite and chlorate to normal healthy adult male volunteers. *J. environ. Pathol. Toxicol. Oncol.*, 5, 229-238

Lubbers, J.R., Chauhan, S., Miller, J.K. & Bianchine, J.R. (1984b) The effects of chronic administration of chlorite to glucose-6-phosphate dehydrogenase deficient healthy adult male volunteers. *J. environ. Pathol. Toxicol. Oncol.*, 5, 239-242

Meier, J.R., Bull, R.J., Stober, J.A. & Cimino, M.C. (1985) Evaluation of chemicals used for drinking water disinfection for production of chromosomal damage and sperm-head abnormalities in mice. *Environ. Mutagenesis*, 7, 201-211

Michael, G.E., Miday, R.K., Bercz, J.P., Miller, R.G., Greathouse, D.G., Kraemer, D.F. & Lucas, J.B. (1981) Chlorine dioxide water disinfection: a prospective epidemiology study. *Arch. environ. Health*, 36, 20-27

Ministry of International Trade and Industry (1989) *Yearbook of Chemical Industries Statistics 1988*, Tokyo, International Trade and Industry Association

Moore, G.S. & Calabrese, E.J. (1980) The effects of chlorine dioxide and sodium chlorite on erythrocytes of A/J and C57L/J mice. *J. environ. Pathol. Toxicol.*, 4, 513-524

Moore, G.S., Calabrese, E.J. & Leonard, D.A. (1980) Effects of chlorite exposure on conception rate and litters of A/J strain mice. *Bull. environ. Contam. Toxicol.*, 25, 689-696

Musil, J., Knotek, Z., Chalupa, J. & Schmidt, P. (1964) Toxicologic aspects of chlorine dioxide application for the treatment of water containing phenols. *Technol. Water*, 8, 327-346

National Academy of Sciences (1980) *Toxicity of Selected Drinking Water Contaminants*, Vol. 3, *Chlorite*, Washington DC, pp. 197-200

National Academy of Sciences (1987) *Chemistry and Toxicity of Selected Disinfectants and By-products*, Vol. 7, *Chlorite, Chlorate*, Washington DC, pp. 99-111

Noack, M.G. & Doerr, R.L. (1979) Chlorine oxygen acids and salts. Chlorine dioxide, chlorous acid, and chlorites. In: Mark, H.F., Othmer, D.F., Overberger, C.G., Seaborg, G.T. & Grayson, M., eds, *Kirk-Othmer Encyclopedia of Chemical Technology*, 3rd ed., Vol. 5, New York, John Wiley & Sons, pp. 612-632

Olin Corporation (1984) *Product Data Sheet, Textone® Sodium Chlorite*, Stamford, CT

Olin Corporation (1989) *Product Data Sheet, Technical Sodium Chlorite for Chlorine Dioxide Generation*, Stamford, CT

Shimoyama, T., Hiasa, Y., Kitahori, Y., Konishi, N. & Murata, Y. (1985) Absence of carcinogenic effect of sodium chlorite in rats. *J. Nara med. Assoc.*, 36, 710-718

Suh, D.H. & Abdel-Rahman, M.S. (1983) Kinetics study of chloride in rat. *J. Toxicol. environ. Health*, 12, 467-473

Suh, D.H., Abdel-Rahman, M.S. & Bull, R.J. (1983) Effect of chlorine dioxide and its metabolites in drinking water on fetal development in rats. *J. appl. Toxicol.*, 3, 75-79

Tuthill, R.W., Giusti, R.A., Moore, G.S. & Calabrese, E.J. (1982) Health effects among newborns after prenatal exposure to ClO_2-disinfected drinking water. *Environ. Health Perspect.*, 46, 39-45

Weast, R.C., ed. (1989) *CRC Handbook of Chemistry and Physics*, 70th ed., Boca Raton, FL, CRC Press, pp. 13-130

White, G.C. (1986) *The Handbook of Chlorination*, 2nd ed., New York, Van Nostrand Reinhold

Wilson, I., Bretscher, K.R., Chea, C.K. & Kelly, H.C. (1983) Heme models of peroxidase enzymes: deuteroferriheme-catalyzed chlorination of monochlorodimedone by sodium chlorite. *J. inorg. Biochem.*, 19, 345-357

Yokose, Y., Uchida, K., Nakae, D., Shiraiwa, K., Yamamoto, K. & Konishi, Y. (1987) Studies of carcinogenicity of sodium chlorite in $B6C3F_1$ mice. *Environ. Health Perspect.*, 76, 205-210

HYPOCHLORITE SALTS

1. Chemical and Physical Data

1.1 Synonyms and molecular formulae

Table 1. Synonyms (Chemical Abstracts Service names are given in bold) and molecular formulae of hypochlorite salts

Chemical name	Chem. Abstr. Services Reg. No.[a]	Synonyms	Formula	Molecular weight
Calcium hypochlorite	7778-54-3	Calcium oxychloride; chlorinated lime; chlorolime chemical; **hypochlorous acid, calcium salt**; lime chloride	$Ca(OCl)_2$	142.98
Dibasic calcium hypochlorite	12394-14-8	**Calcium hydroxide hypochlorite**; lime chloride	$Ca(OCl)_2 \cdot 2Ca(OH)_2$	291.14
Calcium hypochlorite dihydrate	22464-76-2	**Hypochlorous acid, calcium salt, dihydrate**	$Ca(OCl)_2 \cdot 2H_2O$	174.98
Lithium hypochlorite	13840-33-0	**Hypochlorous acid, lithium salt**; lithium chloride oxide; lithium oxychloride	$LiOCl$	58.39
Potassium hypochlorite	7778-66-7	**Hypochlorous acid, potassium salt**; potassium chloride oxide	$KOCl$	90.55
Sodium hypochlorite	7681-52-9	**Hypochlorous acid, sodium salt**; sodium chloride oxide; sodium oxychloride	$NaOCl$	74.44
Sodium hypochlorite heptahydrate	64131-03-9	**Hypochlorous acid, sodium salt, heptahydrate**	$NaOCl \cdot 7H_2O$	200.44
Sodium hypochlorite hydrate (2:5)	55248-17-4	**Hypochlorous acid, sodium salt, hydrate (2:5)**	$NaOCl \cdot 2.5H_2O$	119.48
Sodium hypochlorite pentahydrate	10022-70-5	**Hypochlorous acid, sodium salt, pentahydrate**	$NaOCl \cdot 5H_2O$	164.52
Calcium sodium hypochlorite	53053-57-9	**Hypochlorous acid, calcium sodium salt (3:1:1)**	$Ca(OCl)_2 \cdot NaOCl$	217.42

1.2 Chemical and physical properties of the pure substances

From Weast (1989) unless otherwise specified

Calcium hypochlorite

  (a) *Description:* White powder or flat plates
  (b) *Melting-point:* Decomposes at 100°C
  (c) *Density:* Specific gravity = 2.35
  (d) *Solubility:* Soluble in cold water, 21.4% soluble at 25°C (Wojtowicz, 1979); insoluble in ethanol
  (e) *Stability:* Solid form decomposes exothermically when heated to 175°C, releasing oxygen (Mannsville Chemical Products Corp., 1987). Can react vigorously, and sometimes explosively, with organic and inorganic materials; aqueous solutions subject to decomposition which is influenced by concentration, ionic strength, pH, temperature, light and impurities (Wojtowicz, 1979).
  (f) *Reactivity:* Strong oxidizer of organic and inorganic materials; also acts as a chlorinating agent toward some classes of organic compounds (Wojtowicz, 1979)

Sodium hypochlorite pentahydrate

  (a) *Description:* Colourless crystals
  (b) *Melting-point:* 18°C
  (c) *Solubility:* In water (g/l): 293 at 0°C, 942 at 23°C
  (d) *Stability:* Highly unstable (Budavari, 1989)
  (e) *Reactivity:* Strong oxidizer of organic and inorganic compounds (Wojtowicz, 1979)

Sodium hypochlorite solution (aqueous)

  (a) *Description:* Clear or slightly yellow solution
  (b) *Stability:* Anhydrous hypochlorite is highly explosive; the solution is subject to decomposition, which is influenced by its concentration, ionic strength, pH, temperature, light and impurities; also susceptible to catalysis by trace metal impurities (Wojtowicz, 1979)
  (c) *Reactivity:* Strong oxidizer of many organic and inorganic substances and chlorinates some classes of organic compounds. Contact with acid releases chlorine gas (Jones Chemical, 1989). Reacts violently with ammonium salts, aziridine, methanol and phenylacetonitrile, sometimes resulting in explosion. Reacts with primary aliphatic and aromatic amines to form explosively unstable N-chloramines. Reaction with formic acid becomes explosive at 55°C (Sigma-Aldrich Company, 1989).

Sodium hypochlorite dihydrate

(a) *Description*: Colourless hygroscopic crystals
(b) *Melting-point*: 57.5°C
(c) *Solubility*: Very soluble in cold water

The chemistry of hypochlorite ion in aqueous solutions is discussed in the monograph on chlorinated drinking-water (p. 50). All hypochlorite salts, as well as chlorine itself, in aqueous solution produce equilibrium mixtures of hypochlorous acid, hypochlorite ion and chlorine. In concentrated solutions, hypochlorite ion tends to disproportionate to form chlorate and chloride ions. The reaction is slow at room temperature, but in hot solutions (e.g., 80°C) the reaction is rapid and produces high yields of chlorate ions (Aieta & Roberts, 1986).

1.3 Technical products and impurities

Calcium hypochlorite

Trade names: B-K Powder; Camporit; Chemichlor G; Chloride of lime; Eusol BPC; HTH; HTH (bleaching agent); Lime chloride; Losantin; Pittchlor; Solvox KS; T-Eusol

Calcium hypochlorite (bleach liquor) is produced commercially as a solution of calcium hypochlorite and calcium chloride containing some dissolved lime. Commercial products usually contain 50% or more calcium hypochlorite. The available chlorine content varies but is usually about 30-35 g/l (Wojtowicz, 1979; Budavari, 1989).

Calcium hypochlorite is one of the few metal hypochlorites that is stable enough to be produced as a solid salt. It is produced on a large scale as a 65-70% pure product (dihydrate salt) containing sodium chloride and water as the main impurities. It is also manufactured, to a smaller extent, in the form of bleaching powder (Wojtowicz, 1979), which contains approximately 37% available chlorine in a complex mixture of calcium hydroxide, calcium chloride and various calcium hypochlorite species. Calcium oxide is often blended with bleach powder as a desiccant in order to avoid deliquescence of the powder in hot and humid conditions; this blended product, tropical bleach, contains about 15-30% available chlorine (Baum *et al.*, 1978).

Sodium hypochlorite (liquid bleach)

Trade names: Antiformin; B-K Liquid; Carrel-Dakin solution; Chloros; Clorox; Dakin's solution; Deosan; Hyclorite; Javex; Klorocin; Milton; Neo-cleaner; Neoseptal CL; Parozone; Purin B; Surchlor

Commercial strength sodium hypochlorite is available as a solution that contains 12-15% available chlorine; a weaker solution that is marketed contains approximately 5% available chlorine. The main impurities in these solutions include sodium chlorate, sodium carbonate, sodium chloride and sodium hydroxide. Sodium hypochlorite solution produced on-site for industrial processes generally contains 30-40 g/l of available chlorine (Wojtowicz, 1979).

Lithium hypochlorite

Commercial lithium hypochlorite is a solid product usually containing 35% lithium hypochlorite, 34% sodium chloride and various additional salts (Baum *et al.*, 1978; Wojtowicz, 1979).

2. Production, Use, Occurrence and Analysis

2.1 Production and use

(a) Production

Berthollet first used chlorine in a commercial textile bleaching process in the 1790s; he later discovered that chlorine could be absorbed by caustic potash to form potassium hypochlorite solution (Javel water). Labarraque replaced the expensive potash with caustic soda, and by the early 1800s Labarraque's solution had replaced potassium hypochlorite in the bleaching of textiles. Tennant experimented with a solution of chlorine and milk of lime in 1798 and later discovered that when slaked lime was treated with chlorine a solid bleaching powder (calcium hypochlorite and other salts) was formed, representing the first solid form of chlorine bleach that could be easily transported. Bleaching powder remained the principal textile bleach throughout the 1800s. Tropical bleach, stable in high tropical temperatures, was produced by the addition of quicklime to bleaching powder. After the First World War, technology for shipping liquid chlorine and caustic economically was developed, allowing bleach solutions to be made at the point of use. In 1928, the first dry calcium hypochlorite with 70% available chlorine was produced in the USA and was used widely in the bleaching of textiles and pulp (Baum *et al.*, 1978; Wojtowicz, 1979).

At the inception of the commercial laundry industry in about 1900, sodium hypochlorite solution made with bleaching powder and soda ash was used. When chlorine became more readily available, sodium hypochlorite was produced directly at the point of use. Dry calcium hypochlorite bleaches were introduced in the 1930s. Home bleaching became more common when sodium hypochlorite solutions began to displace bleaching powders in the 1930s; they came into extensive use in the 1940s (Baum *et al.*, 1978). Other hypochlorites, such as lithium hypochlorite, first introduced in 1964, have had limited commercial use (Wojtowicz, 1979).

Calcium hypochlorite solution (bleach liquor) is prepared by adding chlorine to a diluted high quality lime slurry. Solid calcium hypochlorite is generally made by drying a filter cake of neutral calcium hypochlorite dihydrate prepared from hydrated lime, caustic and chlorine. Several industrial processes were developed to eliminate or minimize calcium chloride in the hypochlorite product (Wojtowicz, 1979).

Sodium hypochlorite is usually prepared by chlorinating aqueous sodium hydroxide solution at reduced temperatures to prevent excessive chlorate formation, which can contribute to lower stability. Conversion of sodium hydroxide to hypochlorite is usually limited to 92-94% to prevent overchlorination and to improve stability. Sodium hypochlorite is also prepared electrolytically using small diaphragm-less or membrane cells with a capacity of 1-150 kg per day of equivalent chlorine; these produce dilute hypochlorite solutions of 1-3 and 5-6 g/l from seawater and brine, respectively (Wojtowicz, 1979).

Solid lithium hypochlorite is produced by combining concentrated solutions of sodium hypochlorite and lithium chloride (Baum et al., 1978).

Production capacity of calcium hypochlorite in 1989-90 in countries for which data were available are presented in Table 2. Japanese production of sodium hypochlorite (12% solution) was 947 thousand tonnes in 1984, 954 in 1985, 996 in 1986, 972 in 1987 and 989 in 1988 (Anon., 1985; Ministry of International Trade and Industry, 1989).

Table 2. Production capacity of calcium hypochlorite in selected countries (in thousands of tonnes)[a]

Country	Active chlorine (%)	
	65-70	60
Canada	5.9	None
China	0.5	10.8
India	None	1.0
Italy	0.9	4.1
Japan	34.7	11.0
Republic of Korea	1.0	None
South Africa	12.0	None
USA	105.3	None
Total	160.3	26.9

[a]From PPG Industries (1990)

Sodium hypochlorite is produced by two companies in Africa, three in the Middle East, five in Oceania, 13 in North America, 18 in South America, 30 in Asia and 48 companies in Europe (Chemical Information Services Ltd., 1988). Calcium hypochlorite is produced by 21 companies and potassium chlorite by three companies throughout the world (Anon., 1989).

(b) Use

Calcium hypochlorite is widely used as a sanitizer, oxidizer and bleaching agent. Calcium hypochlorite solutions are used primarily in pulp and textile bleaching, while the solid form is used in less developed countries for textile bleaching and commercial laundering (Baum *et al.*, 1978). The largest use of calcium hypochlorite within the USA is in swimming pools to kill bacteria, control algae and oxidize organic contaminants. It is also used to destroy cyanides in industrial wastes, in disinfection and deodourizing of wastes generated from canneries, dairy plants, beet sugar plants and tanneries, as a biocide in controlling contamination in public, private and industrial water supplies, in sanitizing beverage plants and food processing operations and equipment, in disinfecting sewage disposal plants, in sanitizing fruits and vegetables during growth and following harvest, as a toilet tank sanitizer and in multistage pulp bleaching processes. It can also be reacted with acetone to produce USP chloroform (Mannsville Chemical Products Corp., 1987).

Consumption of calcium hypochlorite in 1989-90 was estimated to be 7.0 thousand tonnes in Japan, 80.1 thousand tonnes in the USA and 40 thousand tonnes in the rest of the world. Use distribution figures for 1989-90 were estimated to be 85% for swimming pools and 15% for other uses in the USA, and 55% for swimming pools and 45% for other uses in the rest of the world (PPG Industries, 1990).

The largest use for sodium hypochlorite solutions (5% concentration) is as a household bleach. More concentrated solutions are used in swimming pool sanitation, in commercial laundry bleaching, in paper and pulp production, in disinfecting municipal water (particularly in small water supplies) and sewage, in the sanitation of dairy plants and food processing operations, to control fungal plugging of oil production equipment, as a desulfurizing agent in oil refineries, and as a disinfectant and sanitizer in health care industries. Large quantities of sodium hypochlorite are used in the chemical industry, primarily in the production of hydrazine (see IARC, 1987) as well as in the synthesis of organic chemicals and the manufacture of chlorinated trisodium phosphate. Sodium hypochlorite solutions produced directly by electrolysis of seawater or brine are used primarily in sewage and wastewater treatment, commercial laundries, large swimming pools and aboard ships (Baum *et al.*, 1978; Wojtowicz, 1979; White, 1986; Mannsville Chemical Products Corp., 1987).

Small quantities of lithium hypochlorite are produced for use in swimming pool sanitation (Mannsville Chemical Products Corp., 1987) and in household laundry detergents (Baum *et al.*, 1978).

2.2 Occurrence

(a) Natural occurrence

Hypochlorous acid is generated in mammalian neutrophils by myeloperoxidase (MP) by the following reaction:

$$MP + H_2O_2 \to MP \cdot H_2O_2$$
$$MP \cdot H_2O_2 + Cl^- \to MP + HOCl + OH^-$$

(Winterbourn, 1985).

(b) Occupational exposure

Due to the wide range of uses of hypochlorite salts, many workers may be exposed to them by dermal (and ocular) contact or inhalation. During routine monitoring in a US calcium hypochlorite manufacturing facility, personal samples contained an 8-h time-weighted average of 0.31 mg/m^3 (geometric mean) and a 15-min short-term exposure level of 0.38 mg/m^3 (geometric mean); work area samples contained an 8-h time-weighted average of 0.13 mg/m^3 and a mean 15-min short-term exposure level of 0.88 mg/m^3 (PPG Industries, 1990). No published data were available on occupational exposures to or the environmental occurrence of hypochlorites. No regulatory standards or guidelines have been established for exposures to hypochlorite.

(c) Water

The chemistry of hypochlorous acid and hypochlorite ion in aqueous solutions is discussed in the monograph on chlorinated drinking-water (p. 50). Residual chlorine in drinking-water is present in part as hypochlorite, indicating widespread exposure of the general population to low levels of hypochlorite in solution.

(d) Other

Another significant potential route of exposure to hypochlorite derives from its widespread use as a household sanitizer and bleach and in swimming pools. No data were found which directly characterize these exposures.

2.3 Analysis

The analysis of hypochlorite and related chlorine species in aqueous media is well documented (White, 1986) and involves a variety of colorimetric and iodometric procedures. No specific method is available for the analysis of occupational exposures to hypochlorite salts in air. General methods for dusts (e.g., NIOSH Method 500; Eller, 1984a) and the appropriate metal (e.g., NIOSH Method 7020; Eller, 1984b) are used.

3. Biological Data Relevant to the Evaluation of Carcinogenic Risk to Humans

3.1 Carcinogenicity studies in animals

(a) *Oral administration*

Mouse: Groups of 50 male and 50 female B6C3F$_1$ mice, four to six weeks old, were given 500 or 1000 mg/l sodium hypochlorite (14% effective chlorine [purity unspecified]) in the drinking-water for 103 weeks. Groups of 73 male and 72 female mice served as controls. Survival at 106 weeks was: males—control, 48/73, low-dose, 39/50; high-dose, 37/50; females—control, 56/72; low-dose, 40/50; high-dose, 39/50. There was no effect upon tumour incidence in either male or female mice (Kurokawa *et al.*, 1986).

Rat: A group of 60 male and female BDII (cPah albino) rats, 100 days old [sex ratio unspecified], was given tap water [organic content not analysed] containing 100 mg/l chlorine prepared with chlorine gas. The animals were mated and the treatment was continued for life through six generations, with the exception of F$_3$ and F$_4$ animals, which were treated during the weaning period only. Altogether, 236 animals in five generations were exposed. Two groups of 20 and 36 rats [sex and age unspecified] from two previous experiments served as controls. There was no difference in survival or in tumour incidence in any generation group as compared to untreated controls (Druckrey, 1968).

Groups of 50 male and 50 female Fischer 344 rats, seven weeks old, were given 0, 500 or 1000 (males) and 0, 1000 or 2000 (females) mg/l sodium hypochlorite (14% effective chlorine [purity unspecified]) in the drinking-water for 104 weeks. Survival at 112 weeks was: males—control, 30/50; low-dose, 26/50; high-dose, 31/50; females—control, 31/50; low-dose, 36/50; high-dose, 35/50. The occurrence of tumours at any site was not significantly greater in rats receiving sodium hypochlorite than in controls. The proportions of low- and high-dose female rats with fibroadenomas of the mammary gland were significantly lower than among controls (control, 8/50; low-dose, 0/50; $p < 0.01$, chi-square test; high-dose, 1/50; $p < 0.01$). Similarly, the proportion of high-dose male rats with nodular hyperplasia of the liver was decreased (control, 23/49; low-dose, 17/50; high-dose, 10/50; $p < 0.01$) (Hasegawa *et al.*, 1986).

(b) Skin application

Mouse: A group of 40 strain ddN female mice, five weeks old, was given 60 topical applications of sodium hypochlorite (10% effective chlorine solution) [purity, vehicle and frequency of application unspecified]. Another group of 40 female mice was given 20 applications of 4-nitroquinoline 1-oxide [dose, purity and frequency unspecified]; and a third group of 40 mice was given 45 applications of sodium hypochlorite following applications of 4-nitroquinoline 1-oxide [number and frequency of applications and dose unspecified]. No skin tumour occurred in mice given applications of sodium hypochlorite alone, whereas skin tumours occurred in 9/32 mice given applications of sodium hypochlorite following initiating doses of 4-nitroquinoline 1-oxide. The skin tumours included one fibrosarcoma, three squamous-cell carcinomas and five papillomas. No skin tumour occurred in mice given applications of 4-nitroquinoline 1-oxide only (Hayatsu *et al.*, 1971). [The Working Group noted the lack of details on dose, frequency of applications, age at termination of the study, and survival.]

A group of 20 female Sencar mice, six weeks old, was given topical applications of 0.2 ml of a solution of 10 g/l sodium hypochlorite [purity unspecified] in acetone twice a week for 51 weeks at which time the study was terminated. A group of 15 female mice given applications of acetone served as controls. All mice survived to the end of the study and no skin tumour was observed in the treated or control groups (Kurokawa *et al.*, 1984). [The Working Group noted the small number of animals used.]

In an initiation/promotion study, a group 20 female Sencar mice, six weeks old, was given a single topical application of 20 nmol [5 µg] 7,12-dimethylbenz[*a*]-anthracene in acetone followed by applications of 0.2 ml of a 10 g/l sodium hypochlorite solution [purity unspecified] in acetone twice a week for 51 weeks. A group of 15 female mice given a single application of 7,12-dimethylbenz[*a*]-anthracene followed by applications of acetone served as controls. The effective number of mice was 20; the number of survivors was not given. A squamous-cell carcinoma of the skin occurred in 1/20 mice treated with 7,12-dimethylbenz[*a*]-anthracene and sodium hypochlorite; none occurred in the initiated controls (Kurokawa *et al.*, 1984).

3.2 Other relevant data

(a) Experimental systems

(i) *Absorption, distribution, excretion and metabolism.*

Radiolabel (^{36}Cl) derived from hypochlorous acid given to male rats was absorbed and appeared in serum at a rate constant of 0.3/h compared to 0.04/h for sodium chloride. The radiolabel was eliminated from plasma with a half-time of 44 h, compared to 52 h for sodium chloride. The distribution of radiolabel from hypochlorous acid and from sodium chloride in tissues was similar (Abdel-Rahman, 1985).

(ii) *Toxic effects*

Blabaum and Nichols (1956) provided mice with concentrations of 100 and 200 mg/l chlorine in the drinking-water (pH 5.9-6.5) for up to 50 days. They reported no effect on body weight or gross morphology.

Chronic treatment of male and female BDII albino rats with drinking-water containing 100 mg/l chlorine [pH unspecified] had no significant toxic effect over seven generations (Druckrey, 1968).

Cunningham (1980) gave sodium hypochlorite in drinking-water at levels of 0, 20, 40 and 80 mg/l as chlorine to male Wistar rats for up to six weeks. In a separate experiment, female rats were administered sodium hypochlorite at doses equivalent to 0, 8, 40 and 200 mg/kg bw available chlorine in milk by gavage twice daily for 14 days. Guinea-pigs were administered sodium hypochlorite at 0 and 50 mg/l as chlorine in drinking-water for five weeks. There were small but significant increases in body weights in rats given drinking-water, and increased kidney weights in rats treated with 200 mg/kg bw by gavage. No significant increase in body weight was seen in guinea-pigs.

Administration of sodium hypochlorite in drinking-water to mice at levels of 25-30 mg/l reduced the number of peritoneal exudate cells recovered by lavage after one to four weeks of treatment (Fidler, 1977). The phagocytic activity of macrophages recovered from treated animals was decreased by approximately 50% relative to control animals during the first two weeks of treatment and was completely absent by the third week. Subsequent experiments demonstrated that the in-vivo phagocytic activity of macrophages recovered from the liver and spleen was also decreased. The treatment also prevented the destruction by injection of macrophage activating factors of spontaneous metastases arising from B16-BL6 melanoma cells implanted subcutaneously (Fidler *et al.*, 1982). Exon *et al.* (1987) observed no decrement in the in-vitro phagocytic activity of peritoneal macrophages recovered by lavage from rats given 5, 15, 30 mg/l sodium hypochlorite; however, decreases in spleen weights, delayed-type hypersensitivity reactions and macrophage oxidative metabolism were observed at the high dose.

In rats, concentrations of sodium hypochlorite of 625 mg/l and above given in drinking-water for 14 days progressively depressed water consumption. In a 92-day study, no significant effect on body weight, organ weights or serum chemistry was observed until concentrations reached 4000 mg/l (Furukawa *et al.*, 1980). In chronic studies in mice and rats (see section 3.1), there was no significant effect on survival of either mice or rats treated with 500-2000 mg/l drinking-water, but dose-related decreases in body weight gain occurred (Hasegawa *et al.*, 1986; Kurokawa *et al.*, 1986).

Robinson *et al.* (1986) examined the effects of hypochlorous acid and hypochlorite on mouse skin. With 10 min of contact per day for four days, hypochlorous acid (pH 6.5) markedly increased skin thickness at concentrations of 300 mg/l and above; a similar but less marked effect was observed with hypochlorite (pH 8.5) at 1000 mg/l.

Cotter *et al.* (1985) applied gauze soaked in 0.1 and 0.5% solutions of sodium hypochlorite (pH 7.49) to the skin of guinea-pigs for two weeks. The 0.1% solution produced no effect on isolated epidermal basal-cell viability, but the 0.5% solution was reported to reduce it.

Male white Carneau pigeons and New Zealand rabbits administered 15 mg/l chlorine in drinking-water (pH 6.5 or 8.5) for three months had increased plasma low-density lipid cholesterol levels and decreased plasma thyroxine levels. The effects were more pronounced in animals fed high-cholesterol diets (Revis *et al.*, 1986). Subsequent experiments by the same authors failed to confirm these observations (Holdsworth *et al.*, 1990).

In a neutrophil migration assay *in vitro*, sodium hypochlorite (0.00025% solution buffered with sodium carbonate) suppressed migration of stimulated and nonstimulated neutrophils (Kozol *et al.*, 1988).

(iii) *Effects on reproduction and prenatal toxicity*

In BDII rats given water containing free chlorine at 100 mg/l in drinking-water daily during seven generations, there was no toxic effect on fertility, growth or survival (Druckrey, 1968).

(iv) *Genetic and related effects* (Table 3)

In one differential toxicity test involving DNA repair-deficient bacteria, a positive result was obtained with sodium hypochlorite. Mutations were induced in *Salmonella typhimurium*.

In a single study, sodium hypochlorite caused chromosomal aberrations in Chinese hamster CHL cells but not in human fibroblasts. [The Working Group noted that lower doses were used in the latter tests.] It was reported in an abstract that sodium hypochlorite did not induce transformation in C3H 10T½ cells (Abernethy *et al.*, 1983). The number of micronuclei was increased in erythrocytes of newt larvae reared in hypochlorite-containing water for 12 days. Neither micronuclei, chromosomal aberrations nor aneuploidy were observed in mice after repeated oral dosing; but abnormal sperm morphology was seen.

Table 3. Genetic and related effects of sodium hypochlorite

Test system	Result		Dose LED/HID	Reference
	Without exogenous metabolic system	With exogenous metabolic system		
ECD, *Escherichia coli pol A*/W3110–P3478, differential toxicity	+	0	0.4000	Rosenkranz (1973)
SA0, *Salmonella typhimurium* TA100, reverse mutation	–	+	2500.0000	Ishidate et al. (1984)
SA3, *Salmonella typhimurium* TA1530, reverse mutation	+	0	0.0100	Wlodkowski & Rosenkranz (1975)[a]
SA3, *Salmonella typhimurium* TA1530, reverse mutation	+	0	5.0000	Wlodkowski & Rosenkranz (1975)
SA5, *Salmonella typhimurium* TA1535, reverse mutation	+	0	0.0100	Wlodkowski & Rosenkranz (1975)
SA8, *Salmonella typhimurium* TA1538, reverse mutation	–	0	5.0000	Wlodkowski & Rosenkranz (1975)
???, Micronucleus test, newt larvae	+	0	0.0000	Gauthier et al. (1989)[b]
CIC, Chromosomal aberrations, Chinese hamster CHL cells	+	(+)	500.0000	Ishidate (1987)
SHF, Sister chromatid exchange, human diploid fibr. lung cell line	+	0	74.0000	Sasaki et al. (1980)
CHF, Chromosomal aberrations, human diploid fibr. lung cell line	–	0	149.0000	Sasaki et al. (1980)
MVM, Micronucleus test, bone–marrow cells of CD–1 mice *in vivo*	–	0	6.0000	Meier et al. (1985)
MVM, Micronucleus test, bone–marrow cells of ddY mice *in vivo*	–	0	1250.0000	Hayashi et al. (1988)
MVM, Micronucleus test, bone–marrow cells of ddY mice *in vivo*	–	0	300.0000	Hayashi et al. (1988)
CBA, Chromosomal aberrations, bone–marrow cells of CD–1 mice *in vivo*	–	0	6.0000	Meier et al. (1985)
AVA, Aneuploidy, bone–marrow cells of CD–1 mice *in vivo*	–	0	6.0000	Meier et al. (1985)
SPM, Sperm morphology, B6C3F$_1$ mice *in vivo*	+	0	3.0000	Meier et al. (1985)

[a]Treatment in liquid medium
[b]Larvae reared in hypochlorite-containing water

In a report lacking details, negative findings were reported in a *Bacillus subtilis* $rec^{+/-}$ assay, in a mutation assay in silkworms and in a test for chromosomal aberrations in rat bone marrow *in vivo* (Kawachi *et al.*, 1980).

(b) *Humans*

(i) *Absorption, distribution, excretion and metabolism*

No data were available to the Working Group.

(ii) *Toxic effects*

Release of chlorine during acidification of sodium hypochlorite solutions (below pH 7.5) is an occasional cause of poisoning (Phillip *et al.*, 1985). The effects are reversible if the exposure is low enough to permit survival from the acute respiratory distress that results (Jones *et al.*, 1986).

The use of hypochlorite solutions to disinfect haemodialysis machines has led to accidental introduction of sodium hypochlorite into the blood. If such exposures are high, they can lead to massive haemolysis (Hoy, 1981).

Skin hypersensitivity to concentrations of 400-600 mg/l sodium hypochlorite was reported in one patient (Eun *et al.*, 1984).

No clinical sign of general or local toxicity was observed following the use of sodium hypochlorite for bladder irrigation in urological patients (Eisen *et al.*, 1976).

In a clinical trial, Wones *et al.* (1990) examined the biochemical effects in 17 healthy men given a daily amount of 1.5 l distilled drinking-water fortified with chlorine at concentrations increasing from 2 to 10 mg/l during 12 weeks. Each person served as his own control. There was a small but significant increase in serum cholesterol levels and total thyroxine concentrations during the exposure period.

(iii) *Effects on reproduction and prenatal toxicity*

No data were available to the Working Group.

(iv) *Genetic and related effects*

No data were available to the Working Group.

3.3 Case reports and epidemiological studies of carcinogenicity to humans

No data were available to the Working Group.

4. Summary of Data Reported and Evaluation

4.1 Exposure data

The principal hypochlorite salts produced commercially are calcium, sodium and lithium hypochlorites. Calcium hypochlorite (solid or aqueous solution) is widely used for disinfection in swimming pools and in industrial applications and for pulp and textile bleaching. Sodium hypochlorite (aqueous solution) is used as a household laundry bleach, in commercial laundering, in pulp and paper manufacture, in industrial chemical synthesis and in the disinfection of drinking-water. Lithium hypochlorite (solid) is used in swimming pools for disinfection and in household detergents.

Hypochlorite salts (principally sodium hypochlorite) are used to disinfect drinking-water at many small treatment works. In the disinfection of drinking-water and wastewater, addition of hypochlorite salts and of chlorine gas gives the same chlorine species in solution—i.e., an equilibrium mixture of mainly hypochlorous acid and hypochlorite anion. In this way, much of the general population is exposed to hypochlorite *via* chlorinated drinking-water (see the monograph on Chlorinated drinking-water).

4.2 Experimental carcinogenicity data

Sodium hypochlorite was tested for carcinogenicity in a two-year study in male and female $B6C3F_1$ mice and Fischer 344 rats by oral administration in drinking-water, in limited studies in female Sencar mice and in female ddN mice by skin application. Sodium hypochlorite was also tested for promoting effects in female Sencar mice following initiation with 7,12-dimethylbenz[*a*]anthracene and in female ddN mice following initiation with 4-nitroquinoline 1-oxide. Sodium hypochlorite administered in the drinking-water did not increase the proportion of rats or mice with tumours. Sodium hypochlorite applied to the skin of Sencar mice or ddN mice did not produce skin tumours. No skin promoting effect was observed in the study with 7,12-dimethylbenz[*a*]anthracene, whereas some effect was seen in the study with 4-nitroquinoline 1-oxide.

Drinking-water containing 100 mg/l chlorine was tested for carcinogenicity in a multigeneration study in male and female BDII rats. No increase in the incidence of tumours was seen in treated animals relative to controls through six generations.

4.3 Human carcinogenicity data

No data were available to the Working Group.

4.4 Other relevant data

Sodium hypochlorite induced genotoxic effects in bacteria. In single studies, chromosomal aberrations were observed in cultured mammalian cells, whereas sister chromatid exchange but no chromosomal aberration was seen in cultured human cells. In a single study, micronuclei were induced in newt larvae. In mice, no induction of micronuclei, aneuploidy or chromosomal aberrations was observed in bone marrow, but abnormal sperm morphology was seen after administration of sodium hypochlorite.

4.5 Evaluation[1]

There is *inadequate evidence* for the carcinogenicity of hypochlorite salts in experimental animals.

No data were available from studies in humans on the carcinogenicity of hypochlorite salts.

Overall evaluation

Hypochlorite salts *are not classifiable as to their carcinogenicity to humans (Group 3)*.

5. References

Abdel-Rahman, M.S. (1985) Pharmacokinetics of chlorine obtained from chlorine dioxide, chlorine, chloramine, and chloride. In: Jolley, R.L., Bull, R.J., Davis, W.P., Katz, S., Roberts, M.H., Jr & Jacobs, V.A., eds, *Water Chlorination: Chemistry, Environmental Impact and Health Effects*, Vol. 5, Ann Arbor, MI, Lewis Publishers, pp. 281-293

Abernethy, D.J., Frazelle, J.H. & Boreiko, C.J. (1983) Relative cytotoxic and transforming potential of respiratory irritants in the C3H/10T1/2 cell transformation system (Abstract No. Cd-20). *Environ. Mutagenesis, 5*, 419

Aieta, E.M. & Roberts, P.V. (1985) The chemistry of oxo-chlorine compounds relevant to chlorine dioxide generation. In: Jolley, R.L., Bull, R.J., Davis, W.P., Katz, S., Roberts, M.H., Jr & Jacobs, V.A., eds, *Water Chlorination: Chemistry, Environmental Impact and Health Effects*, Vol. 5, Ann Arbor, MI, Lewis Publishers, pp. 783-794

Anon. (1985) Chemical industry trends in 1984. In: Japan Chemical Week, ed., *Japan Chemical Annual 1985*, Tokyo, The Chemical Daily Co., pp. 14-18

Anon. (1989) *OPD Chemical Buyers Directory*, 76th annual ed., New York, Schnell Publishing Company, pp. 125, 150, 505, 576

Baum, B.M., Finley, J.H., Blumbergs, J.H., Elliott, E.J., Scholer, F. & Wooten, H.L. (1978) Bleaching agents. In: Mark, H.F., Othmer, D.F., Overberger, C.G., Seaborg, G.T. & Grayson, M., eds, *Kirk-Othmer Encyclopedia of Chemical Technology*, 3rd ed., Vol. 3, New York, John Wiley & Sons, pp. 938-958

---

[1] For description of the italicized terms, see Preamble, pp. 30-33.

Bercz, J.P., Jones, L., Garner, L., Murray, D., Ludwig, D.A. & Boston, J. (1982) Subchronic toxicity of chlorine dioxide and related compounds in drinking water in the nonhuman primate. *Environ. Health Perspect.*, 46, 47-55

Blabaum, C.J. & Nichols, M.S. (1956) Effect of highly chlorinated drinking water on white mice. *J. Am. Water Works Assoc.*, 48, 1503-1506

Budavari, S., ed. (1989) *The Merck Index*, 11th ed., Rahway, NJ, Merck & Co., pp. 254, 1363

Chemical Information Services Ltd (1988) *Directory of World Chemical Producers 1989/90*, Oceanside, NY, p. 522

Cotter, J.L., Fader, R.C., Lilley, C. & Herndon, D.N. (1985) Chemical parameters, antimicrobial activities, and tissue toxicity of 0.1 and 0.5% sodium hypochlorite solutions. *Antimicrob. Agents Chemother.*, 28, 118-122

Cunningham, H.M. (1980) Effect of sodium hypochlorite on the growth of rats and guinea pigs. *Am. J. vet. Res.*, 41, 295-297

Druckrey, H. (1968) Chlorinated drinking water, toxicity tests, involving seven generations of rats (Ger.). *Food Cosmet. Toxicol.*, 6, 147-154

Eisen, M., Jurcovic, K., Pfeiffer, E., Skoluda, D. & Busse, K. (1976) Clinical use of sodium hypochlorite for local treatment and prevention of urinary tract infections and treatment of retracted bladders (Ger.). *Urologe A*, 15, 39-43

Eller, P.M. (1984a) *NIOSH Manual of Analytical Methods*, 3rd ed., Vol. 2, *Nuisance Dust, Total, Method 500*, Cincinnati, OH, National Institute for Occupational Safety and Health, pp. 500-1–500-3

Eller, P.M. (1984b) *NIOSH Manual of Analytical Methods*, 3rd ed., Vol. 1, *Calcium and Compounds, as Ca, Method 7020*, Cincinnati, OH, National Institute for Occupational Safety and Health, pp. 7020-1–7020-3

Eun, H.C., Lee, A.Y. & Lee, Y.S. (1984) Sodium hypochlorite dermatitis. *Contact Derm.*, 11, 45

Exon, J.H., Koller, L.D., O'Reilly, C.A. & Bercz, J.P. (1987) Immunotoxicologic evaluation of chlorine-based drinking water disinfectants, sodium hypochlorite and monochloramine. *Toxicology*, 44, 257-269

Fidler, I.J. (1977) Depression of macrophages in mice drinking hyperchlorinated water. *Nature*, 270, 735-736

Fidler, I.J., Barnes, Z., Fogler, W.E., Kirsh, R., Bugelski, P. & Poste, G. (1982) Involvement of macrophages in the eradication of established metastases following intravenous injection of liposomes containing macrophage activators. *Cancer Res.*, 42, 496-501

Furukawa, F., Kurata, Y., Kokubo, T., Takahashi, M. & Nakadate, M. (1980) Oral acute and subchronic toxicity studies for sodium hypochlorite in F-344 rat. *Bull. natl Inst. Hyg. Sci.*, 98, 62-69

Gauthier, L., Levi, Y. & Jaylet, A. (1989) Evaluation of the clastogenicity of water treated with sodium hypochlorite or monochloramine using a micronucleus test in newt larvae (*Pleurodeles waltl*). *Mutagenesis*, 4, 170-173

Hasegawa, R., Takahashi, M., Kokubo, T., Furukawa, F., Toyoda, K., Sato, H., Kurokawa, Y. & Hayashi, Y. (1986) Carcinogenicity study of sodium hypochlorite in F344 rats. *Food chem. Toxicol.*, 24, 1295-1302

Hayashi, M., Kishi, M., Sofuni, T. & Ishidate, M., Jr (1988) Micronucleus tests in mice on 39 food additives and eight miscellaneous chemicals. *Food chem. Toxicol.*, 26, 487-500

Hayatsu, H., Hoshino, H. & Kawazoe, Y. (1971) Potential cocarcinogenicity of sodium hypochlorite. *Nature*, 233, 495

Holdsworth, G., McCauley, P. & Revis, N.W. (1990) Long-term effects of chlorine-containing disinfectants on plasma levels of cholesterol and thyroxine, in rabbits and pigeons. In: Jolley, R.L., Condie, L.W., Johnson, J.D., Katz, S., Minear, R.A. & Mattice, J.S., eds, *Water Chlorination: Chemistry, Environmental Impact and Health Effects*, Vol. 6, Chelsea, MI, Lewis Publishers, pp. 319-328

Hoy, R.H. (1981) Accidental systemic exposure to sodium hypochlorite (Clorox) during hemodialysis. *Am. J. Hosp. Pharm.*, 38, 1512-1514

IARC (1987) *IARC Monographs on the Evaluation of Carcinogenic Risks to Humans*, Suppl. 7, *Overall Evaluations of Carcinogenicity: An Updating of* IARC Monographs *Volumes 1 to 42*, Lyon, pp. 223-224

Ishidate, M., Jr (1987) *Data Book of Chromosomal Aberration In Vitro*, rev. ed., Tokyo, Life-Science Information Center, p. 383

Ishidate, M., Jr, Sofuni, T., Yoshikawa, K., Hayashi, M., Nohmi, T., Sawada, M. & Matsuoka, A. (1984) Primary mutagenicity screening of food additives currently used in Japan. *Food chem. Toxicol.*, 22, 623-636

Jones Chemical (1989) *Material Data Safety Sheet, Sodium Hypochlorite*, New York

Jones, R.N., Hughes, J.M., Glindmeyer, H. & Weill, H. (1986) Lung function after acute chlorine exposure. *Am. Rev. respir. Dis.*, 134, 1190-1195

Kawachi, T., Komatsu, T., Kada, T., Ishidate, M., Sasaki, M., Sugiyama, T. & Tazima, Y. (1980) Results of recent studies on the relevance of various short-term screening tests in Japan. In: Williams, G.M., Kroes, R., Waaijers, H.W. & van de Poll, K.W., eds, *The Predictive Value of Short-term Screening Tests in Carcinogenicity Evaluation*, Amsterdam, Elsevier/North-Holland Biomedical Press, pp. 253-260

Kozol, R.A., Gillies, C. & Elgebaly, S.A. (1988) Effects of sodium hypochlorite (Dakin's solution) on cells of the wound module. *Arch. Surg.*, 123, 420-423

Kurokawa, Y., Takamura, N., Matsushima, Y., Takayoshi, I. & Hayashi, Y. (1984) Studies on the promoting and complete carcinogenic activities of some oxidizing chemicals in skin carcinogenesis. *Cancer Lett.*, 24, 299-304

Kurokawa, Y., Takayama, S., Konishi, Y., Hiasa, Y., Asahina, S., Takahashi, M., Maekawa, A. & Hayashi, Y. (1986) Long-term in vivo carcinogenicity tests of potassium bromate, sodium hypochlorite, and sodium chlorite conducted in Japan. *Environ. Health Perspect.*, 69, 221-235

Mannsville Chemical Products Corp. (1987) *Chemical Products Synopsis: Calcium Hypochlorite*, Asbury Park, NJ

Meier, J.R., Bull, R.J., Stober, J.A. & Cimino, M.C. (1985) Evaluation of chemicals used for drinking water disinfection for production of chromosomal damage and sperm-head abnormalities in mice. *Environ. Mutagenesis*, 7, 201-211

Ministry of International Trade and Industry (1989) *Yearbook of Chemical Industries Statistics, 1988*, Tokyo, International Trade and Industry Association

Phillip, R., Shepherd, C., Fawthrop, F. & Poulsom, B. (1985) Domestic chlorine poisoning (Letter to the Editor). *Lancet, ii,* 495

PPG Industries (1990) *Report on Calcium Hypochlorite from PPG Chemicals Group,* Pittsburgh, PA

Revis, N.W., McCauley, P. & Holdsworth, G. (1986) Relationship of dietary iodide and drinking water disinfectants to thyroid function in experimental animals. *Environ. Health Perspect., 69,* 243-248

Robinson, M., Bull, R.J., Schamer, M. & Long, R.E. (1986) Epidermal hyperplasia in mouse skin following treatment with alternative drinking water disinfectants. *Environ. Health Perspect., 69,* 293-300

Rosenkranz, H.S. (1973) Sodium hypochlorite and sodium perborate: preferential inhibitors of DNA polymerase-deficient bacteria. *Mutat. Res., 21,* 171-174

Sasaki, M., Sugimura, K., Yoshida, M.A. & Abe, S. (1980) Cytogenetic effects of 60 chemicals on cultured human and Chinese hamster cells. *Kromosomo II, 20,* 574-584

Sigma-Aldrich Company (1989) *Material Safety Data Sheet, Sodium Hypochlorite,* Gillingham, Dorset

Weast, R.C., ed. (1989) *CRC Handbook of Chemistry and Physics,* 70th ed., Boca Raton, FL, CRC Press, pp. B-79, B-117, B-130

White, G.C. (1986) *The Handbook of Chlorination,* 2nd ed., New York, Van Nostrand Reinhold

Winterbourn, C.C. (1985) Comparative reactivities of various biological compounds with myeloperoxidase-hydrogen peroxide-chloride, and similarity of the oxidant to hypochlorite. *Biochim. biophys. Acta, 840,* 204-210

Wlodkowski, T.J. & Rosenkranz, H.S. (1975) Mutagenicity of sodium hypochlorite for *Salmonella typhimurium. Mutat. Res., 31,* 39-42

Wojtowicz, J.A. (1979) Chlorine oxygen acids and salts (chlorine monoxide, hypochlorous acid, and hypochlorites). In: Mark, H.F., Othmer, D.F., Overberger, C.G., Seaborg, G.T. & Grayson, M., eds, *Kirk-Othmer Encyclopedia of Chemical Technology,* 3rd ed., Vol. 5, New York, John Wiley & Sons, pp. 586-611

Wones, R.G, Mieczkowski, L. & Frohman, L.A. (1990) Effects of drinking water chlorine on human lipid and thyroid metabolism. In: Jolley, R.L., Condie, L.W., Johnson, J.D., Katz, S., Minear, R.A. & Mattice, J.S., eds, *Water Chlorination: Chemistry, Environmental Impact and Health Effects,* Vol. 6, Chelsea, MI, Lewis Publishers, pp. 301-308

CHLORINATION BY-PRODUCTS

BROMODICHLOROMETHANE

1. Chemical and Physical Data

1.1 Synonyms

Chem. Abstr. Services Reg. No.: 75-27-4
Chem. Abstr. Name: Bromodichloromethane
IUPAC Systematic Name: Bromodichloromethane
Synonyms: Dichlorobromomethane; dichloromonobromomethane; monobromodichloromethane

1.2 Structural and molecular formulae and molecular weight

```
        Cl
        |
    H - C - Br
        |
        Cl
```

$CHBrCl_2$ Mol. wt: 163.83

1.3 Chemical and physical properties of the pure substance

(a) *Description*: Colourless liquid (Verschueren, 1983)
(b) *Boiling-point*: 90.1°C (Weast, 1989)
(c) *Melting-point*: −57.1°C (Weast, 1989)
(d) *Density*: 1.980 at 20/4°C (Weast, 1989)
(e) *Spectroscopy data*[1]: Infrared (Sadtler Research Laboratories, 1980, prism [1898], grating [18024]; Pouchert, 1981, 1985a,b), nuclear magnetic resonance (Sadtler Research Laboratories, 1980, proton [6709], C-13

---

[1]In square brackets, spectrum number in compilation

[4023]; Pouchert, 1974, 1983) and mass spectral data [1026] (Bunn *et al.*, 1975; Coleman *et al.*, 1984) have been reported.

(*f*) *Solubility*: Soluble in water (4.5 g/l at 20°C) (Mabey *et al.*, 1982), acetone, ethanol, benzene, chloroform and diethyl ether (Weast, 1989)

(*g*) *Volatility*: Vapour pressure, 50 mm Hg at 20°C (Mabey *et al.*, 1982)

(*h*) *Reactivity*: Hydrolysis rate at neutral pH, 25°C, K = 5.76 × 10^{-8} per hour (Mabey *et al.*, 1982)

(*i*) *Octanol/water partition coefficient (P)*: log P, 2.10 (Chemical Information Systems, 1990)

(*j*) *Conversion factor*[1]: mg/m^3 = 6.70 × ppm

1.4 Technical products and impurities

Bromodichloromethane is available at > 97-98% purity, and may be stabilized with potassium carbonate or ethanol (Riedel-de Haën, 1986; American Tokyo Kasei, 1988; Aldrich Chemical Co., 1990).

2. Production, Use, Occurrence and Analysis

2.1 Production and use

(*a*) *Production*

Bromodichloromethane has been prepared by treating a mixture of chloroform (see IARC, 1987) and bromoform (see monograph, p. 213) with triethylbenzylammonium chloride and sodium hydroxide (Fedoryński *et al.*, 1977).

(*b*) *Use*

Bromodichloromethane can be used in organic synthesis, such as in the preparation of phenylbromodichloromethylmercury, which has been widely used for the generation of dichlorocarbene (Fedoryński *et al.*, 1977; Sittig, 1985).

Bromodichloromethane is currently used only as a standard in the analysis of drinking-water (Strobel & Grummt, 1987).

(*c*) *Regulatory status and guidelines*

Standards have been established for trihalomethanes (including bromodichloromethane) in drinking-water (see monograph on Chlorinated drinking-water, p. 59) in several countries.

---

[1]Calculated from: mg/m^3 = (molecular weight/24.45) × ppm, assuming standard temperature (25°C) and pressure (760 mm Hg)

2.2 Occurrence

(a) Natural occurrence

Mean levels of bromodichloromethane in the tissues of a number of temperate marine microalgae (*Ascophyllum nodosum, Fucus vesiculosis, Enteromorpha linza, Ulva lacta, Gigartina stellata*) ranged from 7 to 22 ng/g dry weight; the algae release this compound into seawater, from which it may be released to the atmosphere (Gschwend *et al.*, 1985).

Macroalgae collected near the Bermuda Islands (*Fucales sargassum*) and at the Cape of Good Hope (*Laminariales laminaria*) showed a specific pattern of emissions of volatile organohalides into the surrounding air. The main components were bromoform, bromodichloromethane, chlorodibromomethane; a minor component was bromoethane (Class *et al.*, 1986).

(b) Multimedia exposure assessment

In a study of exposures to volatile organic compounds on two US university campuses in 1980, bromodichloromethane levels in personal breathing-zone air samples ranged from not detected to 3.7 µg/m^3 for 11 students at Lamar University, Texas, and from not detected to 4.3 µg/m^3 for six students at the University of North Carolina. It was detected in the ambient air of 64% of the Lamar University samples at a mean concentration of 1.23 µg/m^3 and in 17% of the University of North Carolina samples at 0.83 µg/m^3. It was not detected in any of the collected breath samples. Tap-water samples contained mean bromodichloromethane levels of 21 ng/ml (range, 13-44) for the Lamar University students and 17 ng/ml (range, 15-20) for those at the University of North Carolina. The authors concluded that drinking-water was an important contributor to total intake of bromodichloromethane from air and water, assuming daily intakes of 10 m^3 of air and 1 litre of drinking-water. The water accounted for 76% of the bromodichloromethane intake (Wallace *et al.*, 1982).

Nine volunteers in New Jersey and three in North Carolina were monitored for exposure to bromodichloromethane in personal air samples, drinking-water, food and breath from July to December 1980. Bromodichloromethane was detected at 0.10-13.40 µg/m^3 in 28 (16 at trace) of the 164 breathing zone air samples in New Jersey and at 0.10-3.66 µg/m^3 in 12 (seven at trace) of the 60 samples in North Carolina. It was detected in none of the exhaled breath samples: 49 in New Jersey (limit of detection, 0.17-0.20 µg/m^3) and 17 in North Carolina (limit of detection, 0.14-2.20 µg/m^3). The drinking-water of the New Jersey volunteers contained bromodichloromethane levels ranging from 4.40 to 42.00 µg/l in home samples and from 0.02 to 37.00 µg/l in workplace samples; it was present in all of the 75 home samples at a median concentration of 18.00 µg/l and in 44 of the 45 workplace samples at a median concentration of 13.0 µg/l. In North Carolina, home drinking-water samples contained bromodichloromethane levels ranging from 12.0 to 20.0 µg/l and those in the workplace from 0.02 to 20.0 µg/l; it was present in all of the 30 home samples at a median concentration of 16 µg/l and in 17 of the 18 workplace samples at a median concentration of 14 µg/l. Bromodichloromethane was detected in composite beverage samples at 1.0 µg/l; it was not detected in composite dairy, meat or fatty food samples. The authors conclude that at least 75% of the volunteers' exposure to bromodichloromethane was contributed by drinking-water and that beverages supplied a significant fraction of the intake (Wallace *et al.*, 1984).

(c) *Air*

The volatilization half-time of bromodichloromethane from rivers and streams has been estimated to range from 33 min to 12 days, depending on turbulence and temperature. A typical half-time, based on actual data, was 35 h (Kaczmar *et al.*, 1984).

Air samples were collected in 1978-79 from eight covered swimming pools in Bremen, Germany, to determine the concentration of bromodichloromethane. The range of means, measured from the surface up to 2 m, in the four pools in which a mixed water source was used was 0.2-22 µg/m^3 (total range, 0.1-38 µg/m^3). The range of means in the four pools in which a groundwater source was used was 0.6-22 µg/m^3 (total range, 0.1-39 µg/m^3). Release of volatile compounds into the air depended on water and air temperature, concentration in the water, and intensity of air circulation (Lahl *et al.*, 1981).

Ullrich (1982) studied the organohalogen compound concentrations in the air of four public indoor swimming pools in western Berlin. Mean concentrations of bromodichloromethane ranged from 36 to 210 µg/m^3.

In the USA, air samples collected 2 cm above the surface of five outdoor pools contained 1-7 µg/m^3 bromodichloromethane; those above four indoor pools contained 1-14 µg/m^3; and those above four spas (whirlpools or hot tubs) contained < 0.1-90 µg/m^3. Samples collected 2 m above the surface contained < 0.1 µg/m^3, < 0.1-10 µg/m^3 and < 0.1-10 µg/m^3, respectively (Armstrong & Golden, 1986).

In a review of data on the presence of volatile organic chemicals in the atmosphere of the USA in 1970-80, a median concentration of 1.2 µg/m^3 bromodichloromethane was reported for the 17 urban/suburban data points examined and 0.11 µg/m^2 for industrial areas (nine data points) (Brodzinsky & Singh, 1983).

Bromodichloromethane levels in air samples collected in 1982-85 over the Atlantic Ocean were 0.1-1 ppt (0.7-6.7 ng/m^3); baseline levels of biogenic bromodichloromethane in air were 0.2-0.6 ppt (1.3-4.0 ng/m^3). Air samples collected in 1985 from a forest area in southern Germany contained 0.5 ppt (3.4 ng/m^3) bromodichloromethane (Class *et al.*, 1986).

According to the Toxic Chemical Release Inventory (National Library of Medicine, 1989), 115 kg of bromodichloromethane were released to ambient air in one US location. No release to media other than air was reported.

(d) Water and sediments

The formation of trihalomethanes in drinking-water and the effects of temperature and pH have been discussed extensively (Williams, 1985; see also the monograph on Chlorinated drinking-water, pp. 56 *et seq.*).

Bromodichloromethane has been measured or detected in many drinking-water systems, both in samples collected at treatment facilities or along the distribution system (Table 1) and in samples collected from natural and untreated water sources (Table 2). Concentrations in treated drinking-water typically ranged from 1 to 50 µg/l (with higher or lower values in some locations), compared with concentrations in untreated (natural) waters, which are typically less than 1 µg/l.

Table 1. Bromodichloromethane concentrations in treated[a] drinking-water, 1973-89

Location, date[b]	Sample site/ raw water source[c]	Concentration (µg/l)[d]	Reference
USA, 1974	NS/surface	2.9-20.8	Bellar *et al.*
USA, 1974	NS/well	1.1-1.9	(1974a,b)
80 US cities (NORS), 1974	Water supplies/ground and surface	ND-116 (mean, 6)	Symons *et al.* (1975)
113 Public water supplies, 1976	Water supplies/NS	Mean, 18; median, 14	Brass *et al.* (1977)
945 US sites, 1981-82	T/ground	1.4-2.2[e] (max., 110)	Westrick *et al.* (1984)
13 US community systems, 1984-85	T and D/ground (3) and surface (10)	< 0.2-58	Reding *et al.* (1989)
10 US utilities, 1985	T/ground and surface	> 10 - < 100 (5 sites) < 10 (5 sites)	Stevens *et al.* (1989)
35 US sites Spring, 1988 Summer, 1988 Autumn, 1988 Winter, 1989	T/ground and surface	Median 6.9 10 5.5 4.1	Krasner *et al.* (1989)

Table 1 (contd)

Location, date[b]	Sample site/ raw water source[c]	Concentration (μg/l)[d]	Reference
Durham, North Carolina, 1975	D/surface	10.99–11.60	McKinney et al. (1976)
15 Kentucky cities, 1977	T/surface	trace–15	Allgeier et al. (1980)
20 Tennessee sites Autumn, 1980 Winter, 1980 Spring, 1981 Summer, 1981	T/ground and surface	0.012–0.287 0.012–0.402 0.051–0.299 0.070–0.480	Minear & Morrow (1983)
Miami, Florida January, 1975 July, 1975	Water supplies ground	78 63	Loy et al. (1976)
Southwestern US city, 1975	Municipal water/ground	0.69–7.76	Henderson et al. (1976)
Lamar University, Texas, 1980	D/NS	13–44 (mean 21)	Wallace et al. (1982)
University of North Carolina, 1980		15–20 (mean, 17)	
East Texas, 1977	NS/surface (14 sites) NS/ground (11 sites)	9.3–89.8 ND–53.4	Glaze & Rawley (1979)
Houston, Texas, Summer 1978 to winter 1980	NS/surface NS/ground	max., 39 max., 5	Cech et al. (1982)
Huron, South Dakota, 1976	T/surface D[f]/surface D[g]/surface D[h]/surface	0–25 22–44 24–57 29–47	Harms & Looyenga (1977)
40 Michigan utilities, NS	T/surface (22 sites) T/ground (18 sites)	2.0–54.2 ND–3.2	Furlong & D'Itri (1986)
Old Love Canal, NY, 1978	D/NS	1.8–10	Barkley et al. (1980)
5 Pennsylvania sites, 1987 5 Virginia sites, 1987	T/surface	12.3–16.3 2.7–10 (range of means)	Smith (1989)
3 Puerto Rican cities, NS	D/ground and surface	0.003–0.011	Rodriguez-Flores (1983)
70 Canadian cities, 1976–77	D and T/ground and surface	0–33 (median, 1.4)	Health & Welfare Canada (1977)

3 Canadian utilities, 1977–78	T/raw water
	  filter effluent
	  plant effluent
	D/surface
		< 0.1
		0.44–0.64
		0.59–0.98
		0.91–1.15
			Otson et al. (1981)

Table 1 (contd)

Location, date[b]	Sample site/ raw water source[c]	Concentration (μg/l)[d]	Reference
30 Canadian sites Aug.-Sep. 1979 Nov.-Dec.. 1979	T/ground and surface	 mean, 3; max., 16 mean, 2; max., 12	Otson et al. (1982)
Burlington, Ontario, Canada, 1981	D/NS	3.8	Comba & Kaiser (1983)
Niagara Falls, Ontario, Canada, 1981		11	
Port Robinson, Ontario, Canada, 1981		160	
Chippawa, Ontario, Canada, 1980–81	D/surface	2.6	Kaiser & Comba (1983)
10 southern Ontario cities, 1981	T/surface D/surface	1.5–9.1 2.2–12	Oliver (1983)
Calgary, Alberta, Canada, 1983	T/surface	0.2–3.0	Hargesheimer (1985)
10 Canadian Great Lakes sites Summer 1982 Winter 1983 Spring 1983	T/ground (1) and surface (9)	Mean 4.4 2.8 4.1	Otson (1987)
Lancashire-Cheshire, UK, 1974	D/NS	1–27	McConnell (1976)
Southeastern UK, NS	D/ground and surface	6.4	Trussell et al. (1980)

Southampton, UK, 1977–78	D/surface
	  pumping station
	  reservoir
		
		mean, 8.66
		mean, 12.58
			Brett & Calverley (1979)

5 Belgian utilities, 1977–78 9 Belgian utilities, 1977–78	T/surface D/surface T/ground D/ground	9.4–56.0 2.4–38.1 0.1–4.1 0.2–5.3	Quaghebeur & De Wulf (1980)
100 German cities, 1977	NS/NS	ND–7.3 (mean, 0.7)	Bauer (1981)
37 German sites, 1976	NS/NS	mean, 0.6	
Bremen and Leverkusen, Germany, NS	NS/NS	ND–2.3	
12 German cities, 1978	NS/NS	ND–19.6 (mean, 3.4)	Eklund et al. (1978)
Tübingen, Germany, 1981	D/NS D/surface	0.10–0.47 (mean, 0.26) 0.78–1.27 (mean, 1.05)	Hagenmaier et al. (1982)
9 German cities, 1978–79	D/ground (1) and surface (8)	0.6–13.1	Lahl et al. (1982)

Table 1 (contd)

Location, date[b]	Sample site/ raw water source[c]	Concentration (μg/l)[d]	Reference
Gothenberg, Sweden	D/surface		Eklund et al. (1978)
1977		2.2	
February 1978		1.4	
Southern Brazil, NS	D/surface	4.4	Trussell et al. (1980)
Eastern Nicaragua, NS	D/surface	ND	
Northern Venezuela, NS	D/surface	10	
Eastern Peru, NS	D/ground and surface	5.7	
Southern China, NS	D/local catchments	7.6	Trussell et al. (1980)
Southern Philippines, NS	D/ground and surface	2.3	
Northern Philippines, NS	D/surface	1.7	
Northern Egypt, NS	D/surface	ND	
Southern Indonesia, NS	D/surface	3.0	
Southeastern Australia, NS	D/surface	4.1	

[a]Treatment not always specified
[b]NORS, National Organics Reconnaissance Survey; NS, not specified
[c]D, distribution system; T, treatment plant
[d]ND, not detected
[e]Range of median values for randomly and nonrandomly selected water supplies serving fewer than and more than 10 000 persons
[f]Short residence time
[g]Medium residence time
[h]Long residence time

Rook (1974) demonstrated that bromodichloromethane, observed at concentrations ranging from 4.2 to 20 μg/l following chlorination of stored surface waters, was a product of chlorination of the humic substances in natural waters.

Bromodichloromethane was detected (but not quantified) by headspace analysis of six of ten Pacific seawater samples collected in 1983; it was not detected in the corresponding marine air samples (Hoyt & Rasmussen, 1985).

Bromodichloromethane was found in water samples collected at various stages of water treatment: not detected (< 0.1 μg/l) in raw river water; 6.3 μg/l in river water treated with chlorine and alum; 18.0 μg/l in three-day-old settled water; 21.9 μg/l in water flowing from settled areas to filters; 18.0 μg/l in the filter effluent; and 20.8 μg/l in finished water (Bellar et al., 1974a,b).

Table 2. Bromodichloromethane concentrations in untreated (natural) water, 1973–89

Location, date[a]	Sample source	Concentration (µg/l)[b]	Reference
80 US cities (NORS), 1975	River, lake and ground	ND–0.8	Symons et al. (1975)
Northern Taiwan, NS	Well	ND	Trussell et al. (1980)
30 Canadian sites, 1979	River, lake and ground	mean, < 1; max., 13	Otson et al. (1982)
Campbellville, Ontario, Canada, 1981	Well	0.004	Comba & Kaiser (1983); Kaiser et al. (1983)
Waterdown, Ontario, 1981	Well	0.027	
Burlington, Ontario, 1981	Well	0.016	
	Well	0.025	
Beamsville, Ontario, 1981	Lake	0.02	
Crawford Lake, Ontario, 1981	Lake	ND–0.02 (mean, 0.002)	
Lake Ontario, Ontario, 1981	Spring	0.006	
Ancaster, Ontario, 1981	River	Trace–0.025 (mean, 0.006)	
Niagara River, Ontario, 1981			
Welland River watershed; Ontario	Surface (river)		Kaiser & Comba (1983)
Summer 1980		0.005–0.02	
Spring 1981		0.015–0.45	
10 Canadian Great Lakes sites	Ground and surface		Otson (1987)
Summer 1982		mean, 0.2	
Winter 1983		0	
Spring 1983		mean, 0.3	
North and South Atlantic, 1985	Seawater	0.0001–0.001 (baseline, 0.0001)	Class et al. (1986)
Ulm, Germany, NS	Rainwater	0.0004	Class et al. (1986)

[a]NORS, National Organics Reconnaissance System; NS, not specified
[b]ND, not detected

According to the US Environmental Protection Agency STORET system, concentrations of bromodichloromethane in 143 samples of surface water in 1970-79 ranged from 0.1 to 1 µg/l in 66% of the samples, 1-10 µg/l in 31% of the samples and 10-100 µg/l in 3% of the samples (Perwak et al., 1980).

The US Environmental Protection Agency estimated that 832 tonnes of bromodichloromethane were generated in the USA in 1978 by water chlorination. On the basis of the 1976 National Organic Monitoring Survey, the general population was estimated to be exposed to 20 μg bromodichloromethane per day from drinking-water, assuming a median concentration of 14 μg/l and a water intake of 1.65 l per day; assuming a maximal concentration of 180 μg/l and an intake of 2.18 l per day, the daily exposure increased to 400 μg per day (Perwak et al., 1980). In a later investigation by the US Environmental Protection Agency STORET data base, analysis of 19 550 water samples revealed a mean bromodichloromethane concentration of 11.14 μg/l (range, 0-10 133 μg/l); analysis of 581 sediment samples revealed a mean of 10.8 μg/kg (range, 0-55 μg/kg) (US Environmental Protection Agency, 1985).

In the 1982 US Nationwide Urban Runoff Program, bromodichloromethane was detected in samples from one of the 15 reporting cities at a concentration of 2 μg/l (Cole et al., 1984).

Samples of finished water collected in 1976 from the clear well storage area at the Huron, South Dakota, USA, water-treatment plant contained 42 μg/l bromodichloromethane. Changing the location of the prechlorination dose (from the presedimentation/chemical addition station to the recarbonation basin) or changing the pH did not substantially affect the bromodichloromethane concentration (Harms & Looyenga, 1977).

Although the levels found were not reported, a detection frequency of 10% for concentrations > 1 μg/l was reported for bromodichloromethane in water samples collected in 1976 from the Delaware, Schuylkill and Lehigh Rivers in the USA. Levels in raw water at treatment plants in Trenton, Torresdale-Philadelphia and Queenslane-Philadelphia increased from < 1 μg/l to 1, 10 and 6 μg/l, respectively, as a result of the chlorination process (DeWalle & Chian, 1978).

Tap-water samples collected between January 1977 and March 1978 in Osaka, Japan, contained bromodichloromethane at levels of 5.8, 7.5 and 14.0 μg/l at seasonal mean water temperatures of 7.4°C, 15.8°C and 25.4°C, respectively. An increase of approximately 0.5 in pH to control pipe corrosion resulted in bromodichloromethane concentrations of 7.8, 9.6 and 14.6 μg/l at similar mean water temperatures (Kajino & Yagi, 1980).

Bromodichloromethane levels in tap water collected at four locations in a Swedish community ranged from 0.84 to 1.2 μg/l; when the treatment facility briefly changed the disinfectant from chlorine to chlorine dioxide, the levels ranged from 0.021 to 0.023 μg/l (detection limit, 0.005 μg/l) (Norin et al., 1981).

Samples of water were collected between August and October 1980 from four supply systems in São Paulo State, Brazil. Mean levels of bromodichloromethane were 6.8-15.8 µg/l in treated water (after treatment), 9.8-23.3 µg/l in treated water from the reservoir and 11.5-27.8 µg/l in treated tap water (de Fernicola & de Azevedo, 1984).

Water samples were collected from five outdoor pools, four indoor pools and four spas (whirlpools or hot tubs) in Lubbock, TX, USA. The concentrations of bromodichloromethane in the outdoor pools, which used chlorine-based materials for disinfection were 1-72 µg/l. Two of the indoor pools in which only chlorination was used had levels of 1.5-90 µg/l; one indoor pool in which only bromination (sodium hypobromite) was used had levels of 1-11 µg/l; and the fourth indoor pool, in which chlorination and bromination were alternated, had levels of 1-8 µg/l. The spa in which only chlorination was used had levels of 1-105 µg/l; the two spas in which only bromination was used had levels of < 0.1-21 µg/l; and the spa in which the combination was used had levels of 0.7-13 µg/l. The average concentration of bromodichloromethane in Lubbock, TX, tap water was 1.0 µg/l (Armstrong & Golden, 1986).

Water samples were collected in 1978-79 from eight covered swimming pools in Bremen, Germany, to determine the concentration of bromodichloromethane. The source of fresh water was mixed river and groundwater for four pools and groundwater for four pools. The level of bromodichloromethane in the pools with mixed sources was 1.5 and that in the pools with groundwater, 1.0 µg/l. The range of means of bromodichloromethane in the four pools with a mixed water source was 15-60 µg/l (total range, 4-150 µg/l); that in the four pools with a groundwater source was 0.1-25 µg/l (total range, 0.1-76 µg/l) (Lahl *et al.*, 1981).

Kaminski and von Loew (1984) found an average concentration of 0.8 µg/l (max, 2.0 µg/l) bromodichloromethane in 26 indoor pools in western Germany. The concentrations of bromodichloromethane in two thermal spas in which the initial bromide concentration was 0.5-0.7 mg/l were 2.3-23.8 µg/l (Weil *et al.*, 1980).

Scotte (1984) studied the concentrations of organohalogens in the water of 10 covered swimming pools in France. The mean concentrations of bromodichloromethane were 4.5 µg/l in the four pools treated with Surchlor GR 60 (anhydrous sodium dichloroisocyanurate); 11.10 µg/l in the two treated with gaseous chlorine, 14.18 µg/l in the two treated with sodium hypochlorite and 0.5 µg/l in the two treated with bromine.

Effluents from a wastewater treatment plant on Boston Harbor, MA, USA, sampled in 1984 and 1985, contained a mean bromodichloromethane concentration of 4.5 µg/l (range, 0.96-10.3) and had an estimated mass input rate of 4.1 kg/day (Kossik et al., 1986).

Heating water to prepare food has been shown to eliminate a large part of trihalomethanes in the water, as a function of temperature and heating time. Levels of bromodichloromethane were reduced from 44.6 µg/l in tap water to 24.1 µg/l after heating at 80°C for 1 min, to 13.5 µg/l after heating to 100°C, to 10.8 µg/l after boiling for 1 min, and to 4.6 µg/l after boiling for 5 min (Lahl et al., 1982).

(e) *Food and beverages*

Entz *et al.* (1982) purchased 39 different food items at retail markets in three geographical areas in the USA and analysed 20 food composites, comprising four groups (dairy, meats, oils-fats, beverages), for bromodichloromethane. It was detected in one dairy composite (milk, ice-cream, cheese and butter) at a concentration of 1.2 µg/l and in two beverage composites at concentrations of 0.3 and 0.6 µg/l. Subsequent analysis of the components of the composites revealed concentrations of 2.3, 3.4 and 3.8 µg/l in three cola soft drinks and 7 µg/kg in butter.

Abdel-Rahman (1982) analysed various US soft drinks for the presence of bromodichloromethane. Cola beverages had mean ranges of 0.9-5.9 µg/l, while other soft drinks had levels of 0.1-3.3 µg/l. Municipal water supplies from which the soft drinks were manufactured were found to contain less than 20 µg/l trihalomethanes.

Uhler and Diachenko (1987) analysed process water and food products from 15 food processing plants located in nine states in the USA, representing 39 food products. Bromodichloromethane was found in seven process waters at levels ranging from < 1 to 14.1 µg/kg, in three soft drinks from one plant at 1.2-2.3 µg/kg and in three ice creams from one plant at 0.6-2.3 µg/kg. The ice-cream plant was the only location at which bromodichloromethane was found in both the process water and the associated food products.

2.3 Analysis

Selected methods for the analysis of bromodichloromethane in air and water are given in Table 3. A variety of analytical methods exist for measuring bromodichloromethane in water. The commonly used methods are based on extraction with solvent followed by gas chromatograph-electron capture detection (US Environmental Protection Agency Method 501-2) (Standing Committee of Analysts, 1980), purge-and-trap, flame ionization detection and microcoulometric gas chromatography (Bellar *et al.*, 1974a,b; Method 501-1), gas chromatography-mass spectrometry and headspace analysis (Otson *et al.*, 1982).

Table 3. Methods for the analysis of bromodichloromethane

Sample matrix	Sample preparation[a]	Assay procedure[b]	Limit of detection[c]	Reference
Air	Collect cryogenically into stainless-steel bottle; inject sample	GC/EC-FI-FPD/GC/MS	NR	Hoyt & Rasmussen (1985)
Seawater	Collect in vacuum extraction flask; pressurize with zero air; inject headspace sample	GC/EC-FI-FPD/GC/MS	NR	Hoyt & Rasmussen (1985)
Water	Purge (inert gas); trap (OV-1 on Chromosorb-W/Tenax/silica gel); desorb as vapour (heat to 180°C, backflush with inert gas) onto packed GC column	GC/ECD	0.10 μg/l	US Environmental Protection Agency (1988a) [Method 601]
		GC/MS	2.2 μg/l	US Environmental Protection Agency (1988b) [Method 624]
Water	Purge (inert gas); trap (OV-1 on Chromosorb-W/Tenax/silica gel); desorb as vapour (heat to 180°C, backflush with inert gas) onto capillary GC column	GC/ED	0.02 μg/l	US Environmental Protection Agency (1988c) [Method 502.2]
		GC/MS	0.08 μg/l	US Environmental Protection Agency (1988d) [Method 524.2]
	Add internal standard (isotope-labelled bromo-dichloromethane); purge; trap and desorb as above	GC/MS	10 μg/l	US Environmental Protection Agency (1988e) [Method 1624]
	Extract in pentane; inject onto GC	GC/EC (>50 μg/l) GC/MS (<50 μg/l)	0.5 μg/l	US Environmental Protection Agency (1988f) [Method 501.2]
Adipose tissue	Purge from liquefied fat at 115°C; trap on silica gel; desorb thermally	GC/HSD	0.8 μg/l	Peoples et al. (1979)
Blood serum	Purge from water-serum mixture containing anti-foam reagent at 115°C; trap on Tenax/silica gel; desorb thermally	GC/HSD	0.8 μg/l	Peoples et al. (1979)

[a]GC, gas chromatograph

[b]GC/EC-FI-FPD, gas chromatography/electron capture-flame ionization-flame photometric detection; GC/MS, gas chromatography/mass spectrometry; GC/ECD, gas chromatography/electrolytic conductivity detection; GC/HSD, gas chromatography/halide selective detection

[c]NR, not reported

The US Environmental Protection Agency methods for analysing water (Methods 8010 and 8240) have also been used for liquid and solid wastes. Volatile components of solid waste samples are first extracted with methanol, prior to purge-and-trap concentration and analysis by gas chromatography-electrolytic conductivity detection (Method 8010) or gas chromatography-mass spectrometry (Method 8240). The detection limit for bromodichloromethane using Method 8010 is 0.10 µg/l, and the practical quantification limit using Method 8240 is 5 µg/l for groundwater and for soil/sediment samples (US Environmental Protection Agency, 1986a,b).

US Environmental Protection Agency Method 624 has also been adapted to the analysis of bromodichloromethane in fish, with an estimated detection limit of 10 µg/kg (Easley *et al.*, 1981).

3. Biological Data Relevant to the Evaluation of Carcinogenic Risk to Humans

3.1 Carcinogenicity studies in animals (Table 4)

(*a*) *Oral administration*

Mouse: Groups of 50 male and 50 female B6C3F$_1$ mice, eight weeks old, were given bromodichloromethane (> 99% pure) in corn oil by gavage at 25 or 50 mg/kg bw (males) and 75 or 150 mg/kg bw (females) on five days per week for 102 weeks. Survival at 104 weeks was: males—vehicle control, 34/50; low-dose, 32/50; high-dose, 42/50; females—control, 26/50; low-dose, 13/50; high-dose, 15/50. Decreased survival in female mice was associated in part with tubo-ovarian abscesses (control, 8/50; low-dose, 19/47; high-dose, 18/49). Tubular-cell adenomas of the kidney occurred in 1/49 vehicle control, 2/50 low-dose and 6/50 high-dose male mice and tubular-cell adenocarcinomas occurred in 4/50 high-dose male mice. The proportion of high-dose male mice with renal tubular-cell neoplasms was significantly greater than that in controls ($p = 0.022$, incidental tumour test). Renal tubular-cell neoplasms were uncommon in control male mice of this strain (historical incidences: study laboratory, 2/299 (0.7%); all National Toxicology Program laboratories, 5/1490 (0.3%)). Hepatocellular adenomas occurred in 1/50 control, 13/48 low-dose and 23/50 high-dose female mice; hepatocellular carcinomas occurred in 2/50 control, 5/48 low-dose and 10/50 high-dose female mice. The proportion of low- and high-dose female mice with hepatocellular neoplasms was significantly greater than that in controls ($p < 0.001$, incidental tumour test; pairwise comparisons and trend test). Adenomas of the anterior pituitary gland occurred at significantly lower incidence in high-dose female mice (control, 17/44; low-dose, 8/43; high-dose, 3/38; $p = 0.006$, pairwise comparison, incidental tumour test). Follicular-cell hyperplasia of the thyroid was significantly more common in treated mice, principally in those given the high dose (males: control, 0/48; low-dose, 3/44; high-dose, 5/49; females: vehicle control, 6/50; low-dose, 18/45; high-dose, 21/48) (National Toxicology Program, 1987).

Groups of male and female CBA × C57Bl/6 hybrid mice [age unspecified] were given bromodichloromethane [purity unspecified] at 0.04 mg/l (50 males, 50 females), 4.0 mg/l (50 males, 50 females) or 400 mg/l (55 males, 55 females) in the drinking-water for 104 weeks. Seventy-five males and 50 females served as controls. The number of animals surviving to the appearance of the first tumour were: males—control, 63; low-dose, 35; medium-dose, 16; and high-dose, 45; females—control, 34; low-dose, 45; medium-dose, 18; and high-dose, 13 [average survival time and numbers of terminal survivors unspecified]. No tumour occurred at increased incidence in treated mice (Voronin *et al.*, 1987). [The Working Group noted the incomplete reporting of the study.]

Rat: Groups of 50 male and 50 female Fischer 344 rats, eight weeks old, were given bromodichloromethane (> 99% pure) in corn oil at 50 or 100 mg/kg bw by gavage on five days per week for 102 weeks. Survival at 104 weeks was: males—vehicle control, 28/50; low-dose, 36/50; high-dose, 28/50; females—vehicle control, 34/50; low-dose, 27/50; high-dose, 41/50. Adenomatous polyps of the large intestine occurred in 0/50 control, 3/50 low-dose and 33/50 high-dose males; adenocarcinomas of the large intestine occurred in 0/50 control, 11/50 low-dose and 38/50 high-dose males. The proportion of male rats receiving 50 or 100 mg/kg bw bromodichloromethane that had neoplasms of the large intestine was significantly greater than in controls ($p < 0.001$, incidental tumour test, pairwise comparisons and trend test). Tubular-cell adenomas of the kidney occurred in 0/50 control, 1/50 low-dose and 3/50 high-dose male rats, whereas tubular-cell adenocarcinomas occurred in 10/50 high-dose males. The proportion of high-dose male rats with renal tubular-cell neoplasms was significantly greater than in controls ($p < 0.001$, incidental tumour test, pairwise comparison and trend test). Renal tubular-cell neoplasms were uncommon in control male rats of this strain (historical incidences: study laboratory, 1/250 (0.4%); all National Toxicology Program laboratories, 8/1448 (0.6%)). Cytomegaly of renal tubular epithelial cells occurred at high incidence in treated male rats but not in control male or treated or control female rats. The proportion of high-dose male rats with adrenal medullary phaeochromocytomas (benign or malignant) was lower than that in controls: control, 18/50; low-dose, 14/50; high-dose, 5/50 ($p = 0.003$, pairwise comparison by incidental tumour test). Adenomatous polyps of the large intestine occurred in 0/46 control, 0/50 low-dose and 7/47 high-dose female rats, while adenocarcinomas occurred in 0/46 control, 0/50 low-dose and 6/47 high-dose rats. The proportion of high-dose female rats with neoplasms of the large intestine was significantly greater than in controls ($p < 0.001$, incidental tumour test, pairwise comparisons and trend test). Renal tubular-cell adenomas occurred in 0/50 control, 1/50 low-dose and 6/50 high-dose females, and tubular-cell adenocarcinomas occurred in 9/50 high-dose female rats. The proportion of high-dose female rats with renal tubular-cell neoplasms was significantly greater than in controls ($p < 0.001$, incidental tumour test, pairwise comparison and trend test). Increased incidences of non-neoplastic lesions of the liver seen in treated females included clear-cell, eosinophilic cytoplasmic, focal cellular and fatty changes. The incidences of neoplasms of the anterior pituitary and of fibroadenomas of the mammary gland were significantly lower in high-dose female rats than in controls (anterior pituitary neoplasms: control, 31/49; low-dose, 20/49; high-dose, 14/49; $p < 0.001$, incidental tumour test and life table test, pairwise comparison and trend; mammary gland fibroadenomas: control, 20/50; low-dose, 15/50; high dose, 1/50; $p < 0.001$, incidental tumour test, pairwise comparison and trend) (National Toxicology Program, 1987).

Groups of 58 male and 58 female weanling Wistar rats received bromodichloromethane [purity unspecified] at 2.4 g/l (approximately the maximal acceptable level) in the drinking-water for 72 weeks followed by 1.2 g/l for the remainder of their life [average survival time not given]. Twenty-six male and 22 female rats received drinking-water without bromodichloromethane and served as controls. Neoplastic nodules of the liver occurred in 0/18 control and 17/53 treated female rats ($p < 0.001$, Fisher's exact test). Adenofibrosis of the liver was also observed in 12/53 treated female and 1/47 treated male rats, but not in controls. The proportion of treated rats with lymphosarcomas in comparison to controls was decreased in treated males and increased in treated females (males: control, 14/22; treated, 9/47; $p < 0.001$, Fisher's exact test; females: control, 2/18; treated, 9/53; $p < 0.01$, Fisher's exact test). The proportion of treated female rats with pituitary gland or mammary gland tumours [types unspecified] was decreased relative to controls (pituitary tumours: controls, 6/18; treated, 5/53; $p < 0.03$, Fisher's exact test; mammary gland tumours: controls, 8/18; treated, 3/53; $p < 0.001$, Fisher's exact test) (Tumasonis *et al.*, 1985). [The Working Group noted that survival-adjusted statistics were not given and the controversial nature of the lesion diagnosed as adenofibrosis.]

(b) *Intraperitoneal administration*

Mouse: In a screening assay based on the enhanced induction of lung tumours, groups of 20 male strain A/St mice, six to eight weeks old, were injected intraperitoneally three times per week with bromodichloromethane ($> 95\%$ pure) in tricaprylin at 20, 40 or 100 mg/kg bw (maximum tolerated dose) for a total of 18 or

24 injections (total doses, 360, 960 or 2400 mg/kg bw). Twenty males receiving tricaprylin only were used as controls. Twenty-four weeks after the first injection, all surviving animals were killed; these were 15/20 of the controls, 15/20 at the low dose, 16/20 at the mid-dose and 13/20 at the high dose. The average numbers of lung adenomas per mouse were 0.27 ± 0.015 (SE) in controls, 0.20 ± 0.11 at the low dose, 0.25 ± 0.11 at the mid-dose and 0.85 ± 0.27 at the high dose [proportion of mice with tumours not given]. In a positive control group given a single intraperitoneal injection of urethane at 1000 mg/kg bw, the average number of lung tumours per mouse was 19.6 ± 2.4 (Theiss *et al.*, 1977).

3.2 Other relevant data

(a) Experimental systems

(i) Absorption, distribution, metabolism and excretion

14C-Bromodichloromethane (0.61 mmol/kg bw; 16 μCi/kg bw; 100 mg/kg bw) administered orally in corn oil to rats by gavage was absorbed and eliminated in the expired air as unchanged bromodichloromethane (42% of dose) or as 14C-carbon dioxide (14% of dose) in 8 h; radiolabel amounting to about 1% of the dose was eliminated in the urine, and about 3% of the dose was retained in body tissues. 14C-Bromodichloromethane (0.92 mmol/kg bw; 32 μCi/kg bw; 150 mg/kg bw) administered similarly to mice was absorbed and eliminated in the expired air as unchanged bromodichloromethane (7% of dose) or as 14C-carbon dioxide (81% of dose) in 8 h; about 2% of the administered radiolabel was eliminated in the urine, and 3% was retained in body tissues (Mink *et al.*, 1986). Bromodichloromethane is also metabolized to carbon monoxide *in vivo* (Anders *et al.*, 1978) and *in vitro* (Ahmed *et al.*, 1977).

Rats given bromodichloromethane at 0.5 or 5 mg by gavage in corn oil once daily for 25 days showed an average serum concentration of 1 or 23 μg/l and an average fat concentration of 51 or 1800 ng/g (Pfaffenberger *et al.*, 1980); three to five days after dosing had ended, the average serum concentration was 1 μg/l and the average fat concentration was 3-4 ng/g fat for both dose levels.

(ii) Toxic effects

The single-dose oral LD_{50} of bromodichloromethane (in Emulphor:ethanol: saline 1:1:8) was 450 mg/kg bw in male and 900 mg/kg bw in female ICR Swiss mice (Bowman *et al.*, 1978). Oral LD_{50} values (in corn oil) of 916 and 969 mg/kg bw were reported in male and female Sprague-Dawley rats (Chu *et al.*, 1980, 1982a), of 651 and 751 mg/kg bw in male and female Fischer 344/N rats and of 300-600 and 651 mg/kg bw in male and female $B6C3F_1$ mice (National Toxicology Program, 1987). Signs of acute toxicity in rats included sedation, prostration, lethargy, laboured

Table 4. Summary of carcinogenicity studies of bromodichloromethane in experimental animals

Reference	Species/strain	Sex	Dose schedule	Experimental parameter/observation	Group 0	1	2	3	Significance	Comments
National Toxicology Program (1987)	Mouse B6C3F$_1$	M	5 d/week, gavage, corn oil, 102 weeks	Dose (mg/kg)	0	25	50	—		
				Survival (104 weeks)	34/50	32/50	42/50			
				Renal tubular-cell						
				Adenoma	1/49	2/50	6/50		} $p = 0.022$	Increase
				Adenocarcinoma	0/50	0/50	4/50			
		F		Dose (mg/kg)	0	75	150	—		Tubo-ovarian abscesses
				Survival (104 weeks)	26/50	13/50	15/50			
				Hepatocellular						
				Adenoma	1/50	13/48	23/50		} $p < 0.001$	Increase
				Carcinoma	2/50	5/48	10/50			
				Anterior pituitary adenoma	17/44	8/43	3/38		$p = 0.006$	Decrease
Voronin et al. (1987)	Mouse CBA × C57Bl/6	M&F	Ad lib. in drinking-water	Dose (mg/l)	0	0.04	4.00	400		
				Survival M	63/75	35/50	16/50	45/55		
				F	34/50	45/50	18/50	13/55		
				Tumours	No increase					
Theiss et al. (1977)	Mouse strain A/St	M	3 d/week, i.p. inj, tricaprylin, 18 or 24 doses	Dose (mg/kg)	0	20	40	100		Screening test in strain in which lung adenomas common; ± SE
				Total dose (mg/kg)	0	360	960	2400		
				Survival (24 weeks)	15/20	15/20	16/20	13/20		
				Lung adenomas per mouse	0.27 ± 0.015	0.20 ± 0.11	0.25 ± 0.11	0.85 ± 0.27		

Table 4 (contd)

Reference	Species/strain	Sex	Dose schedule	Experimental parameter/observation	Group 0	Group 1	Group 2	Group 3	Significance	Comments
National Toxicology Program (1987)	Rat F344	M	5 d/week, gavage, corn oil, 102 weeks	Dose (mg/kg)	0	50	100	—		
				Survival (104 weeks)	28/50	36/50	28/50			
				Large intestine						
				Adenomatous polyp	0/50	3/50	33/50		$p < 0.001$	Increase
				Adenocarcinoma	0/50	11/50	38/50			
				Renal tubular-cell						
				Adenoma	0/50	1/50	3/50		$p < 0.001$	Increase
				Adenocarcinoma	0/50	0/50	10/50			
				Adrenal medullary phaeochromocytoma	18/50	14/50	5/50		$p = 0.003$	Decrease
		F		Dose (mg/kg)	0	50	100	—		
				Survival (104 weeks)	34/50	27/50	41/50			
				Large intestine						
				Adenomatous polyp	0/46	0/50	7/47		$p < 0.001$	Increase
				Adenocarcinoma	0/46	0/50	6/47			
				Renal tubular-cell						
				Adenoma	0/50	1/50	6/50		$p < 0.001$	Increase
				Adenocarcinoma	0/50	0/50	9/50			
				Anterior pituitary adenoma	31/49	20/49	14/49		$p < 0.001$	Decrease
				Mammary gland fibroadenoma	20/50	15/50	1/50		$p < 0.001$	Decrease

Table 4 (contd)

Reference	Species/strain	Sex	Dose schedule	Experimental parameter/observation	Group 0	Group 1	Group 2	Group 3	Significance	Comments
Tumasonis et al. (1985)	Rat Wistar	M	Ad lib. drinking-water; 2.4 g/l for 72 weeks; 1.2 g/l for remainder of life	Dose (g/l)	0	2.4	–	–		
				Survival	NS/26	NS/58				
				Hepatic 'adenofibrosis'	0/22	1/47				
				Neoplastic nodules	5/22	6/47				
				Lymphosarcoma	14/22	9/47			$p < 0.001$	Decrease
		F		Dose (g/l)	0	2.4	–	–		
				Survival	NS/22	NS/58				
				Hepatic 'adenofibrosis'	0/18	12/53			$p < 0.001$	Increase
				Neoplastic nodules	0/18	17/53			$p < 0.01$	Increase
				Lymphosarcoma	2/18	9/53			$p < 0.001$	Decrease
				Mammary tumour	8/18	3/53				

NS, not stated

breathing, ataxia, muscular weakness, anaesthesia and reduction in the number of peripheral lymphocytes (Chu et al., 1980, 1982a; National Toxicology Program, 1987); in mice, sedation and anaesthesia, liver damage including fatty infiltration, kidney lesions and haemorrhage in the adrenal glands, lung and brain were seen (Bowman et al., 1978).

Daily oral treatment of CD-1 mice with 50-250 mg/kg bw bromodichloromethane (in 10% Emulphor in water) for 14 days resulted in decreases in serum glucose in males and increases in transaminases (ALAT and ASAT) and blood urea nitrogen in both males and females treated with 250 mg/kg bw. Decreases in body weight, spleen weight, humoral and cellular immunity and blood fibrinogen as well as increases in liver weight were observed with the highest dose in animals of each sex; some of these effects were also observed with 125 mg/kg bw (Munson et al., 1982). Oral treatment of male CD-1 mice with 37-148 mg/kg bw bromodichloromethane in corn oil for 14 days also led in a dose-dependent fashion to liver damage, as shown by morphological changes, and to increases in serum transaminases (ALAT) at the highest dose level. Morphological and functional changes were observed in the kidney after treatment with the two highest doses (Condie et al., 1983).

Daily oral administration of 150 mg/kg bw bromodichloromethane in corn oil for 14 days was lethal to male but not female $B6C3F_1$ mice, whereas male and female Fischer 344/N rats survived treatment at 600 mg/kg bw. In a 13-week study, male and female Fischer 344/N rats were administered 19-300 mg/kg bw, male $B6C3F_1$ mice were administered 6.25-100 mg/kg bw and female $B6C3F_1$ mice were administered 25-400 mg/kg bw bromodichloromethane in corn oil by gavage on five days per week. In the highest dose groups, treatment was lethal to some rats and resulted in reductions in body weight. Liver and kidney lesions were observed in some of the animals. Atrophy of the thymus, spleen, lymph nodes, seminal vesicles and prostate was observed in rats, the toxicity being less pronounced in females than in males (National Toxicology Program, 1987).

Male and female Sprague-Dawley rats received 5-2500 mg/l bromodichloromethane in drinking-water for 90 days [approximate daily intake, 0.14-49 mg/rat]. The highest concentration resulted in some deaths, mild-to-moderate liver damage, reduction in body weight gain and the number of peripheral lymphocytes and mild changes in the thyroid (Chu et al., 1982b).

Administration of bromodichloromethane in corn oil by gavage for two years (see section 3.1) resulted in kidney damage and fatty changes in the liver in male and female rats (50 or 100 mg/kg bw) and in male (25 or 50 mg/kg bw) but not female (75 or 150 mg/kg bw) mice in all treatment groups. Liver-cell necrosis was observed only in male rats (National Toxicology Program, 1987).

(iii) *Effects on reproduction and prenatal toxicity*

Sprague-Dawley rats were administered bromodichloromethane in corn oil by gavage at daily doses of 0, 50, 100 or 200 mg/kg bw on gestation days 6-15 (Ruddick *et al.*, 1983). Maternal weight gain was significantly decreased at the high-dose level; maternal liver weight was significantly increased at all dose levels, and kidney weight was increased at the highest dose level. There was no difference in the incidence of resorptions, litter size or mean fetal weight. There was no increase in the incidence of external or visceral malformations, but aberrations of the sternum were more prevalent at 100 and 200 mg/kg bw than at 50 mg/kg bw.

(iv) *Genetic and related effects* (Table 5)

Bromodichloromethane induced positive responses in some *Salmonella typhimurium* reverse mutation assays, particularly with strain TA100. Conflicting results were observed in these studies when an exogenous metabolic system was incorporated into the incubation mixture. Generally, the positive results were observed after modifications to the standard assay procedure that resulted in higher doses of the compound reaching the cells, such as exposure in a closed container, a preincubation period or use of a bacterial spot test.

A weak positive response was observed in yeast in a mutation and a gene conversion assay. A mutagenic response was also obtained in L5178Y cells. No sister chromatid exchange was induced in Chinese hamster cells, whereas chromosomal aberrations were observed in two out of three studies. Sister chromatid exchange was observed in one study in human lymphocytes *in vitro*. Sister chromatid exchange but not micronuclei was induced in mouse bone marrow *in vivo*.

(*b*) *Humans*

No data were available to the Working Group.

3.3 Epidemiological studies of carcinogenicity to humans

A single correlation study (Isacson *et al.*, 1983), described in the monograph on chlorinated drinking-water (p. 113), mentioned bromodichloromethane, but the information could not be used to evaluate the carcinogenicity of this chemical individually.

4. Summary of Data Reported and Evaluation

4.1 Exposure data

Bromodichloromethane is found in chlorinated drinking-water as a consequence of the reaction between chlorine, added during water treatment, and

Table 5. Genetic and related effects of bromodichloromethane

Test system	Result		Dose LED/HID	Reference
	Without exogenous metabolic system	With exogenous metabolic system		
SA0, *Salmonella typhimurium* TA100, reverse mutation	+	0	0.0000	Simmon et al. (1977)[a,b]
SA0, *Salmonella typhimurium* TA100, reverse mutation	–	–	500.0000	Mortelmans et al. (1986)
SA0, *Salmonella typhimurium* TA100, reverse mutation	–	–	0.0000	Khudoley et al. (1987)
SA0, *Salmonella typhimurium* TA100, reverse mutation	–	+	5.0000	Strobel & Grummt (1987)
SA0, *Salmonella typhimurium* TA100, reverse mutation	–	–	170.0000	Varma et al. (1988)
SA0, *Salmonella typhimurium* TA100, reverse mutation	–	+	0.0000	Mersch-Sunderman (1989)[c]
SA0, *Salmonella typhimurium* TA100, reverse mutation	0	+	0.0000	Khudoley et al. (1989)[a]
SA2, *Salmonella typhimurium* TA102, reverse mutation	–	–	0.0000	Mersch-Sunderman (1989)
SA4, *Salmonella typhimurium* TA104, reverse mutation	(+)	(+)	125.0000	Strobel & Grummt (1987)
SA5, *Salmonella typhimurium* TA1535, reverse mutation	–	–	500.0000	Mortelmans et al. (1986)
SA5, *Salmonella typhimurium* TA1535, reverse mutation	–	–	170.0000	Varma et al. (1988)
SA7, *Salmonella typhimurium* TA1537, reverse mutation	–	–	500.0000	Mortelmans et al. (1986)
SA7, *Salmonella typhimurium* TA1537, reverse mutation	+	–	130.0000	Varma et al. (1988)
SA9, *Salmonella typhimurium* TA98, reverse mutation	–	–	500.0000	Mortelmans et al. (1986)
SA9, *Salmonella typhimurium* TA98, reverse mutation	–	–	0.0000	Khudoley et al. (1987)
SA9, *Salmonella typhimurium* TA98, reverse mutation	–	(+)	500.0000	Strobel & Grummt (1987)
SA9, *Salmonella typhimurium* TA98, reverse mutation	–	–	170.0000	Varma et al. (1988)
SA9, *Salmonella typhimurium* TA98, reverse mutation	+	+	0.0000	Mersch-Sunderman (1989)[c]
SA9, *Salmonella typhimurium* TA98, reverse mutation	0	+	0.0000	Khudoley et al. (1989)[a]
SAS, *Salmonella typhimurium* TA97, reverse mutation	–	+	5.0000	Strobel & Grummt (1987)
SAS, *Salmonella typhimurium* TA97, reverse mutation	–	+	0.0000	Mersch-Sunderman (1989)[c]
SZG, *Saccharomyces cerevisiae* D7, gene conversion	(+)	–	10.0000	Nestmann & Lee (1985)
SGR, *Saccharomyces cerevisiae* (XV185–14C), reverse mutation	(+)	–	20.0000	Nestmann & Lee (1985)
G5T, Gene mutation, mouse lymphoma L5178Y cells, *tk* locus	–	+	180.0000	McGregor et al. (1988)
SIC, Sister chromatid exchange, Chinese hamster FAF cell line	–	0	8.0000	Strobel & Grummt (1987)

Table 5 (contd)

Test system	Result		Dose LED/HID	Reference
	Without exogenous metabolic system	With exogenous metabolic system		
SIC, Sister chromatid exchange, Chinese hamster CHO cells	–	–	5000.0000	National Toxicology Program (1987)
CIC, Chromosomal aberrations, Chinese hamster FAF cell line	+	0	8.0000	Strobel & Grummt (1987)
CIC, Chromosomal aberrations, Chinese hamster CHO cells	–	–	5000.0000	National Toxicology Program (1987)
CIC, Chromosomal aberrations, Chinese hamster CHO cells	(+)	+	240.0000	Ishidate (1987)
SHL, Sister chromatid exchange, human lymphocytes *in vitro*	+	0	65.0000	Morimoto & Koizumi (1983)
SVA, Sister chromatid exchange, mouse bone–marrow cells *in vivo*	+	0	50.0000	Morimoto & Koizumi (1983)
MVM, Micronucleus test, ddY mice *in vivo*	–	0	500.0000	Hayashi *et al.* (1988)

[a]Closed container
[b]Closed container +, standard –
[c]Spot test +, standard –

natural organic substances in the presence of bromide. The major route of human exposure to bromodichloromethane is *via* drinking-water. It has been detected in chlorinated drinking-water in many parts of the world; it has also been detected in some untreated waters, but at much lower levels. Bromodichloromethane is a major component of the organohalides produced by marine algae.

4.2 Experimental carcinogenicity data

Bromodichloromethane was tested for carcinogenicity in two-year studies in male and female Fischer 344 rats and B6C3F$_1$ mice by oral gavage, in life-span studies in male and female Wistar rats and in CBA × C57Bl/6 hybrid mice by administration in drinking-water. In the gavage studies, bromodichloromethane increased the incidences of adenomatous polyps and adenocarcinomas of the large intestine and of tubular-cell adenomas and adenocarcinomas of the kidney in male and female rats, of tubular-cell adenomas and adenocarcinomas of the kidney in male mice and of hepatocellular adenomas and carcinomas in female mice. In the study by administration in drinking-water, it induced neoplastic nodules and adenofibrosis of the liver in rats; no increase in tumour incidence was seen in mice. In a screening test for lung adenomas by intraperitoneal injection, bromodichloromethane did not increase the incidence of lung tumours in strain A mice.

4.3 Human carcinogenicity data

No relevant data were available to the Working Group.

4.4 Other relevant data

Repeated exposure of rats and mice to bromodichloromethane resulted in toxic effects in several organs, including the liver and kidney.

A study of developmental toxicity in rats given bromodichloromethane throughout the period of major organogenesis showed skeletal variations in the presence of maternal toxicity but no teratogenic effect.

Bromodichloromethane induced mutations in some studies with bacteria and, in a single study, in cultured mammalian cells. Chromosomal aberrations but not sister chromatid exchange were observed in cultured mammalian cells. In single studies, sister chromatid exchange was observed in cultured human cells and in mouse bone marrow *in vivo*. In one study, bromodichloromethane did not induce micronuclei in bone-marrow cells of mice treated *in vivo*.

4.5 Evaluation[1]

There is *inadequate evidence* for the carcinogenicity of bromodichloromethane in humans.

There is *sufficient evidence* for the carcinogenicity of bromodichloromethane in experimental animals.

Overall evaluation

Bromodichloromethane *is possibly carcinogenic to humans (Group 2B).*

5. References

Abdel-Rahman, M.S. (1982) The presence of trihalomethanes in soft drinks. *J. appl. Toxicol.*, 2, 165-166

Ahmed, A.E., Kubic, V.L. & Anders, M.W. (1977) Metabolism of haloforms to carbon monoxide. I. In vitro studies. *Drug Metab. Disposition*, 5, 198-204

Aldrich Chemical Co. (1990) *1990-1991 Aldrich Handbook of Fine Chemicals*, Milwaukee, WI, p. 195

Allgeier, G.D., Mullins, R.L., Jr, Wilding, D.A., Zogorski, J.S. & Hubbs, S.A. (1980) Trihalomethane levels at selected water utilities in Kentucky, USA. In: Afghan, B.K. & Mackay, D., eds, *Hydrocarbons and Halogenated Hydrocarbons in the Aquatic Environment*, New York, Plenum Press, pp. 473-490

American Tokyo Kasei (1988) *TCI American Organic Chemicals 88/89 Catalog*, Portland, OR, p. 184

Anders, M.W., Stevens, J.L., Sprague, R.W., Shaath, Z. & Ahmed, A.E. (1978) Metabolism of haloforms to carbon monoxide. II. In vivo studies. *Drug Metab. Disposition*, 6, 556-560

Armstrong, D.W. & Golden, T. (1986) Determination of distribution and concentration of trihalomethanes in aquatic recreational and therapeutic facilities by electron-capture GC. *LC-GC*, 4, 652-655

Barkley, J., Bunch, J., Bursey, J.T., Castillo, N., Cooper, S.D., Davis, J.M., Erickson, M.D., Harris, B.S.H., III, Kirkpatrick, M., Michael, L.C., Parks, S.P., Pellizzari, E.D., Ray, M., Smith, D., Tomer, K.B., Wagner, R. & Zweidinger, R.A. (1980) Gas chromatography mass spectrometry computer analysis of volatile halogenated hydrocarbons in man and his environment—a multimedia environmental study. *Biomed. Mass Spectrometr.*, 7, 139-147

---

[1] For definition of the italicized terms, see Preamble, pp. 30-33.

Bauer, U. (1981) Human exposure to environmental chemicals—investigations on volatile organic halogenated compounds in water, air, food, and human tissues. III. Communication: results of investigations (Ger.). *Zentralbl. Bakteriol. Mikrobiol. Hyg. Abt. 1 Orig. B, 174*, 200-237

Bellar, T.A., Lichtenberg, J.J. & Kroner, R.C. (1974a) *The Occurrence of Organohalides in Chlorinated Drinking Waters* (EPA-670/4-74-008; NTIS PB-238 589), Cincinnati, OH, US Environmental Protection Agency

Bellar, T.A., Lichtenberg, J.J. & Kroner, R.C. (1974b) The occurrence of organohalides in chlorinated drinking waters. *J. Am. Water Works Assoc., 66*, 703-706

Bowman, F.J., Borzelleca, J.F. & Munson, A.E. (1978) The toxicity of some halomethanes in mice. *Toxicol. appl. Pharmacol., 44*, 213-215

Brass, A.J., Feige, M.A., Halloran, T., Mello, J.W., Munch, D. & Thomas, R.F. (1977) The National Organic Monitoring Survey: samplings and analyses for purgeable organic compounds. In: Pojasek, R.B., ed., *Drinking Water Quality Enhancement Through Source Protection*, Ann Arbor, MI, Ann Arbor Science, pp. 393-416

Brett, R.W. & Calverley, R.A. (1979) A one-year survey of trihalomethane concentration changes within a distribution system. *J. Am. Water Works Assoc., 71*, 515-520

Brodzinsky, R. & Singh, H.B. (1983) *Volatile Organic Chemicals in the Atmosphere: An Assessment of Available Data* (EPA-600/3-83-027a; NTIS PB83-195503), Research Triangle Park, NC, US Environmental Protection Agency

Bunn, W.W., Haas, B.B., Deane, E.R. & Kleopfer, R.D. (1975) Formation of trihalomethanes by chlorination of surface water. *Environ. Lett., 10*, 205-213

Cech, I., Smith, V. & Henry, J. (1982) Spatial and seasonal variations in concentration of trihalomethanes in drinking water. *Pergamon Ser. environ. Sci., 7*, 19-38

Chemical Information Systems (1990) *ISHOW Database*, Baltimore, MD

Chu, I., Secours, V., Marino, I. & Villeneuve, D.C. (1980) The acute toxicity of four trihalomethanes in male and female rats. *Toxicol. appl. Pharmacol., 52*, 351-353

Chu, I., Villeneuve, D.C., Secours, V.E. & Becking, G.C. (1982a) Toxicity of trihalomethanes: I. The acute and subacute toxicity of chloroform, bromodichloromethane, chlorodibromomethane and bromoform in rats. *J. environ. Sci. Health, B17*, 205-224

Chu, I., Villeneuve, D.C., Secours, V.E. & Becking, G.C. (1982b) Trihalomethanes: II. Reversibility of toxicological changes produced by chloroform, bromodichloromethane, chlorodibromomethane and bromoform in rats. *J. environ. Sci. Health, B17*, 225-240

Class, T., Kohnle, R. & Ballschmiter, K. (1986) Chemistry of organic traces in air. VII: Bromo- and bromochloroethanes in air over the Atlantic Ocean. *Chemosphere, 15*, 429-436

Cole, R.H., Frederick, R.E., Healy, R.P. & Rolan, R.G. (1984) Preliminary findings of the Priority Pollutant Monitoring Project of the Nationwide Urban Runoff Program. *J. Water Pollut. Control Fed., 56*, 898-908

Coleman, W.E., Munch, J.W., Kaylor, W.H., Streicher, R.P., Ringhand, H.P. & Meier, J.R. (1984) Gas chromatography/mass spectroscopy analysis of mutagenic extracts of aqueous chlorinated humic acid. A comparison of the byproducts to drinking water contaminants. *Environ. Sci. Technol.*, *18*, 674-681

Comba, M.E. & Kaiser, K.L.E. (1983) Determination of volatile contaminants at the ng/l level in water by capillary gas chromatography with electron capture detection. *Int. J. environ. anal. Chem.*, *16*, 17-31

Condie, L.W., Smallwood, C.L. & Laurie, R.D. (1983) Comparative renal and hepatotoxicity of halomethanes: bromodichloromethane, bromoform, chloroform, dibromochloromethane and methylene chloride. *Drug chem. Toxicol.*, *6*, 563-578

DeWalle, F.B. & Chian, E.S.K. (1978) Presence of trace organics in the Delaware River and their discharge by municipal and industrial sources. In: Bell, J.M., ed., *Proceedings of the Industrial Waste Conference*, Vol. 32, Ann Arbor, MI, Ann Arbor Science, pp. 908-919

Easley, D.M., Kleopfer, R.D. & Carasea, A.M. (1981) Gas chromatographic-mass spectrometric determination of volatile organic compounds in fish. *J. Assoc. off. anal. Chem.*, *64*, 653-656

Eklund, G., Josefsson, B. & Roos, C. (1978) Trace analysis of volatile organic substances in Goeteborg municipal drinking water. *Vatten*, *3*, 195-206

Entz, R.C., Thomas, K.W. & Diachenko, G.W. (1982) Residues of volatile halocarbons in foods using headspace gas chromatography. *J. agric. Food Chem.*, *30*, 846-849

Fedoryński, M., Popławska, M., Nitschke, K., Kowalski, W. & Makosza, M. (1977) A simple method for preparation of dibromochloromethane and bromodichloromethane. *Synth. Commun.*, *7*, 287-292

de Fernicola, N.A.G.G. & de Azevedo, F.A. (1984) Water levels of trihalomethanes (THM) in four different supply systems from Sao Paulo State, (Brazil). *J. environ. Health*, *46*, 187-188

Furlong, E.A.-N. & D'Itri, F.M. (1986) Trihalomethane levels in chlorinated Michigan drinking water. *Ecol. Modelling*, *32*, 215-225

Glaze, W.H. & Rawley, R. (1979) A preliminary survey of trihalomethane levels in selected East Texas water supplies. *J. Am. Water Works Assoc.*, *71*, 509-515

Gschwend, P.M., MacFarlane, J.K. & Newman, K.A. (1985) Volatile halogenated organic compounds released to seawater from temperate marine macroalgae. *Science*, *227*, 1033-1035

Hagenmaier, H., Werner, G. & Jäger, W. (1982) Quantitative determination of volatile halogenated hydrocarbons in water samples by capillary gas chromatography and electron capture detection. *Z. Wasser Abwasser Forsch.*, *15*, 195-198

Hargesheimer, E.E. (1985) Identifying water main leaks with trihalomethane tracers. *J. Am. Water Works Assoc.*, *77*, 71-75

Harms, L.L. & Looyenga, R.W. (1977) *Formation and Removal of Halogenated Hydrocarbons in Drinking Water: A Case Study at Huron, South Dakota* (EPA-908/3-77-001; NTIS PB80-159718), Denver, CO, US Environmental Protection Agency

Hayashi, M., Kishi, M., Sofuni, T. & Ishidate, M., Jr (1988) Micronucleus tests in mice on 39 food additives and eight miscellaneous chemicals. *Food chem. Toxicol.*, 26, 487-500

Health and Welfare Canada (1977) *National Survey for Halomethanes in Drinking Water* (Report 77-EHD-9), Ottawa, Environmental Health Directorate, Health Protection Branch

Henderson, J.E., Peyton, G.R. & Glaze, W.H. (1976) A convenient liquid-liquid extraction method for the determination of halomethanes in water at the parts-per-billion level. In: Keith, L.H. ed., *Identification and Analysis of Organic Pollutants in Water*, Ann Arbor, MI, Ann Arbor Science, pp. 105-111

Hoyt, S.D. & Rasmussen, R.A. (1985) Determining trace gases in air and seawater. *Adv. Chem. Ser. Mapp. Strategies chem. Oceanogr.*, 209, 31-56

IARC (1987) *IARC Monographs on the Evaluation of Carcinogenic Risks to Humans*, Suppl. 7, *Overall Evaluations of Carcinogenicity: An Updating of* IARC Monographs *Volumes 1 to 42*, Lyon, pp. 152-154

Isacson, P., Bean, J.A. & Lynch, C. (1983) Relationship of cancer incidence rates in Iowa municipalities to chlorination status of drinking water. In: Jolley, R.J., Brungs, W.A., Cotruvo, Y.A., Cumming, R.B., Mattice, Y.S. & Jacobs, V.A., eds, *Water Chlorination: Environmental Impact and Health Effects*, Vol. 4, Book 2, *Environment, Health, and Risk*, Ann Arbor, MI, Ann Arbor Sciences, pp. 1353-1364

Ishidate, M., Jr (1987) *Data Book of Chromosomal Aberration Test in vitro*, Tokyo, Life-Science Information Center, pp. 49-50

Kaczmar, S.W., D'Itri, F.M. & Zabik, M.J. (1984) Volatilization rates of selected haloforms from aqueous environments. *Environ. Toxicol. Chem.*, 3, 31-35

Kaiser, K.L.E. & Comba, M.E. (1983) Volatile contaminants in the Welland River watershed. *J. Great Lakes Res.*, 9, 274-280

Kaiser, K.L.E., Comba, M.E. & Huneault, H. (1983) Volatile halocarbon contaminants in the Niagara River and in Lake Ontario. *J. Great Lakes Res.*, 9, 212-223

Kajino, M. & Yagi, M. (1980) Formation of trihalomethanes during chlorination and determination of halogenated hydrocarbons in drinking water. In: Afghan, B.K. & Mackay, D., eds, *Hydrocarbons and Halogenated Hydrocarbons in the Aquatic Environment*, New York, Plenum Press, pp. 491-501

Kaminski, L. & von Loew, E. (1984) Haloforms in swimming pool water (Ger.). *Forum Staedte-Hyg.*, 35, 19-20

Khudoley, V.V., Mizgireuv, I. & Pliss, G.B. (1987) The study of mutagenic activity of carcinogens and other chemical agents with *Salmonella typhimurimum* assays: testing of 126 compounds. *Arch. Geschwulstforsch.*, 57, 453-462

Khudoley, V.V., Gvildis, V. Yu. & Pliss, G.B. (1989) Identification of some trihalomethanes in drinking water and assessment of their toxicity, mutagenicity, and carcinogenicity (Russ.). *Vopr. Onkol.*, 35, 837-842

Kossik, R.F., Gschwend, P.S. & Adams, E.E. (1986) *Tracing and Modeling Pollutant Transport in Boston Harbor* (MITSG-86/16; NTIS PB87-136891), Rockville, MD, National Oceanic and Atmospheric Administration

Krasner, S.W., McGuire, M.J., Jacangelo, J.G., Patania, N.L., Reagan, K.M. & Aieta, E.M. (1989) The occurrence of disinfection by-products in drinking water in a nationwide study. In: *Annual Conference of the American Water Works Association, Los Angeles, CA*, Denver, CO, American Water Works Association

Lahl, U., Bätjer, K., von Düszeln, J., Gabel, B., Stachel, B. & Thiemann, W. (1981) Distribution and balance of volatile halogenated hydrocarbons in the water and air of covered swimming pools using chlorine for water disinfection. *Water Res.*, 15, 803-814

Lahl, U., Cetinkaya, M., von Düszeln, J., Gabel, B., Stachel, B. & Thiemann, W. (1982) Health risks for infants caused by trihalomethane generation during chemical disinfection of feeding utensils. *Ecol. Food Nutr.*, 12, 7-17

Loy, E.W., Jr, Brown, D.W., Stephenson, J.H.M. & Little, J.A. (1976) Uses of wastewater discharge compliance monitoring data. In: Keith, L.H., ed., *Identification and Analysis of Organic Pollutants in Water*, Ann Arbor, MI, Ann Arbor Science, pp. 499-516

Mabey, W.R., Smith, J.H., Podoll, R.T., Johnson, H.L., Mill, T., Chou, T.W., Gates, J., Waight Partridge, I. & Vandenberg, D. (1982) *Aquatic Fate Process Data for Organic Priority Pollutants* (EPA-440/4-81-014; NTIS PB87-169090), Washington DC, Office of Water Regulations and Standards, US Environmental Protection Agency, pp. 179-180

McConnell, G. (1976) Haloorganics in water supplies. *J. Inst. Water Eng. Sci.*, 30, 431-445

McGregor, D.B., Brown, A., Cattanach, P., Edwards, I., McBride, D. & Caspary, W.J. (1988) Responses of the L5178Y tk$^+$/tk$^-$ mouse lymphoma cell forward mutation assay. II: 18 coded chemicals. *Environ. mol. Mutagenesis*, 11, 91-118

McKinney, J.D., Maurer, R.R., Hass, J.R. & Thomas R.O. (1976) Possible factors in the drinking water of laboratory animals causing reproductive failure. In: Keith, L.H., ed., *Identification and Analysis of Organic Pollutants in Water*, Ann Arbor, MI, Ann Arbor Science, pp. 417-432

Mersch-Sunderman, V. (1989) Examination of the mutagenicity of organic microcontaminations in the environment. II. The mutagenicity of halogenated aliphatic hydrocarbons with the *Salmonella*-microsome test (Ames test) in relation to contamination of ground- and drinking-water (Ger.). *Zbl. Bakt. Hyg. B.*, 187, 230-243

Minear, R.A. & Morrow, C.M. (1983) *Raw Water Bromide: Levels and Relationship to Distribution of Trihalomethanes in Finished Drinking Water* (OWRT-A-063-TENN; NTIS PB83-256735), Washington DC, US Environmental Protection Agency

Mink, F.L., Brown, T.J. & Rickabaugh, J. (1986) Absorption, distribution, and excretion of ^{14}C-trihalomethanes in mice and rats. *Bull. environ. Contam. Toxicol.*, 37, 752-758

Morimoto, K. & Koizumi, A. (1983) Trihalomethanes induce sister chromatid exchanges in human lymphocytes *in vitro* and mouse bone marrow cells *in vivo*. *Environ. Res.*, 32, 72-79

Mortelmans, K., Haworth, S., Lawlor, T., Speck, W., Tainer, B. & Zeiger, E. (1986) *Salmonella* mutagenicity tests: II. Results from the testing of 270 chemicals. *Environ. Mutagenesis, 8 (Suppl. 7)*, 1-119

Munson, A.E., Sain, L.E., Sanders, V.M., Kauffmann, B.M., White, K.L., Page, D.G., Barnes, D.W. & Borzelleca, J.F. (1982) Toxicology of organic drinking water contaminants: trichloromethane, bromodichloromethane, dibromochloromethane and tribromomethane. *Environ. Health Perspect.*, 46, 117-126

National Library of Medicine (1989) *TRI (Toxic Chemical Release Inventory Data Bank)*, Bethesda, MD

National Toxicology Program (1987) *Toxicology and Carcinogenesis Studies of Bromodichloromethane (CAS No. 75-27-4) in F344/N Rats and B6C3F$_1$ Mice (Gavage Studies)* (NTP Technical Report Series No. 321), Research Triangle Park, NC, US Department of Health and Human Services

Nestmann, E.R. & Lee, E.G.-H. (1985) Genetic activity in *Saccharomyces cerevisiae* of compounds found in effluents of pulp and paper mills. *Mutat. Res., 155*, 53-60

Norin, H., Renberg, L., Hjort, J. & Lundblad, P.O. (1981) Factors influencing formation of trihalomethanes in drinking water with special reference to Swedish conditions. *Chemosphere, 10*, 1265-1273

Oliver, B.G. (1983) Dihaloacetonitriles in drinking water: algae and fulvic acid as precursors. *Environ. Sci. Technol., 17*, 80-83

Otson, R. (1987) Purgeable organics in Great Lakes raw and treated water. *Int. J. environ. anal. Chem., 31*, 41-53

Otson, R., Williams, D.T., Bothwell, P.D. & Quon, T.K. (1981) Comparison of trihalomethane levels and other water quality parameters for three treatment plants on the Ottawa River. *Environ. Sci. Technol., 15*, 1075-1080

Otson, R., Williams, D.T. & Bothwell, P.D. (1982) Volatile organic compounds in water at thirty Canadian potable water treatment facilities. *J. Assoc. off. anal. Chem., 65*, 1370-1374

Peoples, A.J., Pfaffenberger, C.D., Shafik, T.M. & Enos, H.F. (1979) Determination of volatile purgeable halogenated hydrocarbons in human adipose tissue and blood serum. *Bull. environ. Contam. Toxicol., 23*, 244-249

Perwak, J., Goyer, M., Harris, J., Schimke, G., Scow, K., Wallace, D. & Slimak, M. (1980) *An Exposure and Risk Assessment for Trihalomethanes: Chloroform, Bromoform, Bromodichloromethane, Dibromochloromethane* (EPA-440/4-81-018; NTIS PB85-211977), Washington DC, Office of Water Regulations and Standards, US Environmental Protection Agency

Pfaffenberger, C.D., Peoples, A.J. & Enos, H.F. (1980) Distribution of volatile halogenated organic compounds between rat blood serum and adipose tissue. *Int. J. environ. anal. Chem., 8*, 55-65

Pouchert, C.J., ed. (1974) *The Aldrich Library of NMR Spectra*, Vol. 1, Milwaukee, WI, Aldrich Chemical Co., p. 64C

Pouchert, C.J., ed. (1981) *The Aldrich Library of Infrared Spectra*, 3rd ed., Milwaukee, WI, Aldrich Chemical Co., p. 52A

Pouchert, C.J., ed. (1983) *The Aldrich Library of NMR Spectra*, 2nd ed., Vol. 1, Milwaukee, WI, Aldrich Chemical Co., p. 82B

Pouchert, C.J., ed. (1985a) *The Aldrich Library of FT-IR Spectra*, Vol. 1, Milwaukee, WI, Aldrich Chemical Co., p. 83C

Pouchert, C.J., ed. (1985b) *The Aldrich Library of FT-IR Spectra*, Vol. 3, Milwaukee, WI, Aldrich Chemical Co., p. 116D

Quaghebeur, D. & De Wulf, E. (1980) Volatile halogenated hydrocarbons in Belgian drinking waters. *Sci. total Environ.*, *14*, 43-52

Reding, R., Fair, P.S., Shipp, C.J. & Brass, H.J. (1989) Measurement of dihaloacetonitriles and chloropicrin in US drinking waters. In: *Disinfection By-products: Current Perspectives*, Denver, CO, American Water Works Association, pp. 11-22

Riedel-de Haën (1986) *Laboratory Chemicals 1986*, Seelze 1/Hannover, p. 154

Rodriguez-Flores, M. (1983) *Characterization of Organic Contaminants in Selected Potable Water Supplies in Puerto Rico* (OWRT-A-057-PR; NTIS PB83-178038), Washington DC, US Environmental Protection Agency

Rook, J.J. (1974) Formation of haloforms during chlorination of natural waters. *Water Treat. Exam.*, *23*, 234-243

Ruddick, J.A., Villeneuve, D.C. & Chu, I. (1983) A teratological assessment of four trihalomethanes in the rat. *J. environ. Sci. Health*, *B18*, 333-349

Sadtler Research Laboratories (1980) *The Sadtler Standard Spectra, 1980 Cumulative Index*, Philadelphia, PA

Scotte, P. (1984) Study of halogenated organic compounds in covered swimming pools (Fr). *Eau Ind. Nuisances*, *86*, 37-41

Simmon, V.F., Kauhanen, K. & Tardiff, R.G. (1977) Mutagenic activity of chemicals identified in drinking water. In: Scott, D., Bridges, B.A. & Sobels, F.H., eds, *Progress in Genetic Toxicology*, Amsterdam, Elsevier/North-Holland Biomedical Press, pp. 249-258

Sittig, M. (1985) *Handbook of Toxic and Hazardous Chemicals and Carcinogens*, 2nd ed., Park Ridge, NJ, Noyes Publications, pp. 148-149

Smith, R.L. (1989) A computer assisted, risk-based screening of a mixture of drinking water chemicals. *Trace Subst. environ. Health*, *22*, 215-232

Standing Committee of Analysts (1980) *Chloro and Bromo Trihalogenated Methanes in Water* (Methods for the Examination of Waters and Associated Materials), London, Her Majesty's Stationery Office

Stevens, A.A., Moore, L.A., Slocum, C.J., Smith, B.L., Seeger, D.R. & Ireland, J.C. (1989) By-products of chlorination at ten operating utilities. In: *Disinfection By-products: Current Perspectives*, Denver, CO, American Water Works Association, pp. 23-61

Strobel, K. & Grummt, T. (1987) Aliphatic and aromatic halocarbons as potential mutagens in drinking water. Part I. Halogenated methanes. *Toxicol. environ. Chem.*, *13*, 205-221

Symons, J.M., Bellar, T.A., Carswell, J.K., DeMarco, J., Kropp, K.L., Robeck, G.G., Seeger, D.R., Slocum, C.J., Smith, B.L. & Stevens, A.A. (1975) National Organics Reconnaissance Survey for halogenated organics. *J. Am. Water Works Assoc.*, *67*, 634-647

Theiss, J.C., Stoner, G.D., Shimkin, M.B. & Weisburger, E.K. (1977) Test for carcinogenicity of organic contaminants of United States drinking waters by pulmonary tumor response in strain A mice. *Cancer Res.*, *37*, 2717-2720

Trussell, A.R., Cromer, J.L., Umphres, M.D., Kelley, P.E. & Moncur, J.G. (1980) Monitoring of volatile halogenated organics: a survey of twelve drinking waters from various parts of the world. In: Jolley, R.L., Brungs, W.A., Cumming, R.B. & Jacobs, V.A., eds, *Water Chlorination: Environmental Impact and Health Effects*, Vol. 3, Ann Arbor, MI, Ann Arbor Science, pp. 39-53

Tumasonis, C.F., McMartin, D.N. & Bush, B. (1985) Lifetime toxicity of chloroform and bromodichloromethane when administered over a lifetime in rats. *Ecotoxicol. environ. Saf.*, *9*, 233-240

Uhler, A.D. & Diachenko, G.W. (1987) Volatile halocarbon compounds in process water and processed foods. *Bull. environ. Contam. Toxicol.*, *39*, 601-607

Ullrich, D. (1982) 7. Organohalogen compounds in the air of some public swimming pools in Berlin (Ger.). *WaBoLu-Ber.*, *1*, 50-52

US Environmental Protection Agency (1985) *Health and Environmental Effects Profile for Bromochloromethanes* (EPA-600/X-85-397; NTIS PB88-174610), Cincinnati, OH, Environmental Criteria and Assessment Office

US Environmental Protection Agency (1986a) Method 8010. Halogenated volatile organics. In: *Test Methods for Evaluating Solid Waste—Physical/Chemical Methods*, 3rd ed. (EPA No. SW-846), Washington DC, Office of Solid Waste and Emergency Response

US Environmental Protection Agency (1986b) Method 8240. Gas chromatography/mass spectrometry for volatile organics. In: *Test Methods for Evaluating Solid Waste—Physical/Chemical Methods*, 3rd ed. (EPA No. SW-846), Washington DC, Office of Solid Waste and Emergency Response

US Environmental Protection Agency (1988a) Methods for organic chemical analysis of municipal and industrial wastewater. Method 601. Purgeable halocarbons. *US Code fed. Regul., Title 40*, Part 136, Appendix A, pp. 267-281

US Environmental Protection Agency (1988b) Methods for organic chemical analysis of municipal and industrial wastewater. Method 624. Purgeables. *US Code fed. Regul., Title 40*, Part 136, Appendix A, pp. 432-446

US Environmental Protection Agency (1988c) Method 502.2. Volatile organic compounds in water by purge and trap capillary column gas chromatography with photoionization and electrolytic conductivity detectors in series. Revision 2.0. In: *Methods for the Determination of Organic Compounds in Drinking Water* (PB89-220461 EPA-600/4-88 039), Cincinnati, OH, Environmental Monitoring Systems Laboratory, pp. 31-62

US Environmental Protection Agency (1988d) Method 524.2. Measurement of purgeable organic compounds in water by capillary column gas chromatography/mass spectrometry. Revision 3.0. In: *Methods for the Determination of Organic Compounds in Drinking Water* (PB89-220461, EPA-600/4-88 039), Cincinnati, OH, Environmental Monitoring Systems Laboratory, pp. 285-323

US Environmental Protection Agency (1988e) Methods for organic chemical analysis of municipal and industrial wastewater. Method 1624 Revision B. Volatile organic compounds by isotope dilution GC/MS. *US Code fed. Regul., Title 40*, Part 136, Appendix A, pp. 475-488

US Environmental Protection Agency (1988f) Analysis of trihalomethanes. Part II—Analysis of trihalomethanes in drinking water by liquid/liquid extraction. *US Code fed. Regul., Title 40*, Part 141, Appendix C, pp. 563-571

Varma, M.M., Ampy, F.R., Verma, K. & Talbot, W.W. (1988) In vitro mutagenicity of water contaminants in complex mixtures. *J. appl. Toxicol.*, 8, 243-248

Verschueren, K. (1983) *Handbook of Environmental Data on Organic Chemicals*, 2nd ed., New York, Van Nostrand Reinhold, p. 291

Voronin, V.M., Donchenko, A.I. & Korolev, A.A. (1987) Experimental study of the carcinogenicity of dichlorobromomethane and dibromochloromethane formed during chlorination of water (Russ.). *Gig. Sanit.*, 1, 19-21

Wallace, L., Zweidinger, R., Erickson, M., Cooper, S., Whitaker, D. & Pellizzari, E. (1982) Monitoring individual exposure. Measurements of volatile organic compounds in breathing-zone air, drinking water, and exhaled breath. *Environ. int.*, 8, 269-282

Wallace, L., Pellizzari, E., Hartwell, T., Rosenzweig, M., Erickson, M., Sparacino, C. & Zelon, H. (1984) Personal exposure to volatile organic compounds. I. Direct measurements in breathing-zone air, drinking water, food, and exhaled breath. *Environ. Res.*, 35, 293-319

Weast, R.C., ed. (1989) *CRC Handbook of Chemistry and Physics*, 70th ed., Boca Raton, FL, CRC Press, p. C-348

Weil, L., Jandik, J. & Eichelsdoefer, D. (1980) Organic halogenated compounds in swimming pool water. I. Determination of volatile halogenated hydrocarbons (Ger.). *Z. Wasser Abwasser Forsch.*, 13, 141-145

Westrick, J.J., Mello, J.W. & Thomas, R.F. (1984) The groundwater supply survey. *J. Am. Water Works Assoc.*, 76, 52-59

Williams, D.T. (1985) Formation of trihalomethanes in drinking water. In: Fishbein, L. & O'Neill, I.K., eds, *Environmental Carcinogens. Selected Methods of Analysis*, Vol. 7, *Some Volatile Halogenated Hydrocarbons* (IARC Scientific Publications No. 68), Lyon, IARC, pp. 69-88

BROMOFORM

1. Chemical and Physical Data

1.1 Synonyms

Chem. Abstr. Services Reg. No.: 75-25-2
Chem. Abstr. Name: Tribromomethane
IUPAC Systematic Name: Tribromomethane
Synonyms: Methenyl tribromide; methyl tribromide

1.2 Structural and molecular formulae and molecular weight

$$\begin{array}{c} Br \\ | \\ H-C-Br \\ | \\ Br \end{array}$$

CHBr$_3$ Mol. wt: 252.75

1.3 Chemical and physical properties of the pure substance

(a) *Description*: Colourless liquid with chloroform-like odour (Verschueren, 1983; Budavari, 1989; Weast, 1989)
(b) *Boiling-point*: 149.5°C (Weast, 1989)
(c) *Melting-point*: 8.3°C (Weast, 1989)
(d) *Density*: 2.9035 at 15/4°C (Budavari, 1989)
(e) *Spectroscopy data*[1]: Infrared (Sadtler Research Laboratories, 1980, prism [10], grating [8]; Pouchert, 1981, 1985), nuclear magnetic resonance (Sadtler Research Laboratories, 1980, proton [6375], C-13 [1603];

---

[1] In square brackets, spectrum number in compilation

(f) *Solubility*: Soluble in water (3.0 g/l at 20°C, 3.19 g/l at 30°C), ethanol, benzene, chloroform, diethyl ether and ligroin (Mabey *et al.*, 1982; Verschueren, 1983; Weast, 1989)

(g) *Volatility*: Vapour pressure, 5 mm at 20°C, 5.6 mm Hg at 25°C, 10 mm Hg at 34.0°C; relative vapour density (air = 1), 8.7 (Mabey *et al.*, 1982; Verschueren, 1983; Weast, 1989)

(h) *Octanol/water partition coefficient (P)*: log P, 2.38 (Mabey *et al.*, 1982)

(i) *Stability*: Gradually decomposes in air, acquiring a yellow colour; light accelerates this decomposition (Budavari, 1989)

(j) *Reactivity*: Reacts rapidly with strong caustics and metals such as sodium, potassium, calcium, powdered aluminium, zinc and magnesium (Sittig, 1985); hydrolysis rate at neutral pH and 25°C is 2.5×10^{-9} per hour (Mabey *et al.*, 1982).

(k) *Conversion factor*[2]: mg/m^3 = 10.34 × ppm

1.4 Technical products and impurities

Bromoform is available in various grades from > 95% to > 99% purity; it may be stabilized with ethanol (e.g., 1-3%) or diphenylamine. It is also available in grades specifically for coal flotation and the separation of mineral compounds (Riedel-de Häen, 1986; American Tokyo Kasei, 1988; Aldrich Chemical Co., 1990).

2. Production, Use, Occurrence and Analysis

2.1 Production and use

(a) *Production*

Bromoform can be prepared by reacting chloroform (see IARC, 1987) with aluminium tribromide at less than or equal to 60°C; by reacting ethanol with sodium hypobromite; or by the redistribution reaction between chloroform and ethyl bromide (Harlow & Ross, 1932; Soroos & Hinkamp, 1945; Sherman & Kavasmaneck, 1980).

---

[2]Calculated from: mg/m^3 = (molecular weight/24.45) × ppm, assuming standard temperature (25°C) and pressure (760 mm Hg)

(b) Use

Bromoform has been used as an intermediate in organic synthesis, as an ingredient in fire-resistant chemicals, as a gauge fluid in geological assaying, as a solvent for waxes, greases and oils, and as a sedative (Sittig, 1985; Sax & Lewis, 1987).

(c) Regulatory status and guidelines

Standards have been established for trihalomethanes (including bromoform) in drinking-water (see monograph on chlorinated drinking-water, p. 59) in several countries. Occupational exposure limits and guidelines for bromoform are presented in Table 1.

Table 1. Occupational exposure limits and guidelines for bromoform[a]

Country	Year	Concentration (mg/m^3)	Interpretation[b]
Belgium	1987	5 (skin)	TWA
Brazil	1987	4 (skin)	TWA
Canada (Saskatchewan)	1980	10	STEL
Denmark	1987	5 (skin)	TWA
Finland	1987	5 (skin)	TWA
		15	STEL
Indonesia	1987	5 (skin)	TWA
Mexico	1987	5	TWA
Netherlands	1987	5 (skin)	TWA
Norway	1984	5 (skin)	TWA
Switzerland	1987	5 (skin)	TWA
UK	1987	5 (skin)	TWA
USA			
ACGIH	1989	5 (skin)	TWA
OSHA	1989	5 (skin)	TWA
USSR	1987	5	MAC
Venezuela	1987	5	TWA and ceiling
Yugoslavia	1987	5 (skin)	TWA

[a]From Cook (1987); American Conference of Governmental Industrial Hygienists (ACGIH) (1989); US Occupational Safety and Health Administration (OSHA) (1989)
[b]TWA, time-weighted average; STEL, short-term exposure limit; MAC, maximum allowable concentration

2.2 Occurrence

(a) Natural occurrence

Mean levels of bromoform in the tissues of temperate marine macroalgae (*Ascophyllum nodosum, Fucus vesiculosis, Enteromorpha linza, Ulva lacta, Gigartina stellata*) ranged from 8 to 120 ng/g dry weight. Bromoform was released to seawater at a rate of 0.14-14 µg/g of dry algae per day (Gschwend *et al.*, 1985).

Macroalgae collected near the Bermuda Islands (*Fucales sargassum*) and at the Cape of Good Hope (*Laminariales laminaria*) showed a specific pattern of emissions of volatile organohalides into the air. The main components were bromoform, bromodichloromethane and chlorodibromomethane (Class *et al.*, 1986).

(b) Air

The volatilization half-time of bromoform from rivers and streams has been estimated to range from 63 min to 24 days, depending on turbulence and temperature. A typical half-time, based on actual data, was 66 h (Kaczmar *et al.*, 1984).

Ullrich (1982) studied the organohalogen compound concentrations in the air of four public indoor swimming pools in western Berlin. Mean bromoform concentrations ranged from 0.7 to 6 µg/m^3.

In a review of data on the presence of volatile organic chemicals in the atmosphere of the USA in 1970-80, a median concentration of 0.64 µg/m^3 bromoform was reported for the four urban/suburban data points examined and below the limit of detection for industrial areas (74 data points) (Brodzinsky & Singh, 1983).

Bromoform levels in air samples collected in 1985 over the Atlantic Ocean were 0.6-460 ppt (6.2-4.8 µg/m^3); baseline levels of biogenic bromoform in air were 0.6-6.6 ppt (6.2-68.2 ng/m^3). The highest levels were found in samples taken at beaches with intense algal populations. Air samples collected in 1985 from a forest area in southern Germany contained 1.2 ppt (12.4 ng/m^3) bromoform (Class *et al.*, 1986).

In the USA, air samples collected 2 cm above the surface of five outdoor pools contained < 0.1 µg/m^3 bromoform; those above four indoor pools contained < 0.1-20 µg/m^3 and those above four spas (whirlpools or hot tubs) contained < 0.1-62 µg/m^3. Samples collected 2 m above the surface contained < 0.1 µg/m^3, < 0.1-14 µg/m^3 and < 0.1-14 µg/m^3, respectively (Armstrong & Golden, 1986).

Air samples collected in the USA between 1984 and 1987 showed that the atmospheric concentrations of bromoform averaged 6.3 ng/l at Point Barrow, Alaska, and 3.1 ng/l at Cape Kumukahi, Hawaii. Large pulses of organic bromine (mainly bromoform) enter the atmosphere during the three-month Arctic spring (Fogelqvist, 1985; Cicerone *et al.*, 1988).

(c) Water and sediments

The formation of trihalomethanes in drinking-water (see also the monograph on Chlorinated drinking-water, p. 56 *et seq.*) and the effects of temperature and pH have been discussed extensively (Williams, 1985). The formation of bromoform during drinking-water chlorination depends on the presence of bromide in untreated water.

Bromoform was detected (but not quantified) in headspace analysis of nine of ten Pacific seawater samples collected in 1983 (Hoyt & Rasmussen, 1985).

Bromoform has been measured or detected in many drinking-water systems, both in samples collected at treatment facilities or along the distribution system (Table 2) and in samples collected from natural and untreated water sources (Table 3). Concentrations in treated drinking-water typically ranged from < 1 to 10 µg/l (with higher values in some locations), compared with concentrations in untreated (natural) waters which are typically less than 1 µg/l.

Table 2. Bromoform concentrations in treated[a] drinking-water, 1973-89

Location, date[b]	Sample site/raw water source (treatment)[c]	Concentration (µg/l)[d]	Reference
80 US cities (NORS), 1975	T and D/ground and surface (75% of raw water chlorinated)	ND-92	Symons *et al.* (1975)
113 Public water supplies, 1976	Water supplies/NS	Mean, 12; median, < 0.3	Brass *et al.* (1977)
945 US sites, 1981-82	T/ground	2.4-5.1[e] (max., 110)	Westrick *et al.* (1984)
13 US community systems, 1984-85	T and D/NS	< 0.2-3.1	Reding *et al.* (1989)
10 US utilities, 1985	T/ground and surface (chlorinated)	< 10 (6 sites) ND (4 sites)	Stevens *et al.* (1989)
35 US sites Spring 1988 Summer 1988 Autumn 1988 Winter 1989	T/ground and surface	Median 0.33 0.57 0.88 0.51	Krasner *et al.* (1989)
Southwestern US city, 1975	NS/ground (treated)	43.1-74.2	Henderson *et al.* (1976)

Table 2 (contd)

Location, date[b]	Sample site/raw water source (treatment)[c]	Concentration ($\mu g/l$)[d]	Reference
East Texas, 1977	NS/surface (14 sites) NS/ground (11 sites) (chlorinated)	ND-85.0 ND-258	Glaze & Rawley (1979)
Houston, Texas, summer 1978 to winter 1980	D/surface D/ground (chlorinated and unchlorinated)	max., 2 max., 28 (mean, 5)	Cech et al. (1982)
Miami, Florida January 1975 July 1975	NS/ground (finished water)	 1.5 4.0	Loy et al. (1976)
19 Tennessee sites Autumn 1980 Winter 1980 Spring 1981 Summer 1981	T/ground and surface (13 samples) (2 of 20 samples) (3 of 20 samples) (5 of 20 samples)	 ND ND-0.17 ND-0.012 ND-0.20	Minear & Morrow (1983)
Old Love Canal, NY, 1978	D/NS (5 of 8 sites)	0.02-0.08	Barkley et al. (1980)
New Jersey, 1977-79	NS/surface (197 of 604 samples) NS/ground (235 of 1072 samples)	max., 3.7 max., 34.3	Page (1981)
40 Michigan utilities, NS	T/surface (22 sites) T/ground (18 sites) (chlorinated)	ND ND-1.6	Furlong & D'Itri (1986)
3 Puerto Rican cities, NS	D/ground and surface	ND-0.1	Rodriguez-Flores (1983)
70 Canadian cities, 1976-77	D and T/ground and surface (chlorinated)	0-0.2 (median, <0.01)	Health & Welfare Canada (1977)
30 Canadian sites, 1979	T/ground and surface	mean, <1; max., 1-2	Otson et al. (1982)
10 Canadian Great Lakes sites Summer 1982 Winter 1983 Spring 1983	T/ground and surface	 ND-<0.1 ND-<0.1 mean, 0.1	Otson (1987)
Burlington, Ontario, Canada, 1981 Niagara Falls, Ontario 1981 Port Robinson, Ontario 1981	D/surface, ground and treated	0.150 0.84 0.3	Comba & Kaiser (1983)

Table 2 (contd)

Location, date[b]	Sample site/raw water source (treatment)[c]	Concentration (μg/l)[d]	Reference
Chippawa, Ontario, 1981	D/surface	0.01	Kaiser & Comba (1983)
Lancashire-Cheshire, UK, 1974	D/NS (chlorination)	< 0.01-2.5	McConnell (1976)
Southeastern England/NS	D/ground and surface (chlorinated)	ND	Trussell et al. (1980)
5 Belgian utilities, 1977-78	T/surface	0-3.6	Quaghebeur & De Wulf (1980)
9 Belgian utilities, 1977-78	D/surface (10/14 chlorinated)	0-0.7	
	T/ground	0-4.4	
	D/ground	0-2.7	
12 German cities, 1977	NS/NS	ND-4.9 (mean, 0.9)	Eklund et al. (1978a,b,c)
18 German cities, 1978-80	D/surface (river) (43 samples)	0.3-14 (mean, 4.3)	Gabel et al. (1981)
50 German cities, 1978-80	D/surface (river) (100 samples) (chlorinated)	0.1-14 (mean, 2.7)	
9 German cities, 1978-79	D/NS	0.4-40.4	Lahl et al. (1982)
Bremen and Leverkusen, Germany, 1976-78	NS/NS	ND-28	Bauer (1981)
Tübingen, Germany, 1981	D/NS	0.35-1.15 (mean, 0.77)	Hagenmaier et al. (1982)
	D/surface	0.01-0.05 (mean, 0.04)	
Gothenberg, Sweden, 1977	D/surface (treated)	0.016	Eklund et al. (1978a,b,c)
Southern China, NS	D/local catchments (Cl$_2$)	6.3	Trussell et al. (1980)
Southern Philippines/NS	D/ground and surface (Cl$_2$)	ND	Trussell et al. (1980)
Northern Philippines/NS	D/surface (Cl$_2$)	ND	Trussell et al. (1980)
Northern Egypt/NS	D/surface (Cl$_2$)	ND	Trussell et al. (1980)
Southern Indonesia/NS	D/surface (Cl$_2$)	ND	Trussell et al. (1980)
Southeastern Australia/NS	D/surface (Cl$_2$)	ND	Trussell et al. (1980)
Southern Brazil/NS	D/surface (Cl$_2$)	ND	Trussell et al. (1980)
Eastern Nicaragua/NS	D/surface (Cl$_2$)	ND	Trussell et al. (1980)

Table 2 (contd)

Location, date[b]	Sample site/raw water source (treatment)[c]	Concentration (μg/l)[d]	Reference
Northern Venezuela/NS	D/surface (Cl$_2$)	ND	Trussell et al. (1980)
Eastern Peru/NS	D/ground and surface (Cl$_2$)	ND	Trussell et al. (1980)

[a]Treatment not always specified
[b]NORS, National Organics Reconnaissance Survey; NS, not specified
[c]D, distribution system; T, treatment plant; treatment given as and when described by the author(s)
[d]ND, not detected
[e]Range of median values for randomly and nonrandomly selected water supplies serving fewer than and more than 10 000 people

Table 3. Bromoform concentrations in untreated (natural) water, 1973-79

Location, date[a]	Sample source	Concentration (ng/l)[b]	Reference
Gothenberg, Sweden, 1977	Seawater	27	Eklund et al. (1978b,c)
Northern Taiwan, NS	Well	ND	Trussell et al. (1980)
Lake Ontario, Canada, 1981	Lake (82 sites)	ND-7	Kaiser et al. (1983)
Niagara River, Ontario, Canada, 1981	River (17 sites)	ND-6	
Welland River, Ontario, Canada, watershed, Summer, 1980 Spring 1981	Surface (river)	ND-60 ND-1100	Kaiser & Comba (1983)
North and South Atlantic, 1985	Seawater	0.8->6	Class et al. (1986)
Ulm, Germany, NS	Rainwater	5	Class et al. (1986)

[a]NS, not specified
[b]ND, not detected

Rook (1974) demonstrated that bromoform, observed at concentrations ranging from 0.5 to 10 μg/l following chlorination of stored surface waters, was a product of chlorination of the humic substances in natural waters.

In the 1975 US National Organics Reconnaissance Survey of water supplies in 80 US cities, bromoform was not found in any of the raw water samples (minimum quantifiable concentration, 1-4 µg/l) (Symons *et al.*, 1975).

According to the US Environmental Protection Agency STORET system, concentrations of bromoform in 131 samples of surface water in 1970-79 ranged from 0.1 to 1 µg/l in 1% of the samples, 1-10 µg/l in 97% of the samples, and 10-100 µg/l in 2% of the samples (Perwak *et al.*, 1980).

In the 1982 US Nationwide Urban Runoff Program, bromoform was detected in samples from one of the 15 reporting cities at a concentration of 1 µg/l (Cole *et al.*, 1984).

The US Environmental Protection Agency estimated that 17 tonnes of bromoform were generated in the US in 1978 by water chlorination. On the basis of the 1976 National Organic Monitoring Survey, the general population was estimated to be exposed to 7 µg bromoform per day from drinking-water, assuming a maximal concentration of 280 µg/l and an intake of 2.18 l per day, the daily exposure increased to 600 µg per day (Perwak *et al.*, 1980).

Bromoform levels in tap water collected at four locations in a Swedish community ranged from 0.24 to 0.34 µg/l; when the treatment facility briefly changed the disinfectant from chlorine to chlorine dioxide, bromoform was not detected (detection limit, 0.01 µg/l) (Norin *et al.*, 1981).

Samples of water were collected between August and October 1980 from one supply system in São Paulo State, Brazil. Mean levels of bromoform were 0.1 µg/l in treated water (after chlorination) and 0.06 µg/l in treated water from the reservoir (II) (de Fernicola & de Azevedo, 1984).

Bromoform has been identified as the major halogenated organic compound produced as a result of chlorinating seawater. Potential daily bromoform production by a pilot plant for conversion of ocean thermal energy was estimated to be 1300 g/day; the bromoform concentration contained in the diluted discharge water was estimated to be 0.93 µg/l (Hartwig & Valentine, 1983).

Bromoform concentrations in samples of chlorinated seawater at a desalination test facility on Wrightsville Beach, NC, USA, were 23-209 µg/l. In general, bromoform concentrations increased through the various pretreatment processes; in effluents from activated carbon columns, however, levels ranged from not detected to 36 µg/l (Singer, 1982).

Pretreatment of seawater with chlorine at a pilot desalination facility in the USA produced average concentrations of bromoform ranging from 13 to 110 µg/l during the different steps of the two pretreatment processes (Chang & Singer, 1984).

Samples of chlorinated intake and discharge seawater collected in 1980 from a power plant in Port Everglades, FL, contained bromoform at levels ranging from 75 to 78 µg/l and from 32 to 86 µg/l, respectively. The concentration of bromoform in the unchlorinated intake water was 1 µg/l. Chlorination of the seawater in the laboratory with 1, 2, and 4 mg/l chlorine resulted in corresponding bromoform concentrations of 6.5, 107 and 272 µg/l (Carpenter *et al.*, 1981).

In 1980, the mean concentrations of bromoform in Arctic seawater were 9.8 ng/l in open surface water and 58 ng/l in surface water close to the Svalbard shore (Fogelqvist, 1985).

Water samples were collected from five outdoor pools, four indoor pools, and four spas (whirlpools or hot tubs) in Lubbock, TX, USA. The concentrations of bromoform in the outdoor pools, which used chlorine-based chemicals for chlorination, were < 0.1-1 µg/l. Two of the indoor pools in which only chlorination was used had levels of < 0.1-1 µg/l; one indoor pool in which only bromination (sodium hypobromite) was used had levels of 8-60 µg/l; and the fourth indoor pool, in which chlorination and bromination were alternated had levels of 11-49 µg/l. The spa in which only chlorination was used had levels of < 0.1-1 µg/l; the two spas in which only bromination was used had levels of 3-183 µg/l; and the spa in which the combination was used had levels of 50-89 µg/l. The average concentration of bromoform in Lubbock, TX, tap water was < 0.1 µg/l (Armstrong & Golden, 1986).

Bromination (3 mg bromine/l) of the water in a Swedish public swimming pool resulted in the formation of bromoform at 400 µg/l (Norin & Renberg, 1980).

Water samples were collected in 1978-79 from eight covered swimming pools in Bremen, Germany, to determine the concentration of bromoform. The source of fresh water was mixed river and groundwater for four pools and groundwater for four pools. The level of bromoform in the pools with mixed sources was 45 µg/l, and that for the pools with groundwater, not detectable (ND). The range of means of bromoform in the four pools with a mixed water source was 0.5-12 µg/l (total range, ND-88 µg/l); that in the four pools with a groundwater source was ND-0.1 µg/l (total range, ND-0.2 µg/l) (Lahl *et al.*, 1981).

The concentrations of bromoform in three thermal spas in western Germany in which the initial bromide concentration was 0.5-0.7 mg/l were 2.3-9.6 µg/l (Weil *et al.*, 1980).

Scotte (1984) studied the concentrations of organohalogen compounds in the water of 10 covered swimming pools in France. The mean concentrations of bromoform were 0.1 µg/l in the four pools treated with Surchlor GR 60 (anhydrous sodium dichloroisocyanurate), none in the two treated with gaseous chlorine, 1.0 µg/l in two treated with sodium hypochlorite and 481.2 µg/l in two treated with bromine.

Effluents from a wastewater treatment plant on Boston Harbor, MA, USA, sampled in 1984 and 1985, contained a mean bromoform level of 1.65 µg/l (range, not detected to 6.9 µg/l) and had an estimated mass input rate of 1.39 kg per day (Kossik *et al.*, 1986).

Heating water to prepare food has been shown to eliminate a large part of trihalomethanes in the water, particularly as a function of temperature and heating time. Bromoform levels were reduced from 35.9 µg/l in tap water to 21.3 µg/l after heating at 80°C for 1 min, to 13.9 µg/l after heating to 100°C, to 13.5 µg/l after boiling for 1 min, and to 6.8 µg/l after boiling for 5 min (Lahl *et al.*, 1982).

(d) Animals

Bioaccumulation of bromoform has been studied in five marine species. The uptake and depuration is rapid (equilibrium is reached in 24-48 h), and the concentration factors are relatively low (< 1-10 times the water concentration) (Gibson *et al.*, 1980).

(e) Human tissues and secretions

Analysis of whole blood samples collected from a population of 250 patients with no known exposure to volatile organic compounds revealed a mean bromoform level of 0.6 ng/ml (range, ND-3.4 ng/ml); the number of individuals with detected levels of bromoform was not specified, but 11 had significantly elevated levels (greater than two standard deviations above the mean) (Antoine *et al.*, 1986).

2.3 Analysis

Selected methods for the analysis of bromoform in air, water and other matrices are presented in Table 4. The US Environmental Protection Agency methods for analysing water (Methods 8010 and 8240) have also been used for liquid and solid wastes. Volatile components of solid waste samples are first extracted with methanol, prior to purge-and-trap concentration and analysis by gas chromatography-electrolytic conductivity detection (Method 8010) or gas chromatography-mass spectrometry (Method 8240). The detection limit for bromoform using Method 8010 is 0.20 µg/l, and the practical quantification limit using Method 8240 is 5 µg/l for groundwater and for soil/sediment samples (US Environmental Protection Agency, 1986a,b).

US Environmental Protection Agency Method 624 has also been adapted to the analysis of bromoform in fish, with an estimated detection limit of 10 µg/kg (Easley *et al.*, 1981).

Table 4. Methods for the analysis of bromoform

Sample matrix	Sample preparation[a]	Assay procedure[b]	Limit of detection	Reference
Air	Adsorb on activated charcoal; desorb (carbon disulfide); inject aliquot	GC/FID	0.01 mg per sample	Eller (1987)
	Draw air through Tenax sample tube; heat; desorb into cold trap	GC/MS	20 ng	US Environmental Protection Agency (1988a) [Method TO-1]
	Collect cryogenically into stainless-steel bottle; inject sample	GC/EC-FI-FPD/GC/MS	0.7 ppt	Hoyt & Rasmussen (1985)
Seawater	Collect in vacuum extraction flask; pressurize with zero air; inject headspace sample	GC/EC-FI-FPD/GC/MS	1 ppt	Hoyt & Rasmussen (1985)
Water	Purge (inert gas); trap (OV-1 on Chromosorb-W/Tenax/silica gel); desorb as vapour (heat to 180°C, backflush with inert gas) onto packed GC column	GC/ECD	0.20 μg/l	US Environmental Protection Agency (1988b) [Method 601]
		GC/MS	4.7 μg/l	US Environmental Protection Agency (1988c) [Method 624]
	Purge (inert gas); trap (OV-1 on Chromosorb-W/Tenax/silica gel); desorb as vapour (heat to 180°C, backflush with inert gas) onto capillary GC column	GC/ECD	1.6 μg/l	US Environmental Protection Agency (1988d) [Method 502.2]
		GC/MS	0.12 μg/l	US Environmental Protection Agency (1988e) [Method 524.2]
	Add internal standard (isotope-labelled bromoform); purge; trap and desorb as above	GC/MS	1.0 μg/l	US Environmental Protection Agency (1988f) [Method 1624]
	Extract in pentane; inject onto GC	GC/EC (>50 μg/l) GC/MS (<50 μg/l)	0.5 μg/l	US Environmental Protection Agency (1988g) [Method 501.2]
Adipose tissue	Purge from liquefied fat at 115°C; trap on silica gel; desorb thermally	GC/HSD	2.3 μg/l	Peoples et al. (1979)

Table 4 (contd)

Sample matrix	Sample preparation[a]	Assay procedure[b]	Limit of detection	Reference
Blood serum	Purge from water-serum mixture containing antifoam reagent at 115°C; trap on Tenax/silica gel; desorb thermally	GC/HSD	2.3 µg/l	Peoples et al. (1979)

[a]GC, gas chromatograph

[b]GC/FID, gas chromatography/flame ionization detection; GC/MS, gas chromatography/mass spectrometry; GC/EC-FI-FPD, gas chromatography/electron capture-flame ionization-flame photometric detection; GC/ECD, gas chromatography/electrolytic conductivity detection; GC/HSD, gas chromatography/halide selective detection

Sorbent tube sampling is the approved method of the National Institute for Occupational Health/Occupational Safety and Health Administration (USA) for collecting the most hazardous gases and vapours from the air—in this case bromoform. A sample is collected by drawing air through a tube, where the airborne chemical is trapped in the sorbent. The trapped chemical is solvent-extracted from the tube and analysed using gas chromatography with a flame ionization detector to determine the amount of chemical present. A direct-reading system is available with colour detector tubes to monitor and detect bromoform in air (SKC Inc., 1990).

Bromoform can be determined in water by glass capillary column gas chromatography and electron capture detection, with a detection limit of 1 ng/l (Eklund et al., 1978b,c).

3. Biological Data Relevant to the Evaluation of Carcinogenic Risk to Humans

3.1 Carcinogenicity studies in animals (Table 5)

(a) Oral administration

Mouse: Groups of 50 male and 50 female B6C3F$_1$ mice, eight weeks old, were given bromoform (> 95% pure) in corn oil by gavage at 50 or 100 mg/kg bw (males) and 100 or 200 mg/kg bw (females) on five days per week for 103 weeks. Survival at 105 weeks was: males—vehicle control, 41/50; low-dose, 37/50; high-dose, 36/50; females—control, 25/50; low-dose, 13/50; high-dose, 20/50. Decreased survival in female mice was associated in part with utero-ovarian abscesses. No tumour occurred in a significantly larger proportion of treated than of control mice. The proportion of high-dose male mice with alveolar/bronchiolar neoplasms of the lung was significantly lower than that of controls ($p < 0.015$, incidental tumour test) (National Toxicology Program, 1989).

Rat: Groups of 50 male and 50 female Fischer 344 rats, seven to eight weeks old, were given bromoform (> 95% pure) in corn oil by gavage at 100 or 200 mg/kg bw on five days per week for 103 weeks. Survival at 105 weeks was: males—vehicle control, 34/50; low-dose, 30/50; high-dose, 11/50; females—control, 33/50; low-dose, 28/50; high-dose, 28/50. Adenomatous polyps of the large intestine occurred in 0/50 control, 0/50 low-dose and 2/50 high-dose male rats, and adenocarcinoma of the large intestine occurred in one additional high-dose male rat. Adenomatous polyps occurred in 0/50 control, 1/50 low-dose and 6/50 high-dose female rats, and adenocarcinomas occurred in two additional high-dose animals. The proportion of high-dose female rats with neoplasms of the large intestine was significantly larger than in controls ($p = 0.004$, pairwise comparison; $p < 0.001$, trend test, logistic regression). Neoplasms of the large intestine are uncommon in controls of this strain (males—study laboratory, 0/285; all National Toxicology Program laboratories, 3/1873 (0.2%); females—all National Toxicology Program laboratories, 0/1888). The incidence of preputial gland neoplasms was significantly decreased in high-dose male rats (control, 10/41; low-dose, 5/38; high-dose, 1/34; $p = 0.014$, pairwise comparison, logistic regression). In female rats, the incidences of stromal polyps of the uterus, fibroadenomas of the mammary gland and adenomas of the anterior pituitary gland were significantly decreased in the high-dose group (stromal polyps—vehicle control, 10/49; low-dose, 9/50; high-dose, 2/50; $p = 0.019$, logistic regression; fibroadenomas—vehicle control, 22/50; low-dose, 17/50; high-dose, 6/50; $p < 0.001$, logistic regression; pituitary adenomas—vehicle control, 29/48; low-dose, 12/46; high-dose, 16/48; $p = 0.011$, logistic regression) (National Toxicology Program, 1989).

(b) *Intraperitoneal administration*

Mouse: In a screening assay based on the enhanced induction of lung tumours, groups of 20 male strain A/St mice, six to eight weeks old, were injected intraperitoneally three times per week with bromoform (> 95% pure) in tricaprylin at 4, 48 or 100 mg/kg bw for a total of 18, 23 or 24 injections (total doses, 72, 1100 or 2400 mg/kg bw). Twenty males given tricaprylin only served as controls. All surviving mice were killed 24 weeks after the first injection, at which time survival was 15/20 of controls, 17/20 in the low-dose, 15/20 in the mid-dose and 15/20 in the high-dose group. The average numbers of lung tumours per mouse were 0.27 ± 0.15 (SE) in controls, 0.53 ± 0.21 at the low dose, 1.13 ± 0.36 at the mid-dose and

0.67 ± 0.21 at the high dose [proportion of mice with tumours not given]; the average number of lung tumours in the mid-dose group was significantly higher than that in controls ($p < 0.05$, Student's t test). In a positive control group given a single intraperitoneal injection of urethane at 1000 mg/kg bw, the average number of lung tumours per mouse was 19.6 ± 2.4 (Theiss *et al.*, 1977).

3.2 Other relevant data

(a) *Experimental systems*

(i) *Absorption, distribution, excretion and metabolism*

Bromoform at 0.008-0.04 mmol/kg bw (2-10 mg/kg bw) given by gavage as an aqueous solution to rats was rapidly absorbed and distributed to liver, brain, kidney, blood and fat; the highest concentrations were found in fat 30 min after administration (Parra *et al.*, 1986). ^{14}C-Bromoform at 0.39 mmol/kg bw (16 µCi/kg bw; 100 mg/kg bw) administered orally in corn oil to rats by gavage was absorbed and eliminated in the expired air as unchanged bromoform (67% of dose) or as ^{14}C-carbon dioxide (4% of dose) in 8 h; radiolabel amounting to about 2% of the dose was eliminated in the urine, and about 2% of the dose was retained in body tissues. ^{14}C-Bromoform administered similarly to mice (0.59 mmol/kg bw; 32 µCi/kg bw; 150 mg/kg bw) was absorbed and eliminated in the expired air as unchanged bromoform (6% of dose) or as ^{14}C-carbon dioxide (40% of dose) in 8 h; about 5% of the administered radiolabel was eliminated in the urine, and 12% was retained in body tissues (Mink *et al.*, 1986). Bromoform is metabolized to carbon monoxide *in vivo* (Anders *et al.*, 1978; Stevens & Anders, 1981) and *in vitro*; the latter reaction requires NADPH and oxygen and is stimulated eight fold by glutathione (Ahmed *et al.*, 1977). Bromoform was metabolized to dibromocarbonyl, the bromine analogue of phosgene, by rat hepatic microsomal fractions (Stevens & Anders, 1979; Pohl *et al.*, 1980).

(ii) *Toxic effects*

The LD_{50} of a single subcutaneous injection of bromoform in olive oil was estimated to be 7.2 mmol/kg (1.82 g/kg) bw in male Swiss mice (Kutob & Plaa, 1962); the single intraperitoneal 48-h LD_{50} in corn oil was 414 µl/kg bw in male Sprague-Dawley rats (CR-1 strain; Agarwal & Mehendale, 1983). The single-dose oral LD_{50}s (Emulphor:ethanol:saline 1:1:8 administered by gavage) were 1400 mg/kg bw in male and 1500 mg/kg bw in female ICR Swiss mice (Bowman *et al.*, 1978). Oral LD_{50} values (in corn oil) were 1388 and 1147 mg/kg bw in male and female Sprague-Dawley rats (Chu *et al.*, 1980, 1982a) and 933 mg/kg bw in male and female Fischer 344/N rats (National Toxicology Program, 1989); the values for male

Table 5. Summary of carcinogenicity studies of bromoform in experimental animals

Reference	Species/strain	Sex	Dose schedule	Experimental parameter/observation	Group 0	Group 1	Group 2	Group 3	Significance	Comments
National Toxicology Program (1989)	Mouse B6C3F$_1$	M	5 d/week, gavage, corn oil, 103 weeks	Dose (mg/kg) Survival (105 weeks) Alveolar/bronchiolar neoplasm	0 41/50 11/50	50 37/50 7/50	100 36/50 2/49	—	$p < 0.015$	Decrease
		F		Dose (mg/kg) Survival (105 weeks) Lymphoma	0 25/50 11/49	100 13/50 5/50	200 20/50 3/50	—		Utero-ovarian abscesses
Theiss et al. (1977)	Mouse strain A/St	M	3 d/week, i.p. inject., tricaprylin, 18, 23 or 24 doses	Dose (mg/kg) Total dose (mg/kg) Survival (24 weeks) Lung adenomas per mouse	0 0 15/20 0.27 ± 0.15	4 72 17/20 0.53 ± 0.21	48 1100 15/20 1.13 ± 0.36	100 2400 15/20 0.67 ± 0.21		Screening test in strain in which lung adenomas are common; ± SE

National Toxicology Program (1989)	Rat F344	M	5 d/week, gavage, corn oil, 103 weeks	Dose (mg/kg)
				Survival (105 weeks)
				Large intestine
				  Adenomatous polyp
				  Adenocarcinoma
				  Preputial gland neoplasm
					0
					34/50
					
					0/50
					0/50
					10/41
						100
						30/50
						
						0/50
						0/50
						5/38
							200
							11/50
							
							2/50
							1/50
							1/34
								—	$p = 0.014$	Decrease

		F		Dose (mg/kg)
				Survival (105 weeks)
				Large intestine
				  Adenomatous polyp
				  Adenocarcinoma
				Uterine stromal polyp
				Mammary fibroadenoma
				Pituary adenomas
					0
					33/50
					
					0/50
					0/50
					10/49
					22/50
					29/48
						100
						28/50
						
						1/50
						0.50
						9/50
						17/50
						12/46
							200
							28/50
							
							6/50
							2/50
							2/50
							6/50
							16/48
								—	$p = 0.004$
									
									$p = 0.019$
									$p < 0.001$
									$p = 0.011$
										Increase
										
										Decrease
										Decrease
										Decrease

and female B6C3F$_1$ mice were 707 and 1072 mg/kg bw bromoform, respectively. Signs of acute toxicity included sedation, prostration, lachrymation, lethargy, shallow breathing, reduction in peripheral lymphocyte count and liver damage (Agarwal & Mehendale, 1983; Bowman *et al.*, 1978; Chu *et al.*, 1982a; National Toxicology Program, 1989).

Daily oral treatment of male and female CD-1 mice with 50-250 mg/kg bw bromoform (in 10% Emulphor in water) for 14 days resulted in decreased fibrinogen levels in blood, liver damage (increased ASAT), decreased serum glucose and blood urea nitrogen and decreased cellular and humoral immunity at 250 mg/kg bw (Munson *et al.*, 1982). Similarly, administration of 145 and 289 mg/kg bw bromoform in corn oil per day for 14 days to male CD-1 mice resulted in centrilobular pallor, focal inflammation and increased mitosis in the liver. Tubular hyperplasia and glomerular degeneration and reduced uptake of organic acids were observed in the kidney after administration of the high dose (Condie *et al.*, 1983).

Daily oral administration of 600 and 800 mg/kg bw bromoform by gavage for 14 days induced lethargy, shallow breathing and ataxia and was lethal to all male and female Fischer 344/N rats. Body weight reduction was observed at 400 mg/kg bw; enlargement of the thyroid gland was observed at 400 and 800 mg/kg bw. B6C3F$_1$ mice receiving similar treatment were less sensitive. In a 13-week study, male and female Fischer 344/N rats were administered 12-200 mg/kg bw bromoform and male and female B6C3F$_1$ mice were administered 25-400 mg/kg bw bromoform by gavage on five days per week. Body weight reduction was observed in male mice treated with the high dose. Hepatocellular vacuolization was observed in male rats at all doses and in male mice at 200 and 400 mg/kg bw (National Toxicology Program, 1989).

Male and female Sprague-Dawley rats received 5-2500 mg/l bromoform in drinking-water for 90 days. The highest dose, which corresponded to an approximate daily intake of 29-55 mg/rat, resulted in liver lesions, reduced lymphocyte counts (after a 90-day recovery period) and a reduction in serum lactic dehydrogenase activity (Chu *et al.*, 1982b).

Administration of bromoform for two years (see section 3.1) resulted in hepatic fatty changes, including vacuolization, in male and female rats and in female but not male mice in all treated groups (National Toxicology Program, 1989).

In a screening assay based on the production of γ-glutamyltranspeptidase-positive foci in the liver, nine male rats were given a single dose of bromoform at 1 mmol/kg (250 mg/kg) bw [route of administration unspecified] following a two-thirds hepatectomy. Bromoform did not have a significant effect (Herren-Freund & Pereira, 1986).

(iii) *Effects on reproduction and prenatal toxicity*

Swiss CD-1 mice were given bromoform by gavage in corn oil at daily doses of 0, 50, 100 or 200 mg/kg bw in a reproductive study using a continuous breeding protocol. Treatment was continued for 18 weeks: one week prior to cohabitation, 14 weeks of cohabitation and three weeks thereafter. Toxicity was observed at the two higher levels, as decreased body weight and kidney weight and increased liver weight, but there was no adverse effect on fertility in either the parental or F_1 generation. A decrease in neonatal (F_1) survival was noted in the high-dose group. The body weights of treated F_1 males were significantly lower than the corresponding control values (Gulati *et al.*, 1989).

Sprague-Dawley rats were administered bromoform in corn oil by gavage at daily doses of 0, 50, 100 or 200 mg/kg bw on gestation days 6-15. No maternal toxicity was observed, and bromoform did not alter the incidence of resorptions, litter size or fetal weight. There was a dose-related increase in the incidence of skeletal variations, primarily involving the sternum, the interparietal bones and ribs (Ruddick *et al.*, 1983). [The Working Group noted that too few fetuses (12-16 per dose level) were examined for visceral malformations for a conclusion about lack of malformations in soft tissues to be drawn.]

(iv) *Genetic and related effects* (Table 6)

Bromoform was mutagenic to *Salmonella typhimurium* occasionally, especially when tested in closed containers. In single studies, bromoform caused mitotic arrest (c-mitosis) in *Allium cepa*, mutation in *Drosophila melanogaster* and mouse lymphoma L5178Y cells, and sister chromatid exchange in human lymphocytes and (weakly) in Chinese hamster CHO cells. In two of three studies with cultured Chinese hamster cells, chromosomal aberrations were reported. Sister chromatid exchange was induced in mice *in vivo* in two reports, and induction of micronuclei was seen in one but not in a second study. In one study, chromosomal aberrations were not induced in bone-marrow cells of mice treated *in vivo*. No DNA binding was observed in livers of bromoform-treated rats.

(b) *Humans*

No data were available to the Working Group.

3.3 Case reports and epidemiological studies of carcinogenicity to humans

A single correlation study (Isacson *et al.*, 1983), described in the monograph on chlorinated drinking-water (p. 113), mentioned bromoform, but the information could not be used to evaluate the carcinogenicity of this chemical individually.

Table 6. Genetic and related effects of bromoform

Test system	Result		Dose LED/HID	Reference
	Without exogenous metabolic system	With exogenous metabolic system		
SA0, *Salmonella typhimurium* TA100, reverse mutation	+	0	0.0000	Simmon et al. (1977)[a]
SA0, *Salmonella typhimurium* TA100, reverse mutation	–	0	500.0000	Rapson et al. (1980)
SA0, *Salmonella typhimurium* TA100, reverse mutation	(+)	–	300.0000	Haworth et al. (1983)[b]
SA0, *Salmonella typhimurium* TA100, reverse mutation	(+)	–	250.0000	Varma et al. (1988)
SA0, *Salmonella typhimurium* TA100, reverse mutation	–	–	1667.0000	National Toxicology Program (1989)[c]
SA0, *Salmonella typhimurium* TA100, reverse mutation	–	0	0.0000	Mersch-Sunderman (1989)
SA2, *Salmonella typhimurium* TA102, reverse mutation	–	0	0.0000	Mersch-Sunderman (1989)
SA5, *Salmonella typhimurium* TA1535, reverse mutation	+	0	0.0000	Simmon et al. (1977)[a]
SA5, *Salmonella typhimurium* TA1535, reverse mutation	–	–	1667.0000	National Toxicology Program (1989)
SA5, *Salmonella typhimurium* TA1535, reverse mutation	–	–	300.0000	Haworth et al. (1983)
SA5, *Salmonella typhimurium* TA1535, reverse mutation	–	0	0.0000	Varma et al. (1988)
SA7, *Salmonella typhimurium* TA1537, reverse mutation	–	–	300.0000	Haworth et al. (1983)
SA7, *Salmonella typhimurium* TA1537, reverse mtuation	–	–	1667.0000	National Toxicology Program (1989)[c]
SA7, *Salmonella typhimurium* TA1537, reverse mutation	–	0	0.0000	Varma et al. (1988)
SA9, *Salmonella typhimurium* TA98, reverse mutation	–	–	300.0000	Haworth et al. (1983)
SA9, *Salmonella typhimurium* TA98, reverse mutation	–	–	500.0000	National Toxicology Program (1989)[c]
SA9, *Salmonella typhimurium* TA98, reverse mutation	–	0	0.0000	Varma et al. (1988)
SA9, *Salmonella typhimurium* TA98, reverse mutation	+	–	0.0000	Mersch-Sunderman (1989)
SAS, *Salmonella typhimurium* TA97, reverse mutation	–	–	1667.0000	National Toxicology Program (1989)[c]
SAS, *Salmonella typhimurium* TA97, reverse mutation	+	–	0.0000	Mersch-Sunderman (1989)[d]
ACC, *Allium cepa*, C-mitosis	+	0	0.0000	Östergren (1944)
DMX, *Drosophila melanogaster*, sex-linked recessive lethal mutations	+	0	3000.0000	Woodruff et al. (1985)
DMX, *Drosophila melanogaster*, sex-linked recessive lethal mutations	–	0	1000.0000	Woodruff et al. (1985)
DMH, *Drosophila melanogaster*, reciprocal translocations	–	0	3000.0000	Woodruff et al. (1985)
G5T, Gene mutation, mouse lymphoma L5178Y cells, *tk* locus	+	+	70.0000	National Toxicology Program (1989)
SIC, Sister chromatid exchange, Chinese hamster CHO cells	(+)	–	290.0000	National Toxicology Program (1989)[b]
CIC, Chromosomal aberrations, Chinese hamster CHO cells	+	–	1070.0000	National Toxicology Program (1989)[b]

Table 6 (contd)

Test system	Result		Dose LED/HID	Reference
	Without exogenous metabolic system	With exogenous metabolic system		
CIC, Chromosomal aberrations, Chinese hamster CHL cells	(+)	+	116.0000	Ishidate (1987)
SHL, Sister chromatid exchange, human lymphocytes in vitro	+	0	20.0000	Morimoto & Koizumi (1983)
SVA, Sister chromatid exchange, mouse bone–marrow cells in vivo	+	0	25.0000	Morimoto & Koizumi (1983)
SVA, Sister chromatid exchange, mouse bone–marrow cells in vivo	–	0	800.0000	National Toxicology Program (1989)
CBA, Chromosomal aberrations, mouse bone–marrow cells in vivo	–	0	800.0000	National Toxicology Program (1989)
MVM, Micronucleus test, ddy mice in vivo	–	0	1400.0000	Hayashi et al. (1988)
MVM, Micronucleus test, B6C3F₁ mice in vivo	+	0	800.0000	National Toxicology Program (1989)
BVD, DNA binding to rat liver in vivo	–	0	380.0000	Pereira et al. (1982)

^aClosed container
^bOne of two participating laboratories obtained positive results.
^cResults from SRI laboratories
^dSpot test +, standard –

4. Summary of Data Reported and Evaluation

4.1 Exposure data

Bromoform has a limited number of industrial uses. It is also found in chlorinated drinking-water as a consequence of the reaction between chlorine, added during water treatment, and natural organic substances in the presence of bromide ion. Bromoform has been detected in chlorinated drinking-water in many parts of the world; it has also been detected in untreated water, but at lower levels. Bromoform is the major organohalide produced by chlorination of seawater during desalination. It is a major component of the organohalides produced by marine algae.

The major route of human exposure to bromoform is from drinking-water, although ambient air is also an important source of exposure in some areas.

4.2 Experimental carcinogenicity data

Bromoform was tested for carcinogenicity in a two-year study by oral gavage in male and female $B6C3F_1$ mice and Fischer 344 rats. It induced adenomatous polyps and adenocarcinomas of the large intestine in male and female rats. Bromoform did not increase the proportion of mice with tumours. In a screening test by intraperitoneal injection, there was a slight increase in the average number of lung tumours in strain A mice given the middle dose of bromoform.

4.3 Human carcinogenicity data

No relevant data were available to the Working Group.

4.4 Other relevant data

In experimental animals, bromoform induced liver and kidney damage and decreased the immune response.

There is some evidence of developmental toxicity in the absence of maternal toxicity in rats.

Mutagenic effects of bromoform were observed occasionally in bacteria. In single studies, bromoform induced mitotic arrest in plants, mutations in insects and in cultured mammalian cells and sister chromatid exchange in human lymphocytes.

Chromosomal aberrations were induced in cultured mammalian cells. In single studies in rodents *in vivo*, bromoform did not bind to DNA or cause chromosomal aberrations. Sister chromatid exchange was induced in rodents *in vivo*.

4.5 Evaluation[1]

There is *inadequate evidence* for the carcinogenicity of bromoform in humans.
There is *limited evidence* for the carcinogenicity of bromoform in experimental animals.

Overall evaluation

Bromoform *is not classifiable as to its carcinogenicity to humans (Group 3)*.

5. References

Agarwal, A.K. & Mehendale, H.M. (1983) Absence of potentiation of bromoform hepatotoxicity and lethality by chlordecone. *Toxicol. Lett.*, 15, 251-257

Ahmed, A.E., Kubic, V.L. & Anders, M.W. (1977) Metabolism of haloforms to carbon monoxide. I. In vitro studies. *Drug Metab. Disposition*, 5, 198-204

Aldrich Chemical Co. (1990) *1990-1991 Aldrich Handbook of Fine Chemicals*, Milwaukee, WI, p. 200

American Conference of Governmental Industrial Hygienists (1989) *TLVs Threshold Limit Values and Biological Exposure Indices for 1989-1990*, Cincinnati, OH, p. 14

American Tokyo Kasei (1988) *TCI American Organic Chemicals 88/89 Catalog*, Portland, OR, p. 187

Anders, M.W., Stevens, J.L., Sprague, R.W., Shaath, Z. & Ahmed, A.E. (1978) Metabolism of haloforms to carbon monoxide. II. In vivo studies. *Drug Metab. Disposition*, 6, 556-560

Antoine, S.R., DeLeon, I.R. & O'Dell-Smith, R.M. (1986) Environmentally significant volatile organic pollutants in human blood. *Bull. environ. Contam. Toxicol.*, 36, 364-371

Armstrong, D.W. & Golden, T. (1986) Determination of distribution and concentration of trihalomethanes in aquatic recreational and therapeutic facilities by electron-capture GC. *LC-GC*, 4, 652-655

Barkley, J., Bunch, J., Bursey, J.T., Castillo, N., Cooper, S.D., Davis, J.M., Erickson, M.D., Harris, B.S.H., III, Kirkpatrick, M., Michael, L.C., Parks, S.P., Pellizzari, E.D., Ray, M., Smith, D., Tomer, K.B., Wagner, R. & Zweidinger, R.A. (1980) Gas chromatography mass spectrometry computer analysis of volatile halogenated hydrocarbons in man and his environment—a multimedia environmental study. *Biomed. Mass Spectrometr.*, 7, 139-147

---

[1]For definition of the italicized terms, see Preamble, pp. 30-33.

Bauer, U. (1981) Human exposure to environmental chemicals—Investigations on volatile organic halogenated compounds in water, air, food, and human tissues. III. Communication: results of investigations. *Zentralbl. Bakteriol. Mikrobiol. Hyg. Abt. 1 Orig. B, 174*, 200-237

Bowman, F.J., Borzelleca, J.F. & Munson, A.E. (1978) The toxicity of some halomethanes in mice. *Toxicol. appl. Pharmacol., 44*, 213-215

Brass, A.J., Feige, M.A., Halloran, T., Mello, J.W., Munch, D. & Thomas, R.F. (1977) The National Organic Monitoring Survey: samplings and analyses for purgeable organic compounds. In: Pojasek, R.B., ed., *Drinking Water Quality Enhancement Through Source Protection*, Ann Arbor, MI, Ann Arbor Science, pp. 393-416

Brodzinsky, R. & Singh, H.B. (1983) *Volatile Organic Chemicals in the Atmosphere: An Assessment of Available Data* (EPA-600/3-83-027a; NTIS PB83-195503), Research Triangle Park, NC, US Environmental Protection Agency

Budavari, S., ed. (1989) *The Merck Index*, 11th ed., Rahway, NJ, Merck & Co., p. 215

Bunn, W.W., Haas, B.B., Deane, E.R. & Kleopfer, R.D. (1975) Formation of trihalomethanes by chlorination of surface water. *Environ. Lett., 10*, 205-213

Carpenter, J.H., Smith, C.A. & Zika, R.G. (1981) *Reaction Products from the Chlorination of Seawater* (EPA-600/4-81-010; NTIS PB81-172280), Gulf Breeze, FL, US Environmental Protection Agency

Cech, I., Smith, V. & Henry, J. (1982) Spatial and seasonal variations in concentration of trihalomethanes in drinking water. *Pergamon Ser. environ. Sci., 7*, 19-38

Chang, N.-W. & Singer, P.C. (1984) Formation and fate of bromoform during desalination. *J. environ. Eng., 110*, 1189-1193

Chu, I., Secours, V., Marino, I. & Villeneuve, D.C. (1980) The acute toxicity of four trihalomethanes in male and female rats. *Toxicol. appl. Pharmacol., 52*, 351-353

Chu, I., Villeneuve, D.C., Secours, V.E. & Becking, G.C. (1982a) Toxicity of trihalomethanes: I. The acute and subacute toxicity of chloroform, bromodichloromethane, chlorodibromomethane and bromoform in rats. *J. environ. Sci. Health, B17*, 205-224

Chu, I., Villeneuve, D.C., Secours, V.E. & Becking, G.C. (1982b) Trihalomethanes: II. Reversibility of toxicological changes produced by chloroform, bromodichloromethane, chlorodibromomethane and bromoform in rats. *J. environ. Sci. Health, B17*, 225-240

Cicerone, R.J., Heidt, L.E. & Pollock, W.H. (1988) Measurements of atmospheric methyl bromide and bromoform. *J. geophys. Res., 93*, 3745-3749

Class, T., Kohnle, R. & Ballschmiter, K. (1986) Chemistry of organic traces in air. VII: Bromo- and bromochloroethanes in air over the Atlantic Ocean. *Chemosphere, 15*, 429-436

Cole, R.H., Frederick, R.E., Healy, R.P. & Rolan, R.G. (1984) Preliminary findings of the Priority Pollutant Monitoring Project of the Nationwide Urban Runoff Program. *J. Water Pollut. Control Fed., 56*, 898-908

Coleman, W.E., Munch, J.W., Kaylor, W.H., Streicher, R.P., Ringhand, H.P. & Meier, J.R. (1984) Gas chromatography/mass spectroscopy analysis of mutagenic extracts of aqueous chlorinated humic acid. A comparison of the byproducts to drinking water contaminants. *Environ. Sci. Technol.*, *18*, 674-681

Comba, M.E. & Kaiser, K.L.E. (1983) Determination of volatile contaminants at the ng/l level in water by capillary gas chromatography with electron capture detection. *Int. J. environ. anal. Chem.*, *16*, 17-31

Condie, L.W., Smallwood, C.L. & Laurie, R.D. (1983) Comparative renal and hepatotoxicity of halomethanes: bromodichloromethane, bromoform, chloroform, dibromochloromethane and methylene chloride. *Drug chem. Toxicol.*, *6*, 563-578

Cook, W.A. (1987) *Occupational Exposure Limits—Worldwide*, Cincinnati, OH, American Industrial Hygiene Association, pp. 85, 130, 166

Easley, D.M., Kleopfer, R.D. & Carasea, A.M. (1981) Gas chromatographic-mass spectrometric determination of volatile organic compounds in fish. *J. Assoc. off. anal. Chem.*, *64*, 653-656

Eklund, G., Josefsson, B. & Roos, C. (1978a) Trace analysis of volatile organic substances in Goeteborg municipal drinking water. *Vatten*, *3*, 195-206

Eklund, G., Josefsson, B. & Roos, C. (1978b) Determination of volatile halogenated hydrocarbons in environmental water by glass capillary column gas chromatography and electron capture detection. *Chromatogr. Newsl.*, *6*, 39-41

Eklund, G., Josefsson, B. & Roos, C. (1978c) Determination of volatile halogenated hydrocarbons in tap water, seawater and industrial effluents by glass capillary gas chromatography and electron capture detection. *J. high Resolut. Chromatogr. Chromatogr. Commun.*, *1*, 34-40

Eller, P.M. (1987) *NIOSH Manual of Analytical Methods*, 3rd ed., Vol. 2, rev. 1 (DHHS (NIOSH) Publ. No. 84-100), Washington DC, US Government Printing Office, pp. 1003-1–1003-6

de Fernicola, N.A.G.G. & de Azevedo, F.A. (1984) Water levels of trihalomethanes (THM) in four different supply systems from Sao Paulo State, (Brazil). *J. environ. Health*, *46*, 187-188

Fogelqvist, E. (1985) Carbon tetrachloride, tetrachloroethylene, 1,1,1-trichloroethane and bromoform in Arctic seawater. *J. geophys. Res.*, *90*, 9181-9193

Furlong, E.A.-N. & D'Itri, F.M. (1986) Trihalomethane levels in chlorinated Michigan drinking water. *Ecol. Modelling*, *32*, 215-225

Gabel, B., Lahl, U., Baetjer, K., Cetinkaya, M., von Dueszeln, J., Kozicki, R., Podbielski, A., Stachel, B. & Thiemann, W. (1981) Volatile halogenated compounds (VOHal) in drinking waters of the FRG. *Sci. total Environ.*, *18*, 363-366

Gibson, C.I., Tone, F.C., Schirmer, R.E. & Blaylock, J.W. (1980) Bioaccumulation and depuration of bromoform in five marine species. In: Jolley, R.L., Brungs, W.A., Cumming, R.B. & Jacobs, V.A., eds, *Water Chlorination: Environmental Impact and Health Effects*, Vol. 3, Ann Arbor, MI, Ann Arbor Science, pp. 517-533

Glaze, W.H. & Rawley, R. (1979) A preliminary survey of trihalomethane levels in selected East Texas water supplies. *J. Am. Water Works Assoc.*, *71*, 509-515

Gschwend, P.M., MacFarlane, J.K. & Newman, K.A. (1985) Volatile halogenated organic compounds released to seawater from temperate marine macroalgae. *Science*, 227, 1033-1035

Gulati, D.K., Hope, E., Barnes, L.H., Russell, S. & Poonacha, K.B. (1989) *Bromoform: Reproductive and Fertility Assessment in Swiss CD-1 Mice When Administered by Gavage* (NTP-86-FACB-053), Research Triangle Park, NC, National Institute for Environmental Health Sciences

Hagenmaier, H., Werner, G. & Jäger, W. (1982) Quantitative determination of volatile halogenated hydrocarbons in water samples by capillary gas chromatography and electron capture detection. *Z. Wasser Abwasser Forsch.*, 15, 195-198

Harlow, I.F. & Ross, O.C. (1932) *Preparation of Brominated Hydrocarbons* (US Patent 1,891,415 to Dow Chemical Co)

Hartwig, E.O. & Valentine, R. (1983) Bromoform production in tropical open-ocean waters: ocean thermal energy conversion chlorination. In: Jolley, R.L., Brungs, W.A., Cotruvo, J.A., Cumming, R.B., Mattice, J.S. & Jacobs, V.A., eds, *Water Chlorination: Environmental Impact and Health Effects*, Vol. 4, Book 1, *Chemistry and Water Treatment*, Ann Arbor, MI, Ann Arbor Science, pp. 311-330

Haworth, S., Lawlor, T., Mortelmans, K., Speck, W. & Zeiger, E. (1983) *Salmonella* mutagenicity test results for 250 chemicals. *Environ. Mutagenesis, Suppl. 1*, 3-142

Hayashi, M., Kishi, M., Sofuni, T. & Ishidate, M., Jr (1988) Micronucleus tests in mice on 39 food additives and eight miscellaneous chemicals. *Food chem. Toxicol.*, 26, 487-500

Health and Welfare Canada (1977) *National Survey for Halomethanes in Drinking Water* (Report 77-EHD-9), Ottawa, Environmental Health Directorate, Health Protection Branch

Henderson, J.E., Peyton, G.R. & Glaze, W.H. (1976) A convenient liquid-liquid extraction method for the determination of halomethanes in water at the parts-per-billion level. In: Keith, L.H., ed., *Identification and Analysis of Organic Pollutants in Water*, Ann Arbor, MI, Ann Arbor Science, pp. 105-111

Herren-Freund, S.L. & Pereira, M.A. (1986) Carcinogenicity of by-products of disinfection in mouse and rat liver. *Environ. Health Perspect.*, 69, 59-65

Hoyt, S.D. & Rasmussen, R.A. (1985) Determining trace gases in air and seawater. *Adv. chem. Ser. Mapping Strategies chem. Oceanogr.*, 209, 31-56

IARC (1987) *IARC Monographs on the Evaluation of Carcinogenic Risks to Humans*, Suppl. 7, *Overall Evaluations of Carcinogenicity: An Updating of* IARC Monographs *Volumes 1 to 42*, Lyon, pp. 152-154

Isacson, P., Bean, J.A. & Lynch, C. (1983) Relationship of cancer incidence rates in Iowa municipalities to chlorination status of drinking water. In: Jolley, R.L., Brungs, W.A., Cotruvo, J.A., Cumming, R.B., Mattice, J.S. & Jacobs, V.A., eds, *Water Chlorination: Environmental Impact and Health Effects*, Vol. 4, Book 2, *Environment, Health, and Risk*, Ann Arbor, MI, Ann Arbor Science, pp. 1353-1364

Ishidate, M., Jr (1987) *Data Book of Chromosomal Aberration Test in vitro* rev. ed., Tokyo, Life-Science Information Center, pp. 421-422

Kaczmar, S.W., D'Itri, F.M. & Zabik, M.J. (1984) Volatilization rates of selected haloforms from aqueous environments. *Environ. Toxicol. Chem.*, 3, 31-35

Kaiser, K.L.E. & Comba, M.E. (1983) Volatile contaminants in the Welland River watershed. *J. Great Lakes Res.*, 9, 274-280

Kaiser, K.L.E., Comba, M.E. & Huneault, H. (1983) Volatile halocarbon contaminants in the Niagara River and in Lake Ontario. *J. Great Lakes Res.*, 9, 212-223

Kossik, R.F., Gschwend, P.S. & Adams, E.E. (1986) *Tracing and Modeling Pollutant Transport in Boston Harbor* (Report No. MITSG-86/16; NTIS PB87-136891), Rockville, MD, National Oceanic and Atmospheric Administration

Krasner, S.W., McGuire, M.J., Jacangelo, J.G., Patania, N.L., Reagan, K.M. & Aieta, E.M. (1989) The occurrence of disinfection by-products in drinking water in a nationwide study. In: *Annual Conference of the American Water Works Association, Los Angeles, CA*, Denver, CO, American Water Works Association

Kutob, S.D. & Plaa, G.L. (1962) A procedure for estimating the hepatotoxic potential of certain industrial solvents. *Toxicol. appl. Pharmacol.*, 4, 354-361

Lahl, U., Bätjer, K., von Düszeln, J., Gabel, B., Stachel, B. & Thiemann, W. (1981) Distribution and balance of volatile halogenated hydrocarbons in the water and air of covered swimming pools using chlorine for water disinfection. *Water Res.*, 15, 803-814

Lahl, U., Cetinkaya, M., von Düszeln, J., Gabel, B., Stachel, B. & Thiemann, W. (1982) Health risks for infants caused by trihalomethane generation during chemical disinfection of feeding utensils. *Ecol. Food Nutr.*, 12, 7-17

Loy, E.W., Jr, Brown, D.W., Stephenson, J.H.M. & Little, J.A. (1976) Uses of wastewater discharge compliance monitoring data. In: Keith, L.H., ed., *Identification and Analysis of Organic Pollutants in Water*, Ann Arbor, MI, Ann Arbor Science, pp. 499-516

Mabey, W.R., Smith, J.H., Podoll, R.T., Johnson, H.L., Mill, T., Chou, T.W., Gates, J., Waight Partridge, I. & Vandenberg, D. (1982) *Aquatic Fate Process Data for Organic Priority Pollutants* (EPA-440/4-81-014; NTIS PB87-169090), Washington DC, Office of Water Regulations and Standards, US Environmental Protection Agency, pp. 183-184

McConnell, G. (1976) Haloorganics in water supplies. *J. Inst. Water Eng. Sci.*, 30, 431-445

Mersch-Sunderman, V. (1989) Examination of the mutagenicity of organic microcontaminations inn the environment. II. The mutagenicity of halogenated aliphatic hydrocarbons with the *Salmonella*-microsome test (Ames test) in relation to contamination of ground- and drinking-water (Ger.). *Zbl. Bakt. Hyg. B.*, 187, 230-243

Minear, R.A. & Morrow, C.M. (1983) *Raw Water Bromide: Levels and Relationship to Distribution of Trihalomethanes in Finished Drinking Water* (OWRT-A-063-TENN; NTIS PB83-256735), Washington DC, US Environmental Protection Agency

Mink, F.L., Brown, T.J. & Rickabaugh, J. (1986) Absorption, distribution, and excretion of ^{14}C-trihalomethanes in mice and rats. *Bull. environ. Contam. Toxicol.*, 37, 752-758

Morimoto, K. & Koizumi, A. (1983) Trihalomethanes induce sister chromatid exchanges in human lymphocytes *in vitro* and mouse bone marrow cells *in vivo*. *Environ. Res.*, 32, 72-79

Munson, A.E., Sain, L.E., Sanders, V.M., Kauffmann, B.M., White, K.L., Jr, Page, D.G., Barnes, D.W. & Borzelleca, J.F. (1982) Toxicology of organic drinking water contaminants: trichloromethane, bromodichloromethane, dibromochloromethane and tribromomethane. *Environ. Health Perspect.*, 46, 117-126

National Toxicology Program (1989) *Toxicology and Carcinogenesis Studies of Tribromomethane (Bromoform) (CAS No. 75-25-2) in F344/N Rats B6C3F$_1$ Mice (Gavage Studies)* (NTP Technical Report No. 350), Research Triangle Park, NC

Norin, H. & Renberg, L. (1980) Determination of trihalomethanes (THM) in water using high efficiency solvent extraction. *Water Res.*, 14, 1397-1402

Norin, H., Renberg, L., Hjort, J. & Lundblad, P.O. (1981) Factors influencing formation of trihalomethanes in drinking water with special reference to Swedish conditions. *Chemosphere*, 10, 1265-1273

Östergren, C. (1944) Colchicine mitosis chromosome contraction, narcosis and protein chain folding. *Hereditas*, 30, 429-467

Otson, R. (1987) Purgeable organics in Great Lakes raw and treated water. *Int. J. environ. anal. Chem.*, 31, 41-53

Otson, R., Williams, D.T. & Bothwell, P.D. (1982) Volatile organic compounds in water at thirty Canadian potable water treatment facilities. *J. Assoc. off. anal. Chem.*, 65, 1370-1374

Page, G.W. (1981) Comparison of groundwater and surface water for patterns and levels of contamination by toxic substances. *Environ. Sci. Technol.*, 15, 1475-1481

Parra, P., Martinez, E., Suñol, C., Artigas, F., Tusell, J.M., Gelpi, E. & Albaigés, J. (1986) Analysis, accumulation and central effects of trihalomethanes. I. Bromoform. *Toxicol. environ. Chem.*, 24, 79-91

Peoples, A.J., Pfaffenberger, C.D., Shafik, T.M. & Enos, H.F. (1979) Determination of volatile purgeable halogenated hydrocarbons in human adipose tissue and blood serum. *Bull. environ. Contam. Toxicol.*, 23, 244-249

Pereira, M.A., Lin, L.-H.C., Lippitt, J.M. & Herren, S.L. (1982) Trihalomethanes as initiators and promoters of carcinogenesis. *Environ. Health Perspect.*, 46, 151-156

Perwak, J., Goyer, M., Harris, J., Schimke, G., Scow, K., Wallace, D. & Slimak, M. (1980) *An Exposure and Risk Assessment for Trihalomethanes: Chloroform, Bromoform, Bromodichloromethane, Dibromochloromethane* (EPA-440/4-81-018; NTIS PB85-211977), Washington DC, Office of Water Regulations and Standards, US Environmental Protection Agency

Pohl, L.R., Martin, J.L., Taburet, A.M. & George, J.W. (1980) Oxidative bioactivation of haloforms into hepatotoxins. In: Coon, M.J., Conney, A.H., Estabrook, R.W., Gelboin, H.V., Gillette, J.R. & O'Brien, P.J., eds, *Microsomes, Drug Oxidations, and Chemical Carcinogenesis*, Vol. II, New York, Academic Press, pp. 881-884

Pouchert, C.J., ed. (1974) *The Aldrich Library of NMR Spectra*, Vol. 1, Milwaukee, WI, Aldrich Chemical Co., p. 64B

Pouchert, C.J., ed. (1981) *The Aldrich Library of Infrared Spectra*, 3rd ed., Milwaukee, WI, Aldrich Chemical Co., p. 51H

Pouchert, C.J., ed. (1985) *The Aldrich Library of FT-IR Spectra*, Vol. 3, Milwaukee, WI, Aldrich Chemical Co., p. 117B

Quaghebeur, D. & De Wulf, E. (1980) Volatile halogenated hydrocarbons in Belgian drinking waters. *Sci. total Environ.*, 14, 43-52

Rapson, W.H., Nazar, M.A. & Butsky, V.V. (1980) Mutagenicity produced by aqueous chlorination of organic compounds. *Bull. environ. Contam. Toxicol.*, 24, 590-596

Reding, R., Fair, P.S., Shipp, C.J. & Brass, H.J. (1989) Measurement of dihaloacetonitriles and chloropicrin in US drinking waters. In: *Disinfection By-products: Current Perspectives*, Denver, CO, American Water Works Association, pp. 11-22

Riedel-de Haën (1986) *Laboratory Chemicals 1986*, Seelze 1/Hannover, pp. 161-162

Rodriguez-Flores, M. (1983) *Characterization of Organic Contaminants in Selected Potable Water Supplies in Puerto Rico* (OWRT-A-057-PR; NTIS PB83-178038), Washington DC, US Environmental Protection Agency

Rook, J.J. (1974) Formation of haloforms during chlorination of natural waters. *Water Treat. Exam.*, 23, 234-243

Ruddick, J.A., Villeneuve, D.C. & Chu, I. (1983) A teratological assessment of four trihalomethanes in the rat. *J. environ. Sci. Health*, B18, 333-349

Sadtler Research Laboratories (1980) *The Sadtler Standard Spectra, 1980 Cumulative Index*, Philadelphia, PA

Sax, N.I. & Lewis, R.J., Sr (1987) *Hawley's Condensed Chemical Dictionary*, 11th ed., New York, Van Nostrand Reinhold, p. 172

Scotte, P. (1984) Study of halogenated organic compounds in covered swimming pools (Fr.). *Eau Ind. Nuisances*, 86, 37-41

Sherman, P.D., Jr & Kavasmaneck, P.R. (1980) Ethanol. In: Mark, H.F., Othmer, D.F., Overberger, C.G., Seaborg, G.T. & Grayson, M., eds, *Kirk-Othmer Encyclopedia of Chemical Technology*, 3rd ed., Vol. 9, New York, John Wiley & Sons, pp. 341-342

Simmon, V.F., Kauhanen, K. & Tardiff, R.G. (1977) Mutagenic activity of chemicals identified in drinking water. In: Scott, D., Bridges, B.A. & Sobels, F.H., eds, *Progress in Genetic Toxicology*, Amsterdam, Elsevier/North-Holland Biomedical Press, pp. 249-258

Singer, P.C. (1982) *Formation and Removal of Bromoform During Desalination* (OWRT-C-90013-S(9507); NTIS PB87-197646), Washington DC, US Environmental Protection Agency

Sittig, M. (1985) *Handbook of Toxic and Hazardous Chemicals and Carcinogens*, 2nd ed., Park Ridge, NJ, Noyes Publications, pp. 149-150

SKC Inc. (1990) *Comprehensive Catalog and Guide. Air Sampling Products for Worker Monitoring, Chemical Hazard Detection and Industrial Hygiene*, Eighty Four, PA

Soroos, H. & Hinkamp, J.B. (1945) The redistribution reaction. XI. Application to the preparation of carbon tetraiodide and related halides. *J. Am. chem. Soc.*, 67, 1642-1643

Stevens, A.A., Moore, L.A., Slocum, C.J., Smith, B.L., Seeger, D.R. & Ireland, J.C. (1989) By-products of chlorination at ten operating utilities. In: *Disinfection By-Products: Current Perspectives*, Denver, CO, American Water Works Association, pp. 23-61

Stevens, J.L. & Anders, M.W. (1979) Metabolism of haloforms to carbon monoxide. III. Studies on the mechanism of the reaction. *Biochem. Pharmacol.*, 28, 3189-3194

Stevens, J.L. & Anders, M.W. (1981) Metabolism of haloforms to carbon monoxide. IV. Studies on the reaction mechanism in vivo. *Chem.-biol. Interact.*, *37*, 365-374

Symons, J.M., Bellar, T.A., Carswell, J.K., DeMarco, J., Kropp, K.L., Robeck, G.G., Seeger, D.R., Slocum, C.J., Smith, B.L. & Stevens, A.A. (1975) National Organics Reconnaissance Survey for halogenated organics. *J. Am. Water Works Assoc.*, *67*, 634-647

Theiss, J.C., Stoner, G.D., Shimkin, M.B. & Weisburger, E.K. (1977) Test for carcinogenicity of organic contaminants of United States drinking waters by pulmonary tumor response in strain A mice. *Cancer Res.*, *37*, 2717-2720

Trussell, A.R., Cromer, J.L., Umphres, M.D., Kelley, P.E. & Moncur, J.G. (1980) Monitoring of volatile halogenated organics: a survey of twelve drinking waters from various parts of the world. In: Jolley, R.L., Brungs, W.A., Cumming, R.B. & Jacobs, V.A., eds, *Water Chlorination: Environmental Impact and Health Effects*, Vol. 3, Ann Arbor, MI, Ann Arbor Science, pp. 39-53

Ullrich, D. (1982) 7. Organohalogenated compounds in the air of some public swimming pools of Berlin (Ger.). *WaBoLu-Ber.*, *1*, 50-52

US Environmental Protection Agency (1986a) Method 8010. Halogenated volatile organics. In: *Test Methods for Evaluating Solid Waste—Physical/Chemical Methods*, 3rd ed. (EPA No. SW-846), Washington DC, Office of Solid Waste and Emergency Response

US Environmental Protection Agency (1986b) Method 8240. Gas chromatography/mass spectrometry for volatile organics. In: *Test Methods for Evaluating Solid Waste—Physical/Chemical Methods*, 3rd ed. (EPA No. SW-846), Washington DC, Office of Solid Waste and Emergency Response

US Environmental Protection Agency (1988a) Method TO-1. Method for the determination of volatile organic compounds in ambient air using Tenax(R) absorption and gas chromatography/mass spectrometry (GC/MS). In: *Compendium of Methods for the Determination of Toxic Organic Compounds in Ambient Air* (EPA-600/4-89-017; NTIS PB90-116989), Research Triangle Park, NC, Atmospheric Research and Exposure Laboratory, Office of Research and Development

US Environmental Protection Agency (1988b) Methods for organic chemical analysis of municipal and industrial wastewater. Method 601. Purgeable halocarbons. *US Code fed. Regul.*, *Title 40*, Part 136, Appendix A, pp. 267-281

US Environmental Protection Agency (1988c) Methods for organic chemical analysis of municipal and industrial wastewater. Method 624. Purgeables. *US Code fed. Regul.*, *Title 40*, Part 136, Appendix A, pp. 432-446

US Environmental Protection Agency (1988d) Method 502.2. Volatile organic compounds in water by purge and trap capillary column gas chromatography with photoionization and electrolytic conductivity detectors in series. Revision 2.0. In: *Methods for the Determination of Organic Compounds in Drinking Water* (EPA-600/4-88-039; PB89-220461), Cincinnati, OH, Environmental Monitoring Systems Laboratory, pp. 31-62

US Environmental Protection Agency (1988e) Method 524.2. Measurement of purgeable organic compounds in water by capillary column gas chromatography/mass spectrometry. Revision 3.0. In: *Methods for the Determination of Organic Compounds in Drinking Water* (EPA-600/4-88-039; PB98-220461), Cincinnati, OH, Environmental Monitoring Systems Laboratory, pp. 285-323

US Environmental Protection Agency (1988f) Methods for organic chemical analysis of municipal and industrial wastewater. Method 1624 Revision B. Volatile organic compounds by isotope dilution GC/MS. *US Code fed. Regul., Title 40*, Part 136, Appendix A, pp. 475-488

US Environmental Protection Agency (1988g) Analysis of trihalomethanes. Part II—Analysis of trihalomethanes in drinking water by liquid/liquid extraction. *US Code fed. Regul., Title 40*, Part 141, Appendix C, pp. 563-571

US Occupational Safety and Health Administration (1989) Air contaminants—permissible exposure limits (Report No. OSHA 3112). *US Code fed. Regul., Title 29*, Part 1910.1000

Varma, M.M., Ampy, F.R., Verma, K. & Talbot, W.W. (1988) In vitro mutagenicity of water contaminants in complex mixtures. *J. appl. Toxicol.*, 8, 243-248

Verschueren, K. (1983) *Handbook of Environmental Data on Organic Chemicals*, 2nd ed., New York, Van Nostrand Reinhold, pp. 291-292

Weast, R.C., ed. (1989) *CRC Handbook of Chemistry and Physics*, 70th ed., Boca Raton, FL, CRC Press, pp. C-350, D-198

Weil, L., Jandik, J. & Eichelsdoefer, D. (1980) Organic halogenated compounds in swimming pool water. I. Determination of volatile halogenated hydrocarbons (Ger.). *Z. Wasser Abwasser Forsch.*, 13, 141-145

Westrick, J.J., Mello, J.W. & Thomas, R.F. (1984) The groundwater supply survey. *J. Am. Water Works Assoc.*, 76, 52-59

Williams, D.T. (1985) Formation of trihalomethanes in drinking water. In: Fishbein, L. & O'Neill, I.K., eds, *Environmental Carcinogens. Selected Methods of Analysis*, Vol. 7, *Some Volatile Halogenated Hydrocarbons* (IARC Scientific Publications No. 68), Lyon, IARC, pp. 69-88

Woodruff, R.C., Mason, J.M., Valencia, R. & Zimmering, S. (1985) Chemical mutagenesis testing in *Drosophila*. V. Results of 53 coded compounds tested for the National Toxicology Program. *Environ. Mutagenesis*, 7, 677-702

CHLORODIBROMOMETHANE

1. Chemical and Physical Data

1.1 Synonyms

Chem. Abstr. Services Reg. No.: 124-48-1
Chem. Abstr. Name: Chlorodibromomethane
IUPAC Systematic Name: Chlorodibromomethane
Synonyms: Dibromomethanechloro; dibromomonochloromethane; monochlorodibromomethane

1.2 Structural and molecular formulae and molecular weight

$$\begin{array}{c} Br \\ | \\ H - C - Cl \\ | \\ Br \end{array}$$

$CHBr_2Cl$ Mol. wt: 208.29

1.3 Chemical and physical properties of the pure substance

(a) *Description*: Clear, colourless to pale-yellow liquid (Verschueren, 1983)
(b) *Boiling-point*: 119-120°C at 748 mm Hg (Weast, 1989)
(c) *Melting-point*: < -20°C (Verschueren, 1983)
(d) *Density*: 2.451 at 20/4°C (Weast, 1989)
(e) *Spectroscopy data*[1]: Infrared (Sadtler Research Laboratories, 1980, prism [1896], grating [284]; Pouchert, 1981, 1985a,b), nuclear magnetic resonance (Sadtler Research Laboratories, 1980, proton [6708], C-13 [2846]; Pouchert, 1983) and mass spectral data [1458] (Bunn *et al.*, 1975; Coleman *et al.*, 1984) have been reported.

---

[1]In square brackets, spectrum number in compilation

(f) *Solubility:* Soluble in acetone, ethanol, benzene and diethyl ether (Weast, 1989)
(g) *Volatility:* Vapour pressure, 76 mm Hg at 20°C (Mabey *et al.*, 1982)
(h) *Reactivity:* Hydrolysis constant at 25°C and neutral pH, K = 2.88 × 10^{-8} per hour (Mabey *et al.*, 1982)
(i) *Octanol/water partition coefficient (P)*: log P, 2.24 (Mabey *et al.*, 1982)
(j) *Conversion factor*[1]: mg/m^3 = 8.52 × ppm

1.4 Technical products and impurities

Chlorodibromomethane is available at > 95-98% purity (American Tokyo Kasei, 1988; Aldrich Chemical Co., 1990).

2. Production, Use, Occurrence and Analysis

2.1 Production and use

(a) Production

Chlorodibromomethane has been synthesized by the addition of dibromochloroacetone to 5N sodium hydroxide, followed by drying and distillation of the combined oil and ether extract (Barrett *et al.*, 1971). It is also formed in the reaction of a mixture of chloroform (see IARC, 1987) and bromoform (see monograph, p. 213) with triethylbenzylammonium chloride and sodium hydroxide (Fedorynski *et al.*, 1977).

(b) Use

Chlorodibromomethane has been used as a chemical intermediate in the manufacture of fire extinguishing agents, aerosol propellants, refrigerants and pesticides (Sittig, 1985). It is also used for the preparation of phenyl-dibromochloromethylmercury, which has been used for the generation of bromochlorocarbene (Fedoryński *et al.*, 1977).

(c) Regulatory status and guidelines

Standards for trihalomethanes (including chlorodibromomethane) have been established in drinking-water (see monograph on chlorination of drinking-water, p. 59) in several countries.

---

[1]Calculated from: mg/m^3 = (molecular weight/24.45) × ppm, assuming standard temperature (25°C) and pressure (760 mm Hg)

2.2 Occurrence

(a) Natural occurrence

Levels of chlorodibromomethane in tissues of temperate marine macroalgae (*Ascophyllum nodosum, Fucus vesiculosis, Enteromorpha linza, Ulva lacta, Gigartina stellata*) ranged from 150 to 590 ng/g dry weight. Chlorodibromomethane was released to seawater at rates of 150-4300 ng/g of dry algae per day (Gschwend *et al.*, 1985).

Macroalgae collected near the Bermuda Islands (*Fucales sargassum*) and at the Cape of Good Hope (*Laminariales laminaria*) showed a specific pattern of emission of volatile organohalides into the surrounding air. The main components were bromoform, bromodichloromethane and chlorodibromomethane; a minor component was bromoethane (Class *et al.*, 1986).

(b) Air

The volatilization half-time of chlorodibromomethane from rivers and streams has been estimated to range from 43 min to 17 days, depending on turbulence and temperature. A typical half-time, based on actual data, was 46 h (Kaczmar *et al.*, 1984).

Chlorodibromomethane was detected in seven of 22 air samples collected in 1978 from Bochum University and Bochum-Kemnade, Germany, at a concentration of 0.1 µg/m^3; and in five of 12 air samples from Bochum-City, at not detected to 0.9 µg/m^3 (Bauer, 1981).

Ullrich (1982) studied the organohalogen concentrations in the air of four public indoor swimming pools in western Berlin. Mean chlorodibromomethane concentrations ranged from 30 to 170 µg/m^3 in three of them.

In a review of data on the presence of volatile organic chemicals in the atmosphere of the USA in 1970-80, a median concentration of 120 ng/m^3 chlorodibromomethane was reported for the 21 urban/suburban data points examined and below the limit of detection for industrial areas (68 data points) (Brodzinsky & Singh, 1983).

In the USA, air samples collected 2 cm above the surface of five outdoor pools contained 0.4-2 µg/m^3 chlorodibromomethane; those above indoor pools contained 0.5-17 µg/m^3 and those above four spas (whirlpools or hot tubs) contained < 0.1-31 µg/m^3. Samples collected 2 m above the surface contained < 0.1 µg/m^3, < 0.1-5 µg/m^3 and < 0.1-5 µg/m^3, respectively (Armstrong & Golden, 1986).

Levels of chlorodibromomethane in air samples collected in 1982-85 over the Atlantic Ocean were 0.06-10 ppt (0.5-85.2 ng/m^3); baseline levels of biogenic chlorodibromomethane in air were 0.1-0.5 ppt (0.9-4.3 ng/m^3). Air samples collected in 1985 from a forest area in southern Germany contained 0.4 ppt (3.4 ng/m^3) chlorodibromomethane (Class *et al.*, 1986).

(c) Water and sediments

The formation of trihalomethanes in drinking-water (see also the monograph on chlorinated drinking-water, p. 56 *et seq.*) and the effects of temperature and pH have been discussed extensively (Williams, 1985). The formation of chlorodibromomethane during drinking-water chlorination depends on the presence of bromine in untreated water.

Chlorodibromomethane has been measured or detected in many drinking-water systems, both in samples collected at treatment facilities or along the distribution system (Table 1) and in samples collected from natural and untreated water sources (Table 2). Concentrations in treated drinking-water typically ranged from 1 to 20 µg/l (with higher or lower values in some locations), compared with concentrations in untreated (natural) waters which are typically less than 1 µg/l.

Table 1. Chlorodibromomethane concentrations in treated[a] drinking-water, 1973-89

Location, date[b]	Sample site/raw water source[c]	Concentration (µg/l)[d]	Reference
NS, 1973, 1974	NS/surface	<0.1-2.0	Bellar *et al.*
NS, 1974	NS/well	0.8-0.9	(1974a,b)
80 US cities (NORS), 1975	NS/ground and surface	ND-100	Symons *et al.* (1975)
113 Public water supplies, 1976	Water supplies/NS	Mean, 14; median, 3.5	Brass *et al.* (1977)
945 US sites, 1981-82	T/ground	2.1-4.6[e] (max., 63)	Westrick *et al.* (1984)
13 US community systems, 1984-85	T and D/NS	<0.2-27	Reding *et al.* (1989)
10 US utilities, 1985	T/ground and surface	>10-<100 (2 sites) <10 (8 sites)	Stevens *et al.* (1989)

35 US sites
  Spring 1988
  Summer 1988
  Autumn 1988
  Winter 1989
	T/ground and surface	Median
		2.6
		4.5
		3.8
		2.7
			Krasner *et al.* (1989)

Durham, North Carolina, 1975	D/surface	2.16-3.40	McKinney *et al.* (1976)
15 Kentucky cities, 1977	T/surface	ND-trace	Allgeier *et al.* (1980)

Table 1 (contd)

Location, date[b]	Sample site/raw water source[c]	Concentration (μg/l)[d]	Reference
20 Tennessee sites	T/ground and surface		Minear & Morrow (1983)
Autumn 1980		1–96 ng/l	
Winter 1980		7–291 ng/l	
Spring 1981		10–152 ng/l	
Summer 1981		19–373 ng/l	
Miami, Florida	NS/ground		Loy et al. (1976)
January 1975		35	
July 1975		37	
Southwestern US city, 1975	NS/ground	7.53–42.8	Henderson et al. (1976)
East Texas, 1977	NS/surface (14 sites)	trace–183.7	Glaze & Rawley (1979)
	NS/ground (11 sites)	ND–173.1	
Houston, Texas, summer 1978 to winter 1980	NS/surface	max., 46	Cech et al. (1982)
	NS/ground	max., 7	
40 Michigan utilities, NS	T/surface (22 sites)	ND–39.6	Furlong & D'Itri (1986)
	T/ground (18 sites)	ND–3.1	
Old Love Canal, NY, 1978	D/NS	ND–6.8	Barkley et al. (1980)
5 Pennsylvania sites, 1987	T surface	3.2–4.7	Smith (1989)
5 Virginia sites, 1987		0.3–1.49 (range of means)	
3 Puerto Rican cities, NS	D/ground and surface	6.1–11 ng/l	Rodriguez-Flores (1983)
70 Canadian cities, 1976–77	D and T/ground and surface	0–6.2 (median, 0.1)	Health & Wefare Canada (1977)
30 Canadian sites,	T/ground and surface		Otson et al. (1982)
Aug.–Sep. 1979		mean, < 1; max., 4	
Nov.–Dec. 1979		mean, < 1; max., 3	
Burlington, Ontario, Canada, 1981	D/NS	0.91	Comba & Kaiser (1983)
Niagara Falls, Ontario, 1981		1	
Port Robinson, Ontario, 1981		4	
Chippawa, Ontario, 1981	D/surface	2.1	Kaiser & Comba (1983)
10 Ontario cities, 1981	T/surface	0.1–2.9	Oliver (1983)
	D/surface	0.1–4.8	
10 Canadian Great Lakes sites, Ontario	T/ground and surface		Otson (1987)
Summer 1982		mean, 1.8	
Winter 1983		mean, 1.4	
Spring 1983		mean, 1.9	

Table 1 (contd)

Location, date[b]	Sample site/raw water source[c]	Concentration ($\mu g/l$)[d]	Reference
Lancashire-Cheshire, UK, 1974	D/NS	<0.01-3 $\mu g/kg$	McConnell (1976)
Southeastern UK, NS	D/ground and surface	2.3	Trussell et al. (1980)
Southampton, UK, 1977-78	D/surface	pumping station mean, 1.50 reservoir mean, 2.29	Brett & Calverley (1979)
5 Belgian utilities, 1977-78	T/surface D/surface	0.3-14.6 0.4-12.8	Quaghebeur & De Wulf (1980)
9 Belgian utilities, 1977-78	T/ground D/ground	0-10.4 0-5.5	
100 German cities, 1977	NS/NS	<0.1-17.1 (mean, 0.4)	Bauer (1981)
37 German sites, 1976	NS/NS	mean, 0.3	
Bremen and Leverkusen, Germany, NS	NS/NS	0.1-1.5	
12 German cities, 1978	NS/NS	ND-19.8 (mean, 2.2)	Eklund et al. (1978)
9 German cities, 1978-79	D/NS	1.2-9.6	Lahl et al. (1982)
Gothenberg, Sweden, 1977	D/surface	0.6	Eklund et al. (1978)
Southern China, NS	D/local catchments	13	Trussell et al. (1980)
Southern Philippines, NS	D/ground and surface	1.2	Trussell et al. (1980)
Northern Philippines, NS	D/surface	1.5	Trussell et al. (1980)
Northern Egypt, NS	D/surface	ND	Trussell et al. (1980)
Southern Indonesia, NS	D/surface	0.7	Trussell et al. (1980)
Southeastern Australia, NS	D/surface	ND	Trussell et al. (1980)
Southern Brazil, NS	D/surface	ND	Trussell et al. (1980)
Eastern Nicaragua, ND	D/surface	1.1	Trussell et al. (1980)
Northern Venezuela, NS	D/surface	ND	Trussell et al. (1980)
Eastern Peru, NS	D/ground and surface	2.7	Trussell et al. (1980)

[a]Treatment not always specified
[b]NORS, National Organics Reconnaissance Survey; NS, not specified
[c]D, distribution system; T, treatment plant
[d]Unless otherwise noted; ND, not detected
[e]Range of median values for randomly and nonrandomly selected water supplies serving fewer than and more than 10 000 people

Table 2. Chlorodibromomethane concentrations in untreated (natural) water, 1973-89

Location, date	Sample source	Concentration (ng/l)[a]	Reference
Northern Taiwan, NS	Well	ND	Trussell et al. (1980)
Rhine River, Germany, 1976	River	< 0.1–0.3	Bauer (1981)
New Jersey, USA, 1977–79	Ground Surface	2.4 µg/l 8.2 µg/l	Page (1981)
30 Canadian sites, 1979	River, lake and ground	mean, < 1 µg/l	Otson et al. (1982)
Lake Ontario, Canada, 1981	Lake	ND–630	Kaiser et al. (1983)
Niagara River, Ontario, Canada, 1981	River	trace–15 (mean, 4)	
Welland River watershed, Ontario, Canada	Surface (river)		Kaiser & Comba (1983)
Summer 1980		ND–15	
Spring 1981		ND–800	
10 Canadian Great Lakes sites	Ground and surface		Otson (1987)
Summer 1982		mean, < 100	
Winter 1983		0	
Spring 1983		mean, 100	
North and South Atlantic 1985	Seawater	0.1–2.2 (baseline, 0.12)	Class et al. (1986)
Ulm, Germany, NS	Rainwater	0.7	Class et al. (1986)

[a]ND, not detected

Rook (1974) demonstrated that chlorodibromomethane, observed at concentrations ranging from 1.7 to 13.3 µg/l following chlorination of stored surface waters, was a product of chlorination of the humic substances in natural waters.

Chlorodibromomethane was detected (but not quantified) by headspace analysis in two of ten Pacific seawater headspace samples collected in 1983 (Hoyt & Rasmussen, 1985).

Chlorodibromomethane was found in water samples collected at various stages of water treatment: not detected (< 0.1 µg/l) in raw river water; 0.7 µg/l in river water treated with chlorine and alum; 1.1 µg/l in three-day-old settled water; 2.4 µg/l in water flowing from settled areas to filters; 1.7 µg/l in the filter effluent; and 2.0 µg/l in finished water (Bellar et al., 1974a,b).

In the 1975 US National Organics Reconnaissance Survey of water supplies from 80 US cities, chlorodibromomethane was not found in any of the raw water samples (minimum quantifiable concentration, 0.4 µg/l) (Symons *et al.*, 1975).

According to the US Environmental Protection Agency STORET system, chlorodibromomethane concentrations in 130 samples of surface water in 1970-79 ranged from 0.1 to 1.0 µg/l in 80% of the samples and 1 to 10 µg/l in 20% of the samples (Perwak *et al.*, 1980).

The US Environmental Protection Agency estimated that 204 tonnes of chlorodibromomethane were generated in the USA in 1978 by water chlorination. On the basis of the 1976 National Organic Monitoring Survey, the general population was estimated to be exposed to 7 µg chlorodibromomethane per day from drinking-water, assuming a median concentration of 4 µg/l and a water intake of 1.65 l per day; assuming a maximal concentration of 290 µg/l and an intake of 2.18 l per day, the daily exposure increased to 600 µg per day (Perwak *et al.*, 1980). In a later investigation by the US Environmental Protection Agency STORET data base, analysis of 18 616 water samples revealed a mean chlorodibromomethane concentration of 10.09 µg/l (range, 0-10133 µg/l); analysis of 590 sediment samples revealed a mean of 11.3 µg/kg (range, 0-237 µg/kg) (US Environmental Protection Agency, 1985).

In the 1982 US Nationwide Urban Runoff Program, chlorodibromomethane was detected in samples from one of the 15 reporting cities at a concentration of 2 µg/l (Cole *et al.*, 1984).

Tap-water samples collected between January 1977 and March 1978 in Osaka, Japan, contained chlorodibromomethane at levels of 1.6, 2.6 and 4.3 µg/l at seasonal mean water temperatures of 7.4°C, 15.8°C, and 25.4°C, respectively. An increase of approximately 0.5 in pH to control pipe corrosion resulted in chlorodibromomethane concentrations of 2.4, 3.4 and 4.9 µg/l at similar mean water temperatures (Kajino & Yagi, 1980).

Chlorodibromomethane levels in tap water collected at four locations in a Swedish community ranged from 0.87 to 1.3 µg/l; when the treatment facility briefly changed the disinfectant from chlorine to chlorine dioxide, chlorodibromomethane was not detected (detection limit, 0.006 µg/l) (Norin *et al.*, 1981).

In a survey of large American water utilities, chlorodibromomethane was identified as one of 36 unregulated chemicals that was detected at greatest frequency during routine monitoring of drinking-water (Anon., 1983).

Samples of water were collected between August and October 1980 from four supply systems in São Paulo State, Brazil. Mean levels of chlorodibromomethane were 1.3-3.4 µg/l in treated water (after treatment), 1.8-6.4 µg/l in treated water from the reservoir and 2.2-80 µg/l in treated tap water (de Fernicola & de Azevedo, 1984).

Water samples were collected from five outdoor pools, four indoor pools, and four spas (whirlpools or hot tubs) in Lubbock, TX, USA. The concentration of chlorodibromomethane in the outdoor pools, which used chlorine-based materials for chlorination, were < 0.1-8 µg/l. Two of the indoor pools in which only chlorination was used had chlorodibromomethane levels of 0.3-11 µg/l; one indoor pool in which only bromination (sodium hypobromite) was used had levels of 8-30 µg/l; and the fourth indoor pool, in which chlorination and bromination were alternated, had levels of 9-26 µg/l. The spa in which only chlorination was used had levels of < 0.1-5 µg/l; the two spas in which only bromination was used had levels of 0.6-48 µg/l; and the spa in which the combination was used had levels of 11-26 µg/l. The average concentration of chlorodibromomethane in Lubbock, TX, tap water was 0.43 µg/l (Armstrong & Golden, 1986).

Water samples were collected in 1978-79 from eight covered swimming pools in Bremen, Germany, to determine the concentration of chlorodibromomethane. The source of fresh water was mixed river and groundwater for four pools and groundwater for four pools. The level of chlorodibromomethane in the pools with mixed sources was 7.0 and for the pools with groundwater, 0.4 µg/l. The range of means of chlorodibromomethane in the four pools with a mixed water source was 6-28 µg/l (total range, 0.1-140 µg/l); that in the four pools with a groundwater source was 4-10 µg/l (total range, 0.2-140 µg/l) (Lahl *et al.*, 1981).

The concentrations of chlorodibromomethane in three thermal spas in western Germany in which the initial bromide concentration was 0.5-0.7 mg/l were 13.6-32.2 µg/l (Weil *et al.*, 1980).

Scotte (1984) studied the concentrations of organohalogen compounds in the water of 10 covered swimming pools in France. The mean concentrations of chlorodibromomethane were 2.4 µg/l in the four pools treated with Surchlor GR 60 (anhydrous sodium dichloroisocyanurate), 2.5 µg/l in the two treated with gaseous chlorine and 4.5 µg/l in the two treated with sodium hypochlorite and the two treated with bromine.

Kaminski and von Loew (1984) found an average concentration of 2.3 µg/l (max, 6.7 µg/l) chlorodibromomethane in 26 indoor pools in western Germany.

Effluents from a wastewater treatment plant on Boston Harbor, MA, USA, sampled in 1984 and 1985, contained a mean chlorodibromomethane level of 3.34 µg/l (range, 0.20-12.16) and had an estimated mass input rate of 2.98 kg per day (Kossik *et al.*, 1986).

Heating water to prepare food has been shown to eliminate a large part of trihalomethanes in the water, as a function of temperature and heating time. Chlorodibromomethane levels were reduced from 42.3 µg/l in tap water to 24.1 µg/l after heating at 80°C for 1 min, to 14.4 µg/l after heating to 100°C, to 12.3 µg/l after boiling for 1 min and to 5.5 µg/l after boiling for 5 min (Lahl *et al.*, 1982).

(d) Food and beverages

The mean concentration of chlorodibromomethane in 12 milk products (ice cream, yoghurt, curds, buttermilk) was 0.1 µg/kg fresh weight. The maximal level found was 0.3 µg/kg (Bauer, 1981).

(e) Tissues and secretions

The mean concentrations (micrograms per kilogram fresh weight) of chlorodibromomethane in human tissues samples collected in 1978 from the Ruhr District of Germany were: kidney capsule fat, 5.1 (max, 40.1); hypodermis fat, 0.9; lung, 0.7; liver, 0.4; and muscle, 1.3 (Bauer, 1981).

In a gas chromatography-mass spectrometry screening study of human milk samples collected in four urban areas of the USA, traces of chlorodibromomethane were detected in one out of eight samples (Pellizzari *et al.*, 1982).

Chlorodibromomethane was detected at concentrations of 0.6 ± 0.02 (SD) in human kidney and 0.06 ± 0.01 µg/kg in human lung tissues collected at surgery or during pathological examination (Kroneld, 1989).

(f) Other

Chlorodibromomethane was detected in cosmetic products (face wash and shaving lotion) bought in shops in Bochum, Germany, at a concentration of 0.2 µg/l; it was detected in one of seven cough mixtures at a concentration of 0.9 µg/kg (Bauer, 1981).

2.3 Analysis

Selected methods for the analysis of chlorodibromomethane in air, water and other media are given in Table 3. A variety of analytical methods exist for measuring chlorodibromomethane in water. The commonly used methods are based on extraction with solvent, followed by gas chromatography-electron capture detection (Standing Committee of Analysts, 1980; US Environmental Protection Agency, 1988f), purge-and-trap and flame ionization detection and microcoulometric gas chromatography (Bellar *et al.*, 1974a,b; US Environmental Protection Agency, 1988g) and gas chromatography-mass spectrometry and headspace analysis (Otson *et al.*, 1982).

Table 3. Methods for the analysis of chlorodibromomethane

Sample matrix	Sample preparation[a]	Assay procedure[b]	Limit of detection[c]	Reference
Air	Collect cryogenically into stainless-steel bottle; inject sample	GC/EC-FI-FPD/GC/MS	NR	Hoyt & Rasmussen (1985)
Seawater	Collect in vacuum extraction flask; pressurize with zero air; inject headspace sample	GC/EC-FI-FPD/GC/MS	NR	Hoyt & Rasmussen (1985)
Water	Purge (inert gas); trap (OV-1 on Chromosorb-W/Tenax/silica gel); desorb as vapour (heat to 180°C, backflush with inert gas) onto packed GC column	GC/ECD	0.09 µg/l	US Environmental Protection Agency (1988a) [Method 601]
		GC/MS	3.1 µg/l	US Environmental Protection Agency (1988b) [Method 624]
	Purge (inert gas); trap (OV-1 on Chromosorb-W/Tenax/silica gel); desorb as vapour (heat to 180°C, backflush with inert gas) onto capillary GC column	GC/ECD	0.3 µg/l	US Environmental Protection Agency (1988c) [Method 502.2]
		GC/MS	0.05 µg/l	US Environmental Protection Agency (1988d) [Method 524.2]
	Add internal standard (isotope-labelled chlorodibromomethane); purge; trap and desorb as above	GC/MS	10 µg/l	US Environmental Protection Agency (1988e) [Method 1624]
	Extract in pentane; inject onto GC	GC/EC (>50 µg/l) GC/MS (<50 µg/l)	0.5 µg/l	US Environmental Protection Agency (1988f) [Method 501.2]
Adipose tissue	Purge from liquefied fat at 115°C; trap on silica gel; desorb thermally	GC/HSD	2.1 µg/l	Peoples et al. (1979)
Blood serum	Purge from water-serum mixture containing anti-foam reagent at 115°C; trap on Tenax/silica gel; thermally desorb	GC/ECD	2.1 µg/l	Peoples et al. (1979)

[a]GC, gas chromatograph

[b]GC/EC-FI-FPD, gas chromatography/electron capture-flame ionization-flame photometric detection; GC/ECD, gas chromatography/electrolytic conductivity detection; GC/MS, gas chromatography/mass spectrometry; GC/HSD, gas chromatography/halide selective detection

[c]NR, not reported

The US Environmental Protection Agency methods for analysing water (Methods 8010 and 8240) have also been used for liquid and solid wastes. Volatile components of solid waste samples are first extracted with methanol prior to purge-and-trap concentration and analysis by gas chromatography-electrolytic conductivity detection (Method 8010) or gas chromatography-mass spectrometry (Method 8240). The method detection limit using Method 8010 is 0.09 µg/l, and the practical quantification limit using Method 8240 is 5 µg/l for groundwater and for soil/sediment samples (US Environmental Protection Agency, 1986a,b).

US Environmental Protection Agency Method 624 has also been adapted to the analysis of chlorodibromomethane in fish, with an estimated detection limit of 10 µg/kg (Easley *et al.*, 1981).

3. Biological Data Relevant to the Evaluation of Carcinogenic Risk to Humans

3.1 Carcinogenicity studies in animals (Table 4)

Oral administration

Mouse: Groups of 50 male and 50 female B6C3F$_1$ mice, eight weeks old, were given chlorodibromomethane (> 98% pure) in corn oil by gavage at 50 or 100 mg/kg bw on five days per week for 105 weeks. Survival at 107 weeks was: males—vehicle control, 44/50; low-dose, 7/50; high-dose, 29/50; females—control, 32/50; low-dose, 27/50; high-dose, 36/50. The poor survival of low-dose males was caused by an accidental overdose which killed 35 rats in weeks 58-59. Hepatocellular adenomas occurred in 14/50 control and 10/50 high-dose male mice, whereas hepatocellular carcinomas occurred in 10/50 vehicle control and 19/50 high-dose males. Although the proportion of high-dose male mice with carcinomas was significantly increased relative to controls ($p = 0.03$, incidental tumour test), the overall proportion with hepatocellular adenomas or carcinomas combined was not. In female mice, hepatocellular adenomas occurred in 2/50 control, 4/49 low-dose and 11/50 high-dose animals; hepatocellular carcinomas occurred in 4/50 control, 6/49 low-dose and 8/50 high-dose females. The proportion of high-dose female mice with hepatocellular adenoma or carcinoma was significantly greater than in controls ($p = 0.004$, pairwise comparison; $p = 0.003$, trend test, incidental tumour test). The proportion of high-dose male mice with lymphomas was significantly smaller than in controls (vehicle control, 9/50; high dose, 0/50; $p = 0.006$, incidental tumour test) (National Toxicology Program, 1985).

Groups of male and female CBA × C57Bl/6 hybrid mice [age unspecified] were given chlorodibromomethane [purity unspecified] at 0.04 mg/l (50 males, 50 females), 4 mg/l (50 males, 50 females) or 400 mg/l (55 males, 55 females) in the drinking-water for 104 weeks. Seventy-five males and 50 females served as controls. The numbers of animals surviving to the appearance of the first tumour were: males—control, 63; low-dose, 13; medium-dose, 33; and high-dose, 39; females—control, 34; low-dose, 27; medium-dose, 40; and high-dose, 32 [average survival time and number of terminal survivors unspecified]. No tumour occurred at increased incidence in treated mice (Voronin et al., 1987). [The Working Group noted the incomplete reporting of the study.]

Rat: Groups of 50 male and 50 female Fischer 344 rats, eight weeks old, were given chlorodibromomethane (> 98% pure) in corn oil by gavage at 40 or 80 mg/kg bw on five days per week for 104 weeks. Survival at 106 weeks was: males—vehicle control, 34/50; low-dose, 38/50; high-dose, 43/50; females—control, 39/50; low-dose, 37/50; high-dose, 41/50. No tumour occurred in a significantly greater proportion of rats receiving chlorodibromomethane than in controls. The incidences of stromal polyps of the uterus and fibroadenomas of the mammary gland were significantly lower in high-dose female rats than in controls (stromal polyps—control, 14/50; low-dose, 8/50; high-dose, 5/50; $p = 0.021$, incidental tumour test; fibroadenomas—control, 18/50; low-dose, 12/50; high-dose, 4/50; $p < 0.001$, incidental tumour test) (National Toxicology Program, 1985).

3.2 Other relevant data

(a) *Experimental systems*

(i) *Absorption, distribution, excretion and metabolism*

The blood/gas partition coefficient (at 37°C) of chlorodibromomethane in rats was 116 (Gargas et al., 1989), which predicts that it will be readily absorbed by inhalation.

^{14}C-Chlorodibromomethane at 0.48 mmol/kg (16 µCi/kg; 100 mg/kg bw) administered orally in corn oil to rats by gavage was absorbed and eliminated in the expired air as bromodichloromethane (48% of dose) or as ^{14}C-carbon dioxide (18% of dose) in 8 h; radiolabel amounting to about 1% of the dose was eliminated in the urine, and about 1% of the dose was retained in body tissues. ^{14}C-Chlorodibromomethane (0.72 mmol/kg; 32 µCi/kg; 150 mg/kg bw) administered similarly to mice was absorbed and eliminated in the expired air as unchanged chlorodibromomethane (12% of dose) or as ^{14}C-carbon dioxide (72% of dose) in 8 h; about 2% of the administered radiolabel was eliminated in the urine, and 5% was retained in body tissues (Mink et al., 1986). Chlorodibromomethane is metabolized to carbon monoxide *in vivo* (Anders et al., 1978) and *in vitro* (Ahmed et al., 1977).

Table 4. Summary of carcinogenicity studies of chlorodibromomethane in experimental animals

Reference	Species/strain	Sex	Dose schedule	Experimental parameter/observation	Group 0	1	2	3	Significance	Comments
National Toxicology Program (1985)	Mouse B6C3F$_1$	M	5 d/week, gavage, corn oil, 105 weeks	Dose (mg/kg)	0	50	100	—		Accidental overdose
				Survival (107 weeks)	44/50	7/50	29/50			
				Hepatocellular						
				Adenoma	14/50	5/50	10/50		$p = 0.03$	Increase (but combined tumours NS)
				Carcinoma	10/50	9/50	19/50		$p = 0.06$	
				Lymphoma	9/50	4/50	0/50			Decrease
		F		Dose (mg/kg)	0	50	100	—		
				Survival (107 weeks)	32/50	27/50	36/50			
				Hepatocellular						
				Adenoma	2/50	4/49	11/50		} $p = 0.003$	Increase
				Carcinoma	4/50	6/49	8/50			
Voronin et al. (1987)	Mouse CBA × C57Bl/6	M/F	Ad lib. in drinking-water	Dose (mg/l)	0	0.04	4.0	400		
				Survival M	63/75	13/50	33/50	39/55		
				F	34/50	27/50	40/50	32/55		
				Tumours	—	—	—	—		
National Toxicology Program (1985)	Rat F344	M	5 d/week, gavage, corn oil, 104 weeks	Dose (mg/kg)	0	40	80	—		
				Survival (106 weeks)	34/50	38/50	43/50			
		F		Dose (mg/kg)	0	40	80	—		
				Survival (106 weeks)	39/50	37/50	41/50			
				Uterine stromal polyp	14/50	8/50	5/50		$p = 0.021$	Decrease
				Mammary fibroadenoma	18/50	12/50	4/50		$p < 0.001$	Decrease

(ii) *Toxic effects*

The single-dose oral LD_{50} of chlorodibromomethane (Emulphor:ethanol:saline, 1:1:8) was 800 mg/kg bw in male and 1200 mg/kg bw in female ICR Swiss mice (Bowman *et al.*, 1978). Oral LD_{50} values (corn oil) of 1186 and 848 mg/kg bw were reported in male and female Sprague-Dawley rats, respectively, in one study (Chu *et al.*, 1980, 1982a), and of 1.08 ml/kg bw in males in another (Hewitt *et al.*, 1983). Signs of acute toxicity included sedation, prostration, anaesthesia, ataxia, muscular weakness, increased blood cholesterol and reduction in the number of peripheral lymphocytes.

Daily oral treatment by gavage of male CD-1 mice with 250 mg/kg bw chlorodibromomethane (Emulphor:ethanol:saline, 1:1:8) for 14 days resulted in body weight reduction, liver damage, depression of humoral and cellular immunity and reduced spleen and thymus weights (Munson *et al.*, 1982). Oral treatment by gavage of male CD-1 mice with 147 mg/kg bw chlorodibromomethane in corn oil for 14 days led to morphological and functional kidney lesions (Condie *et al.*, 1983).

Daily oral administration by gavage of 500 and 1000 mg/kg bw chlorodibromomethane in corn oil was lethal after a few days to 3/5 male and to all female Fischer 344/N rats, whereas a few male and female $B6C3F_1$ mice survived treatment with 500 mg/kg bw for 14 days. Mottled livers and renal alterations were observed at necropsy (National Toxicology Program, 1985). In a 13-week study, male and female Fischer 344/N rats and $B6C3F_1$ mice were administered 15-250 mg/kg bw chlorodibromomethane in corn oil by gavage on five days per week. In the highest dose groups, treatment was lethal to 9/10 male and 9/10 female rats and resulted in fatty changes and centrilobular necrosis in the liver and proximal tubular-cell degeneration and regeneration in the kidney. Lesions of the salivary gland were inflammation and squamous metaplasia. $B6C3F_1$ mice were less sensitive. Fatty metamorphosis of the liver and tubular degeneration of the kidney were observed in 5/10 of the males but not the females in the high-dose groups (Dunnick *et al.*, 1985; National Toxicology Program, 1985).

Male and female Sprague-Dawley rats received 5-2500 mg/l chlorodibromomethane in drinking-water for 90 days. The highest dose, which corresponded to an approximate daily intake of 55 mg/rat, resulted in some deaths, moderate liver damage, reduction in the number of blood lymphocytes and mild changes in the thyroid (Chu *et al.*, 1982b).

Administration of chlorodibromomethane in corn oil for two years (see section 3.1) resulted in fatty changes of the liver in rats and mice of each sex in all treatment groups; cytomegaly and necrosis were observed only in livers of high-dose male mice. Nephrosis was observed in female rats and male mice at low and high doses. Furthermore, follicular-cell hyperplasia of the thyroid gland was observed in female but not male mice in all treatment groups (Dunnick *et al.*, 1985; National Toxicology Program, 1985).

(iii) *Effects on reproduction and prenatal toxicity*

As reported in an abstract, multigeneration reproduction and dominant lethal studies were conducted in ICR Swiss mice administered chlorodibromomethane in drinking-water at 0, 0.1, 1.0 or 4.0 g/l (Borzelleca *et al.*, 1980). There appeared to be dose-dependent effects on survival and weight gain of pups and on viability and lactation indices in the multigeneration study. [The Working Group noted that the data presented were insufficient to allow assessment of the design of this study.]

As reported in an abstract, no apparent teratogenic effect was observed in ICR Swiss mice given chlorodibromomethane at 0, 0.1, 1.0 or 4.0 g/l in drinking-water (Borzelleca *et al.*, 1980). [The Working Group noted that the data presented were insufficient to allow assessment of the design of this study.]

Sprague-Dawley rats were given chlorodibromomethane daily in corn oil by gavage at 0, 50, 100 or 200 mg/kg bw on gestation days 6-15. Group sizes were 10-12 pregnant dams at the time of caesarean section. Maternal weight gain was significantly decreased from control values at the highest dose level. There was no increase in the incidence of resorptions or change in litter size or fetal body weight or in the incidence of external, visceral or skeletal malformations (Ruddick *et al.*, 1983). [The Working Group noted that the selection of dose levels appears to be appropriate but that the small number of animals per group gives this study a low power for detecting changes in the proportion of malformed fetuses.]

(iv) *Genetic and related effects* (Table 5)

Chlorodibromomethane was mutagenic to *Salmonella typhimurium* in some reported studies; experiments performed in closed containers generally gave positive results. In *Saccharomyces cerevisiae*, there was an increase in the frequency of gene conversion, but no mutation was observed. Chromosomal aberrations were induced in Chinese hamster CHL cells and sister chromatid exchange in human lymphocytes *in vitro*. Sister chromatid exchange but not micronuclei was induced in mouse bone marrow *in vivo*. As reported in an abstract, dominant lethal effects were not induced in male mice (Borzelleca *et al.*, 1980).

(b) *Humans*

No data were available to the Working Group.

3.3 Case reports and epidemiological studies of carcinogenicity to humans

A single correlation study (Isacson *et al.*, 1983), described in the monograph on chlorinated drinking-water (p. 113), mentioned chlorodibromomethane, but the information could not be used to evaluate the carcinogenicity of this compound individually.

Table 5. Genetic and related effects of chlorodibromomethane

Test system	Result		Dose LED/HID	Reference
	Without exogenous metabolic system	With exogenous metabolic system		
SA0, *Salmonella typhimurium* TA100, reverse mutation	+	0	0.0000	Simmon et al. (1977)[a]
SA0, *Salmonella typhimurium* TA100, reverse mutation	–	–	500.0000	National Toxicology Program (1985)
SA0, *Salmonella typhimurium* TA100, reverse mutation	–	–	0.0000	Khudoley et al. (1987)
SA0, *Salmonella typhimurium* TA100, reverse mutation	–	+	8.0000	Varma et al. (1988)
SA0, *Salmonella typhimurium* TA100, reverse mutation	0	+	0.0000	Khudoley et al. (1989)[a]
SA0, *Salmonella typhimurium* TA100, reverse mutation	–	–	0.0000	Mersch-Sunderman (1989)[b]
SA2, *Salmonella typhimurium* TA102, reverse mutation	–	–	0.0000	Mersch-Sunderman (1989)[c]
SA5, *Salmonella typhimurium* TA1535, reverse mutation	–	–	5000.0000	National Toxicology Program (1985)
SA5, *Salmonella typhimurium* TA1535, reverse mutation	+	+	8000.0000	Varma et al. (1988)
SA7, *Salmonella typhimurium* TA1537, reverse mutation	–	–	1667.0000	National Toxicology Program (1985)
SA7, *Salmonella typhimurium* TA1537, reverse mutation	+	+	8.0000	Varma et al. (1988)
SA9, *Salmonella typhimurium* TA98, reverse mutation	–	–	1667.0000	National Toxicology Program (1985)
SA9, *Salmonella typhimurium* TA98, reverse mutation	–	–	0.0000	Khudoley et al. (1987)
SA9, *Salmonella typhimurium* TA98, reverse mutation	–	+	8.0000	Varma et al. (1988)
SA9, *Salmonella typhimurium* TA98, reverse mutation	0	+	0.0000	Khudoley et al. (1989)[a]
SA9, *Salmonella typhimurium* TA98, reverse mutation	–	–	0.0000	Mersch-Sunderman (1989)[b]
SAS, *Salmonella typhimurium* TA97, reverse mutation	–	–	0.0000	Mersch-Sunderman (1989)[b]
SCG, *Saccharomyces cerevisiae* D7, gene conversion	+	(+)	245.0000	Nestmann & Lee (1985)
SCR, *Saccharomyces cerevisiae* XVI85–14C, reverse mutation	–	–	1225.0000	Nestmann & Lee (1985)
CIC, Chromosomal aberrations in Chinese hamster CHL cells	–	+	100.0000	Ishidate (1987)
SHL, Sister chromatid exchange, human lymphocytes *in vitro*	+	0	80.0000	Morimoto & Koizumi (1983)
SVA, Sister chromatid exchange, ICR/SJ mouse bone-marrow cells	+	0	25.0000	Morimoto & Koizumi (1983)
MVM, Micronucleus test, ddY mice *in vivo*	–		1000.0000	Hayashi et al. (1988)

[a]Closed container
[b]Standard assay, closed container or spot test
[c]Standard assay or spot test

4. Summary of Data Reported and Evaluation

4.1 Exposure data

Chlorodibromomethane has limited commercial use but is used industrially as a chemical intermediate. It is found in chlorinated drinking-water as a consequence of the reaction between chlorine, added during drinking-water treatment, and natural organic substances in the presence of bromide. The major route of human exposure is *via* drinking-water. Chlorodibromomethane has been detected in chlorinated drinking-water in many parts of the world; it is not normally present in untreated water. It is a major component of organohalide emissions from marine algae.

4.2 Experimental carcinogenicity data

Chlorodibromomethane was tested for carcinogenicity in two-year studies by oral gavage in male and female $B6C3F_1$ mice and Fischer 344 rats and in a lifetime study in CBA × C57Bl/6 hybrid mice by administration in drinking-water. In $B6C3F_1$ mice, it produced a significant increase in the incidence of hepatocellular neoplasms in females and a marginal increase in males. Chlorodibromomethane did not increase the proportion of rats with tumours at any site relative to that in controls. There was no increase in tumour incidence in CBA × C57Bl/6 hybrid mice given chlorodibromomethane in drinking-water.

4.3 Human carcinogenicity data

No relevant data were available to the Working Group.

4.4 Other relevant data

Chlorodibromomethane was mutagenic to bacteria. In single studies, it induced mitotic recombination, but not mutation, in yeast, chromosomal aberrations in cultured mammalian cells and sister chromatid exchange in cultured human cells. Sister chromatid exchange but not micronuclei were observed in rodents treated *in vivo*.

4.5 Evaluation[1]

There is *inadequate evidence* for the carcinogenicity of chlorodibromomethane in humans.

---

[1]For definition of the italicized terms, see Preamble, pp. 30-33.

There is *limited evidence* for the carcinogenicity of chlorodibromomethane in experimental animals.

Overall evaluation

Chlorodibromomethane *is not classifiable as to its carcinogenicity to humans (Group 3).*

5. References

Ahmed, A.E., Kubic, V.L. & Anders, M.W. (1977) Metabolism of haloforms to carbon monoxide. I. In vitro studies. *Drug Metab. Disposition*, 5, 198-204

Aldrich Chemical Co. (1990) *1990-1991 Aldrich Handbook of Fine Chemicals*, Milwaukee, WI, p. 286

Allgeier, G.D., Mullins, R.L., Jr, Wilding, D.A., Zogorski, J.S. & Hubbs, S.A. (1980) Trihalomethane levels at selected water utilities in Kentucky, USA. In: Afghan, B.K. & Mackay, D., eds, *Hydrocarbons and Halogenated Hydrocarbons in the Aquatic Environment*, New York, Plenum Press, pp. 473-490

American Tokyo Kasei (1988) *TCI American Organic Chemicals 88/89 Catalog*, Portland, OR, p. 275

Anders, M.W., Stevens, J.L., Sprague, R.W., Shaath, Z. & Ahmed, A.E. (1978) Metabolism of haloforms to carbon monoxide. II. In vivo studies. *Drug Metab. Disposition*, 6, 556-560

Anon. (1983) Water substances spur pleas for regulation. *J. Commerce*, 30 September, 22b1

Armstrong, D.W. & Golden, T. (1986) Determination of distribution and concentration of trihalomethanes in aquatic recreational and therapeutic facilities by electron-capture GC. *LC-GC*, 4, 652-655

Barkley, J., Bunch, J., Bursey, J.T., Castillo, N., Cooper, S.D., Davis, J.M., Erickson, M.D., Harris, B.S.H., III, Kirkpatrick, M., Michael, L.C., Parks, S.P., Pellizzari, E.D., Ray, M., Smith, D., Tomer, K.B., Wagner, R. & Zweidinger, R.A. (1980) Gas chromatography mass spectrometry computer analysis of volatile halogenated hydrocarbons in man and his environment—a multimedia environmental study. *Biomed. Mass Spectrometr.*, 7, 139-147

Barrett, G.C., Hall, D.M., Hargreaves, M.K. & Modarai, B. (1971) Preparation of some halogeno-acetones including 1-bromo-1-chloro-1-fluoroacetone. *J. chem. Soc. (C)*, 279-282

Bauer, U. (1981) Human exposure to environmental chemicals—investigations on volatile organic halogenated compounds in water, air, food, and human tissues. III. Communication: results of investigations. *Zentralbl. Bakteriol. Mikrobiol. Hyg. Abt. 1 Orig. B*, 174, 200-237

Bellar, T.A., Lichtenberg, J.J. & Kroner, R.C. (1974a) *The Occurrence of Organohalides in Chlorinated Drinking Waters* (EPA-670/4-74-008; NTIS PB-238 589), Cincinnati, OH, US Environmental Protection Agency

Bellar, T.A., Lichtenberg, J.J. & Kroner, R.C. (1974b) The occurrence of organohalides in chlorinated drinking waters. *J. Am. Water Works Assoc.*, *66*, 703-706

Borzelleca, J.F., Skalsky, H.L. & Riddle, B.L. (1980) Effects of dibromochloromethane and trichloromethane in drinking water on reproduction and development in mice (Abstract). *Fed. Proc.*, *39*, 999

Bowman, F.J., Borzelleca, J.F. & Munson, A.E. (1978) The toxicity of some halomethanes in mice. *Toxicol. appl. Pharmacol.*, *44*, 213-215

Brass, A.J., Feige, M.A., Halloran, T., Mello, J.W., Munch, D. & Thomas, R.F. (1977) The National Organic Monitoring Survey: samplings and analyses for purgeable organic compounds. In: Pojasek, R.B., ed., *Drinking Water Quality Enhancement Through Source Protection*, Ann Arbor, MI, Ann Arbor Science, pp. 393-416

Brett, R.W. & Calverley, R.A. (1979) A one-year survey of trihalomethane concentration changes within a distribution system. *J. Am. Water Works Assoc.*, *71*, 515-520

Brodzinsky, R. & Singh, H.B. (1983) *Volatile Organic Chemicals in the Atmosphere: An Assessment of Available Data* (EPA-600/3-83-027a; NTIS PB83-195503), Research Triangle Park, NC, US Environmental Protection Agency

Bunn, W.W., Haas, B.B., Deane, E.R. & Kleopfer, R.D. (1975) Formation of trihalomethanes by chlorination of surface water. *Environ. Lett.*, *10*, 205-213

Cech, I., Smith, V. & Henry, J. (1982) Spatial and seasonal variations in concentration of trihalomethanes in drinking water. *Pergamon Ser. environ. Sci.*, *7*, 19-38

Chu, I., Secours, V., Marino, I. & Villeneuve, D.C. (1980) The acute toxicity of four trihalomethanes in male and female rats. *Toxicol. appl. Pharmacol.*, *52*, 351-353

Chu, I., Villeneuve, D.C., Secours, V.E. & Becking, G.C. (1982a) Toxicity of trihalomethanes: I. The acute and subacute toxicity of chloroform, bromodichloromethane, chlorodibromomethane and bromoform in rats. *J. environ. Sci. Health*, *B17*, 205-224

Chu, I., Villeneuve, D.C., Secours, V.E. & Becking, G.C. (1982b) Trihalomethanes: II. Reversibility of toxicological changes produced by chloroform, bromodichloromethane, chlorodibromomethane and bromoform in rats. *J. environ. Sci. Health*, *B17*, 225-240

Class, T., Kohnle, R. & Ballschmiter, K. (1986) Chemistry of organic traces in air. VII. Bromo- and bromochloromethanes in air over the Atlantic Ocean. *Chemosphere*, *15*, 429-436

Cole, R.H., Frederick, R.E., Healy, R.P. & Rolan, R.G. (1984) Preliminary findings of the Priority Pollutant Monitoring Project of the Nationwide Urban Runoff Program. *J. Water Pollut. Control Fed.*, *56*, 898-908

Coleman, W.E., Munch, J.W., Kaylor, W.H., Streicher, R.P., Ringhand, H.P. & Meier, J.R. (1984) Gas chromatography/mass spectroscopy analysis of mutagenic extracts of aqueous chlorinated humic acid. A comparison of the byproducts to drinking water contaminants. *Environ. Sci. Technol.*, *18*, 674-681

Comba, M.E. & Kaiser, K.L.E. (1983) Determination of volatile contaminants at the ng/l level in water by capillary gas chromatography with electron capture detection. *Int. J. environ. anal. Chem.*, *16*, 17-31

Condie, L.W., Smallwood, C.L. & Laurie, R.D. (1983) Comparative renal and hepatotoxicity of halomethanes: bromodichloromethane, bromoform, chloroform, dibromochloromethane and methylene chloride. *Drug chem. Toxicol.*, *6*, 563-578

Dunnick, J.K., Haseman, J.K., Lilja, H.S. & Wyand, S. (1985) Toxicity and carcinogenicity of chlorodibromomethane in Fischer 344/N rats and B6C3F$_1$ mice. *Fundam. appl. Toxicol.*, *5*, 1128-1136

Easley, D.M., Kleopfer, R.D. & Carasea, A.M. (1981) Gas chromatographic-mass spectrometric determination of volatile organic compounds in fish. *J. Assoc. off. anal. Chem.*, *64*, 653-656

Eklund, G., Josefsson, B. & Roos, C. (1978) Trace analysis of volatile organic substances in Goeteborg municipal drinking water. *Vatten*, *3*, 195-206

Fedoryński, M., Popławska, M., Nitschke, K., Kowalski, W. & Makosza, M. (1977) A simple method for preparation of dibromochloromethane and bromodichloromethane. *Synth. Commun.*, *7*, 287-292

de Fernicola, N.A.G.G. & de Azevedo, F.A. (1984) Water levels of trihalomethanes (THM) in four different supply systems from Sao Paulo State (Brazil). *J. environ. Health*, *46*, 187-188

Furlong, E.A.-N. & D'Itri, F.M. (1986) Trihalomethane levels in chlorinated Michigan drinking water. *Ecol. Modelling*, *32*, 215-225

Gargas, M.L., Burgess, R.J., Voisard, D.E., Cason, G.H. & Andersen, M.E. (1989) Partition coefficients of low-molecular weight volatile chemicals in various liquids and tissues. *Toxicol. appl. Pharmacol.*, *98*, 87-99

Glaze, W.H. & Rawley, R. (1979) A preliminary survey of trihalomethane levels in selected East Texas water supplies. *J. Am. Water Works Assoc.*, *71*, 509-515

Gschwend, P.M., MacFarlane, J.K. & Newman, K.A. (1985) Volatile halogenated organic compounds released to seawater from temperate marine macroalgae. *Science*, *227*, 1033-1035

Hagenmaier, H., Werner, G. & Jäger, W. (1982) Quantitative determination of volatile halogenated hydrocarbons in water samples by capillary gas chromatography and electron capture detection. *Z. Wasser Abwasser Forsch.*, *15*, 195-198

Hayashi, M., Kishi, M., Sofuni, T. & Ishidate, M. Jr (1988) Micronucleus tests in mice on 39 food additives and eight miscellaneous chemicals. *Food chem. Toxicol.*, *26*, 487-500

Health and Welfare Canada (1977) *National Survey for Halomethanes in Drinking Water* (Report 77-EHD-9), Ottawa, Environmental Health Directorate, Health Protection Branch

Henderson, J.E., Peyton, G.R. & Glaze, W.H. (1976) A convenient liquid-liquid extraction method for the determination of halomethanes in water at the parts-per-billion level. In: Keith, L.H. ed., *Identification and Analysis of Organic Pollutants in Water*, Ann Arbor, MI, Ann Arbor Science, pp. 105-111

Hewitt, W.R., Brown, E.M. & Plaa, G.L. (1983) Acetone-induced potentiation of trihalomethane toxicity in male rats. *Toxicol. Lett.*, *16*, 285-296

Hoyt, S.D. & Rasmussen, R.A. (1985) Determining trace gases in air and seawater. *Adv. chem. Ser. Mapping Strategies chem. Oceanogr.*, *209*, 31-56

IARC (1987) *IARC Monographs on the Evaluation of Carcinogenic Risks to Humans*, Suppl. 7, *Overall Evaluations of Carcinogenicity: An Updating of* IARC Monographs *Volumes 1 to 42*, Lyon, pp. 152-154

Isacson, P., Bean, J.A. & Lynch, C. (1983) Relationship of cancer incidence rates in Iowa municipalities to chlorination status of drinking water. In: Jolley, R.J., Brungs, W.A., Cotruvo, J.A., Cumming, R.B., Mattice, J.S. & Jacobs, V.A., eds, *Water Chlorination: Environmental Impact and Health Effects*, Vol. 4, Book 2, *Environment, Health, and Risk*, Ann Arbor, MI, Ann Arbor Science, pp. 1353-1363

Ishidate, M., Jr (1987) *Data Book of Chromosomal Aberration Test in vitro*, rev. ed., Tokyo, Life-Science Information Center, pp. 118-119

Kaczmar, S.W., D'Itri, F.M. & Zabik, M.J. (1984) Volatilization rates of selected haloforms from aqueous environments. *Environ. Toxicol. Chem.*, 3, 31-35

Kaiser, K.L.E. & Comba, M.E. (1983) Volatile contaminants in the Welland River watershed. *J. Great Lakes Res.*, 9, 274-280

Kaiser, K.L.E., Comba, M.E. & Huneault, H. (1983) Volatile halocarbon contaminants in the Niagara River and in Lake Ontario. *J. Great Lakes Res.*, 9, 212-223

Kajino, M. & Yagi, M. (1980) Formation of trihalomethanes during chlorination and determination of halogenated hydrocarbons in drinking water. In: Afghan, B.K. & Mackay, D., eds, *Hydrocarbons and Halogenated Hydrocarbons in the Aquatic Environment*, New York, Plenum Press, pp. 491-501

Kaminski, L. & von Loew, E. (1984) Haloforms in swimming pool water (Ger.). *Forum Staedte-Hyg.*, 35, 19-20

Khudoley, V.V., Mizgireuv, I. & Pliss, G.B. (1987) The study of mutagenic activity of carcinogens and other chemical agents with *Salmonella* typhimurium assays: testing of 126 compounds. *Arch. Geschwulstforsch.*, 57, 453-462

Khudoley, V.V., Gvildis, V.Y. & Pliss, G.B. (1989) Identification of some trihalomethanes in drinking water and assessment of their toxicity, mutagenicity, and carcinogenicity (Russ.). *Vopr. Onkol.*, 35, 837-842

Kossik, R.F., Gschwend, P.S. & Adams, E.E. (1986) *Tracing and Modeling Pollutant Transport in Boston Harbor* (MITSG-86/16; NTIS PB87-136891), Rockville, MD, National Oceanic and Atmospheric Administration

Krasner, S.W., McGuire, M.J., Jacangelo, J.G., Patania, N.L., Reagan, K.M. & Aieta, E.M. (1989) The occurrence of disinfection by-products in drinking water in a nationwide study. In: *Annual Conference of the American Water Works Association, Los Angeles, CA*, Denver, CO, American Water Works Association, pp. 1-32

Kroneld, R. (1989) Volatile pollutants in the environment and human tissues. *Bull. environ. Contam. Toxicol.*, 42, 873-877

Lahl, U., Bätjer, K., von Düszeln, J., Gabel, B., Stachel, B. & Thiemann, W. (1981) Distribution and balance of volatile halogenated hydrocarbons in the water and air of covered swimming pools using chlorine for water disinfection. *Water Res.*, 15, 803-814

Lahl, U., Cetinkaya, M., von Düszeln, J., Gabel, B., Stachel, B. & Thiemann, W. (1982) Health risks for infants caused by trihalomethane generation during chemical disinfection of feeding utensils. *Ecol. Food Nutr.*, 12, 7-17

Loy, E.W., Jr, Brown, D.W., Stephenson, J.H.M. & Little, J.A. (1976) Uses of wastewater discharge compliance monitoring data. In: Keith, L.H., ed., *Identification and Analysis of Organic Pollutants in Water*, Ann Arbor, MI, Ann Arbor Science, pp. 499-516

Mabey, W.R., Smith, J.H., Podoll, R.T., Johnson, H.L., Mill, T., Chou, T.W., Gates, J., Waight Partridge, I. & Vandenberg, D. (1982) *Aquatic Fate Process Data for Organic Priority Pollutants. Final Report* (EPA-440/4-81-014; NTIS PB87-169090), Washington, DC, Office of Water Regulations and Standards, US Environmental Protection Agency, pp. 181-182

McConnell, G. (1976) Haloorganics in water supplies. *J. Inst. Water Eng. Sci.*, 30, 431-445

McKinney, J.D., Maurer, R.R., Hass, J.R. & Thomas R.O. (1976) Possible factors in the drinking water of laboratory animals causing reproductive failure. In: Keith, L.H., ed., *Identification and Analysis of Organic Pollutants in Water*, Ann Arbor, MI, Ann Arbor Science, pp. 417-432

Mersch-Sunderman, V. (1989) Examination of the mutagenicity of organic microcontaminations in the environment. II. The mutagenicity of halogenated aliphatic hydrocarbons with the *Salmonella*-microsome test (Ames test) in relation to contamination of ground- and drinking-water (Ger.). *Zbl. Bakt. Hyg. B.*, 187, 230-243

Minear, R.A. & Morrow, C.M. (1983) *Raw Water Bromide: Levels and Relationship to Distribution of Trihalomethanes in Finished Drinking Water* (OWRT-A-063-TENN; NTIS PB83-256735), Washington DC, US Environmental Protection Agency

Mink, F.L., Brown, T.J. & Rickabaugh, J. (1986) Absorption, distribution, and excretion of 14C-trihalomethanes in mice and rats. *Bull. environ. Contam. Toxicol.*, 37, 752-758

Morimoto, K. & Koizumi, A. (1983) Trihalomethanes induce sister chromatid exchanges in human lymphocytes *in vitro* and mouse bone marrow cells *in vivo*. *Environ. Res.*, 32, 72-79

Munson, A.E., Sain, L.E., Sanders, V.M., Kauffmann, B.M., White, K.L., Page, D.G., Barnes, D.W. & Borzelleca, J.F. (1982) Toxicology of organic drinking water contaminants: trichloromethane, bromodichloromethane, dibromochloromethane and tribromomethane. *Environ. Health Perspect.*, 46, 117-126

National Toxicology Program (1985) *Toxicology and Carcinogenesis Studies of Chlorodibromomethane (CAS No. 124-48-1) in F344/N Rats and B6C3F1 Mice (Gavage Studies)* (NTP Technical Report Series No. 282), Research Triangle Park, NC, US Department of Health and Human Services

Nestmann, E.R. & Lee, E.G.-H. (1985) Genetic activity in *Saccharomyces cerevisiae* of compounds found in effluents of pulp and paper mills. *Mutat. Res.*, 155, 53-60

Norin, H., Renberg, L., Hjort, J. & Lundblad, P.O. (1981) Factors influencing formation of trihalomethanes in drinking water with special reference to Swedish conditions. *Chemosphere*, 10, 1265-1273

Oliver, B.G. (1983) Dihaloacetonitriles in drinking water: algae and fulvic acid as precursors. *Environ. Sci. Technol.*, 17, 80-83

Otson, R. (1987) Purgeable organics in Great Lakes raw and treated water. *Int. J. environ. anal. Chem.*, 31, 41-53

Otson, R., Williams, D.T. & Bothwell, P.D. (1982) Volatile organic compounds in water at thirty Canadian potable water treatment facilities. *J. Assoc. off. anal. Chem.*, 65, 1370-1374

Page, G.W. (1981) Comparison of groundwater and surface water for patterns and levels of contamination by toxic substances. *Environ. Sci. Technol.*, 15, 1475-1481

Pellizzari, E.D., Whitaker, D.A. & Erickson, M.D. (1982) Purgeable organic compounds in mother's milk. *Bull. environ. Contam. Toxicol.*, 28, 322-328

Peoples, A.J., Pfaffenberger, C.D., Shafik, T.M. & Enos, H.F. (1979) Determination of volatile purgeable halogenated hydrocarbons in human adipose tissue and blood serum. *Bull. environ. Contam. Toxicol.*, 23, 244-249

Perwak, J., Goyer, M., Harris, J., Schimke, G., Scow, K., Wallace, D. & Slimak, M. (1980) *An Exposure and Risk Assessment for Trihalomethanes: Chloroform, Bromoform, Bromodichloromethane, Dibromochloromethane* (EPA-440/4-81-018; NTIS PB85-211977), Washington DC, Office of Water Regulations and Standards, US Environmental Protection Agency

Pouchert, C.J., ed. (1981) *The Aldrich Library of Infrared Spectra*, 3rd ed., Milwaukee, WI, Aldrich Chemical Co., p. 52B

Pouchert, C.J., ed. (1983) *The Aldrich Library of NMR Spectra*, 2nd ed., Vol. 1, Milwaukee, WI, Aldrich Chemical Co., p. 82A

Pouchert, C.J., ed. (1985a) *The Aldrich Library of FT-IR Spectra*, Vol. 1, Milwaukee, WI, Aldrich Chemical Co., p. 83D

Pouchert, C.J., ed. (1985b) *The Aldrich Library of FT-IR Spectra*, Vol. 3, Milwaukee, WI, Aldrich Chemical Co., p. 117A

Quaghebeur, D. & De Wulf, E. (1980) Volatile halogenated hydrocarbons in Belgian drinking waters. *Sci. total Environ.*, 14, 43-52

Reding, R., Fair, P.S., Shipp, C.J. & Brass, H.J. (1989) Measurement of dihaloacetonitriles and chloropicrin in US drinking waters. In: *Disinfection By-products: Current Perspectives*, Denver, CO, American Water Works Association, pp. 11-22

Rodriguez-Flores, M. (1983) *Characterization of Organic Contaminants in Selected Potable Water Supplies in Puerto Rico* (OWRT-A-057-PR; NTIS PB83-178038), Washington DC, US Environmental Protection Agency

Rook, J.J. (1974) Formation of haloforms during chlorination of natural waters. *Water Treat. Exam.*, 23, 234-243

Ruddick, J.A., Villeneuve, D.C. & Chu, I. (1983) A teratological assessment of four trihalomethanes in the rat. *J. environ. Sci. Health*, B18, 333-349

Sadtler Research Laboratories (1980) *The Sadtler Standard Spectra, 1980 Cumulative Index*, Philadelphia, PA

Scotte, P. (1984) Study of halogenated organic compounds in covered swimming pools (Fr.). *Eau ind. Nuisances*, 86, 37-41

Simmon, V.F., Kauhanen, K. & Tardiff, R.G. (1977) Mutagenic activity of chemicals identified in drinking water. In: Scott, D., Bridges, B.A. & Sobels, F.H., eds, *Progress in Genetic Toxicology*, Amsterdam, Elsevier/North-Holland Biomedical Press, pp. 249-258

Sittig, M. (1985) *Handbook of Toxic and Hazardous Chemicals and Carcinogens*, 2nd ed., Park Ridge, NJ, Noyes Publications, p. 306

Smith, R.L. (1989) A computer assisted, risk-based screening of a mixture of drinking water chemicals. *Trace Subst. environ. Health*, 22, 215-232

Standing Committee of Analysts (1980) *Chloro and Bromo Trihalogenated Methans in Water* (Methods for the Examination of Waters and Associated Materials), London, Her Majesty's Stationery Office

Stevens, A.A., Moore, L.A., Slocum, C.J., Smith, B.L., Seeger, D.R. & Ireland, J.C. (1989) By-products of chlorination at ten operating utilities. In: *Disinfection By-products: Current Perspectives*, Denver, CO, American Water Works Association, pp. 23-61

Symons, J.M., Bellar, T.A., Carswell, J.K., DeMarco, J., Kropp, K.L., Robeck, G.G., Seeger, D.R., Slocum, C.J., Smith, B.L. & Stevens, A.A. (1975) National Organics Reconnaissance Survey for halogenated organics. *J. Am. Water Works Assoc.*, 67, 634-647

Trussell, A.R., Cromer, J.L., Umphres, M.D., Kelley, P.E. & Moncur, J.G. (1980) Monitoring of volatile halogenated organics: a survey of twelve drinking waters from various parts of the world. In: Jolley, R.L., Brungs, W.A., Cumming, R.B. & Jacobs, V.A., eds, *Water Chlorination: Environmental Impact and Health Effects*, Vol. 3, Ann Arbor, MI, Ann Arbor Science, pp. 39-53

Ullrich, D. (1982) 7. Organohalogens in the air of some swimming pools of Berlin (Ger.). *WaBoLu-Ber.*, 1, 50-52

US Environmental Protection Agency (1985) *Health and Environmental Effects Profile for Bromochloromethanes* (EPA-600/X-85-397; NTIS PB88-174610), Cincinnati, OH, Environmental Criteria and Assessment Office

US Environmental Protection Agency (1986a) Method 8010. Halogenated volatile organics. In: *Test Methods for Evaluating Solid Waste—Physical/Chemical Methods*, 3rd ed. (EPA No. SW-846), Washington DC, Office of Solid Waste and Emergency Response

US Environmental Protection Agency (1986b) Method 8240. Gas chromatography/mass spectrometry for volatile organics. In: *Test Methods for Evaluating Solid Waste—Physical/Chemical Methods*, 3rd ed. (EPA No. SW-846), Washington DC, Office of Solid Waste and Emergency Response

US Environmental Protection Agency (1988a) Methods for organic chemical analysis of municipal and industrial wastewater. Method 601. Purgeable halocarbons. *US Code fed. Regul.*, Title 40, Part 136, Appendix A, pp. 267-281

US Environmental Protection Agency (1988b) Methods for organic chemical analysis of municipal and industrial wastewater. Method 624. Purgeables. *US Code fed. Regul.*, Title 40, Part 136, Appendix A, pp. 432-446

US Environmental Protection Agency (1988c) Method 502.2. Volatile organic compounds in water by purge and trap capillary column gas chromatography with photoionization and electrolytic conductivity detection in series. Revision 2.0. In: *Methods for the Determination of Organic Compounds in Drinking Water* (PEA-600/4-88-039; PB89-220461), Cincinnati, OH, Environmental Monitoring Systems Laboratory, pp. 285-323

US Environmental Protection Agency (1988d) Method 524.2. Measurement of purgeable organic compounds in water by capillary column gas chromatography/mass spectrometry. Revision 3.0. In: *Methods for the Determination of Organic Compounds in Drinking Water* (PEA-600/4-88-039; PB89-220461), Cincinnati, OH, Environmental Monitoring Systems Laboratory, pp. 255-284

US Environmental Protection Agency (1988e) Methods for organic chemical analysis of municipal and industrial wastewater. Method 1624 Revision B. Volatile organic compounds by isotope dilution GC/MS. *US Code fed. Regul., Title 40*, Part 136, Appendix A, pp. 475-488

US Environmental Protection Agency (1988f) Analysis of trihalomethanes. Part II—Analysis of trihalomethanes in drinking water by liquid/liquid extraction. *US Code fed. Regul., Title 40*, Part 141, Appendix C, pp. 563-571

US Environmental Protection Agency (1988g) Analysis of trihalomethanes. Part I—The analysis of trihalomethanes in drinking water by the purge and trap method. *US Code fed. Regul., Title 40*, Part 141, Appendix C, pp. 549-562

Varma, M.M., Ampy, F.R., Verma, K. & Talbot, W.W. (1988) In vitro mutagenicity of water contaminants in complex mixtures. *J. appl. Toxicol., 8*, 243-248

Verschueren, K. (1983) *Handbook of Environmental Data on Organic Chemicals*, 2nd ed., New York, Van Nostrand Reinhold, p. 464

Voronin, V.M., Donchenko, A.I. & Korolev, A.A. (1987) Experimental study of the carcinogenicity of dichlorobromomethane and dibromochloromethane formed during chlorination of water (Russ.). *Gig. Sanit., 1*, 19-21

Weast, R.C., ed. (1989) *CRC Handbook of Chemistry and Physics*, 70th ed., Boca Raton, FL, CRC Press, p. C-349

Weil, L., Jandik, J. & Eichelsdoefer, D. (1980) Organic halogenated compounds in swimming pool water. I. Determination of volatile halogenated hydrocarbons (Ger.). *Z. Wasser Abwasser Forsch., 13*, 141-145

Westrick, J.J., Mello, J.W. & Thomas, R.F. (1984) The groundwater supply survey. *J. Am. Water Works Assoc., 76*, 52-59

Williams, D.T. (1985) Formation of trihalomethanes in drinking water. In: Fishbein, L. & O'Neill, I.K., eds, *Environmental Carcinogens. Selected Methods of Analysis*, Vol. 7, *Some Volatile Halogenated Hydrocarbons* (IARC Scientific Publications No. 68), Lyon, IARC, pp. 69-88

HALOGENATED ACETONITRILES

1. Chemical and Physical Data

Bromochloroacetonitrile

1.1 Synonyms

Chem. Abstr. Services Reg. No.: 83463-62-1
Chem. Abstr. Name: Bromochloroacetonitrile
IUPAC Systematic Name: Bromochloroacetonitrile

1.2 Structural and molecular formulae and molecular weight

$$\begin{array}{c} \text{Br} \\ | \\ \text{H} - \text{C} - \text{C} \equiv \text{N} \\ | \\ \text{Cl} \end{array}$$

$C_2HBrClN$ Mol. wt: 154.39

1.3 Chemical and physical properties of the pure substance

(a) *Description*: Liquid
(b) *Boiling-point*: 138-140°C (Oliver, 1983)
(c) *Density*: 1.68 (Trehy & Bieber, 1981)
(d) *Spectroscopy data*: Mass spectroscopy data have been reported (Coleman *et al.*, 1984).
(e) *Octanol/water partition coefficient (P)*: log P, 0.28 (calculated by the method of Leo *et al.*, 1971)
(f) *Half-time in water at 25°C*: 55 h at pH 8.32 (Trehy & Bieber, 1981)
(g) *Conversion factor*[1]: $mg/m^3 = 6.32 \times ppm$

---

[1] Calculated from: mg/m^3 = (molecular weight/24.45) × ppm, assuming standard temperature (25°C) and pressure (760 mm Hg)

Chloroacetonitrile

1.1 Synonyms

Chem. Abstr. Services Reg. No.: 107-14-2
Chem. Abstr. Name: Chloroacetonitrile
IUPAC Systematic Name: Chloroacetonitrile
Synonyms: Chloracetonitrile; 2-chloroacetonitrile; alpha-chloroacetonitrile; chloromethyl cyanide; monochloroacetonitrile; monochloromethyl cyanide

1.2 Structural and molecular formulae and molecular weight

$$\text{Cl}-\overset{\overset{\text{H}}{|}}{\underset{\underset{\text{H}}{|}}{\text{C}}}-\text{C} \equiv \text{N}$$

C_2H_2ClN Mol. wt: 75.50

1.3 Chemical and physical properties of the pure substance

(a) *Description*: Colourless liquid (Eastman Kodak Co., 1987; Sax & Lewis, 1987)
(b) *Boiling-point*: 126-127°C (Weast, 1989)
(c) *Density*: 1.193 at 20°C (Weast, 1989)
(d) *Spectroscopy data*[1]: Infrared, raman [7886]; Pouchert, 1981, 1985a,b), nuclear magnetic resonance (Sadtler Research Laboratories, 1980, proton [6783]; Pouchert, 1974, 1983) and mass spectral data [23] have been reported (STN International, 1989a).
(e) *Solubility*: Very soluble in ethanol and diethyl ether (STN International, 1989a; Weast, 1989)
(f) *Octanol/water partition coefficient (P)*: log P, 0.23 (Chemical Information Systems, 1990)
(g) *Conversion factor*[2]: $mg/m^3 = 3.09 \times ppm$

---

[1] In square brackets, spectrum number in compilation
[2] Calculated from: mg/m^3 = (molecular weight/24.45) x ppm, assuming standard temperature (25°C) and pressure (760 mm Hg)

Dibromoacetonitrile

1.1 Synonyms

Chem. Abstr. Services Reg. No.: 3252-43-5
Chem. Abstr. Name: Dibromoacetonitrile
IUPAC Systematic Name: Dibromoacetonitrile

1.2 Structural and molecular formulae and molecular weight

$$\begin{array}{c} Br \\ | \\ H-C-C \equiv N \\ | \\ Br \end{array}$$

C_2HBr_2N Mol. wt: 198.84

1.3 Chemical and physical properties of the pure substance

(a) *Description*: Liquid
(b) *Boiling-point*: 169-170°C (Oliver, 1983)
(c) *Density*: 2.369 (Trehy & Bieber, 1981)
(d) *Spectroscopy data*[1]: Infrared, raman [6434]; Pouchert, 1981, 1985a,b), nuclear magnetic resonance (Sadtler Research Laboratories, 1980, proton [2567], grating [883]; Pouchert, 1974, 1983) and mass spectral data (STN International, 1989a [162C]) have been reported.
(e) *Octanol/water partition coefficient (P)*: log P, 0.42 (calculated by the method of Leo *et al.*, 1971)
(f) *Half-time in water at 25°C*: 500 h at pH 7.4; 85 h at pH 8.3; 19 h at pH 9.0 (Bieber & Trehy, 1983)
(g) *Conversion factor*[2]: $mg/m^3 = 8.13 \times ppm$

Dichloroacetonitrile

1.1 Synonyms

Chem. Abstr. Services Reg. No.: 3018-12-0
Chem. Abstr. Name: Dichloroacetonitrile

---

[1]In square brackets, spectrum number in compilation
[2]Calculated from: mg/m^3 = (molecular weight/24.45) × ppm, assuming standard temperature (25°C) and pressure (760 mm Hg)

IUPAC Systematic Name: Dichloroacetonitrile
Synonyms: Dichloromethyl cyanide

1.2 Structural and molecular formulae and molecular weight

$$\begin{array}{c} Cl \\ | \\ H-C-C \equiv N \\ | \\ Cl \end{array}$$

C_2HCl_2N Mol. wt: 109.94

1.3 Chemical and physical properties of the pure substance

(a) *Description*: Liquid
(b) *Boiling-point*: 112-113°C (Weast, 1989)
(c) *Density*: 1.369 at 20°C (Weast, 1989)
(d) *Spectroscopy data*[1]: Infrared, raman [4637, 4639]), nuclear magnetic resonance (Sadtler Research Laboratories, 1980, proton [23934]) and mass spectral data (Coleman *et al.*, 1984) have been reported.
(e) *Solubility*: Soluble in ethanol (Weast, 1989) and methanol (STN International, 1989a)
(f) *Octanol/water partition coefficient (P)*: log P, 0.14 (Chemical Information Systems, 1990)
(g) *Half-time in water at 25°C*: 30 h at pH 8.3; 0.75 h at pH 9.77 (Bieber & Trehy, 1983)
(h) *Conversion factor*[2]: mg/m^3 = 4.50 × ppm

Trichloroacetonitrile

1.1 Synonyms

Chem. Abstr. Services Reg. No.: 545-06-2
Chem. Abstr. Name: Trichloroacetonitrile
IUPAC Systematic Name: Trichloroacetonitrile

---

[1] In square brackets, spectrum number in compilation

[2] Calculated from: mg/m^3 = (molecular weight/24.45) × ppm, assuming standard temperature (25°C) and pressure (760 mm Hg)

Synonyms: Cyanotrichloromethane; 2,2,2-trichloroacetonitrile; trichloromethyl cyanide; trichloromethylnitrile

1.2 Structural and molecular formulae and molecular weight

$$\begin{array}{c} Cl \\ | \\ Cl-C-C\equiv N \\ | \\ Cl \end{array}$$

C_2Cl_3N Mol. wt: 144.39

1.3 Chemical and physical properties of the pure substance

(a) *Description*: Colourless liquid (Budavari, 1989; Sax & Lewis, 1987)
(b) *Boiling-point*: 85.7°C (Budavari, 1989)
(c) *Melting-point*: –42°C (Weast, 1989)
(d) *Density*: 1.4403 at 25/4°C (Weast, 1989)
(e) *Spectroscopy data*[1]: Infrared (Sadtler Research Laboratories, 1980, prism [38232, 42996]; grating [17218, 23996]; Pouchert, 1981, 1985a,b) and mass spectral data [383]; Coleman *et al.*, 1984) have been reported.
(f) *Solubility*: Insoluble in water (STN International, 1989a)
(g) *Conversion factor*[2]: $mg/m^3 = 5.91 \times ppm$

1.4 Technical products and impurities

Chloroacetonitrile is available at 98-> 99% purity, dibromoacetonitrile at 95-97% purity, dichloroacetonitrile at 95-98% purity and trichloroacetonitrile at ≥ 98% purity (Riedel-de Haën, 1986; Eastman Kodak Co., 1987; American Tokyo Kasei, 1988; Fairfield Chemical Co., 1988; Pfaltz & Bauer, 1988; Aldrich Chemical Co., 1990).

The trade name for trichloroacetonitrile is Tritox.

---

[1] In square brackets, spectrum number in compilation
[2] Calculated from: mg/m^3 = (molecular weight/24.45) × ppm, assuming standard temperature (25°C) and pressure (760 mm Hg)

2. Production, Use, Occurrence and Analysis

2.1 Production and use

(a) Production

Chloroacetonitrile was synthesized by Bisschopinck in 1873 by the reaction of chloroacetamide with phosphoric anhydride (STN International, 1989b). It has been produced commercially by the high-temperature chlorination of acetonitrile (Movsum-Zade *et al.*, 1977).

Dibromoacetonitrile and dichloroacetonitrile have been synthesized by the reaction of *N*-bromosuccinimide (or *N*-chlorosuccinimide) with cyanoacetic acid (Trehy & Bieber, 1981). Bromochloroacetonitrile has been synthesized by the reaction of a mixture of *N*-chlorosuccinimide and *N*-bromosuccinimide with cyanoacetic acid, and fractional distillation of the resulting mixture of haloacetonitriles (Pereira *et al.*, 1984a; Bull *et al.*, 1985). Bromochloroacetonitrile has also been synthesized by the bromination of chloroacetonitrile with bromine (Trehy & Bieber, 1981).

Trichloroacetonitrile was synthesized by Bisschopinck in 1873 by the reaction of trichloroacetamide with phosphoric anhydride (STN International, 1989b). It has been produced commercially by reaction of acetonitrile with chlorine or sulfonyl chloride in the presence of a catalyst at elevated temperatures (Dow Chemical Co., 1975). Trichloroacetonitrile has also been synthesized from ethyl trichloroacetate and aqueous ammonia and by reaction of methylnitrile, hydrochloric acid and chlorine gas (Budavari, 1989).

(b) Use

Chloroacetonitrile has reportedly been used as a fumigant and as a chemical intermediate; trichloroacetonitrile has reportedly been used as an insecticide (Sax & Lewis, 1987).

(c) Regulatory status and guidelines

No regulatory standards or guidelines have been published for exposures to halogenated acetonitriles.

2.2 Occurrence

Halogenated acetonitriles have been identified in the environment only as by-products of the chlorination of ground- and surface waters for disinfection of drinking-water supplies. Therefore, the only known route of environmental release of halogenated acetonitriles is as a constituent of potable water supplies. Potential precursors for the formation of these compounds during chlorination are algae, humic substances and proteinaceous material—all naturally occurring water components. The brominated acetonitriles are formed when bromide is present in the water during chlorination. Halogenated acetonitriles have not been detected in raw (untreated) water sources (Trehy & Bieber, 1981; Keith *et al.*, 1982; Oliver, 1983).

(a) Natural occurrence

None of the halogenated acetonitriles is known to occur as a natural product.

(b) Water and sediments

Dichloroacetonitrile was detected in tap-water samples taken from two buildings in Durham, NC, USA (McKinney *et al.*, 1976).

In a two-year study of trace organic compounds in drinking-water in Philadelphia, PA, Suffet *et al.* (1980) tested samples collected at three treatment plants and one hotel, which originated from two surface water (river) sources. Dichloroacetonitrile was identified in four of nine samples originating from the Delaware River but in neither of the two samples originating from the Schuylkill River. No quantitative analysis was performed.

Three dihaloacetonitriles (bromochloroacetonitrile, dibromoacetonitrile and dichloroacetonitrile) were identified in chlorinated ground- and surface water supplies in southern Florida (USA). Although individual concentrations were not determined, the highest total concentration of dihaloacetonitriles found was 42 μg/l in a chlorinated well-water supply. The relative concentrations of the three dihaloacetonitriles varied considerably from supply to supply, and none was found in water supplies that had been lime-softened. Experimental chlorination of lakewater samples confirmed that dihaloacetonitriles were formed in the chlorination process (Trehy & Bieber, 1980, 1981).

Bromochloroacetonitrile and dibromoacetonitrile were identified in stored, chlorinated Rhine water in The Netherlands. The concentrations of bromochloroacetonitrile before and after chlorination were < 0.1 and 3 μg/l, respectively, and those of dibromoacetonitrile, < 0.1 and 1 μg/l (Zoeteman *et al.*, 1982).

Analysis of chlorinated drinking-water samples from ten cities in Ontario, Canada, showed dichloroacetonitrile at 0.3-8.1 μg/l and bromochloroacetonitrile at not detected (< 0.1 μg/l) to 1.8 μg/l. Neither was detected in any of the raw water samples tested; dibromoacetonitrile was not detected in either the raw-water or drinking-water samples. When chlorine was reacted with fulvic acid (a major component of humic substances found in natural waters) for 4-120 h, dichloroacetonitrile was produced in concentrations of 3.5-6.5 µg/l, and bromochloroacetonitrile was not detected (< 0.1 µg/l); when chlorine was reacted with a blue-green alga, dichloroacetonitrile occurred at 0.5-3.5 µg/l and bromochloroacetonitrile at not detected to 0.6 µg/l; after reaction with a green alga, dichloroacetonitrile occurred at 0.3-1.0 µg/l and the bromine compound at not detected to 1.1 µg, indicating that dihaloacetonitriles can be produced by chlorinating aquatic humic substances and algae under conditions used for water treatment (Oliver, 1983).

Samples collected following prechlorination of raw water in 1984 from a drinking-water treatment plant at Cholet, France, contained dichloroacetonitrile at 0.14 µg/l and trichloroacetonitrile at 0.11 µg/l (Bruchet et al., 1985).

Samples of chlorinated wastewater from an extended aeration treatment plant in the USA contained 7-14 µg/l dichloroacetonitrile. Chlorinated lakewater samples from West Palm Beach, FL, contained 5-19 µg/l dichloroacetonitrile and those from Gainsville, FL, 10-17 µg/l (chlorine contact time, 5-70 min). Experimental studies confirmed that dichloroacetonitrile is formed when amino acids, such as aspartic acid, tyrosine and tryptophan, are chlorinated (Trehy et al., 1986, 1987).

Treated water at 10 Canadian sites along the Great Lakes in the spring contained dichloroacetonitrile at a mean concentration of 0.3 µg/l. It was not detected (detection limit, 0.1 µg/l) in treated water samples during the summer or winter or in any of the raw-water samples (Otson, 1987).

Treatment plant samples collected from 29 US community water systems (using free chlorine disinfection) in 1984 and 1985 contained bromochloroacetonitrile at 0.2-9.4 µg/l (25 of 29 sites), dibromoacetonitrile at 0.3-11 µg/l (14 of 29 sites) and dichloroacetonitrile at 0.2-21 µg/l (27 of 29 sites). Distribution samples from the system contained bromochloroacetonitrile at 0.4-10 µg/l (21 of 26 sites), dibromoacetonitrile at 0.2-2.5 µg/l (11 of 26 sites) and dichloroacetonitrile at 0.3-24 µg/l (21 of 26 sites) (quantification limits, 0.2-0.3 µg/l) (Reding et al., 1989).

Water samples collected from 10 US utilities in 1985 (using free chlorine disinfection, in one of which ammonia was added before distribution) contained bromochloroacetonitrile at < 10 µg/l at all of seven sites for which data were available, dibromoacetonitrile at < 10 µg/l at three of the seven sites and dichloroacetonitrile at < 10 µg/l at all of 10 sites; trichloroacetonitrile was not found at any of the eight sites for which data were available (Stevens et al., 1989).

Samples of finished water were taken from 35 US utilities in spring, summer, autumn and winter, 1988-89. The median concentrations (µg/l) of halogenated acetonitriles measured were: bromochloroacetonitrile, 0.50-0.70;

dibromoacetonitrile, 0.48-0.54; dichloroacetonitrile, 1.1-1.2; and trichloroacetonitrile, not detected (detection limits, < 0.012 and < 0.029, according to season) (Krasner *et al.*, 1989).

2.3 Analysis

Difficulties in analysising halogenated acetonitriles in environmental samples have been attributed to their instability under the analytical conditions used (Oliver, 1983). Halogenated acetonitriles hydrolyse in water to haloacetamides and then to haloacetic acids (Bieber & Trehy, 1983). Destruction by hydrolysis is appreciable at the pH (8.0-8.5) frequently found in water treatment plants. The rate of hydrolysis increases at elevated temperature and pH (Trehy & Bieber, 1981; Bieber & Trehy, 1983; Oliver, 1983).

Dichloroacetonitrile has been determined in drinking-water by concentration on an adsorption column, followed by gas chromatography-flame ionization detection or gas chromatography-electron capture detection (Chriswell *et al.*, 1983).

Dihaloacetonitriles can be measured in water by solvent (pentane) extraction, usually after salting out using sodium chloride or sodium sulfate, followed by gas chromatography-electron capture detection (Oliver, 1983; Reckhow & Singh, 1984; Krasner *et al.*, 1989; Reding *et al.*, 1989.

3. Biological Data Relevant to the Evaluation of Carcinogenic Risk to Humans

3.1 Carcinogenicity studies in animals (Tables 1 and 2)

(a) Oral administration

Mouse: In a screening assay based on the enhanced induction of lung tumours, groups of 40 female strain A/J mice, 10 weeks old, were given 10 mg/kg bw chloroacetonitrile, dichloroacetonitrile, trichloroacetonitrile, bromochloroacetonitrile or dibromoacetonitrile [purity unspecified] in 10% Emulphor by oral gavage three times per week for eight weeks. A group of 40 animals given 10% Emulphor only served as controls. Survival at the end of the study (at nine months of age) was: control, 31/40; chloroacetonitrile-treated, 28/40; dichloroacetonitrile-treated, 30/40; trichloroacetonitrile-treated, 32/40; bromochloroacetonitrile-treated, 32/40; and dibromoacetonitrile-treated, 31/40. The numbers of animals with lung tumours and the average numbers of tumours per animal were: control, 3/31 and 0.1; chloroacetonitrile-treated, 9/28 and 0.43 ($p < 0.05$); dichloroacetonitrile-treated, 7/30 and 0.23; trichloroacetonitrile-treated, 9/32 and 0.38 ($p < 0.05$); bromochloroacetonitrile-treated, 10/32 and 0.34 ($p < 0.05$); dibromoacetonitrile-treated, 5/31 and 0.19 (Bull & Robinson, 1985).

Table 1. Results of skin application of halogenated acetonitriles to female Sencar mice[a]

Expt. no.	Treatment	Dose (mg/kg)	No. of animals			
			Effective	With tumours	With papilloma	With carcinoma
1	Chloroacetonitrile	200×6	38	11**	7	4
2		400×6	37	11**	9	4
2		800×6	38	6	1	6
1	Dichloroacetonitrile	200×6	39	4	4	0
2		400×6	35	4	2	3
2		800×6	35	1	1	0
1	Trichloroacetonitrile	200×6	34	2	1	1
2		400×6	36	11**	5	6
3		400×6	38	1	0	1
2		800×6	36	3	2	1
3		800×6	29	2	1	1
2	Bromochloroacetonitrile	200×6	35	1	0	1
2		400×6	37	7	0	7
2		800×6	37	8*	3	6
1	Dibromoacetonitrile	200×6	36	8*	6	2
2		400×6	35	17**	9	8
3		400×6	35	16**	7	9
2		800×6	37	7	5	2
3		800×6	37	3	1	2
1	Acetone	0.2 ml×6	34	1	0	1
2		0.2 ml×6	37	3	3	0
3		0.2 ml×6	34	5	1	4

[a]From Bull et al. (1985)
*$p < 0.05$, Fisher's exact test
**$p < 0.01$, Fisher's exact test

Table 2. Summary of carcinogenicity studies of halogenated acetonitriles in experimental animals

Reference	Species/strain	Sex	Dose schedule	Experimental parameter/observation	Group[a] 0	1	2	3	4	5	Significance	Comment
Bull & Robinson (1985)	Mouse A/J	F	3 d/week, orally, Emulphor, 24 doses	Dose (mg/kg) Survival (9 months) Lung adenoma Lung adenomas per mouse	0 31/40 3/31 0.10	10 28/40 9/28* 0.43	10 30/40 7/30 0.23	10 32/40 9/32* 0.38	10 32/40 10/32* 0.34	10 31/40 5/31 0.19	*$p < 0.05$	Low doses
Bull et al. (1985)	Mouse Sencar	F	3 d/week, skin appl., acetone, 24 weeks	Dose (mg/kg)	0	800	800	800	800	400		No tumour; few details

[a]Group 0, vehicle control; group 1, chloroacetonitrile; group 2, dichloroacetonitrile; group 3, trichloroacetonitrile; group 4, bromochloroacetonitrile; group 5, dibromoacetonitrile

(b) Skin application

Mouse: In a series of three initiation-promotion experiments, groups of 40 female Sencar mice [age unspecified] were given topical applications of 200, 400 or 800 mg/kg bw chloroacetonitrile (> 99% pure), dichloroacetonitrile (> 99% pure), trichloroacetonitrile (98% pure), bromochloroacetonitrile (93% pure) or dibromoacetonitrile (96% pure) in 0.2 ml acetone three times per week for two weeks. Two weeks following the last application of the halogenated acetonitrile, 1.0 µg 12-*O*-tetradecanoylphorbol 13-acetate (TPA) was applied topically three times per week for 20 weeks. Animals were observed for one year. The numbers of animals with skin tumours and the numbers of squamous-cell papillomas and carcinomas are given in Table 1 (Bull *et al.*, 1985). [The Working Group noted that the results were variable and were derived from three separate studies not conducted simultaneously.]

In another experiment, groups of 40 female Sencar mice [age unspecified] were given topical applications of 800 mg/kg bw chloroacetonitrile, dichloroacetonitrile, trichloroacetonitrile or bromochloroacetonitrile or 400 mg/kg bw dibromoacetonitrile at the same purity described above in 0.2 ml acetone three times per week for 24 weeks. A group of 40 female mice given topical applications of acetone on the same schedule served as controls. No skin tumour occured (Bull *et al.*, 1985). [The Working Group noted that the duration of the study was not specified.]

3.2 Other relevant data

(a) Experimental systems

(i) Absorption, distribution, excretion and metabolism

Rats given 0.75 mmol/kg bw bromochloroacetonitrile (116 mg/kg bw), chloroacetonitrile (57 mg/kg bw), dibromoacetonitrile (149 mg/kg bw), dichloroacetonitrile (82.5 mg/kg bw) or trichloroacetonitrile (108 mg/kg bw) dissolved in tricaprylin by gavage excreted 12.8, 14.2, 7.7, 9.3 or 2.3%, respectively, of the dose in the urine as thiocyanate within 24 h (Pereira *et al.*, 1984b). The halonitriles are apparently metabolized to the corresponding cyanohydrins, which eliminate cyanide; the cyanide is metabolized to thiocyanate by rhodanese.

Rats given 0.002 mmol/kg bw (0.20 mg/kg bw), 0.02 (2.0 mg/kg bw) or 0.14 (15 mg/kg bw) [1-^{14}C]dichloroacetonitrile by gavage in water excreted 62-73% of the dose by six days; most of the radiolabel was excreted in the urine (42-45%) and faeces (14-20%) (Roby *et al.*, 1986). Exhalation of ^{14}C-carbon dioxide amounted to 3.2-8.1% of the administered dose of dichloroacetonitrile. After six days, 19.3% of the dose was retained in the tissues; the largest amounts were found in blood (4-8% of dose), muscle (4-8%), skin (3-6%) and liver (about 2%). In rats given

0.002 mmol/kg bw (0.21 mg/kg bw), 0.02 mmol/kg bw (2.0 mg/kg bw) or 0.14 mmol/kg bw (15 mg/kg bw) [2-^{14}C]-dichloroacetonitrile by gavage in water, a total of 82-86% of the dose was eliminated in the urine, faeces and expired air (as ^{14}C-carbon dioxide) within 48 h; urinary radiolabel accounted for 35-40% and ^{14}C-carbon dioxide for 33-34% of the dose. After two days, 12-17% of the dose was retained in the tissues; and the largest amounts were found in liver (about 5% of dose), muscle (3-5%), blood (2-5%) and skin (about 1%).

After oral administration of 0.02 mmol/kg bw (2.0 mg/kg bw) or 0.14 mmol/kg bw (15 mg/kg bw) [1-^{14}C]-dichloroacetonitrile to mice by gavage in water, 85 and 83%, respectively, of the dose was eliminated in the urine, faeces and expired air (as ^{14}C-carbon dioxide) by 24 h (Roby *et al.*, 1986). The urine contained 64-70% of the dose, the faeces contained 9-13% and about 5% was eliminated as carbon dioxide; 11-12% was retained in the tissues. The largest amount of radiolabel was found in the liver (about 4% of dose), muscle and skin (about 2%) and blood and fat (about 1%). When mice were given 0.02 mmol/kg bw (2.0 mg/kg) or 0.14 mmol/kg bw (15 mg/kg) [2-^{14}C]-dichloroacetonitrile by gavage in water, 84-88% of the dose was eliminated in the urine, faeces and expired air (as ^{14}C-carbon dioxide) within 24 h; the urine (42-43% of dose) and expired air (31-37%) contained the most radiolabel. Nine percent of the administered radiolabel was retained in the tissues after 24 h; most was found in the liver (about 5% of dose), and 0.5-1% of the dose was present in muscle, kidney and skin.

Chloroacetonitrile is metabolized to hydrogen cyanide by mouse hepatic microsomal fractions (Tanii & Hashimoto, 1984).

Except trichloroacetonitrile, the haloacetonitriles considered here were shown to be electrophiles on the basis of their reactivity with *para*-nitrobenzylpyridine and DNA, in the ranking dibromoacetonitrile > bromochloroacetonitrile > chloroacetonitrile > dichloroacetonitrile (Lin *et al.*, 1986).

(ii) *Toxic effects*

Chloroacetonitrile

The acute oral LD$_{50}$ of chloroacetonitrile was estimated to be 1.8 mmol/kg bw [136 mg/kg] in male ddY mice (Tanii & Hashimoto, 1984). Chloroacetonitrile was inactive in a bioassay for γ-glutamyltranspeptidase-positive foci in a bioassay for rat liver (Herren-Freund & Pereira, 1986).

Dichloroacetonitrile

The acute oral (by gavage) LD$_{50}$s of dichloroacetonitrile were 339 and 330 mg/kg bw for male and female CD rats and 270 and 279 mg/kg bw for male and female CD-1 ICR mice, respectively (Hayes *et al.*, 1986). Signs of toxicity included ataxia, depressed respiration and prostration but no gross pathological change.

Daily administration of 12-90 mg/kg bw dichloroacetonitrile in corn oil by gavage to male and female CD rats for 14 days induced no toxicity, as measured by mortality rates, body weights and clinical chemistry and haematological parameters, although the relative weight of some organs was slightly reduced (Hayes *et al.*, 1986).

Daily treatment of male and female CD rats by gavage with 8, 33 or 65 mg/kg bw dichloroacetonitrile in corn oil for 90 days resulted in 25-50% mortality in the 65-mg/kg bw group and 5-10% in the 8- and 33-mg/kg bw groups. Significant signs of toxicity were reduced body weights in males and females, reduced weights of most organs in males and reduced serum cholesterol levels at the highest dose; clinical chemistry and haematological parameters were generally unchanged. Livers appeared to be larger in treated females (Hayes *et al.*, 1986).

Dichloroacetonitrile was inactive in a bioassay for γ-glutamyltranspeptidase-positive foci in rat liver (Herren-Freund & Pereira, 1986).

Trichloroacetonitrile

Trichloroacetonitrile was inactive in a bioassay for γ-glutamyltranspeptidase-positive foci in rat liver (Herren-Freund & Pereira, 1986).

Dibromoacetonitrile

The acute oral LD_{50}s of dibromoacetonitrile were 245 and 361 mg/kg bw in male and female CD rats and 289 and 303 mg/kg bw in male and female CD-1 ICR mice, respectively (Hayes *et al.*, 1986). Signs of toxicity included ataxia, depressed respiration and prostration, but no gross pathological change was observed.

Daily administration of 23-180 mg/kg bw dibromoacetonitrile to male and female CD rats in corn oil by gavage for 14 days resulted in 100% mortality in the 180-mg/kg bw group and 20-40% in the 90-mg/kg bw group. Besides a dose-dependent depression in weight gain, the only significant signs of toxicity were decreases in relative spleen and thymus weights (males) and an increase in relative liver weight (females) in the 90-mg/kg bw group (Hayes *et al.*, 1986).

Daily treatment of male and female CD rats by gavage with 6, 23 or 45 mg/kg bw dibromoacetonitrile in corn oil for 90 days resulted in 5-10% mortality in the 23- and 45-mg/kg bw groups. Significant signs of toxicity were reduced body weights in males in the 45-mg/kg bw group. Relative organ weights, clinical chemistry and haematological parameters were generally unchanged (Hayes *et al.*, 1986).

Dibromoacetonitrile was inactive in a bioassay for γ-glutamyltranspeptidase-positive foci in rat liver (Herren-Freund & Pereira, 1986).

No data on acute toxicity were available for bromochloroacetonitrile or trichloroacetonitrile. No data on subacute or subchronic toxicity were available for chloroacetonitrile, bromochloroacetonitrile or trichloroacetonitrile. None of the halogenated acetonitriles has been examined for chronic toxicity.

(iii) *Effects on reproduction and prenatal toxicity*

Chloroacetonitrile

In a screening study (Smith *et al.*, 1987), Long-Evans rats were administered chloroacetonitrile in tricaprylin by gavage at 55 mg/kg bw daily on gestation days 7-21. The litters were culled on postnatal day 6 (to six to eight pups) and again at weaning, when litters were reduced to four pups, which were retained until 41 or 42 days of age. One of the 30 treated dams died. There was a significant decrease in maternal weight gain during the period of treatment. No effect on pregnancy, proportion of resorptions, pup survival or growth after birth was observed. Litter weight at the time of birth was significantly lower than in the controls.

Bromochloroacetonitrile

In the screening study described above (Smith *et al.*, 1987), one of the 20 dams treated with bromochloroacetonitrile died. Litter weight at the time of birth was slightly lower than in the controls, and weight gain to day 4 was also decreased. There was no significant effect on the percentages of pregnant females or resorptions and no effect on neonatal survival.

Dibromoacetonitrile

In the screening study described above but using a dose of 50 mg/kg bw (Smith *et al.*, 1987), four of the 26 dams treated with dibromoacetonitrile died. There was a significant decrease in maternal weight gain during the period of treatment. Litter weight at the time of birth was lower than in the controls, and weight gain to lactation day 4 was significantly decreased. There was no significant effect on the percentages of pregnant females or resorptions and no effect on neonatal survival after birth.

Dichloroacetonitrile

In two screening studies performed as described above (Smith *et al.*, 1987), no mortality was seen in a group of 20 females treated with 55 mg/kg bw/day dichloroacetonitrile in one study, but two of the 23 treated dams died in a second study. Maternal weight gain was significantly decreased in both studies. There was a significant decrease in the percentage of dams delivering viable litters, an increase in resorption of litters and fetuses and decreased survival of neonates after birth. Litter weight at the time of birth was significantly decreased, and weight gain to lactation day 4 was lower in the treated group than in controls.

Groups of 22-24 mated Long-Evans rats were administered dichloroacetonitrile in tricaprylin by gavage at 0, 5, 15, 25 or 45 mg/kg bw daily on gestation days 6-18 (Smith *et al.*, 1989). Maternal deaths (9%) occurred at the high-dose levels, and significant increases in postimplantation losses were observed in the groups given 25 mg/kg bw (35%) and 45 mg/kg bw (80%). At the high-dose level, resorption of entire litters occurred in 60% of the survivors. Maternal weight gain and fetal body weight were decreased at the high-dose level. The frequency of malformations of the soft tissues, particularly cardiovascular and urogenital organs, was significantly increased at the high-dose level, at which there was also an increase in the percentage of skeletal malformations. Thus, dichloroacetonitrile induced malformations, but only at dose levels at which there was both severe embryolethality and maternal toxicity.

Trichloroacetonitrile

In two screening studies (Smith *et al.*, 1987), performed as described above, five of 25 females treated with 55 mg/kg bw/day trichloroacetonitrile died in one study while none of 20 died in the second. Maternal weight gain was significantly decreased in both studies. There was a significant decrease in the percentage of pregnant females, an increase in the number of litters resorbed and a decrease in neonatal survival. Litter weight was decreased at birth and postnatally.

Groups of 22-24 mated Long-Evans rats were administered trichloroacetonitrile gavage in tricaprylin by gavage at 0, 1, 7.5, 15, 35 or 55 mg/kg bw daily on gestation days 6-18 (Smith *et al.*, 1988). Maternal mortality occurred in the groups given 35 and 55 mg/kg bw, and maternal weight gain was significantly decreased at the highest dose level. Whole litters were resorbed at levels of 7.5 mg/kg bw and above, and a significant increase in the incidence of fetal resorptions was observed at 15 mg/kg bw (49%), 35 mg/kg bw (62%) and 55 mg/kg bw (78%). Fetal body weights were significantly decreased at 35 mg/kg bw, but not at other dose levels. The incidences of malformations, primarily cardiovascular and urogenital, were significantly increased in the groups given 15, 35 and 55 mg/kg bw. Thus, trichloroacetonitrile induced malformations in rats at embryolethal dose levels. Embryolethality occurred at dose levels below those which caused maternal toxicity and malformations.

(iv) *Genetic and related effects* (Table 3)

The genetic and related effects of the haloacetonitriles considered have been reviewed (Bull, 1985).

Bromochloroacetonitrile

Mutation was induced in *Salmonella typhimurium*. Sister chromatid exchange was induced in Chinese hamster ovary (CHO) cells and DNA strand breaks in human lymphoblast cell lines. In mice dosed for five days, neither micronuclei in bone marrow nor abnormal sperm morphology was induced.

Chloroacetonitrile

Chloroacetonitrile did not induce mutation in a single study with *S. typhimurium*. Sister chromatid exchange was induced in one study using Chinese hamster ovary (CHO) cells, and DNA strand breaks (weakly) were induced in another using a human lymphoblast cell line. In mice dosed for five days, neither micronuclei in bone marrow nor abnormal sperm morphology was induced.

Dibromoacetonitrile

Dibromoacetonitrile was not mutagenic in either *S. typhimurium* or *Drosophila melanogaster*. Sister chromatid exchange was induced in Chinese hamster ovary (CHO) cells and DNA strand breaks in human lymphoblast cell lines. In mice dosed for five days, neither micronuclei in bone marrow nor abnormal sperm morphology was induced.

Dichloroacetonitrile

Dichloroacetonitrile induced mutation in *S. typhimurium* and in one study in *D. melanogaster*. Mitotic recombination was not induced in yeast. Dichloroacetonitrile induced sister chromatid exchange in Chinese hamster ovary (CHO) cells and DNA strand breaks (weakly) in a human cell line. In mice dosed for five days, neither micronuclei in bone marrow nor abnormal sperm morphology was induced.

Trichloroacetonitrile

Trichloroacetonitrile did not induce mutation in *S. typhimurium*. Sister chromatid exchange was induced in Chinese hamster ovary (CHO) cells and DNA strand breaks in human cell lines. In mice dosed for five days, neither micronuclei in bone marrow nor abnormal sperm morphology was induced.

Table 3. Genetic and related effects of bromochloroacetonitrile

Test system	Result		Dose LED/HID	Reference
	Without exogenous metabolic system	With exogenous metabolic system		
SA0, *Salmonella typhimurium* TA100, reverse mutation	−	+	13.0000	Bull et al. (1985)
SA5, *Salmonella typhimurium* TA1535, reverse mutation	+	+	26.0000	Bull et al. (1985)
SA7, *Salmonella typhimurium* TA1537, reverse mutation	−	−	0.0000	Bull et al. (1985)
SA8, *Salmonella typhimurium* TA1538, reverse mutation	−	−	0.0000	Bull et al. (1985)
SA9, *Salmonella typhimurium* TA98, reverse mutation	−	−	105.0000	Bull et al. (1985)
SIC, Sister chromatid exchange, Chinese hamster CHO cells *in vitro*	+	+	0.6500	Bull et al. (1985)
DIH, DNA strand breaks, human lymphoblast cell line	+	0	0.0000	Daniel et al. (1986)
MVM, Micronucleus test, CD–1 mice *in vivo*	−	0	50.0000	Bull et al. (1985)
SPF, Sperm morphology, B6C3F$_1$ mice *in vivo*	−	0	50.0000	Meier et al. (1985)

Table 3 (contd). Genetic and related effects of chloroacetonitrile

Test system	Result		Dose LED/HID	Reference
	Without exogenous metabolic system	With exogenous metabolic system		
SA0, *Salmonella typhimurium* TA100, reverse mutation	–	–	1500.0000	Bull et al. (1985)
SA5, *Salmonella typhimurium* TA1535, reverse mutation	–	–	1500.0000	Bull et al. (1985)
SA7, *Salmonella typhimurium* TA1537, reverse mtuation	–	–	0.0000	Bull et al. (1985)
SA8, *Salmonella typhimurium* TA1538, reverse mutation	–	–	0.0000	Bull et al. (1985)
SA9, *Salmonella typhimurium* TA98, reverse mutation	–	–	1500.0000	Bull et al. (1985)
SIC, Sister chromatid exchange, Chinese hamster CHO cells *in vitro*	+	+	4.0000	Bull et al. (1985)
DIH, DNA strand breaks, human lymphoblast cell line	(+)	0	225.0000	Daniel et al. (1986)
MVM, Micronucleus test, CD-1 mice *in vivo*	–	0	50.0000	Bull et al. (1985)
SPF, Sperm morphology, B6C3F$_1$ mice *in vivo*	–	0	50.0000	Meier et al. (1985)

Table 3 (contd). Genetic and related effects of dibromoacetonitrile

Test system	Result		Dose LED/HID	Reference
	Without exogenous metabolic system	With exogenous metabolic system		
SA0, *Salmonella typhimurium* TA100, reverse mutation	–	–	115.0000	Bull et al. (1985)
SA5, *Salmonella typhimurium* TA1535, reverse mutation	–	–	115.0000	Bull et al. (1985)
SA7, *Salmonella typhimurium* TA1537, reverse mtuation	–	–	0.0000	Bull et al. (1985)
SA8, *Salmonella typhimurium* TA1538, reverse mutation	–	–	0.0000	Bull et al. (1985)
SA9, *Salmonella typhimurium* TA98, reverse mutation	–	–	115.0000	Bull et al. (1985)
DMX, *Drosophila melanogaster*, sex-linked recessive lethal mutation	–	0	200.0000	Valencia et al. (1985)
SIC, Sister chromatid exchange, Chinese hamster CHO cells *in vitro*	+	+	0.0300	Bull et al. (1985)
DIH, DNA strand breaks, human lymphoblast cell line	+	0	0.0000	Daniel et al. (1986)
MVM, Micronucleus test, CD-1 mice *in vivo*	–	0	50.0000	Bull et al. (1985)
SPF, Sperm morphology, B6C3F$_1$ mice *in vivo*	–	0	50.0000	Meier et al. (1985)

Table 3 (contd). Genetic and related effects of dichloroacetonitrile

Test system	Result		Dose LED/HID	Reference
	Without exogenous metabolic system	With exogenous metabolic system		
SA0, *Salmonella typhimurium* TA100, reverse mutation	+	0	250.0000	Simmon *et al.* (1977)
SA0, *Salmonella typhimurium* TA100, reverse mutation	+	+	175.0000	Bull *et al.* (1985)
SA5, *Salmonella typhimurium* TA1535, reverse mutation	+	+	22.0000	Bull *et al.* (1985)
SA7, *Salmonella typhimurium* TA1537, reverse mtuation	–	–	0.0000	Bull *et al.* (1985)
SA8, *Salmonella typhimurium* TA1538, reverse mutation	–	–	0.0000	Bull *et al.* (1985)
SA9, *Salmonella typhimurium* TA98, reverse mutation	+	+	680.0000	Bull *et al.* (1985)
SCG, *Saccharomyces cerevisiae*, mitotic recombination	–	0	0.0000	Simmon *et al.* (1977)
DMX, *Drosophila melanogaster*, sex-linked recessive lethal mutation	+	0	200.0000	Valencia *et al.* (1985)
SIC, Sister chromatid exchange, Chinese hamster CHO cells *in vitro*	(+)	+	1.0000	Bull *et al.* (1985)
DIH, DNA strand breaks, human lymphoblast cell line	(+)	0	220.0000	Daniel *et al.* (1986)
MVM, Micronucleus test, CD-1 mice *in vivo*	–	0	50.0000	Bull *et al.* (1985)
SPF, Sperm morphology, B6C3F$_1$ mice *in vivo*	–	0	50.0000	Meier *et al.* (1985)

Table 3 (contd). Genetic and related effects of trichloroacetonitrile

Test system	Result		Dose LED/HID	Reference
	Without exogenous metabolic system	With exogenous metabolic system		
SA0, *Salmonella typhimurium* TA100, reverse mutation	–	–	840.0000	Bull et al. (1985)
SA5, *Salmonella typhimurium* TA1535, reverse mutation	–	–	840.0000	Bull et al. (1985)
SA7, *Salmonella typhimurium* TA1537, reverse mutation	–	–	0.0000	Bull et al. (1985)
SA8, *Salmonella typhimurium* TA1538, reverse mutation	–	–	0.0000	Bull et al. (1985)
SA9, *Salmonella typhimurium* TA98, reverse mutation	–	–	840.0000	Bull et al. (1985)
SIC, Sister chromatid exchange, Chinese hamster CHO cells *in vitro*	+	+	2.0000	Bull et al. (1985)
DIH, DNA strand breaks, human lymphoblast cell line	+	0	0.0000	Daniel et al. (1986)
MVM, Micronucleus test, CD-1 mice *in vivo*	–	0	50.0000	Bull et al. (1985)
SPF, Sperm morphology, B6C3F$_1$ mice *in vivo*	–	0	50.0000	Meier et al. (1985)

(b) Humans

No data were available to the Working Group.

3.3 Case reports and epidemiological studies of carcinogenicity to humans

No data were available to the Working Group.

4. Summary of Data Reported and Evaluation

4.1 Exposure data

Halogenated acetonitriles are not produced on an industrial scale. Chloro- and trichloroacetonitriles have been used on a limited basis in the past as pesticides.

Several halogenated acetonitriles have been detected in chlorinated drinking-water in a number of countries as a consequence of the reaction of chlorine with natural organic substances (and bromine in the case of brominated acetonitriles) present in untreated water. The only known route of human exposure is through chlorinated drinking-water.

4.2 Experimental carcinogenicity data

Halogenated acetonitriles (chloroacetonitrile, dichloroacetonitrile, trichloroacetonitrile, bromochloroacetonitrile and dibromoacetonitrile) were tested in a limited carcinogenicity study in female Sencar mice by skin application, in an initiation/promotion study in female Sencar mice by skin application and in a screening assay for lung tumours in female strain A mice by oral administration. No skin tumour was produced by any of the haloacetonitriles after skin application in mice. In the initiation/promotion study, reproducible, significant increases in the numbers of animals with skin tumours were seen only with dibromoacetonitrile; no dose-related increase in the incidence of skin tumours was observed. Marginal increases in the proportion of mice with lung tumours occurred with all of the halogenated acetonitriles, but the increases were significant only with chloroacetonitrile, trichloroacetonitrile and bromochloroacetonitrile.

4.3 Human carcinogenicity data

No data were available to the Working Group.

4.4 Other relevant data

In short-term screening studies *in vivo*, chloroacetonitrile, bromochloroacetonitrile and dibromoacetonitrile caused minimal developmental toxicity in the presence of significant maternal toxicity. In developmental toxicity studies, dichloroacetonitrile and trichloroacetonitrile caused malformations and embryolethality in the presence of maternal toxicity; with trichloroacetonitrile, embryolethality was also observed at lower dose levels in the absence of maternal toxicity.

Mutations were induced in bacteria by bromochloroacetonitrile and dichloroacetonitrile but not by chloroacetonitrile, dibromoacetonitrile or trichloroacetonitrile. Mutations were induced in insects by dichloroacetonitrile but not by dibromoacetonitrile.

Sister chromatid exchange was induced in cultured mammalian cells by all five halogenated acetonitriles. DNA strand breaks were induced in human lymphocytes *in vitro* by bromochloroacetonitrile, dibromoacetonitrile and trichloroacetonitrile.

In orally treated mice, neither micronuclei in bone-marrow cells nor sperm-head abnormalities were induced by any of the five halogenated acetonitriles.

4.5 Evaluations[1]

There is *inadequate evidence* for the carcinogenicity of bromochloroacetonitrile in experimental animals.

There is *inadequate evidence* for the carcinogenicity of chloroacetonitrile in experimental animals.

There is *inadequate evidence* for the carcinogenicity of dibromoacetonitrile in experimental animals.

There is *inadequate evidence* for the carcinogenicity of dichloroacetonitrile in experimental animals.

There is *inadequate evidence* for the carcinogenicity of trichloroacetonitrile in experimental animals.

No data were available from studies in humans on the carcinogenicity of halogenated acetonitriles.

---

[1]For definition of the italicized terms, see Preamble, pp. 30-33.

Overall evaluations

Bromochloroacetonitrile *is not classifiable as to its carcinogenicity to humans (Group 3).*

Chloroacetonitrile *is not classifiable as to its carcinogenicity to humans (Group 3).*

Dibromoacetonitrile *is not classifiable as to its carcinogenicity to humans (Group 3).*

Dichloroacetonitrile *is not classifiable as to its carcinogenicity to humans (Group 3).*

Trichloroacetonitrile *is not classifiable as to its carcinogenicity to humans (Group 3).*

5. References

Aldrich Chemical Co. (1990) *1990-1991 Aldrich Handbook of Fine Chemicals*, Milwaukee, WI, pp. 275, 406, 1267

American Tokyo Kasei (1988) *TCI American Organic Chemicals 88/89 Catalog*, Portland, OR, pp. 262, 1217

Bieber, T.I. & Trehy, M.L. (1983) Dihaloacetonitriles in chlorinated natural waters. In: Jolley, R.L., Brungs, W.A., Cotruvo, J.A., Cumming, R.B., Mattice, J.S. & Jacobs, V.A., eds, *Water Chlorination: Environmental Impact and Health Effects*, Vol. 4, Book 1, *Chemistry and Water Treatment*, Ann Arbor, MI, Ann Arbor Science, pp. 85-96

Bruchet, A., Tsutsumi, Y., Duguet, J.P. & Mallevialle, J. (1985) Characterization of total halogenated compounds during water treatment processes. In: Jolley, R.L., Bull, R.J., Davis, W.P., Katz, S., Roberts, M.H., Jr & Jacobs, V.A., eds, *Water Chlorination: Chemistry, Environmental Impact and Health Effects*, Vol. 5, Chelsea, MI, Lewis Publishers, pp. 1165-1184

Budavari, S., ed. (1989) *The Merck Index*, 11th ed., Rahway, NJ, Merck & Co., p. 1515

Bull, R.J. (1985) Carcinogenic and mutagenic properties of chemicals in drinking water. *Sci. total Environ.*, 47, 385-413

Bull, R.J. & Robinson, M. (1985) Carcinogenic activity of haloacetonitrile and haloacetone derivatives in the mouse skin and lung. In: Jolley, R.L., Bull, R.J., Davis, W.P., Katz, S., Roberts, M.H., Jr & Jacobs, V.A., eds, *Water Chlorination: Chemistry, Environmental Impact and Health Effects*, Vol. 5, Chelsea, MI, Lewis Publishers, pp. 221-227

Bull, R.J., Meier, J.R., Robinson, M., Ringhand, H.P., Laurie, R.D. & Stober, J.A. (1985) Evaluation of mutagenic and carcinogenic properties of brominated and chlorinated acetonitriles: by-products of chlorination. *Fundam. appl. Toxicol.*, 5, 1065-1074

Chemical Information Systems (1990) *ISHOW Database*, Baltimore, MD

Chriswell, C.D., Gjerde, D.T., Schultz-Sibbel, G., Fritz, J.S. & Ogawa, I. (1983) *An Evaluation of the Adsorption Properties of Silicalite for Potential Application to Isolating Polar Low-molecular-weight Organics from Drinking Water* (EPA-600/1-83-001; NTIS PB83-148502), Research Triangle Park, NC, US Environmental Protection Agency

Coleman, W.E., Munch, J.W., Kaylor, W.H., Streicher, R.P., Ringhand, H.P. & Meier, J.R. (1984) Gas chromatography/mass spectroscopy analysis of mutagenic extracts of aqueous chlorinated humic acid. A comparison of the byproducts to drinking water contaminants. *Environ. Sci. Technol.*, *18*, 674-681

Daniel, F.B., Schenck, K.M., Mattox, J.K., Lin, E.L.C., Haas, D.L. & Pereira, M.A. (1986) Genotoxic properties of haloacetonitriles: drinking water by-products of chlorine disinfection. *Fundam. appl. Toxicol.*, *6*, 447-453

Dow Chemical Co. (1975) Vapor phase chlorination of acetonitrile in the presence of water. US Patent 3923860 [CA84(9): 58685c]

Eastman Kodak Co. (1987) *Kodak Laboratory and Research Products* (Catalog No. 53), Rochester, NY, pp. 88, 420

Fairfield Chemical Co. (1988) *General Catalog No. 9: Organic Research Chemicals, Custom Synthesis, Bulk Chemicals 1989-1990*, Blythewood, SC, pp. 17, 30, 48, 50

Hayes, J.R., Condie, L.W., Jr & Borzelleca, J.F. (1986) Toxicology of haloacetonitriles. *Environ. Health Perspect.*, *69*, 183-202

Herren-Freund, S.L. & Pereira, M.A. (1986) Carcinogenicity of by-products of disinfection in mouse and rat liver. *Environ. Health Perspect.*, *69*, 59-65

Keith, L.H., Hall, R.C., Hanisch, R.C., Landolt, R.G. & Henderson, J.E. (1982) New methods for analyzing water pollutants. *Water Sci. Technol.*, *14*, 59-71

Krasner, S.W., McGuire, M.J., Jacangelo, J.G., Patania, N.L., Reagan, K.M. & Aieta, E.M. (1989) The occurrence of disinfection by-products in drinking water in a nationwide study. In: *Annual Conference of the American Water Works Association, Los Angeles, CA*, Denver, CO, American Water Works Association, pp. 1-32

Leo, A., Hansch, C. & Elkins, D. (1971) Partition coefficients and their uses. *Chem. Rev.*, *71*, 525-558

Lin, E.L.C., Daniel, F.B., Herren-Freund, S.L. & Pereira, M.A. (1986) Haloacetonitriles: metabolism, genotoxicity, and tumor-initiating activity. *Environ. Health Perspect.*, *69*, 67-71

McKinney, J.D., Maurer, R.R., Hass, J.R. & Thomas, R.O. (1976) Possible factors in the drinking water of laboratory animals causing reproductive failure. In: Keith, L.H., ed., *Identification and Analysis of Organic Pollutants in Water*, Ann Arbor, MI, Ann Arbor Science, pp. 417-432

Meier, J.R., Bull, R.J., Stober, J.A. & Cimino, M.C. (1985) Evaluation of chemicals used for drinking water disinfection for production of chromosomal damage and sperm-head abnormalities in mice. *Environ. Mutagenesis*, *7*, 201-211

Movsum-Zade, E.M., Mamedov, M.G. & Shikhiev, I.A. (1977) Chlorination of acetonitrile (Russ.). *Khim. Prom-st.*, *9*, 662-663

Oliver, B.G. (1983) Dihaloacetonitriles in drinking water: algae and fulvic acid as precursors. *Environ. Sci. Technol.*, *17*, 80-83

Otson, R. (1987) Purgeable organics in Great Lakes raw and treated water. *Int. J. environ. anal. Chem.*, *31*, 41-53

Pereira, M.A., Daniel, F.B. & Lin, E.L.C. (1984a) *Relationship Between the Metabolism of Haloacetonitriles and Chloroform and Their Carcinogenic Activity* (EPA-600/D-84-203; NTIS PB84-246230), Cincinnati, OH, US Environmental Protection Agency

Pereira, M.A., Lin, L.-H.C. & Mattox, J.K. (1984b) Haloacetonitrile excretion as thiocyanate and inhibition of dimethylnitrosamine demethylase: a proposed metabolic scheme. *J. Toxicol. environ. Health*, *13*, 633-641

Pfaltz & Bauer (1988) *Organic and Inorganic Chemicals for Research*, Waterbury, CT, pp. 83, 120, 124

Pouchert, C.J., ed. (1974) *The Aldrich Library of NMR Spectra*, Vol. 3, Milwaukee, WI, Aldrich Chemical Co., pp. 162A, 162C

Pouchert, C.J., ed. (1981) *The Aldrich Library of Infrared Spectra*, 3rd ed., Milwaukee, WI, Aldrich Chemical Co., pp. 506D, 506E, 506G

Pouchert, C.J., ed. (1983) *The Aldrich Library of NMR Spectra*, 2nd ed., Vol. 1, Milwaukee, WI, Aldrich Chemical Co., pp. 705D, 706C

Pouchert, C.J., ed. (1985a) *The Aldrich Library of FT-IR Spectra*, Vol. 1, Milwaukee, WI, Aldrich Chemical Co., pp. 848A, 848C, 849A

Pouchert, C.J., ed. (1985b) *The Aldrich Library of FT-IR Spectra*, Vol. 3, Milwaukee, WI, Aldrich Chemical Co., pp. 806D, 807A, 807B

Reckhow, D.A. & Singer, P.C. (1984) The removal of organic halide precursors by preozonation and alum coagulation. *J. Am. Water Works Assoc.*, *76*, 151-157

Reding, R., Fair, P.S., Shipp, C.J. & Brass, H.J. (1989) Measurement of dihaloacetonitriles and chloropicrin in US drinking waters. In: *Disinfection By-products: Current Perspectives*, Denver, CO, American Water Works Association, pp. 11-22

Riedel-de Haën (1986) *Laboratory Chemicals 1986*, Seelze 1/Hannover, pp. 243, 1070

Roby, M.R., Carle, S., Pereira, M.A. & Carter, D.E. (1986) Excretion and tissue disposition of dichloroacetonitrile in rats and mice. *Environ. Health Perspect.*, *69*, 215-220

Sadtler Research Laboratories (1980) *The Sadtler Standard Spectra, 1980 Cumulative Index*, Philadelphia, PA

Sax, N.I. & Lewis, R.J. (1987) *Hawley's Condensed Chemical Dictionary*, 11th ed., New York, NY, Van Nostrand Reinhold, pp. 261, 1175

Simmon, V.F., Kauhanen, K. & Tardiff, R.G. (1977) Mutagenic activity of chemicals identified in drinking water. In: Scott, D., Bridges, B.A. & Sobels, F.H., eds, *Progress in Genetic Toxicology*, Amsterdam, Elsevier/North-Holland Biomedical Press, pp. 249-258

Smith, M.K., George, E.L., Zenick, H., Manson, J.M. & Stober, J.A. (1987) Developmental toxicity of halogenated acetonitriles: drinking water by-products of chlorine disinfection. *Toxicology*, *46*, 83-93

Smith, M.K., Randall, J.L., Tocco, D.R., York, R.G., Stober, J.A. & Read, E.J. (1988) Teratogenic effects of trichloroacetonitrile in the Long-Evans rat. *Teratology*, *38*, 113-120

Smith, M.K., Randall, J.L., Stober, J.A. & Read, E.J. (1989) Developmental toxicity of dichloroacetonitrile: a by-product of drinking water disinfection. *Fundam. appl. Toxicol.*, *12*, 765-772

Stevens, A.A., Moore, L.A., Slocum, C.J., Smith, B.L., Seeger, D.R. & Ireland, J.C. (1989) By-products of chlorination at ten operating utilities. In: *Disinfection By-products: Current Perspectives*, Denver, CO, American Water Works Association, pp. 23-61

STN International (1989a) *HODOC (Handbook of Data on Organic Chemicals) Database*, Columbus, OH

STN International (1989b) *Beilstein Database*, Columbus, OH

Suffet, I.H., Brenner, L. & Cairo, P.R. (1980) GC/MS identification of trace organics in Philadelphia drinking waters during a 2-year period. *Water Res.*, *14*, 853-867

Tanii, H. & Hashimoto, K. (1984) Studies on the mechanism of acute toxicity of nitriles in mice. *Arch. Toxicol.*, *55*, 47-54

Trehy, M.L. & Bieber, T.I. (1980) Effects of commonly used water treatment processes on the formation of THMs and DHANs. In: *Proceedings of the American Water Works Association 1980 Annual Conference*, Part 1, Denver, CO, American Water Works Association, pp. 125-138

Trehy, M.L. & Bieber, T.I. (1981) Detection, identification and quantitative analysis of dihaloacetonitriles in chlorinated natural waters. In: Keith, L.H., ed., *Advances in the Identification and Analysis of Organic Pollutants in Water*, Vol. 2, Ann Arbor, MI, Ann Arbor Science, pp. 941-975

Trehy, M.L., Yost, R.A. & Miles, C.J. (1986) Chlorination byproducts of amino acids in natural waters. *Environ. Sci. Technol.*, *20*, 1117-1122

Trehy, M.L., Yost, R.A. & Miles, C.J. (1987) Amino acids as potential precursors for halogenated byproducts formed on chlorination of natural waters. In: *194th American Chemical Society Meeting, New Orleans, LA, August 30-Sept 4, 1987* (Extended Abstracts), Washington DC, American Chemical Society, pp. 582-584

Valencia, R., Mason, J.M., Woodruff, R.C. & Zimmering, S. (1985) Chemical mutagenesis testing in *Drosophila*. III. Results of 48 coded compounds tested for the National Toxicology Program. *Environ. Mutagenesis*, *7*, 325-348

Weast, R.C., ed. (1989) *CRC Handbook of Chemistry and Physics*, 70th ed., Boca Raton, FL, CRC Press, pp. C-52, C-204, C-530

Zoeteman, B.C.J., Hrubec, J., de Greef, E. & Kool, H.J. (1982) Mutagenic activity associated with by-products of drinking water disinfection by chlorine, chlorine dioxide, ozone and UV-irradiation. *Environ. Health Perspect.*, *46*, 197-205

SOME OTHER HALOGENATED CHEMICALS

BROMOETHANE

1. Chemical and Physical Data

1.1 Synonyms

Chem. Abstr. Services Reg. No.: 74-96-4
Chem. Abstr. Name: Bromoethane
IUPAC Systematic Name: Bromoethane
Synonyms: Bromic ether; ethyl bromide; hydrobromic ether; monobromoethane

1.2 Structural and molecular formulae and molecular weight

$$\begin{array}{c} \text{H} \quad \text{H} \\ | \quad | \\ \text{H} - \text{C} - \text{C} - \text{Br} \\ | \quad | \\ \text{H} \quad \text{H} \end{array}$$

C_2H_5Br Mol. wt: 108.97

1.3 Chemical and physical properties of the pure substance

(a) *Description*: Clear, colourless liquid with ethereal odour and burning taste (Great Lakes Chemical Corp., 1981; Budavari, 1989)
(b) *Boiling-point*: 38.4°C (Weast, 1989)
(c) *Melting-point*: -118.6°C (Weast, 1989)
(d) *Density*: 1.4604 at 20/4°C (Weast, 1989)
(e) *Spectroscopy data*[1]: Infrared (Sadtler Research Laboratories, 1980, prism [4631, 4632], grating [10951]; Pouchert, 1981, 1985), nuclear magnetic

---

[1]In square brackets, spectrum number in compilation

resonance (Sadtler Research Laboratories, 1980, proton [225, V10], C-13 [616]; Pouchert, 1974, 1983) and mass spectral data [331]) have been reported.

(f) *Solubility*: Soluble in water (1.067 g/100 g at 0°C, 0.914 g/100 g at 20°C), ethanol, chloroform and diethyl ether (Budavari, 1989; Weast, 1989)

(g) *Volatility*: Vapour pressure, 400 mm Hg at 21.0°C (Weast, 1989); relative vapour density (air = 1), 3.75 (Great Lakes Chemical Corp., 1989a)

(h) *Stability*: Turns yellow on exposure to air and light (Budavari, 1989)

(i) *Reactivity*: Reacts rapidly with metals such as sodium, potassium, calcium, powdered aluminium, zinc and magnesium (Sittig, 1985)

(j) *Octanol/water partition coefficient (P)*: log P, 1.61 (Chemical Information Systems, 1990)

(k) *Conversion factor*[1]: mg/m^3 = 4.46 × ppm

1.4 Technical products and impurities

Bromoethane is available as a commercial-grade liquid with a minimum purity of 99% and maximum acidity of 5.0 ppm (as HBr) (Great Lakes Chemical Corp., 1989b).

2. Production, Use, Occurrence and Analysis

2.1 Production and use

(a) *Production*

Bromoethane was first synthesized in France in 1827 by Serullas from alcohol and bromine reacted with phosphorus (STN International, 1989). It can be produced commercially by the following methods: reaction of ethanol with hydrogen bromide; distillation of a mixture of hydrogen bromide, ethanol and sulfuric acid; reaction of ethanol with phosphorus and bromine; reaction of ethane with sulfur trioxide and potassium bromide at 300-325°C; and reaction of ethylene with hydrogen bromide initiated by gamma radiation (Stenger, 1978; Budavari, 1989).

Bromoethane is currently produced by three companies in France, three in the UK, two in Germany, two in Japan, two in the USA and one each in Israel, Italy and the Netherlands (Chemical Information Services, 1988).

---

[1]Calculated from: mg/m^3 = (molecular weight/24.45) × ppm, assuming standard temperature (25°C) and pressure (760 mm Hg)

(b) Use

Bromoethane has been used as an ethylating agent in organic synthesis and gasoline, as a refrigerant and as an extraction solvent; it has had limited use as a local anaesthetic (Sittig, 1985; Strobel & Grummt, 1987). It has been investigated as a possible substitute for chlorofluorocarbons in compression heat pumps (Narodoslawsky & Moser, 1988).

(c) Regulatory status and guidelines

Occupational exposure limits and guidelines for bromoethane are presented in Table 1.

Table 1. Occupational exposure limits and guidelines for bromoethane[a]

Country		Year	Concentration (mg/m^3)	Interpretation[b]
Argentina		1985	1100	TWA and STEL
Australia		1985	890	TWA
			1110	STEL
Austria		1987	890	TWA
Belgium		1989	890	TWA
			1110	STEL
Brazil		1987	695	TWA
Canada		1986	890	TWA
			1110	STEL
Denmark		1987	890	TWA
Finland		1987	890	TWA
			1115	STEL
Germany		1989	890	TWA
Hungary		1985	50	TWA
			250	STEL (30 min)
Indonesia		1987	890	TWA
Italy		1987	145	TWA
Netherlands		1986	890	TWA
Poland		1985	50	TWA
Romania		1985	400	TWA
			500	STEL
Switzerland		1987	890	TWA
UK		1987	890	TWA
		1987	1110	STEL (10 min)
USA	ACGIH	1989	891	TWA
			1110	STEL
	OSHA	1989	890	TWA
			1110	STEL
USSR		1987	5	TWA

Table 1 (contd)

Country	Year	Concentration (mg/m^3)	Interpretation[b]
Venezuela	1987	800	TWA
		1110	STEL
Yugoslavia	1987	890	TWA

[a]From Cook (1987); American Conference of Governmental Industrial Hygienists (ACGIH) (1989); US Occupational Safety and Health Administration (OSHA) (1989); United Nations Environment Programme (1990)
[b]TWA, time-weighted average; STEL, short-term exposure limit

2.2 Occurrence

(a) Natural occurrence

Macroalgae collected near the Bermuda Islands (*Fucales sargassum*) and at the Cape of Good Hope (*Laminariales laminaria*) showed a specific pattern of emission of volatile organohalides into the air. The main components were bromoform, bromodichloromethane and chlorodibromomethane; a minor component was bromoethane (Class *et al.*, 1986).

(b) Air

Bromoethane was not found (detection limit, < 1 ppt) in six air samples taken near the surface of the Pacific Ocean at several sites in the northern hemisphere (Hoyt & Rasmussen, 1985).

(c) Water and sediments

Bromoethane was detected (but not quantified) in headspace analysis of seven out of ten seawater samples collected in 1983 (Hoyt & Rasmussen, 1985).

In a two-year study of trace organic compounds in drinking-water in Philadelphia, PA, Suffet *et al.* (1980) tested samples derived from two surface water (river) sources and collected at four sites. Bromoethane was identified in one of 13 samples collected at the two sites originating from the Delaware River but in neither of the three samples originating from the Schuylkill River. No quantitative analysis was performed.

2.3 Analysis

Selected methods for the analysis of bromoethane in air, breath and water are presented in Table 2.

Table 2. Methods for the analysis of bromoethane

Sample matrix	Sample preparation	Assay procedure[a]	Limit of detection[b]	Reference
Air	Adsorb on activated charcoal; desorb (2-propanol); inject aliquot	GC/FID	0.02 mg/m^3	Eller (1985)
	Draw through tube; compare reaction with standard chart	Colorimetric	NR	Lodge (1989a); SKC Inc. (1989)
	Collect cryogenically into stainless-steel bottle; inject sample	GC/EC-FI-FPD/ GC/MS	1 ppt (4.5 ng/m^3)	Hoyt & Rasmussen (1985)
Seawater	Collect in a vacuum extraction flask; pressurize with zero air; inject headspace sample	GC/EC-FI-FPD/ GC/MS	1 ppt (1 ng/l)	Hoyt & Rasmussen (1985)
Breath	Collect in plastic bag; evacuate cell; draw sample in and scan	FT-IR	10 ppm (45 mg/m^3)	Lodge (1989b)
Water	Put in sample vessel; place probe into headspace; measure peak intensity of fragment ions	MIMS	1 ppb (μg/l)	Wenhu et al. (1987)

[a]GC/FID, gas chromatography/flame ionization detection; GC/EC-FI-FPD/GC-MS, gas chromatography/electron capture-flame ionization-flame photometric detection/gas chromatography-mass spectrometry; FT-IR, Fourier transform-infrared spectroscopy; MIMS, membrane introduction mass spectroscopy

3. Biological Data Relevant to the Evaluation of Carcinogenic Risk to Humans

3.1 Carcinogenicity studies in animals (Table 3)

(a) Inhalation

Mouse: Groups of 50 male and 50 female B6C3F$_1$ mice, nine weeks old, were exposed to 100, 200 or 400 ppm bromoethane (446, 890 or 1780 mg/m3; > 98% pure) by whole-body inhalation for 6 h per day on five days per week for 103 weeks. Survival at 105 weeks was: males—control, 35/50; low-dose, 37/50; mid-dose, 30/50; high-dose, 34/50; females—control, 36/50; low-dose, 37/50; mid-dose, 36/50; high-dose, 22/50. Uterine neoplasms reduced the survival in high-dose female mice. Adenomas of the uterine endometrium occurred in 0/50 control, 1/50 low-dose, 1/47 mid-dose and 6/48 high-dose female mice, and adenocarcinomas occurred in 0/50 control, 2/50 low-dose, 3/47 mid-dose and 19/48 high-dose females; squamous-cell carcinomas of the uterine endometrium occurred in 0/50 control, 1/50 low-dose, 1/47 mid-dose and 3/48 high-dose female mice. The proportion of female mice with uterine endometrium neoplasms (0/50 control, 4/50 low-dose, 5/47 mid-dose, and 27/48 high-dose) was significantly increased over that in controls at all exposure concentrations (low-dose, $p = 0.017$; mid-dose, $p = 0.035$; high-dose, $p < 0.001$, trend test; $p < 0.001$, incidental tumour test). Alveolar/bronchiolar adenomas occurred in 5/50 control, 6/50 low-dose, 8/50 mid-dose and 9/50 high-dose males, and alveolar/bronchiolar carcinomas occurred in 2/50 control, 0/50 low-dose, 5/50 mid-dose and 6/50 high-dose male mice. The proportion of high-dose male mice with alveolar/bronchiolar neoplasms was significantly increased relative to that in controls ($p = 0.049$, pairwise comparison; $p = 0.012$, trend test, incidental tumour test) (National Toxicology Program, 1989).

Rat: Groups of 50 male and 50 female Fischer 344 rats, eight to ten weeks old, were exposed to 100, 200 or 400 ppm bromoethane (446, 890 or 1780 mg/m^3 bromoethane; > 98% pure) by whole-body inhalation 6 h per day on five days per week for 104 weeks. Survival at 106 weeks was: males—control, 17/50; low-dose, 26/50; mid-dose, 26/50; high-dose, 21/50; females—control, 19/50; low-dose, 29/50; mid-dose, 24/50; high-dose, 22/50. Adrenal medullary phaeochromocytomas (benign and malignant combined) occurred in 8/48 control, 23/47 low-dose, 18/50 mid-dose and 21/49 high-dose male rats. The proportion of low and high-dose male rats with phaeochromocytoma was significantly greater than that in controls (low-dose, $p = 0.013$; high-dose, $p = 0.007$, incidental tumour test); however, there was disproportionate sampling of the adrenal medulla among control and exposure groups (numbers of adrenal medullas examined microscopically: control, 66; low-dose, 82; mid-dose, 85; high-dose, 86). When statistical analyses were performed using as denominators the number of medullas examined microscopically [to reduce observation bias], the p values were 0.022 (low-dose) and 0.027 (high-dose). Granular-cell tumours of the brain occurred in 0/49 control, 3/50 low-dose, 1/50 mid-dose and 1/50 high-dose male rats; these incidences are not significant. The incidences of these tumours in historical controls were 0/297 in the study laboratory and 4/1928 in all National Toxicology Program laboratories. Glial-cell tumours of the brain (glioma, astrocytoma or oligodendroglioma) occurred in 0/49 control and 3/50 low-dose male rats and 0/50 control, 1/50 low-dose, 1/48 mid-dose and 3/50 high-dose female rats ($p = 0.045$, trend test); the incidences of these tumours in historical controls were: males—study laboratory, 3/297; all National Toxicology Program studies, 13/1928; females—study laboratory, 1/297; all National Toxicology Program laboratories, 23/1969). Gliosis was reported in one rat in each of the low- and high-dose male groups and control and high-dose female groups. The incidence of mammary gland neoplasms (all histological types) was significantly decreased in high-dose female rats (control, 18/50; low-dose, 15/50; mid-dose, 10/48; high-dose, 7/50; $p = 0.011$, pairwise comparison; $p = 0.004$, trend test, incidental tumour test). Inflammation, epithelial hyperplasia and squamous metaplasia of the nasal cavity mucosa were increased in frequency in exposed rats (National Toxicology Program, 1989).

(b) *Subcutaneous administration*

Rat: Groups of 20 female CB hooded rats, six weeks old, were given a single subcutaneous injection of bromoethane [purity unspecified] in trioctanoin at 1.25, 4.2 or 12.5 mmol/kg bw (136, 460 or 1362 mg/kg bw). Twenty female rats given trioctanoin alone were used as controls. Survival at 96 weeks after injection was: low-dose, 18/20; mid-dose, 17/20; high-dose, 17/20 [survival of controls not given]. No sarcoma was seen at the injection sites in either treated or control rats (Dipple *et al.*, 1981). [The Working Group noted that only single injections were given and the small number of animals.]

(c) *Intraperitoneal administration*

Mouse: In a screening assay based on the enhanced induction of lung tumours, groups of 10 male and 10 female strain A/He mice, six to eight weeks old, were injected intraperitoneally three times per week with three dose levels of bromoethane [purity unspecified] in tricaprylin for a total of 24 injections (total doses, 11, 27.5 or 55 mmol/kg bw [1200, 3000 or 6000 mg/kg bw]). A group of 160 mice given tricaprylin only were used as controls. All surviving animals were killed 24 weeks after the first injection. Survival at that time was 154/160 in the tricaprylin vehicle control, 19/20 in the low-dose, 16/20 in the mid-dose and 20/20 in the high-dose groups. The proportions of mice with lung tumours were 34/154, 4/19, 4/16 and 6/20 in the four groups, respectively; the average numbers of lung tumours per mouse were 0.22 ± 0.07 (SE), 0.21 ± 0.05, 0.31 ± 0.08 and 0.35 ± 0.08. In positive control groups given a single intraperitoneal injection of 10 or 20 mg urethane, the numbers of mice with lung tumours were 19/19 and 18/18, respectively and the average numbers of lung tumours per animal were 8.1 ± 2.3 and 17.8 ± 4.32 (Poirier *et al.*, 1975).

3.2 Other relevant data

(a) *Experimental systems*

(i) *Absorption, distribution, metabolism and excretion*

Rats given 2.3-11 mmol/kg bw (250-1200 mg/kg bw) bromoethane in olive oil by gavage eliminated 66.7-74.5% of the dose in the expired air within 5-11 h. When rats

Table 3. Summary of carcinogenicity studies of bromoethane in experimental animals

Reference	Species/ strain	Sex	Dose schedule	Experimental parameter/ observation	Group 0	Group 1	Group 2	Group 3	Significance	Comments
National Toxicology Program (1989)	Mouse B6C3F$_1$	M	6 h/day, 5 d/week, inhalation, 103 weeks	Dose (ppm) Survival (105 weeks) Alveolar/bronchiolar Adenoma Carcinoma	0 35/50 5/50 2/50	100 37/50 6/50 0/50	200 30/50 8/50 5/50	400 34/50 9/50 6/50	$p = 0.049$	Increase
		F		Dose (ppm) Survival (105 weeks) Uterine Adenoma Adenocarcinoma Squamous-cell carcinoma	0 36/50 0/50 0/50 0/50	100 37/50 1/50 2/50 1/50	200 36/50 1/47 3/47 1/47	400 22/50 6/48 19/48 3/48	$p = 0.017$ low-dose $p = 0.035$ mid-dose $p < 0.001$ high-dose	Increases, also causing death at high dose
Poirier et al. (1975)	Mouse strain A/He	M F	3 d/week, i.p. inj., tricaprylin, 24 doses	Total dose (mmol/kg) Survival (24 weeks) Lung adenomas Lung adenomas per mouse	0 154/160 34/154 0.22 ±0.07	11 19/20 4/19 0.21 ±0.05	27.5 16/20 4/16 0.31 ±0.08	55 20/20 6/20 0.35 ±0.08		Sexes pooled for analysis; screening test in strain in which lung adenomas are common (±SE)
National Toxicology Program (1989)	Rat F344	M	6 h/d, 5 d/week, inhalation, 104 weeks	Dose (ppm) Survival (106 weeks) Adrenal medullary phaeochromocytoma	0 17/50 8/48	100 26/50 23/47	200 26/50 18/50	400 21/50 21/49	$p = 0.013$ low-dose $p = 0.112$ mid-dose $p = 0.007$ high-dose	Increases
				Granular-cell tumour of the brain	0/49	3/50	1/50	1/50		
				Glial-cell tumour of the brain	0/49	3/50	0/50	0/50		
				Alveolar/bronchiolar neoplasm	0/48	0/49	4/48	1/48		

BROMOETHANE

Table 3 (contd)

Reference	Species/ strain	Sex	Dose schedule	Experimental parameter observation	Group 0	1	2	3	Significance	Comments
National Toxicology Program (1989) (contd)		F		Dose (ppm)	0	100	200	400		
				Survival (106 weeks)	19/50	29/50	24/50	22/50		
				Glial-cell tumour of the brain	0/50	1/50	1/48	3/50	$p = 0.045$	Increases
				Alveolar/bronchiolar adenoma	0/50	0/48	0/47	3/49		
				Mammary neoplasms combined	18/50	15/50	10/48	7/50	$p = 0.011$	Decreases
Dipple et al. (1981)	Rat CB hooded	F	Single s.c. inj. in trioctanoin	Dose (mmol/kg)	0	1.25	4.2	12.5		No injection-site sarcoma
				Survival (90 weeks)	Not given	18/20	17/20	17/20		

were given 1.4 mmol/kg bw in five doses of 25 mg/kg at hourly intervals (total dose, 150 mg/kg) or single doses of 4.6 mmol/kg bw (500 mg/kg) or 7.3 mmol/kg bw (800 mg/kg) bromoethane by intraperitoneal injection in olive oil, 73-89% of the dose was eliminated in the expired air by 6 h (Miller & Haggard, 1943).

Williams (1959) reported that 73-89% of a dose of bromoethane injected into rats was eliminated unchanged in the expired air. When bromoethane was given orally in oil at 0.25-1.0 g/kg, 67-76% was eliminated unchanged in the expired air and 34-38% was converted to inorganic bromide.

Enzymatic dehalogenation of bromoethane in the presence of glutathione or cysteine was demonstrated in rat liver extracts (Heppel & Porterfield, 1948).

Bromoethane bound to rat liver cytochrome P450 and inhibited its activity by 27% in microsomes of phenobarbital-induced rats (Ivanetich *et al.*, 1978).

(ii) *Toxic effects*

The toxicology of bromoethane has been reviewed (Torkelson & Rowe, 1981; National Toxicology Program, 1989).

Intraperitoneal LD_{50}s of 2850 mg/kg bw for mice and 1750 mg/kg bw for rats were reported [vehicle unspecified] (Torkelson & Rowe, 1981). The 1-h LC_{50}s of bromoethane were estimated to be 27 000 ppm [120.42 g/m^3] in male Sprague-Dawley rats and 16 200 ppm (72.25 g/m^3) in male CF-1 mice (Vernot *et al.*, 1977). The 4-h LC_{50}s in female Fischer 344/N rats and female B6C3F$_1$ mice were 4681 and 2723 ppm (20.88 and 12.1 g/m^3), respectively. All male and female Fischer 344/N rats died during or after a 4-h exposure to 10 000 ppm (44.6 g/m^3), and exposure concentrations of 10 000 and 5000 ppm (22.3 g/m^3) were lethal to all male and female B6C3F$_1$ mice, respectively (National Toxicology Program, 1989). Guinea-pigs died after inhalation of 14% by volume bromoethane for 10 min and 2.4% for 90 min, whereas all survived inhalation of 1.2% for 90 min (Sayers *et al.*, 1929). Signs of toxicity were increased respiration rate, hyperactivity, loss of coordination, dyspnoea, loss of consciousness and lung and liver congestion.

All male and female Fischer 344/N rats and B6C3F$_1$ mice exposed by inhalation to 4000 ppm (17.8 g/m^3) and 2000 ppm (8.9 g/m^3) bromoethane for 6 h per day on five days per week for two weeks died. Signs of toxicity were prostration, dyspnoea, lachrymation, haemorrhage and congestion in the respiratory tract. Some of these symptoms were also observed after exposure to 1000 ppm (4.46 g/m^3). In 14-week studies, male and female Fischer 344/N rats and B6C3F$_1$ mice were exposed to 100-1600 ppm (0.45-7.14 g/m^3) bromoethane by inhalation for 6 h per day on five days per week. The high dose resulted in some deaths and reduced body weights of the surviving animals; signs of toxicity in both species included ataxia and atrophy of thigh muscles and uterus. In rats, tremors, paresis, mineralization and degeneration in the brain, atrophy of the testis, haemosiderosis of the spleen and some depletion of bone-marrow haematopoietic cells were also observed. Involution of the ovary was observed in mice at the high and mid-doses (see below) (National Toxicology Program, 1989).

(iii) *Effects on reproduction and prenatal toxicity*

In the 14-week inhalation studies in B6C3F$_1$ mice and Fischer 344 rats described above, severe testicular atrophy was observed in all rats, but not in mice, at 1600 ppm (7.14 g/m^3) bromoethane but not at lower concentrations. Four of ten male rats in the 1600-ppm group died. In female mice, but not in rats, the size and number of corpora lutea in the ovary were decreased at 1600 ppm (7/10 animals) and at 800 ppm (3.57 g/m^3, 3/9 animals) (National Toxicology Program, 1989).

No data on reproductive or developmental toxicity were available to the Working Group.

(iv) *Genetic and related effects* (Table 4)

Bromoethane was mutagenic to *Salmonella typhimurium* strains TA100 and TA1535 when tested in closed containers. In single studies, mutations were not induced in *Drosophila melanogaster* and chromosomal aberrations were not induced in cultured mammalian cells. One study showed an increased incidence of sister chromatid exchange in cultured Chinese hamster ovary (CHO) cells. In a study reported as an abstract, bromoethane tested in a closed container enhanced viral transformation in cultured Syrian hamster embryo cells (Hatch *et al.*, 1983).

(b) *Humans*

No data were available to the Working Group.

3.3 Case reports and epidemiological studies of carcinogenicity to humans

No data were available to the Working Group.

4. Summary of Data Reported and Evaluation

4.1 Exposure data

Bromoethane has limited commercial use, including that as an ethylating agent. It has been detected in ocean air as a result of emissions by marine algae.

4.2 Experimental carcinogenicity data

Bromoethane was tested for carcinogenicity in a two-year study in male and female Fischer 344 rats and B6C3F$_1$ mice by inhalation. In male rats, there was a

Table 4. Genetic and related effects of bromoethane

Test system	Result		Dose LED/HID	Reference
	Without exogenous metabolic system	With exogenous metabolic system		
SA0, *Salmonella typhimurium* TA100, reverse mutation	+	0	0.0000	Simmon (1981)[a]
SA0, *Salmonella typhimurium* TA100, reverse mutation	+	+	580.0000	Barber et al. (1981)[a]
SA0, *Salmonella typhimurium* TA100, reverse mutation	+	0	3500.0000	Barber et al. (1983)[a]
SA0, *Salmonella typhimurium* TA100, reverse mutation	–	–	5000.0000	Haworth et al. (1983)
SA5, *Salmonella typhimurium* TA1535, reverse mutation	+	+	580.0000	Barber et al. (1981)[a]
SA5, *Salmonella typhimurium* TA1535, reverse mutation	+	0	3500.0000	Barber et al. (1983)[a]
SA5, *Salmonella typhimurium* TA1535, reverse mutation	–	–	5000.0000	Haworth et al. (1983)
SA7, *Salmonella typhimurium* TA1537, reverse mutation	–	–	5000.0000	Haworth et al. (1983)
SA9, *Salmonella typhimurium* TA98, reverse mutation	–	–	3700.0000	Barber et al. (1981)[a]
SA9, *Salmonella typhimurium* TA98, reverse mutation	–	–	5000.0000	Haworth et al. (1983)
DMX, *Drosophila melanogaster*, sex-linked recessive lethal mutation	–	0	900.0000	Vogel & Chandler (1974)
SIC, Sister chromatid exchange, Chinese hamster CHO cells	+	+	100.0000	Loveday et al. (1989)
CIC, Chromosomal aberrations, Chinese hamster CHO cells	–	–	1000.0000	Loveday et al. (1989)

[a]Closed container

significant increase in the incidence of adrenal phaeochromocytomas, which was not dose-related. A marginal increase in the incidence of uncommon brain tumours occurred in treated females. In mice, bromoethane induced neoplasms of the uterine endometrium; a marginal increase in the incidence of lung tumours was observed in males. In a screening study by intraperitoneal injection, bromoethane did not increase the incidence of lung tumours in strain A mice.

4.3 Human carcinogenicity data

No data were available to the Working Group.

4.4 Other relevant data

Bromoethane was mutagenic in bacteria but not in insects in a single study. In other single studies, bromoethane caused sister chromatid exchange but not chromosomal aberrations in mammalian cells.

4.5 Evaluation[1]

There is *limited evidence* for the carcinogenicity of bromoethane in experimental animals.

No data were available from studies in humans on the carcinogenicity of bromoethane.

Overall evaluation

Bromoethane *is not classifiable as to its carcinogenicity to humans (Group 3)*.

5. References

American Conference of Governmental Industrial Hygienists (1989) *TLVs Threshold Limit Values and Biological Exposure Indices for 1989-1990*, Cincinnati, OH, p. 23

Barber, E.D., Donish, W.H. & Mueller, K.R. (1981) A procedure for the quantitative measurement of the mutagenicity of volatile liquids in the Ames *Salmonella*/microsome assay. *Mutat. Res.*, *90*, 31-48

Barber, E.D., Donish, W.H. & Mueller, K.R. (1983) The relationship between growth and reversion in the Ames *Salmonella* plate incorporation assay. *Mutat. Res.*, *113*, 89-101

---

[1]For definition of the italicized terms, see preamble, pp. 30-33.

Budavari, S., ed. (1989) *The Merck Index*, 11th ed., Rahway, NJ, Merck & Co., p. 596

Chemical Information Services (1988) *Directory of World Chemical Producers 989/90 Edition*, Oceanside, NY

Chemical Information Systems (1990) *ENVIROFATE Database*, Baltimore, MD

Class, T., Kohnle, R. & Ballschmiter, K. (1986) Chemistry of organic traces in air. VII: Bromo- and bromochloromethanes in air over the Atlantic Ocean. *Chemosphere*, 15, 429-436

Cook, W.A. (1987) *Occupational Exposure Limits—Worldwide*, Cincinnati, OH, American Industrial Hygiene Association, pp. 130, 139, 187

Dipple, A., Levy, L.S. & Lawley, P.D. (1981) Comparative carcinogenicity of alkylating agents: comparisons of a series of alkyl and aralkyl bromides of differing chemical reactivities as inducers of sarcoma at the site of a single injection in the rat. *Carcinogenesis*, 2, 103-107

Eller, P.M. (1985) *NIOSH Manual of Analytical Methods*, 3rd ed., Vol. 2 (DHHS (NIOSH) Publ. No. 84-100), Washington DC, US Government Printing Office, pp. 1011-1—1011-3

Great Lakes Chemical Corp. (1981) *Product Information: Ethyl Bromide (Bromoethane)*, West Lafayette, IN

Great Lakes Chemical Corp. (1989a) *Material Safety Data Sheet: Ethyl Bromide*, West Lafayette, IN

Great Lakes Chemical Corp. (1989b) *Product Information: Ethyl Bromide (Bromoethane)*, West Lafayette, IN

Hatch, G., Anderson, T., Elmore, E. & Nesnow, S. (1983) Status of enhancement of DNA viral transformation for determination of mutagenic and carcinogenic potential of gaseous and volatile compounds (Abstract No. Cd-26). *Environ. Mutagenesis*, 5, 442

Haworth, S., Lawlor, T., Mortelmans, K., Speck, W. & Zeiger, E. (1983) Salmonella mutagenicity test results for 250 chemicals. *Environ. Mutagenesis*, Suppl. 1, 3-142

Heppel, L.A. & Porterfield, V.T. (1948) Enzymatic dehalogenation of certain brominated and chlorinated compounds. *J. biol. Chem.*, 76, 763-769

Hoyt, S.D. & Rasmussen, R.A. (1985) Determining trace gases in air and seawater. *Adv. chem. Ser. Mapping Strategies chem. Oceanogr.*, 209, 31-56

Ivanetich, K.M., Lucas, S., Marsh, J.A., Ziman, M.R., Katz, I.D. & Bradshaw, J.J. (1978) Organic compounds: their interaction with and degradation of hepatic microsomal drug-metabolizing enzymes *in vitro*. *Drug Metab. Disposition*, 6, 218-225

Lodge, J.P., Jr, ed. (1989a) *Methods of Air Sampling and Analysis*, 3rd ed., Chelsea, MI, Lewis Publishers, pp. 171-187

Lodge, J.P., Jr, ed. (1989b) *Methods of Air Sampling and Analysis*, 3rd ed., Chelsea, MI, Lewis Publishers, Inc., pp. 78-83

Loveday, K.S., Lugo, M.H., Resnick, M.A., Anderson, B.E. & Zeiger, E. (1989) Chromosome aberration and sister chromatid exchange tests in Chinese hamster ovary cells *in vitro*: II. Results with 20 chemicals. *Environ. mol. Mutagenesis*, 13, 60-94

Miller, D.P. & Haggard, H.W. (1943) Intracellular penetration of bromide as a feature in the toxicity of alkyl bromides. *J. ind. Hyg. Toxicol.*, 25, 423-433

Narodoslawsky, M. & Moser, F. (1988) New compression heat pump media as replacements for CFCs. *Int. J. Refrig.*, *11*, 264-268

National Toxicology Program (1989) *Toxicology and Carcinogenesis Studies of Bromoethane (Ethyl Bromide) (CAS No. 74-96-4) in F344/N Rats and B6C3F$_1$ Mice (Inhalation Studies)* (NTP Technical Report Series No. 363), Research Triangle Park, NC, US Department of Health and Human Services

Poirier, L.A., Stoner, G.D. & Shimkin, M.B. (1975) Bioassay of alkyl halides and nucleotide base analogs by pulmonary tumor response in strain A mice. *Cancer Res.*, *35*, 1411-1415

Pouchert, C.J., ed. (1974) *The Aldrich Library of NMR Spectra*, Vol. 1, Milwaukee, WI, Aldrich Chemical Co., p. 49A

Pouchert, C.J., ed. (1981) *The Aldrich Library of Infrared Spectra*, 3rd ed., Milwaukee, WI, Aldrich Chemical Co., p. 42B

Pouchert, C.J., ed. (1983) *The Aldrich Library of NMR Spectra*, 2nd ed., Vol. 1, Milwaukee, WI, Aldrich Chemical Co., p. 59A

Pouchert, C.J., ed. (1985) *The Aldrich Library of FT-IR Spectra*, Vol. 3, Milwaukee, WI, Aldrich Chemical Co., p. 82D

Sadtler Research Laboratories (1980) *The Sadtler Standard Spectra, 1980 Cumulative Index*, Philadelphia, PA

Sayers, R.R., Yant, W.P., Thomas, B.G.H. & Berger, L.B. (1929) *Physiological Response Attending Exposure to Vapors of Methyl Bromide, Methyl Chloride, Ethyl Bromide and Ethyl Chloride* (Public Health Bulletin No. 185), Washington DC, Treasury Department, US Public Health Service

Simmon, V.F. (1981) Applications of the *Salmonella* microsome assay. In: Stich, H.F. & San, R.H.C., eds, *Short Term Tests for Chemical Carcinogens*, New York, Springer, pp. 120-126

Sittig, M. (1985) *Handbook of Toxic and Hazardous Chemicals and Carcinogens*, 2nd ed., Park Ridge, NJ, Noyes Publications, pp. 414-415

SKC Inc. (1989) *Comprehensive Catalog and Guide. Air Sampling Products for Worker Monitoring, Chemical Hazard Detection and Industrial Hygiene*, Eighty Four, PA

Stenger, V.A. (1978) Bromine compounds. In: Mark, H.F., Othmer, D.F., Overberger, C.G., Seaborg, G.T. & Grayson, M., eds, *Kirk-Othmer Encyclopedia of Chemical Technology*, 3rd ed., Vol. 4, New York, John Wiley & Sons, pp. 243-263

STN International (1989) *Beilstein Database*, Columbus, OH

Strobel, K. & Grummt, T. (1987) Aliphatic and aromatic halocarbons as potential mutagens in drinking water. III. Halogenated ethanes and ethenes. *Toxicol. environ. Chem.*, *15*, 101-128

Suffet, I.H., Brenner, L. & Cairo, P.R. (1980) GC/MS identification of trace organics in Philadelphia drinking waters during a 2-year period. *Water Res.*, *14*, 853-867

Torkelson, T.R. & Rowe, V.K. (1981) Ethyl bromide. In: Clayton, G. & Clayton, F.E., eds, *Patty's Industrial Hygiene and Toxicology*, New York, John Wiley & Sons, pp. 3483-3486

United Nations Environment Programme (1990) *International Register of Potentially Toxic Chemicals, Recommendations—Legal Mechanisms*, Geneva

US Occupational Safety and Health Administration (1989) Air contaminants—permissible exposure limits (Report No. OSHA 3112). *US Code fed. Regul.*, Part 1910.1000

Vernot, E.H., MacEwen, J.D., Haun, C.C. & Kinkead, E.R. (1977) Acute toxicity and skin corrosion data for some organic and inorganic compounds and aqueous solutions. *Toxicol. appl. Pharmacol.*, *42*, 417-423

Vogel, E. & Chandler, J.L.R. (1974) Mutagenicity testing of cyclamate and some pesticides in *Drosophila melanogaster*. *Experientia*, *30*, 621-623

Weast, R.C., ed. (1989) *CRC Handbook of Chemistry and Physics*, 70th ed., Boca Raton, FL, CRC Press, pp. C-264, D-199

Wenhu, D., Kuangnan, C., Jianli, L. & Zhenying, D. (1987) Determination of trace volatile organic compounds in water samples by membrane introduction mass spectrometry. *Shitsuryo Bunseki*, *35*, 122-132

Williams, R.T. (1959) *Detoxication Mechanisms*, 2nd ed., New York, John Wiley & Sons, p. 28

CHLOROETHANE

1. Chemical and Physical Data

1.1 Synonyms

Chem. Abstr. Services Reg. No.: 75-00-3
Chem. Abstr. Name: Chloroethane
IUPAC Systematic Name: Chloroethane
Synonyms: Aethylis; aethylis chloridum; chlorethyl; ether chloratus; ether hydrochloric; ether muriatic; ethyl chloride; hydrochloric ether; monochlorethane; monochloroethane; muriatic ether

1.2 Structural and molecular formulae and molecular weight

$$\begin{array}{c} H \quad H \\ | \quad | \\ H - C - C - Cl \\ | \quad | \\ H \quad H \end{array}$$

C_2H_5Cl Mol. wt: 64.52

1.3 Chemical and physical properties of the pure substance

(a) *Description*: Gas at room temperature and pressure, with characteristic ethereal odour and burning taste (Budavari, 1989)
(b) *Boiling-point*: 12.3°C (Weast, 1989)
(c) *Melting-point*: −136.4°C (Weast, 1989)
(d) *Density*: 0.8978 at 20/4°C (Weast, 1989)
(e) *Spectroscopy data*[1]: Infrared (Sadtler Research Laboratories, 1980, prism [533], grating [36755]; Pouchert, 1985), nuclear magnetic resonance

---

[1] In square brackets, spectrum number in compilation

(Sadtler Research Laboratories, 1980, proton [V11]), ultraviolet (Hubrich & Stuhl, 1980) and mass spectral data [45] have been reported.

(f) *Solubility*: Soluble in water (5.74 g/l at 20°C), ethanol and diethyl ether (Budavari, 1989; Weast, 1989)

(g) *Volatility*: Vapour pressure, 1000 mm Hg at 20°C; relative vapour density (air = 1), 2.22 (Verschueren, 1983; Sax & Lewis, 1987; Budavari, 1989)

(h) *Reactivity*: Reacts rapidly with metals such as sodium, potassium, calcium, powdered aluminium, zinc and magnesium (Sittig, 1985)

(i) *Octanol/water partition coefficient (P)*: log P, 1.54 (Verschueren, 1983)

(j) *Conversion factor*[1]: mg/m^3 = 2.64 × ppm

1.4 Technical products and impurities

Trade names: Anodynon; Chelen; Chlorene; Chloridum; Chloryl; Chloryl anesthetic; Cloretilo; Dublofix; Kelene; Narcotile

Chloroethane is available at greater than 99% purity. One technical-grade product contains water, 0.02% max; nonvolatile residue, 0.01% max; acidity (as HCl), 0.002% max; and total impurities, 0.5% max. It is also available as an anhydrous 2.0M solution in *tert*-butyl methyl ether or diethyl ether (PPG Industries, 1986; American Tokyo Kasei, 1988; Aldrich Chemical Co., 1990).

2. Production, Use, Occurrence and Analysis

2.1 Production and use

(a) *Production*

The dominant process for production of chloroethane in the USA involves the addition of anhydrous hydrogen chloride to ethylene (see IARC, 1987a) in the presence of an aluminium chloride catalyst. The hydrochlorination is a liquid-phase reaction, carried out at about 40°C. Reacted products are fed into a flash evaporator column, where chloroethane is separated from less volatile compounds and then purified by fractionation. Hydrochlorination of ethanol has not been used for US chloroethane production since 1980, and chlorination of ethane

---

[1]Calculated from: mg/m^3 = (molecular weight/24.45) × ppm, assuming standard temperature (25°C) and pressure (760 mm Hg)

(catalytically, electrolytically, thermally or photochemically) has not been used at any production facility in the USA since 1974. Chloroethane is also obtained as a by-product from the production of vinyl chloride (see IARC, 1987b) or chlorofluorocarbon, although this method accounts for only a small amount (Mannsville Chemical Products Corp., 1984; Hume, 1988).

Production of chloroethane in the USA by year is presented in Table 1. Worldwide exports and imports of chloroethane are presented in Tables 2-4.

Table 1. US production of chloroethane, 1960-88 (in thousand tonnes)[a]

Year	Quantity
1960	247
1965	311
1970	308
1975	261
1980	180
1981	147
1982	154
1983	128
1984	132
1985	77
1986	74
1987	70
1988	69

[a]From Dialog Information Services (1989); US International Trade Commission (1989)

(b) Use

Chloroethane is used in the manufacture of tetraethyllead (see IARC, 1987c) and as an alkylating agent in the production of ethylcellulose (which is used in paper coatings, printing inks, films, adhesives and moulded plastics), ethylhydroxyethylcellulose, some pharmaceuticals and as a foam-blowing agent in the manufacture of polystyrene. It is used as a local anaesthetic because of its rapid cooling effect as it vaporizes (Reynolds, 1989). Historical and minor uses include use in organic synthesis, as an alkylating agent in the production of aluminium alkyls and other metal alkyls and as a solvent for phosphorus, sulfur, fats, oils, resins and waxes (Sax & Lewis, 1987; Hume, 1988; Dow Chemical Co., 1989).

Table 2. Worldwide imports of chloroethane, 1979-88 (in tonnes)[a]

Country	1979	1980	1981	1982	1983	1984	1985	1986	1987	1988
Belgium/Luxembourg[b]	4377	5739	5684	15 774	6168	6266	6475	5733	6166	7221
Brazil	NR[c]	168	84	127	56	15	14	158	317	NR
Canada	2029	497	936	76	10	1955	87	524	1215	3965[b]
Denmark[b]	286	435	372	361	408	396	319	326	223	NR
France	4974	509	4	415	121	1	372	123	371	4084
Germany[b]	8932	8227	8581	10 179	11 721	14 511	12 663	11 533	11 730	13 754
India	NR	5	1	NR	NR	NR	7381	3	NR	NR
Ireland[b]	509	302	277	362	478	103	161	100	243	116
Italy	3037	8923	7777	8486	8886	8877	9880	10 174	8812	NR
Japan	NR	NR	NR	NR	NR	NR	NR	NR	NR	24
Mexico	9646	10 620	10 084	10 634	8538	8739	10 496	NR	10 433	NR
Netherlands[b]	1766	2354	1725	1934	2605	3153	4002	2414	2382	2599
Pakistan	31	NR	17	2	51	2	3	3	28	NR
Portugal[b]	NR	0	0	0	54	1	22	38	0	3
Republic of Korea	55	34	2	31	5	8	11	20	19	NR
Spain	NR	NR	235	225	321	315	394	409	411	NR
UK[b]	971	1447	NR	NR	NR	NR	NR	NR	NR	NR
USA[d]	NR	1270	5030	2325	0.16	NR	1.5	NR	0.4	NR
Venezuela	0	NR	NR	0	6	36	83	51	NR	NR
Yugoslavia[e]	8732	7690	2451	1028	833	9430	51 800	36 868	117 563	1[b]

[a]From Dialog Information Services (1990)
[b]Chloromethane and chloroethane
[c]NR, not reported
[d]From US Department of Commerce (1981, 1982, 1983, 1984, 1986, 1988)
[e]Chloroethane and dichloroethane

Table 3. Countries of origin for imports of chloroethane[a]

Importing country	Year	Country of origin
Belgium/Luxembourg	1987	Federal Republic of Germany (49%), France (45%), USA (3%), Netherlands (2%), UK (1%)
	1988	Federal Republic of Germany (54%), France (42%), UK (4%)
Brazil	1986, 1987	USA (100%)
Canada	1987	USA (100%)
	1988	USA (99%), others (1%)
Denmark	1986	Federal Republic of Germany (43%), Switzerland (36%), Belgium/Luxembourg (20%), others (1%)
	1987	Federal Republic of Germany (79%), Netherlands (6%), others (15%)
Germany, Federal Republic of	1987	France (37%), UK (21%), Belgium/Luxembourg (14%), Netherlands (14%), Switzerland (14%), others (1%)
	1988	France (46%), Switzerland (20%), UK (19%), Belgium/Luxembourg (12%), Netherlands (2%), others (1%)
France	1987	Federal Republic of Germany (100%)
	1988	USSR (90%), Federal Republic of Germany (10%)
India	1985	Saudia Arabia (100%)
	1986	France (100%)
Ireland	1987	UK (55%), Federal Republic of Germany (45%)
	1988	Federal Republic of Germany (46%), Belgium/Luxembourg (38%), Netherlands (16%)
Italy	1986	Federal Republic of Germany (56%), France (44%)
	1987	Federal Republic of Germany (57%), France (43%)
Japan	1988	USA (66%), Singapore (34%)
Mexico	1987	USA (100%)
Netherlands	1987	Federal Republic of Germany (89%), France (6%), UK (3%), others (1%)
	1988	Federal Republic of Germany (87%), Belgium/Luxembourg (5%), France (4%), UK (4%)
Pakistan	1986	UK (100%)
	1987	Federal Republic of Germany (77%), UK (23%)
Republic of Korea	1986	Japan (100%)
	1987	Japan (97%), USA (3%)
Spain	1986	France (98%), Federal Republic of Germany (2%)
	1987	France (88%), Federal Republic of Germany (12%)
Yugoslavia	1987	Italy (63%), Greece (16%), Saudi Arabia (12%), USA (6%), others (3%)

[a]From Dialog Information Services (1990)

Table 4. Worldwide exports of chloroethane, 1979–88 (in tonnes)[a]

Country	1979	1980	1981	1982	1983	1984	1985	1986	1987	1988
Belgium/Luxembourg[b]	2598	2562	2192	2073	1223	1488	1333	1598	1547	1754
Brazil	NR[c]	NR	NR	NR	NR	5	NR	NR	NR	NR
Canada[b]	NR	NR	NR	NR	NR	NR	NR	NR	NR	8118
Denmark[b]	NR	20	44	23	0	1	0	NR	0	NR
France	4994	9076	9148	9035	7603	6008	NR	NR	NR	NR
Germany, Federal Republic of[b]	16 612	19 832	15 812	18 668	20 268	20 640	24 339	33 594	46 928	34 613
India	NR	NR	1	NR	NR	NR	3	NR	NR	NR
Ireland[b]	59	51	20	45	56	60	35	49	40	104
Netherlands[b]	6	567	1268	293	201	100	463	691	896	151
Spain	NR	NR	10	7	9	1	1	NR	NR	NR
UK[b]	4197	1209	1029	1852	1963	3968	4066	3590	4362	4193
USA	12 718	11 876	12 186	12 000	9697	9112	9756	13 868	11 227	8562
Yugoslavia	NR	196[d]	NR	0	NR	NR	0	NR	0	NR

[a]From Dialog Information Services (1990)
[b]Chloromethane and chloroethane
[c]NR, not reported
[d]Chloroethane and dichloroethane

The estimated use pattern for chloroethane in the USA in 1979 was 90% as an ethylating agent in the synthesis of tetraethyllead, 5% as exports and 5% for miscellaneous uses (includes ethyl cellulose plastics, dyes and pharmaceuticals) (Anon., 1979). In 1984, the estimated use pattern for chloroethane was 80% as an ethylating agent in the synthesis of tetraethyllead, 15% in the synthesis of ethyl cellulose and 5% for miscellaneous uses (Mannsville Chemical Products Corp., 1984).

Chloroethane has been investigated as a possible substitute for chlorofluorocarbons in compression heat pumps (Narodoslawsky & Moser, 1988).

(c) *Regulatory status and guidelines*

Occupational exposure limits and guidelines for chloroethane are presented in Table 5.

Table 5. Occupational exposure limits and guidelines for chloroethane[a]

Country	Year	Concentration (mg/m^3)	Interpretation[b]
Australia	1985	2600	TWA
		3250	STEL
Austria	1987	2600	TWA
Belgium	1989	2600	TWA
Brazil	1987	2030	TWA
Canada	1986	2600	TWA
Denmark	1987	2600	TWA
Finland	1989	1300	TWA
		1625	STEL
Germany	1989	2600	TWA
Hungary	1985	50	TWA
		250	STEL
Italy	1987	1000	TWA
Japan	1987	2600	TWA
Mexico	1987	2600	TWA
Netherlands	1987	2600	TWA
Norway	1984	2600	TWA
Poland	1985	200	TWA
Romania	1985	1500	TWA
		2000	STEL
Sweden	1988	1300	TWA
		1900	STEL (15 min)
Switzerland	1987	2600	TWA

Table 5 (contd)

Country	Year	Concentration (mg/m^3)	Interpretation[b]
UK	1987	2600	TWA
		3250	STEL (10 min)
USA			
ACGIH	1989	2640	TWA
OSHA	1989	890	TWA
USSR	1982	50	MAC
Venezuela	1987	2600	TWA
		3250	Ceiling
Yugoslavia	1985	260	TWA

[a]From Cook (1987); American Conference of Governmental Industrial Hygienists (ACGIH) (1989); US Occupational Safety and Health Administration (OSHA) (1989); United Nations Environment Programme (1990)

[b]TWA, time-weighted average; STEL, short-term exposure limit; MAC, maximum allowable concentration

2.2 Occurrence

(a) Natural occurrence

Chloroethane is not known to occur as a natural product.

(b) Occupational exposure

The National Occupational Hazard Survey estimated in 1972-74 that 113 000 workers were exposed to chloroethane (National Institute for Occupational Safety and Health, 1978)

A survey at the Ethyl Corp., Pasadena, TX, in 1980 to monitor worker exposure to chloroethane during the manufacture of tetraethyllead found concentrations in personal and area air samples ranging from 274 to 1143 µg/m^3 (Ringenburg, 1983).

(c) Air

According to the US Toxic Chemical Release Inventory, total emissions of chloroethane into the air in 1987 were approximately 2000 tonnes from 42 locations. Industrial releases to other media were estimated to be 0.9 tonnes to ambient water from six locations and 0.9 kg to the land from one location (National Library of Medicine, 1989).

Estimated emissions of chloroethane in the USA in 1970 were 4800 tonnes, with hydrochlorination of ethylene contributing 3700 tonnes, hydrochlorination of ethanol, 600 tonnes, and chlorination of ethane, 500 tonnes (Processes Research, 1972).

In a later study, it was estimated that about 4500 tonnes of chloroethane are released into the atmosphere every year in the USA; the average background concentration of chloroethane at 40°N in 1981 was 26 ng/m^3. Air samples collected in seven US cities in 1980-81 contained mean chloroethane concentrations of 108-598 ng/m^3 (Singh *et al.*, 1983).

Emissions of chloroethane in 1988 from five major US sources were estimated to be 290 tonnes per year from production (process, equipment leaks, transportation and storage; four producers); 339 tonnes per year from tetraethyllead production (one producer); 982 tonnes per year from ethylene dichloride production (18 producers); 60 tonnes per year from ethyl cellulose production (two producers); and 1170 tonnes per year from its use in polystyrene foam blowing (six facilities) (Hume, 1988).

In a review of data on the presence of volatile organic chemicals in the US atmosphere between 1970 and 1980, a median concentration of 160 ng/m^3 chloroethane was reported for the 348 data points examined, 160 ng/m^3 for urban/suburban areas (337 data points) and 120 ng/m^3 (11 data points) close to industrial sources (Brodzinsky & Singh, 1983).

(*d*) *Water and sediments*

In a survey of large US water utilities, chloroethane was identified as one of 36 unregulated chemicals that were detected at the greatest frequency during routine monitoring of drinking-water (Anon., 1983).

Tap-water samples taken from two buildings in Durham, NC, USA, contained chloroethane (levels not specified); chloroethane was also identified in a purified water sample (McKinney *et al.*, 1976).

Sediment samples collected in 1980 from Lake Pontchartrain, LA, USA, contained a mean chloroethane concentration of 0.2 ng/g wet weight (Ferrario *et al.*, 1985).

(*e*) *Tissues and secretions*

In a gas chromatography-mass spectrometry screening study of human milk samples collected in four urban areas of the USA, traces of chloroethane were found in two of eight samples (Pellizzari *et al.*, 1982).

Oyster samples collected in 1980 from Lake Pontchartrain, LA, USA, contained a mean chloroethane concentration of 7.6 ng/g wet weight (Ferrario *et al.*, 1985).

2.3 Analysis

Selected methods for the analysis of chloroethane in air, breath and water are identified in Table 6. The US Environmental Protection Agency methods for analysing water (Methods 8010 and 8240) have also been applied to liquid and solid wastes. Volatile components of solid-waste samples are first extracted with methanol prior to purge-and-trap concentration and analysis by gas chromatography-electrolytic conductivity detection (Method 8010) or gas chromatography-mass spectrometry (Method 8240). The detection limit using Method 8010 is 0.5 µg/l and the practical quantification limit using Method 8240 is 10 µg/l for groundwater and soil/sediment samples (US Environmental Protection Agency, 1986a,b).

US Environmental Protection Agency Method 624 has also been adapted to the analysis of chloroethane in fish, with an estimated detection limit of 250 µg/kg (Easley *et al.*, 1981).

Table 6. Methods for the analysis of chloroethane

Sample matrix	Sample preparation[a]	Assay procedure[b]	Limit of detection[c]	Reference
Air	Adsorb on activated charcoal; desorb (carbon disulfide); inject aliquot	GC/FID	0.01 mg per sample	Eller (1985); SKC Inc. (1989)
	Collect whole air sample in neutralized stainless-steel canister	GC/FID	NR	US Environmental Protection Agency (1988a) [Method TO-14]
	Draw air through tube; compare reaction with standard chart	Colorimetric	NR	Lodge (1989a)
Breath	Collect sample in plastic bag; evacuate cell; draw sample in and scan	FT-IR	10 ppm (26.4 mg/m^3)	Lodge (1989b)
Water	Purge (inert gas); trap (OV-1 on Chromosorb-W/Tenax/silica gel); desorb as vapour (heat to 180°C, backflush with inert gas) onto packed GC column	GC/ECD	0.5 µg/l	US Environmental Protection Agency (1988b) [Method 601]
		GC/MS	NR	US Environmental Protection Agency (1988c) [Method 624]

Table 6 (contd)

Sample matrix	Sample preparation[a]	Assay procedure[b]	Limit of detection[c]	Reference
Water	Purge (inert gas); trap (OV-1 on Chromosorb-W/Tenax/silica gel); desorb as vapour (heat to 180°C, backflush with inert gas) onto capillary GC column	GC/ECD	0.1 µg/l	US Environmental Protection Agency (1988d) [Method 502.2]
		GC/MS	0.1 µg/l	US Environmental Protection Agency (1988e) [Method 524.2]

[a]GC, gas chromatograph

[b]GC/FID, gas chromatography/flame ionization detection; FT-IR, Fourier transform/infrared spectroscopy; GC/ECD, gas chromatography/electrolytic conductivity detection; GC/MS, gas chromatography/mass spectrometry

[c]NR, not reported

3. Biological Data Relevant to the Evaluation of Carcinogenic Risk to Humans

3.1 Carcinogenicity studies in animals (Table 7)

Inhalation

Mouse: Groups of 50 male and 50 female B6C3F$_1$ mice, nine weeks old, were exposed by inhalation to 15 000 ppm (39 600 mg/m^3) chloroethane (99.5% pure) by whole-body exposure for 6 h per day on five days per week for 100 weeks. Survival at 109 weeks of age was: males—control, 28/50; treated, 11/50; females—control, 32/50; treated, 2/50. Uterine neoplasms reduced the survival of treated female mice. Carcinomas of the uterus occurred in 0/49 control and 43/50 treated female mice ($p < 0.001$, logistic regression test), and the carcinomas in many female mice metastasized to a variety of organs. Hepatocellular tumours, primarily carcinomas, occurred in 3/49 control female and 8/48 treated female mice ($p = 0.025$, logistic regression test); the incidence of liver tumours in treated males was not increased (10/47 *versus* 15/50 controls). In males, alveolar/bronchiolar adenomas occurred in 3/50 control and 8/48 treated mice, and alveolar/bronchiolar carcinomas were seen in 2/50 control and 2/48 treated male mice. The proportion of treated male mice with alveolar/bronchiolar tumours was increased relative to that in controls ($p = 0.008$, logistic regression test). The incidence of lung tumours was not increased in females: controls, 5/49; treated, 4/50 (National Toxicology Program, 1989).

Rat: Groups of 50 male and 50 female Fischer 344 rats, eight weeks old, were exposed by inhalation to 15 000 ppm (39 600 mg/m^3) chloroethane (99.5% pure) by whole-body exposure for 6 h per day on five days per week for 102 weeks. Survival at 112 weeks of age was: males—control, 16/50; treated, 8/50; females—control, 31/50; treated, 22/50. Tumours of the skin occurred in 4/50 control male and 9/50 treated male rats. The distribution of tumour types was as follows: trichoepithelioma, 1/50 treated; sebaceous gland adenoma, 1/50 treated; basal-cell carcinoma, 3/50 treated; squamous-cell carcinoma, 2/50 treated; keratoacanthoma, 4/50 control and 2/50 treated. The incidences of all epithelial skin neoplasms in historical controls were 19/300 (6.3%) at the study laboratory and 100/1936 (5.2%) at all National Toxicology Program laboratories. Brain glial-cell tumours (astrocytomas) occurred in 3/50 treated female rats but not in controls; the historical incidence of these tumours in female rats at the study laboratory was 1/297 and that at all National Toxicology Program laboratories was 23/1969 (1.2%) (National Toxicology Program, 1989).

3.2 Other relevant data

(a) Experimental systems

(i) *Absorption, distribution, metabolism and excretion*

The human blood serum/gas partition coefficient (K_D at 25°C) of chloroethane is 2.3 (Morgan *et al.*, 1972).

^{36}Cl-Chloroethane undergoes little (< 0.5%) dechlorination when incubated with rat hepatic microsomal fractions in the presence of NADPH and oxygen (Van Dyke & Wineman, 1971).

(ii) *Toxic effects*

The toxicology of chloroethane has been reviewed (Torkelson & Rowe, 1981).

Acute exposure of guinea-pigs by inhalation to 23-24% chloroethane vapour in air for 5-10 min and to 15.3% for 40 min resulted in some deaths, whereas all animals exposed to 9.1% for 30 min and 1% for 810 min survived (Torkelson & Rowe, 1981). A 2-h inhalation LC$_{50}$ of 152 mg/l (57 600 ppm) chloroethane was reported for white rats [strain unspecified] (Troshina, 1964); signs of toxicity included anaesthesia, liver congestion, haemorrhage and lung oedema. The narcotic concentration of chloroethane for mice, rabbits, dogs and cats was 3.4-4.5% (Henderson & Kennedy, 1930). Cardiac arrhythmia (due to vagus stimulation and sensitization to adrenalin) were observed in dogs under chloroethane anaesthesia (Bush *et al.*, 1952; Morris *et al.*, 1953; Haid *et al.*, 1954). Exposure of male and female Fischer 344/N rats and B6C3F$_1$ mice to 4000 and 10 000 ppm (10.56 and 26.4 g/m^3) chloroethane for 6 h resulted in decreased non-protein sulfhydryl concentrations in the liver 30 min after exposure (Landry *et al.*, 1982).

Table 7. Summary of carcinogenicity studies of chloroethane in experimental animals

Reference	Species/strain	Sex	Dose schedule	Experimental parameter/observation	Group 0	Group 1	Group 2	Group 3	Significance	Comments
National Toxicology Program (1989)	Mouse B6C3F$_1$	M	6 h/d, 5 d/week, inhalation, 100 weeks	Dose (ppm) Survival (109 weeks of age) Alveolar/bronchiolar 　Adenoma 　Carcinoma	0 28/50 3/50 2/50	15 000 11/50 8/48 2/48	– 	– 	$p = 0.008$	Increase
		F		Dose (ppm) Survival (109 weeks of age) Uterine carcinoma Hepatocellular 　Adenoma 　Carcinoma	0 32/50 0/49 0/49 3/49	15 000 2/50 43/50 1/48 7/48	– 	– 	$p < 0.001$ $p = 0.025$	Increase
	Rat F344	M	6 h/d, 5 d/week, inhalation, 102 weeks	Dose (ppm) Survival (112 weeks of age) Skin 　Trichoepithelioma 　Sebaceous adenoma 　Basal-cell carcinoma 　Squamous-cell carcinoma 　Keratoacanthoma	0 16/50 0/50 0/50 0/50 0/50 4/50	15 000 8/50 1/50 1/50 3/50 2/50 2/50	– 	– 	$p = 0.016$	Increase
		F		Dose (ppm) Survival (112 weeks of age) Astrocytoma	0 31/50 0/50	15 000 22/50 3/50	– 	– 		Uncommon tumour

Exposure of male and female Fischer 344/N rats and male beagle dogs to 1600, 4000 or 10 000 ppm (4.2, 10.56 or 26.4 g/m^3) chloroethane for 6 h per day on five days per week for two weeks had no toxic effect except for slight increases in relative liver weights of male rats exposed to 4000 or 10 000 ppm (Landry *et al.*, 1982). Similar results were obtained when male and female B6C3F$_1$ mice were exposed to 5000 ppm (13.2 g/m^3) chloroethane for 23 h per day for 11 consecutive days; exposure to 250 or 1250 ppm (0.66 or 3.3 g/m^3) had no effect on relative liver weights (Landry *et al.*, 1989).

Exposure of male and female Fischer 344/N rats and B6C3F$_1$ mice to 2500-19 000 ppm (6.6-50.2 g/m^3) chloroethane for 6 h per day on five days per week for 13 weeks by whole-body inhalation induced no clinical sign of toxicity, and no gross pathological or histological change was observed except for reduced body weight. Increased relative liver weights were observed in male rats and female mice exposed to 19 000 ppm (National Toxicology Program, 1989). Conversely, chloroethane [purity nonspecified] at 0.57 g/m^3 (216 ppm) for 4 h per day for six months was reported to cause liver and lung damage, decreased blood pressure and reduced phagocytic activity of leukocytes in rats [sex and strain were not specified] (Troshina, 1966).

(iii) *Effects on reproduction and prenatal toxicity*

No data were available to the Working Group.

(iv) *Genetic and related effects* (Table 8)

Chloroethane was mutagenic to *Salmonella typhimurium*, but no response was observed in a cell transformation assay using cultured mammalian cells. A study reporting a positive response in *S. typhimurium* and negative responses in tests for unscheduled DNA synthesis in mouse hepatocytes and for transformation in BALB/c 3T3 cells could not be evaluated [details not given] (Milman *et al.*, 1988).

(b) *Humans*

(i) *Absorption, distribution, excretion and metabolism*

When ^{38}Cl-chloroethane was administered by inhalation at about 5 mg/subject in a single breath [subject weight unspecified] to human volunteers, about 30% of the administered radioactivity was eliminated in the breath within 1 h. Urinary excretion of ^{38}Cl amounted to < 0.01% of the dose/min (Morgan *et al.*, 1970).

(ii) *Toxic effects*

Davidson (1926) exposed volunteers to 1.3-3.36% chloroethane vapour in air. No adverse effect was seen after exposure to 1.3% for 21 min, whereas exposure to 3.36% led to incoordination, unconsciousness and cyanosis within minutes. Decreased reaction times were observed with exposure to concentrations of 2.5%.

Table 8. Genetic and related effects of chloroethane

Test system	Result		Dose LED/HID	Reference
	Without exogenous metabolic system	With exogenous metabolic system		
SA0, *Salmonella typhimurium* TA100, reverse mutation	–	+	0.0000	National Toxicology Program (1989)[a]
SA5, *Salmonella typhimurium* TA1535, reverse mutation	+	+	0.0000	National Toxicology Program (1989)[a]
SA9, *Salmonella typhimurium* TA98, reverse mutation	–	–	0.0000	National Toxicology Program (1989)[a]
TBM, Cell transformation, BALB/c3T3 mouse cells	–	0	467.0000	Tu et al. (1985)[a]

[a]Closed container

The anaesthetic concentration of chloroethane in humans has been estimated to be 4% (Lawson, 1965). During anaesthesia, vagal inhibition may occur (Bush *et al.*, 1952). Deaths that occurred under anaesthesia were due mainly to very high chloroethane concentrations, which caused respiratory depression (Henderson & Kennedy, 1930; Dobkin & Byles, 1971).

Chloroethane has been shown to elicit allergic contact dermatitis (van Ketel, 1976).

(iii) *Effects on reproduction and prenatal toxicity*

No data were available to the Working Group.

(iv) *Genetic and related effects*

No data were available to the Working Group.

3.3 Case reports and epidemiological studies of carcinogenicity to humans

No data were available to the Working Group.

4. Summary of Data Reported and Evaluation

4.1 Exposure data

Chloroethane is produced by the hydrochlorination of ethylene. It is used in the manufacture of tetraethyllead, as an industrial ethylating agent, as a blowing agent in the production of polystyrene foam and as a local anaesthetic. Occupational exposure occurs during the production of tetraethyllead, and industrial emissions have led to detectable levels of chloroethane in ambient air.

4.2 Experimental carcinogenicity data

Chloroethane was tested for carcinogenicity in a two-year study in male and female $B6C3F_1$ mice and Fischer 344 rats by inhalation. It induced uterine carcinomas in mice; marginal increases occurred in the incidence of hepatocellular tumours in female mice and in the incidence of alveolar/bronchiolar tumours in male mice. There was a marginal increase in the incidence of skin tumours in male rats, and a few uncommon glial-cell tumours occurred in female rats.

4.3 Human carcinogenicity data

No data were available to the Working Group.

4.4 Other relevant data

In single studies, chloroethane was mutagenic to bacteria but did not induce transformation in cultured mammalian cells.

4.5 Evaluation[1]

There is *limited evidence* for the carcinogenicity of chloroethane in experimental animals.

No data were available from studies in humans on the carcinogenicity of chloroethane.

Overall evaluation

Chloroethane *is not classifiable as to its carcinogenicity to humans (Group 3)*.

5. References

Aldrich Chemical Co. (1990) *1990-1991 Aldrich Handbook of Fine Chemicals*, Milwaukee, WI, p. 291

American Conference of Governmental Industrial Hygienists (1989) *TLVs Threshold Limit Values and Biological Exposure Indices for 1989-1990*, Cincinnati, OH, p. 23

American Tokyo Kasei (1988) *TCI American Organic Chemicals 88/89 Catalog*, Portland, OR, p. 278

Anon. (1979) Chemical profile: ethyl chloride. *Chem. Mark. Rep.*, 27 August, pp. 9, 14

Anon. (1983) Water substances spur pleas for regulation. *J. Commerce*, 30 September, p. 22b1

Brodzinsky, R. & Singh, H.B. (1983) *Volatile Organic Chemicals in the Atmosphere: An Assessment of Available Data* (EPA-600/3-83-027a; NTIS PB83-195503), Research Triangle Park, NC, US Environmental Protection Agency

Budavari, S., ed. (1989) *The Merck Index*, 11th ed., Rahway, NJ, Merck & Co., p. 597

Bush, O.F., Bittenbender, G. & Adriani, J. (1952) Electrocardiographic changes during ethyl chloride and vinyl ether anesthesia in the dog and man. *Anesthesiology*, 13, 197-202

Cook, W.A. (1987) *Occupational Exposure Limits—Worldwide*, Cincinnati, OH, American Industrial Hygiene Association, pp. 121, 139, 187

Davidson, B.M. (1926) Studies of intoxication. V. The action of ethyl chloride. *J. Pharmacol. exp. Ther.*, 26, 37-42

---

[1]For definition of the italicized terms, see Preamble, pp. 30-33.

Dialog Information Services (1989) *PTS US Time Series* (Database No. 82), Palo Alto, CA

Dialog Information Services (1990) *Chem-Intell* (Database No. 318), Palo Alto, CA

Dobkin, A.B. & Byles, P.H. (1971) The pharmacodynamics of divinyl ether, ethyl chloride, fluroxene, nitrous oxide, and trichloroethylene. In: Soma, L.R., ed., *Textbook of Veterinary Anesthesia*, Baltimore, MD, Williams & Wilkins, pp. 94-104

Dow Chemical Co. (1989) *Technical Information Sheet: Ethyl Chloride*, Midland, MI

Easley, D.M., Kleopfer, R.D. & Carasea, A.M. (1981) Gas chromatographic-mass spectrometric determination of volatile organic compounds in fish. *J. Assoc. off. anal. Chem.*, 64, 653-656

Eller, P.M. (1985) *NIOSH Manual of Analytical Methods*, 3rd ed., Vol. 2 (DHHS (NIOSH) Publ. No. 84-100), Washington DC, US Government Printing Office, pp. 2519-1-2519-4

Ferrario, J.B., Lawler, G.C., DeLeon, I.R. & Laseter, J.L. (1985) Volatile organic pollutants in biota and sediments of Lake Pontchartrain. *Bull. environ. Contam. Toxicol.*, 34, 246-255

Haid, B., White, J.M., Jr & Morris, L.E. (1954) Observations of cardiac rhythm during ethyl chloride anesthesia in the dog. *Curr. Res. Anesth. Analg.*, 33, 318-325

Henderson, V.E. & Kennedy, A.S. (1930) Ethyl chloride. *Can. med. Assoc. J.*, 23, 226-231

Hubrich, C. & Stuhl, F. (1980) The ultraviolet absorption of some halogenated methanes and ethanes of atmospheric interest. *J. Photochem.*, 12, 93-107

Hume, G.L. (1988) *Summary of Emissions Associated with Sources of Ethyl Chloride* (EPA-450/3-88-005; NTIS PB88-240247), Research Triangle Park, NC, US Environmental Protection Agency

IARC (1987a) *IARC Monographs on the Evaluation of Carcinogenic Risks to Humans*, Suppl. 7, *Overall Evaluations of Carcinogenicity: An Updating of* IARC Monographs *Volumes 1 to 42*, Lyon, p. 63

IARC (1987b) *IARC Monographs on the Evaluation of Carcinogenic Risks to Humans*, Suppl. 7, *Overall Evaluations of Carcinogenicity: An Updating of* IARC Monographs *Volumes 1 to 42*, Lyon, pp. 373-376

IARC (1987c) *IARC Monographs on the Evaluation of Carcinogenic Risks to Humans*, Suppl. 7, *Overall Evaluations of Carcinogenicity: An Updating of* IARC Monographs *Volumes 1 to 42*, Lyon, pp. 152-154

van Ketel, W.G. (1976) Allergic contact dermatitis from propellants in deodorant sprays in combination with allergy to ethyl chloride. *Contact Derm.*, 2, 115-119

Landry, T.D., Ayres, J.A., Johnson, K.A. & Wall, J.M. (1982) Ethyl chloride: a two-week inhalation toxicity study and effects on liver non-protein sulfhydryl concentrations. *Fundam. appl. Toxicol.*, 2, 230-234

Landry, T.D., Johnson, K.A., Phillips, J.E. & Weiss, S.K. (1989) Ethyl chloride: 11-day continuous exposure inhalation toxicity study in B6C3F$_1$ mice. *Fundam. appl. Toxicol.*, 13, 516-422

Lawson, J.I.M. (1965) Ethyl chloride. *Br. J. Anaesth.*, 37, 667-670

Lodge, J.P., Jr, ed. (1989a) *Methods of Air Sampling and Analysis*, 3rd ed., Chelsea, MI, Lewis Publishers, Inc., pp. 171-187

Lodge, J.P., Jr, ed. (1989b) *Methods of Air Sampling and Analysis*, 3rd ed., Chelsea, MI, Lewis Publishers, Inc., pp. 78-83

Mannsville Chemical Products Corp. (1984) *Chemical Products Synopsis: Ethyl Chloride*, Cortland, NY

McKinney, J.D., Maurer, R.R., Hass, J.R. & Thomas, R.O. (1976) Possible factors in the drinking water of laboratory animals causing reproductive failure. In: Keith, L.H., ed., *Identification and Analysis of Organic Pollutants in Water*, Ann Arbor, MI, Ann Arbor Science, pp. 417-432

Milman, H.A., Story, D.L., Riccio, E.S., Sivak, A., Tu, A.S., Williams, G.M., Tong, C. & Tyson, C.A. (1988) Rat liver foci and in vitro assays to detect initiating and promoting effects of chlorinated ethanes and ethylenes. *Ann. N.Y. Acad. Sci.*, 534, 521-530

Morgan, A., Black, A. & Belcher, D.R. (1970) The excretion in breath of some aliphatic halogenated hydrocarbons following administration by inhalation. *Ann. occup. Hyg.*, 13, 219-233

Morgan, A., Black, A. & Belcher, D.R. (1972) Studies on the absorption of halogenated hydrocarbons and their excretion in breath using ^{38}Cl tracer techniques. *Ann. occup. Hyg.*, 15, 273-282

Morris, L.E., Noltensmeyer, M.H. & White, J.M., Jr (1953) Epinephrine induced cardiac irregularities in the dog during anesthesia with trichloroethylene, cyclopropane, ethyl chloride and chloroform. *Anesthesiology*, 14, 153-158

Narodoslawsky, M. & Moser, F. (1988) New compression heat pump media as replacements for CFCs. *Int. J. Refrig.*, 11, 264-268

National Institute for Occupational Safety and Health (1978) *Chloroethanes: Review of Toxicity. Current Intelligence Bulletin 27* (DHEW (NIOSH) Publ. No. 78-181), Cincinnati, OH, US Department of Health, Education, and Welfare

National Library of Medicine (1989) *Toxic Release Inventory Data Bank*, Bethesda, MD

National Toxicology Program (1989) *Toxicology and Carcinogenesis Studies of Chloroethane (Ethyl Chloride) (CAS No. 75-00-3) in F344/N Rats and B6C3F$_1$ Mice (Inhalation Studies)* (NTP Technical Report No. 346), Research Triangle Park, NC

Pellizzari, E.D., Hartwell, T.D., Harris, B.S.H., III, Waddell, R.D., Whitaker, D.A. & Erickson, M.D. (1982) Purgeable organic compounds in mother's milk. *Bull. environ. Contam. Toxicol.*, 28, 322-328

Pouchert, C.J., ed. (1985) *The Aldrich Library of FT-IR Spectra*, Vol. 3, Milwaukee, WI, Aldrich Chemical Co., p. 82C

PPG Industries (1986) *Ethyl Chloride Data Sheet*, Pittsburgh, PA

Processes Research (1972) *Air Pollution from Chlorination Processes* (APTD-1110; NTIS PB-218048), Research Triangle Park, NC, US Environmental Protection Agency

Reynolds, J.E.F. (1989) *Martindale. The Extra Pharmacopoeia*, 29th ed., London, The Pharmaceutical Press, p. 1215

Ringenburg, V.L. (1983) *Survey Report of Ethyl Corporation, Pasadena, Texas* (NIOSH Report No. IWS-115.10; NTIS PB83-232876), Cincinnati, OH, National Institute for Occupational Safety and Health

Sadtler Research Laboratories (1980) *The Sadtler Standard Spectra, 1980 Cumulative Index*, Philadelphia, PA

Sax, N.I. & Lewis, R.J. (1987) *Hawley's Condensed Chemical Dictionary*, 11th ed., New York, Van Nostrand Reinhold Co., p. 482

Singh, H.B., Salas, L.J., Stiles, R. & Shigeishi, H. (1983) *Measurements of Hazardous Organic Chemicals in the Ambient Atmosphere* (EPA-600/3-83-002; NTIS PB83-156935), Research Triangle Park, NC, US Environmental Protection Agency

Sittig, M. (1985) *Handbook of Toxic and Hazardous Chemicals and Carcinogens*, 2nd ed., Park Ridge, NJ, Noyes Publications, pp. 417-418

SKC Inc. (1989) *Comprehensive Catalog and Guide. Air Sampling Products for Worker Monitoring, Chemical Hazard Detection and Industrial Hygiene*, Eighty Four, PA

Torkelson, T.R. & Rowe, V.K. (1981) Ethyl chloride. In: Clayton, G.D. & Clayton, F.E., eds, *Patty's Industrial Hygiene and Toxicology*, New York, John Wiley & Sons, pp. 3480-3483

Troshina, M.M. (1964) Toxicology of ethyl chloride (Russ.). *Prom. Khim. Veshchestv.*, 6, 45-55

Troshina, M.M. (1966) Some materials for substantiating maximum permissible concentration of ethyl chloride in the atmosphere of work premises (Russ.). *Gig. Tr. prof. Zabol.*, 10, 37-42

Tu, A.S., Murray, T.A., Hatch, K.M., Sivak, A. & Milman, H.A. (1985) In vitro transformation of BALB/c-3T3 cells by chlorinated ethanes and ethylenes. *Cancer Lett.*, 28, 85-92

United Nations Environment Programme (1990) *International Register of Potentially Toxic Chemicals, Recommendations—Legal Mechanisms*, Geneva

US Department of Commerce (1981) *US Imports for Consumption and General Imports—TSUSA Commodity by Country of Origin* (FT246/Annual 1980), Washington DC, Bureau of the Census, p. 1-266

US Department of Commerce (1982) *US Imports for Consumption and General Imports—TSUSA Commodity by Country of Origin* (FT246/Annual 1981), Washington DC, Bureau of the Census, p. 1-275

US Department of Commerce (1983) *US Imports for Consumption and General Imports—TSUSA Commodity by Country of Origin* (FT246/Annual 1982), Washington DC, Bureau of the Census, p. 1-292

US Department of Commerce (1984) *US Imports for Consumption and General Imports—TSUSA Commodity by Country of Origin* (FT246/Annual 1983), Washington DC, Bureau of the Census, p. 1-309

US Department of Commerce (1986) *US Imports for Consumption and General Imports—TSUSA Commodity by Country of Origin* (FT246/Annual 1985), Washington DC, Bureau of the Census, p. 1-583

US Department of Commerce (1988) *US Imports for Consumption and General Imports—TSUSA Commodity by Country of Origin* (FT246/Annual 1987), Washington DC, Bureau of the Census, p. 1-524

US Environmental Protection Agency (1986a) Method 8010. Halogenated volatile organics. In: *Test Methods for Evaluating Solid Waste—Physical/Chemical Methods*, 3rd ed. (EPA Report No. SW-846), Washington DC, Office of Solid Waste and Emergency Response

US Environmental Protection Agency (1986b) Method 8240. Gas chromatography/mass spectrometry for volatile organics. In: *Test Methods for Evaluating Solid Waste—Physical/Chemical Methods*, 3rd ed. (EPA Report No. SW-846), Washington DC, Office of Solid Waste and Emergency Response

US Environmental Protection Agency (1988a) Method TO-14. The determination of volatile organic compounds (VOCs) in ambient air using SUMMA(R) passivated canister sampling and gas chromatographic analysis. In: *Compendium of Methods for the Determination of Toxic Organic Compounds in Ambient Air* (EPA-600/4-89-017; NTIS PB90-116989), Research Triangle Park, NC, Atmospheric Research and Exposure Laboratory, Office of Research and Development

US Environmental Protection Agency (1988b) Methods for organic chemical analysis of municipal and industrial wastewater. Method 601. Purgeable halocarbons. *US Code fed. Regul.*, Title 40, Part 136, Appendix A, pp. 267-281

US Environmental Protection Agency (1988c) Methods for organic chemical analysis of municipal and industrial wastewater. Method 624. Purgeables. *US Code fed. Regul.*, Title 40, Part 136, Appendix A, pp. 432-446

US Environmental Protection Agency (1988d) Method 502.2. Volatile organic compounds in water by purge and trap capillary column gas chromatography with photoionization and electrolytic conductivity detectors in series. Revision 2.0 (PB89-220461). In: *Methods for the Determination of Organic Compounds in Drinking Water* (EPA-600/4-88/039), Cincinnati, OH, pp. 31-62

US Environmental Protection Agency (1988e) Method 524.2. Measurements of purgeable organic compounds in water by capillary column gas chromatography/mass spectrometry. Revision 3.0 (PB89-220461). In: *Methods for the Determination of Organic Compounds in Drinking Water* (EPA-600/4-88/039), Cincinnati, OH, pp. 285-323

US International Trade Commission (1989) *Synthetic Organic Chemicals, US Production and Sales, 1988* (USITC Publ. 2219), Washington DC, US Government Printing Office, p. 15-7

US Occupational Safety and Health Administration (1989) Air contaminants—Permissible exposure limits (Report No. OSHA 3112). *US Code fed. Regul.*, Part 1910.1000

Van Dyke, R.A. & Wineman, C.G. (1971) Enzymatic dechlorination: dechlorination of chloroethanes and propanes *in vitro*. *Biochem. Pharmacol.*, 20, 463-470

Verschueren, K. (1983) *Handbook of Environmental Data on Organic Chemicals*, 2nd ed., New York, Van Nostrand Reinhold, p. 631

Weast, R.C., ed. (1989) *CRC Handbook of Chemistry and Physics*, 70th ed., Boca Raton, FL, CRC Press, p. C-264

1,1,2-TRICHLOROETHANE

This substance was considered by a previous Working Group, in 1978 (IARC, 1979). Since that time, new data have become available, and these have been incorporated into the monograph and taken into consideration in the evaluation.

1. Chemical and Physical Data

1.1 Synonyms

Chem. Abstr. Services Reg. No.: 79-00-5
Chem. Abstr. Name: 1,1,2-Trichloroethane
IUPAC Systematic Name: 1,1,2-Trichloroethane
Synonyms: Ethane trichloride; trichloroethane; *beta*-trichloroethane; 1,2,2-trichloroethane; vinyl trichloride

1.2 Structural and molecular formulae and molecular weight

$$\begin{array}{c} Cl \quad H \\ | \quad\; | \\ H - C - C - Cl \\ | \quad\; | \\ Cl \quad H \end{array}$$

$C_2H_3Cl_3$ 　　　　　　　　　　　　　　　　　　　Mol. wt: 133.41

1.3 Chemical and physical properties of the pure substance

(a) *Description*: Clear, colourless liquid with a sweet odour (Verschueren, 1983; Sax & Lewis, 1987; Budavari, 1989)
(b) *Boiling-point*: 113.8°C (Weast, 1989)
(c) *Melting-point*: –36.5°C (Weast, 1989)
(d) *Density*: 1.4397 at 20/4°C (Weast, 1989)

(e) *Spectroscopy data*[1]: Infrared (Sadtler Research Laboratories, 1980, prism [9721], grating [29465]; Pouchert, 1981, 1985a,b), nuclear magnetic resonance (Sadtler Research Laboratories, 1980, proton [16882, V2], C-13 [344]; Pouchert, 1974, 1983) and mass spectral data [617] have been reported.

(f) *Solubility*: Slightly soluble in water (4.50 g/l at 20°C); soluble in ethanol, chloroform and diethyl ether (Verschueren, 1983; Weast, 1989)

(g) *Volatility*: Vapour pressure, 19 mm Hg at 20°C, 40 mm Hg at 35°C; relative vapour density (air = 1), 4.63 (Verschueren, 1983)

(h) *Reactivity*: Reacts with strong oxidizers, strong caustics and metals such as sodium, potassium, powdered aluminium and magnesium (Sittig, 1985)

(i) *Octanol/water partition coefficient (P)*: log P, 2.07 (Chemical Information Systems, Inc., 1990)

(j) *Conversion factor*[2]: mg/m^3 = 5.46 × ppm

1.4 Technical products and impurities

1,1,2-Trichloroethane is available in the USA as a commercial-grade product, either stabilized or unstabilized, at a purity of > 99%. The stabilized product contains *sec*-butanol (0.5%) and 1,2-butylene oxide (0.25%) (Dow Chemical Co., 1990). 1,1,2-Trichloroethane is also available in research quantities at 95-99.9% purity for calibration of liquid density meters and as a gas chromatography standard (Riedel-de Haën, 1986; Eastman Kodak Co., 1987; American Tokyo Kasei, 1988; Pfaltz & Bauer, 1988; Aldrich Chemical Co., 1990).

2. Production, Use, Occurrence and Analysis

2.1 Production and use

(a) *Production*

1,1,2-Trichloroethane is produced in the USA by the chlorination of ethylene (see IARC, 1987a). In a two-stage manufacturing process, this initially yields 1,2-dichloroethane (see IARC, 1987b); subsequent chlorination yields

---

[1]In square brackets, spectrum number in compilation

[2]Calculated from: mg/m^3 = (molecular weight/24.45) × ppm, assuming standard temperature (25°C) and pressure (760 mm Hg)

1,1,2-trichloroethane and hydrochloric acid. In an alternative production method, ethylene is combined with hydrochloric acid and oxygen at 280-370°C on a catalyst of cupric chloride and potassium chloride to yield 1,1,2-trichloroethane and other chlorinated ethanes (Reed *et al.*, 1988). 1,1,2-Trichloroethane has also been made by the chlorination of vinyl chloride (see IARC, 1987c) in a liquid phase at 300-320°C (Thomas *et al.*, 1982).

There is only one producer in the USA, with an annual production estimated to be 186 000 tonnes (Reed *et al.*, 1988). 1,1,2-Trichloroethane is also produced by one company each in France and the UK (Chemical Information Services Ltd, 1988).

(b) Use

Over 95% of the 1,1,2-trichloroethane produced in the USA is used as an intermediate in the production of vinylidene chloride (see IARC, 1987d). Vinylidene chloride is made by dehydrochlorination of 1,1,2-trichloroethane with an aqueous suspension of calcium hydroxide at about 320°C or with lime or sodium hydroxide (Wiseman, 1972; Reed *et al.*, 1988). 1,1,2-Trichloroethane has two minor uses: as a solvent for the coating laid down on films and as a chemical intermediate or process solvent in pharmaceutical manufacture. It has been used as a solvent for fats, oils, waxes and resins and, in small amounts, as a solvent for chlorinated rubber and adhesives (US Environmental Protection Agency, 1980; Sax & Lewis, 1987; Strobel & Grummt, 1987).

(c) Regulatory status and guidelines

Occupational exposure limits and guidelines for 1,1,2-trichloroethane are presented in Table 1.

Table 1. Occupational exposure limits and guidelines for 1,1,2-trichloroethane[a]

Country	Year	Concentration (mg/m^3)	Interpretation[b]
Australia	1985	45 (skin)[c]	TWA
		90	STEL
Austria	1987	55 (skin)[d]	TWA
Belgium	1989	45 (skin)	TWA
Brazil	1982	35	TWA
Canada	1986	45 (skin)	TWA
Chile	1987	36 (skin)	TWA
Denmark	1987	45 (skin)	TWA
Finland	1987	45	TWA
		90 (skin)	STEL

Table 1 (contd)

Country	Year	Concentration (mg/m^3)	Interpretation[b]
Germany	1988	55[d]	TWA
		110[d]	STEL (30 min)
Hungary	1985	10	TWA
		50	STEL
Indonesia	1987	45 (skin)	TWA
Japan	1988	45 (skin)	TWA
Mexico	1987	45 (skin)	TWA
Netherlands	1986	45 (skin)	TWA
Poland	1985	100	TWA
Switzerland	1987	55 (skin)	TWA
UK	1987	45 (skin)	TWA
		90 (skin)	STEL (10 min)
USA			
ACGIH	1989	55 (skin)	TWA
OSHA	1989	45 (skin)	TWA
Venezuela	1987	35 (skin)	TWA
		90	STEL

[a]From Cook (1987); Health and Safety Executive (1987); American Conference of Governmental Industrial Hygienists (ACGIH) (1989); US Occupational Safety and Health Administration (OSHA) (1989); United Nations Environment Programme (1990)
[b]TWA, time-weighted average; STEL, short-term exposure limit
[c]With a notation of skin absorption
[d]Suspected of carcinogenic potential

2.2 Occurrence

(a) *Natural occurrence*

1,1,2-Trichloroethane is not known to occur as a natural product.

(b) *Occupational exposure*

The US National Occupational Hazard Survey estimated that in 1972-74 112 000 workers were potentially exposed to 1,1,2-trichloroethane (National Institute for Occupational Safety and Health, 1978)

(c) *Multimedia exposure assessment*

Nine volunteers in New Jersey (four chemical and oil company workers) and three in North Carolina (none considered to be occupationally exposed) were monitored for exposure to 1,1,2-trichloroethane between July and December 1980. Breathing-zone air, drinking-water, food and breath were monitored. 1,1,2-Trichloroethane was detected in 10 (seven at trace) of the 161 breathing-zone air samples in New Jersey at 0.14-34.70 µg/m^3 and in one of the 60 samples in North Carolina at about 0.54 µg/m^3. It was detected in five (four at trace) exhaled breath samples in New Jersey at 0.07-5.13 µg/m^3, with a median concentration of 0.2 µg/m^3, and in none of the 17 exhaled breath samples in North Carolina (limit of detection, 0.20 µg/m^3). It was not detected in any of the 13 home drinking-water samples for the volunteers in New Jersey (limit of detection, 0.01 µg/l) [no information was provided on home samples for the volunteers in North Carolina] (Wallace *et al.*, 1984).

(*d*) *Air*

According to the Toxic Chemical Release Inventory, total emissions of 1,1,2-trichloroethane into the air in the USA in 1987 were approximately 935 tonnes from 49 locations. Industrial releases to other media were 5.5 tonnes to ambient water from 12 locations and 4 kg to the land from two locations (National Library of Medicine, 1989).

Estimated levels of 1,1,2-trichloroethane in ambient air collected in 1977 from four sites in Iberville Parish, LA, USA, ranged from trace to 1.8 µg/m^3 (Pellizzari, 1982).

1,1,2-Trichloroethane was detected in one of ten air samples collected in 1978 at Bochum-University, Germany, at a concentration of 0.6 µg/m^3, in three of 12 air samples from Bochum-Kemnade at 0.3-0.8 µg/m^3, and in three of 12 air samples from Bochum City at 0.8-17 µg/m^3; the mean concentration of 1,1,2-trichloroethane in air samples from Bochum and vicinity was 0.4 µg/m^3 (Bauer, 1981).

In a review of data on the presence of volatile organic chemicals in the US atmosphere in 1970-80, a median concentration of 0.082 µg/m^3 1,1,2-trichloroethane was reported for the 760 data points examined, 0.076 µg/m^3 (667 data points) for urban/suburban areas, and 0.5 µg/m^3 (91 data points) close to industrial sources (Brodzinsky & Singh, 1983).

Air samples collected in 10 US cities in 1979-81 contained mean concentrations of 0.033-0.022 µg/m^3 1,1,2-trichloroethane (Singh *et al.*, 1983).

The mean concentration of 1,1,2-trichloroethane in indoor winter air samples collected in 1982 from 11 homes in Knoxville, TN, USA, was 14.1 µg/m^3; the compound was not detected (< 1 µg/m^3) in any outdoor air sample (Gupta *et al.*, 1984).

(e) *Water and sediments*

Of 10 water supplies surveyed by the US Environmental Protection Agency (1980), only one contained 1,1,2-trichloroethane; a study of finished water from a metropolitan area reported concentrations of 0.1-8.5 µg/l. The highest concentrations of 1,1,2-trichloroethane in samples collected in 1977-79 in ground- and surface water in New Jersey, USA, were 31.1 µg/l and 18.7 µg/l, respectively; the compound was detected in 72 of 1069 groundwater samples and in 53 of 603 surface water samples (Page, 1981).

Samples of Rhine River water collected near Lobith, Germany, between April and December 1976 contained levels of 1,1,2-trichloroethane ranging from not detected to 1.0 µg/l, with a mean of 0.2 µg/l. Levels in drinking-water samples collected from 100 cities in western Germany in 1977 ranged from not detected to 5.8 µg/l (Bauer, 1981).

1,1,2-Trichloroethane was not detected in raw water samples collected in 1979 from 30 Canadian water treatment facilities but was found in two samples of treated water at a mean concentration of < 1 µg/l (max, 7 µg/l) (Otson *et al.*, 1982).

Tap-water samples (30 each) collected in 1981 in Tübingen, Germany, from the local water supply contained an average concentration of 1,1,2-trichloroethane of 0.05 µg/l (range, 0.03-0.09 µg/l); and samples from the Lake Constance water supply (sampled in Tübingen) contained an average of 0.01 µg/l (range, 0.00-0.02 µg/l) (Hagenmaier *et al.*, 1982).

1,1,2-Trichloroethane was detected at concentrations greater than 10 µg/l in 58 of 1982 US industrial wastewater samples. The minimum and maximum concentrations found were 12 µg/l and 3400 µg/l, respectively. The industrial wastewaters in which 1,1,2-trichloroethane was identified include those from factories producing adhesives/sealants, iron/steel foundries, laundries, factories producing mechanical products, organics/plastics and paint/ink, petroleum refineries, plants for pharmaceuticals and phosphates, printing/publishing establishments and timber mills (Thomas *et al.*, 1982).

According to the US Environmental Protection Agency, the entrance of 1,1,2-trichloroethane into groundwater and surface water most probably occurs through the direct discharge of chemical wastes into water and the disposal of waste into landfills, with subsequent leaching of the chemical to groundwater. 1,1,2-Trichloroethane was found in 66% of 399 drinking-water samples from New Jersey, USA, with groundwater sources. Concentrations were distributed as follows: 203 samples, < 1 µg/l; 141 samples, 1-10 µg/l; 55 samples, 10-100 µg/l; one sample, > 100 µg/l. 1,1,2-Trichloroethane was also detected in 50 of 372 well-water samples (maximum, 300 µg/l) from Long Island, NY, in 1978 (Thomas *et al.*, 1982).

Tap-water samples collected in 1981 from Port Robinson, Niagara Falls and Chippawa, Ontario, Canada, contained concentrations of 0.48, 0.045-0.12 and 0.15 µg/l 1,1,2-trichloroethane, respectively. Surface water samples collected from 17 sites in 1980 and 19 sites in 1981 in the same vicinity (Welland River watershed) contained concentrations ranging from not detected to trace (< 0.04 µg/l) (Kaiser & Comba, 1983).

In the 1982 Nationwide Urban Runoff Program, 1,1,2-trichloroethane was detected in samples from one of the 15 reporting cities at a range of 2-3 µg/l (Cole et al., 1984).

Of approximately 20 000 groundwater samples collected in Suffolk County, NY, USA, 5.9% from community water systems and 4.5% from other water systems (non-community and private wells) contained 1,1,2-trichloroethane. The maximal concentration was 13 µg/l in community water systems and 1 µg/l in non-community water systems; none was detected in private wells (Zaki, 1986).

In 1985, the California Department of Health Services carried out a survey in which the water samples from 2949 wells throughout California were analysed for organic contaminants; 1,1,2-trichloroethane was found at concentrations ranging from 0.7 to 1.1 µg/l in four wells (Reed et al., 1988).

(f) Human tissues

The mean concentrations (micrograms per kilogram fresh weight) of 1,1,2-trichloroethane in human tissues samples collected in 1978 from the general population in the Ruhr District of Germany were: kidney capsule fat, 6.1 (max., 35.7); hypodermal fat, 13.9 (max., 158.6); lung, 2.3 (max., 34.0); liver, 2.6 (max., 25.3); and muscle, 17.2 (Bauer, 1981).

(g) Other

1,1,2-Trichloroethane has been identified as a by-product in the production of polyvinyl chloride plastics; it constituted 40% of the volatile components of a by-product known as EDC-tar collected from a Swedish factory (Rosenberg et al., 1975).

1,1,2-Trichloroethane was identified as an impurity in nine of 22 samples of commercial 1,1,1-trichloroethane, at concentrations ranging from 300 to 3015 µg/ml (Henschler et al., 1980).

2.3 Analysis

Selected methods for the analysis of 1,1,2-trichloroethane in air and water are identified in Table 2. The US Environmental Protection Agency methods for analysing water (Methods 8010 and 8240) have also been applied to liquid and solid wastes. Volatile components of solid-waste samples are first extracted with methanol prior to purge-and-trap concentration and analysis by gas chromatography-electrolytic conductivity detection (Method 8010) or gas chromatography-

Table 2. Methods for the analysis of 1,1,2-trichloroethane

Sample matrix	Sample preparation[a]	Assay procedure[b]	Limit of detection[c]	Reference
Air	Adsorb on activated charcoal; desorb (carbon disulfide); inject aliquot	GC/FID	0.01 mg per sample	Eller (1987); Lodge (1989a)
	Collect whole air sample in passivated stainless steel canister	GC/FID	10 ppm (54.6 mg/m^3)	US Environmental Protection Agency (1988a) [Method TO-14]; Lodge (1989b)
	Draw air through tube; compare reaction with standard chart	Colorimetric	NR	SKC Inc. (1990)
Breath	Collect in plastic bag; evacuate cell; draw sample in and scan	FT-IR	10 ppm (54.6 mg/m^3)	Lodge (1989c)
Water	Purge (inert gas); trap (OV-1 on Chromosorb-W/Tenax/silica gel); desorb as vapour (heat to 180°C, backflush with inert gas) onto packed GC column	GC/ECD	0.02 µg/l	US Environmental Protection Agency (1988b) [Method 601]
		GC/MS	5.0 µg/l	US Environmental Protection Agency (1988c) [Method 624]
Water	Purge (inert gas); trap (OV-1 on Chromosorb-W/Tenax/silica gel); desorb as vapour (heat to 180°C, backflush with inert gas) onto capillary GC column	GC/ECD	0.04 µg/l	US Environmental Protection Agency (1988d) [Method 502.2]
		GC/MS	0.10 µg/l	US Environmental Protection Agency (1988e) [Method 524.2]
	Add internal standard (isotope-labelled 1,1,2-trichloroethane); purge; trap and desorb as above	GC/MS	10 µg/l	US Environmental Protection Agency (1988f) [Method 1624]

[a]GC, gas chromatograph
[b]GC/FID, gas chromatography/flame ionization detection; FT-IR, Fourier transform/infrared spectroscopy; GC/ECD, gas chromatography/electrolytic conductivity detection; GC/MS, gas chromatography/mass spectrometry
[c]NR, not reported

mass spectrometry (Method 8240). The detection limit using Method 8010 is 0.02 µg/l, and the practical quantification limit using Method 8240 is 5 µg/l for groundwater and 5 µg/kg for soil/sediment samples (US Environmental Protection Agency, 1986a,b). Method 624 has also been adapted to the analysis of 1,1,2-trichloroethane in fish, with an estimated detection limit of 10 µg/kg (Easley *et al.*, 1981).

3. Biological Data Relevant to the Evaluation of Carcinogenic Risk to Humans

3.1 Carcinogenicity studies in animals (Table 3)

(a) Oral administration

Mouse: Groups of 50 male and 50 female $B6C3F_1$ mice, five weeks of age, received technical-grade 1,1,2-trichloroethane (one batch; purity, 99, 91 and 95% in three analyses over a one-year period [impurities unspecified]) in corn oil by gavage on five consecutive days per week for 78 weeks. Low-dose and high-dose animals received, respectively, 150 and 300 mg/kg bw per day for eight weeks and then 200 and 400 mg/kg bw per day for 70 weeks, followed by 12-13 weeks without treatment, after which the experiment was terminated. The time-weighted average doses were 195 and 390 mg/kg bw per day, respectively [calculated over seven days per week]. Groups of 20 male and 20 female mice either received corn oil alone and served as matched vehicle controls or remained untreated and served as matched untreated controls. At least 50% of the male mice in each group were alive at week 86; 50% of the female mice were still alive after 90, 89, 58 and 81 weeks in the untreated control, vehicle control, low-dose and high-dose groups, respectively. The incidence of hepatocellular neoplasms [reported as carcinomas] was increased significantly ($p < 0.01$) in all treated groups: males—2/17 (untreated controls, 2/20 vehicle controls, 18/49 low-dose animals and 37/49 high-dose animals; in females—2/20 untreated controls, 0/20 vehicle controls, 16/48 low-dose animals and 40/45 high-dose animals. Adrenal phaeochromocytomas were present in 8/48 high-dose males and in 12/43 high-dose females, but in no other group (National Cancer Institute, 1978).

Rat: Groups of 50 male and 50 female Osborne-Mendel rats, six weeks of age, received technical-grade 1,1,2-trichloroethane (see above) in corn oil by gavage on five consecutive days a week for 78 weeks. Low-dose and high-dose groups received, respectively, 35 and 70 mg/kg bw per day for 20 weeks, then 50 and 100 mg/kg bw per day for 58 weeks and were left untreated for the subsequent 34-35 weeks. The time-weighted average doses were 46 and 92 mg/kg bw per day [calculated over seven days per week]. Groups of 20 male and 20 female rats either received corn oil alone and served as matched vehicle controls or remained untreated and served as matched untreated controls. At least 50% of the male rats in untreated, low-dose and high-dose control groups survived more than 96 weeks; 50% of the females in the untreated control, low-dose and high-dose groups survived more than 105 weeks. Vehicle control groups had unexpectedly poor survival, with only 5% (1/20) of males and 20% (4/20) of females still alive at the end of the study; the authors did not, therefore, include them in statistical comparisons. No statistically significant increase in tumour incidence was found, either in males or in females (National Cancer Institute, 1978).

In a screening assay based on the production of γ-glutamyltranspeptidase-positive foci in rat liver, 10 male Osborne-Mendel rats (weighing 180-230 g when received and observed for a further five days before the start of the experiment) were given a single dose of 0.52 mmol/kg bw ([69.4 mg] maximum tolerated dose) 1,1,2-trichloroethane (98% pure) in corn oil by gavage 24 h following a two-thirds partial hepatectomy. Ten male rats given 2 ml/kg bw corn oil following partial hepatectomy served as vehicle controls. Six days after partial hepatectomy, the rats were given 0.05% (w/w) phenobarbital in the diet for seven weeks; they were then transferred to their regular diet for seven more days, at which time they were sacrificed. The number of foci/cm^2 liver in rats given 1,1,2-trichloroethane was not greater than that in the vehicle controls [numbers of foci not reported]. In further studies, groups of 10 male rats were given a single initiating dose of 30 mg/kg bw N-nitrosodiethylamine in 5 ml water or water alone by intraperitoneal injection 24 h after a two-thirds partial hepatectomy. Six days later, the rats were given 0.52 mmol/kg bw 1,1,2-trichloroethane in corn oil or corn oil alone by gavage on five days per week for seven weeks. In rats initiated with N-nitrosodiethylamine, 1,1,2-trichloroethane significantly increased the incidence of γ-glutamyltranspeptidase-positive foci/cm2 liver: control (N-nitrosodiethylamine plus corn oil), 1.6 ± 0.3 (SD); treated, 6.3 ± 2.2 ($p < 0.05$, Student's t test). In rats not initiated with N-nitrosodiethylamine, 1,1,2-trichloroethane also produced a significant increase in the number of foci/cm2 liver: control (water plus corn oil), 0.4 ± 0.2; treated, 4.4 ± 1.3 ($p < 0.05$) (Story et al., 1986).

(b) *Subcutaneous injection*

Rat: Groups of 50 male and 50 female Sprague-Dawley rats [200-250 g] were given 15.37 or 46.77 μmol [2.05 or 6.24 mg] 1,1,2-trichloroethane (> 99% pure) in 0.25 ml dimethylsulfoxide by subcutaneous injection once a week for two years. Groups of 35 male and 50 female rats given 0.25 ml dimethylsulfoxide on the same dosing schedule served as vehicle controls; groups of the same size without treatment served as untreated controls. The median survival time was: males—untreated control, 100 weeks; vehicle control, 87 weeks; low-dose, 90 weeks; high-dose, 85 weeks; females—untreated control, 91 weeks; vehicle control, 95 weeks; low-dose, 86 weeks; high-dose, 83 weeks. Sarcomas occurred at various sites in none of the untreated controls, in 2/35 and 3/50 vehicle control, in 4/50 and 3/50 low-dose and in 8/50 and 5/50 high-dose rats. The proportion of low- or high-dose rats with sarcomas was not significantly larger than that of vehicle controls (Norpoth et al., 1988).

3.2 Other relevant data

(a) Experimental systems

(i) Absorption, distribution, excretion and metabolism

The blood/gas partition coefficient (at 37°C) of 1,1,2-trichloroethane in rats was 58 (Gargas et al., 1989), and the blood serum/gas partition coefficient (at 25°C) in humans was 56 (Morgan et al., 1972); these findings predict that this compound is readily absorbed by inhalation.

1,1,2-Trichloroethane is absorbed through mouse skin (Tsuruta, 1975): when 5.4 mmol (0.5 ml) of undiluted 1,1,2-trichloroethane were applied to a 2.92-cm^2 area of mouse skin, 5.7 μmol (763 μg) were absorbed after 15 min, 0.015 μmol (2 μg) was eliminated in expired air, and the balance was retained in the body. The percutaneous absorption rate was 130 nmol/min per cm^2 of skin. Skin absorption was confirmed in guinea-pigs (Jakobson et al., 1977).

In mice exposed to 1000 ppm (5445 mg/m^3) 1,1,2-trichloroethane by whole-body inhalation for 1 h, the highest concentrations of the halocarbon immediately after exposure were found in fat, followed by liver and kidney (Takahara, 1986).

In dogs given undiluted 1,1,2-trichloroethane (0.375 or 0.75 mmol/kg, 50 or 100 mg/kg) intravenously, about 20 or 30%, respectively, was eliminated in the expired air within 60 min (Hobara et al., 1981).

The metabolic kinetic constants for 1,1,2-trichloroethane in rats exposed *in vivo* by inhalation to 501 ppm [2.7 g/m^3] for 6 h were K_m = 0.75 mg/l (5.63 μM) and V_{max} = 7.70 mg/kg per h (57.7 μmol/kg per h) (Gargas & Andersen, 1989).

^{14}C-1,1,2-Trichloroethane (0.75-1.5 mmol/kg, 100-200 mg/kg) was given in olive oil to mice by intraperitoneal injection, and the elimination of radiolabel was followed for three days (Yllner, 1971); 73-87% of the dose was eliminated in the urine, 0.1-2% in the feces and 16-22% in expired air, largely (60%) as ^{14}C-carbon dioxide. Several urinary metabolites were identified: chloroacetic acid (6-31% of urinary radioactivity), free *S*-carboxymethylcysteine (29-46%), conjugated *S*-carboxymethylcysteine (3-10%), thiodiacetic acid (38-42%), 2,2-dichloroethanol

Table 3. Summary of carcinogenicity studies of 1,1,2-trichloroethane in experimental animals

Reference	Species/strain	Sex	Dose schedule	Experimental parameter/observation[a]	Group				Significance
					0	1	2	3	
National Cancer Institute (1978)	Mouse B6C3F$_1$	M	5 d/week, gavage, corn oil, 78 weeks	Dose (mg/kg TWA)	0*	0*	195	390	
				Survival (91 weeks)	11/20	10/20	25/50	18/50	
				Hepatocellular carcinoma	2/17	2/20	18/49	37/49	$p < 0.001$
				Adrenal phaeochromocytoma	0/18	0/20	0/49	8/48	$p = 0.003$
		F		Dose (mg/kg TWA)	0*	0*	195	390	
				Survival (91 weeks)	11/20	18/20	21/50	15/50	
				Hepatocellular carcinoma	2/20	0/20	16/48	40/45	$p < 0.001$
				Adrenal phaeochromocytoma	0/20	0/20	0/48	12/43	$p = 0.006$
National Cancer Institute (1978)	Rat Osborne-Mendel	M	5 d/week, gavage, corn oil, 78 weeks	Dose (mg/kg TWA)	0*	0*	46	92	
				Survival (111 weeks)	6/20	1/20	11/50	15/50	
		F		Dose (mg/kg TWA)	0*	0*	46	92	
				Survival (111 weeks)	8/20	4/20	32/50	21/50	
Norpoth et al. (1988)	Rat Sprague-Dawley	M	1/week, s.c. inj., dimethyl sulfoxide, 2 years	Dose (µmol/rat)	0**	0**	15.37	46.77	
				Survivl (median days)	696	605	633	594	
				Sarcoma (not necessarily at injection site)	0/35	2/35	4/50	8/50	
		F		Dose (µmol/rat)	0**	0**	15.37	46.77	
				Survival (median days)	639	668	602	584	
				Sarcoma (not necessarily at injection site)	0/50	3/50	3/50	5/50	

[a]TWA, time-weighted average
*Group 0, untreated; group 1, corn oil
**Group 0, untreated; group 1, 0.25 ml dimethylsulfoxide

(1-2%), oxalic acid (about 0.5%) and trichloroacetic acid (1.4-2.3%). The latter compound may be formed from an impurity or may be a metabolite of 1,1,2-trichloroethane.

Thiodiglycolic acid (about 20% of the dose) was identified as a urinary metabolite of 1,1,2-trichloroethane given intraperitoneally (40 or 160 μmol [5.3 or 21.3 mg]/rat) in corn oil to rats (Norpoth et al., 1988).

Daily oral doses of 1,1,2-trichloroethane (0.52 mmol/kg [70 mg/kg] to rats and 2.24 μmol/kg [300 mg/kg] to mice) on five days per week for four weeks followed by a single dose of ^{14}C-1,1,2-trichloroethane in corn oil resulted in elimination of about 7-9 or 3-5% of the dose in the expired air as 1,1,2-trichloroethane and volatile metabolites or as ^{14}C-carbon dioxide, respectively, by 48 h (Mitoma et al., 1985). The excreta contained about 75% of the administered radiolabel, and 2-4% was retained in the carcass. S-Carboxymethylcysteine, thiodiacetic acid and chloroacetic acid were identified as urinary metabolites in both rats and mice.

1,1,2-Trichloroethane is metabolized by rat hepatic cytochromes P450 to chloroacetic acid (Ivanetich & Van den Honert, 1981) and to inorganic chloride (Van Dyke & Wineman, 1971; Van Dyke & Gandolfi, 1975).

When ^{14}C-1,1,2-trichloroethane was incubated in air with primary cultures of rat hepatocytes isolated from phenobarbital-treated rats, covalent binding of 1,1,2-trichloroethane metabolites to protein (about 1.6 nmol/mg protein per 30 min) and to lipid (about 5 nmol/mg lipid phosphorus) was detected (DiRenzo et al., 1984).

(ii) *Toxic effects*

The single-dose oral LD_{50}s (in 10% Emulphor in water) of 1,1,2-trichloroethane in male and female CD-1 mice were estimated to be 378 and 491 mg/kg bw, respectively (White et al., 1985). Six-hour LC_{50} values of 1654 (9 g/m^3) and 416 ppm (2.3 g/m^3) 1,1,2-trichloroethane by whole-body inhalation have been reported for male Sprague-Dawley rats and for mice [sex and strain not specified], respectively (Bonnet et al., 1980). Percutaneous applications of undiluted 1,1,2-trichloroethane at doses of 0.5 or 2.0 ml/3.1 cm^2 skin to male and female guinea-pigs resulted in 100% mortality; 0.25 ml was lethal for 25% of the animals (Wahlberg, 1976). Intraperitoneal administration of 0.25-2.0 ml/animal 1,1,2-trichloroethane to male and female guinea-pigs was also lethal to all animals (Wahlberg & Boman, 1979). Signs of toxicity included sedation, gastric irritation, lung haemorrhage and liver and kidney damage (White et al., 1985).

Daily administration of 3.8 or 38 mg/kg bw 1,1,2-trichloroethane by gavage (in 10% Emulphor in water) to male and female CD-1 mice for 14 days had no significant toxic effect (Sanders et al., 1985; White et al., 1985).

Administration of 0.02, 0.2 or 2 g/l 1,1,2-trichloroethane in the drinking-water of male and female CD-1 mice for 90 days [calculated daily doses: males—4.4, 46 or 305 mg/kg bw; females—3.9, 44 or 384 mg/kg bw] resulted in body weight reduction, depressed peritoneal macrophage function and decreased liver glutathione levels in males; decreased haematocrit and haemoglobin levels in females; and increased serum alkaline phosphatase activities and decreased haemagglutination titres in males and females at some doses. No other significant toxic effect was observed (Sanders *et al.*, 1985; White *et al.*, 1985).

1,1,2-Trichloroethane significantly increased the total number of enzyme-altered foci in livers of Osborne-Mendel rats when administered after treatment with *N*-nitrosodiethylamine (Milman *et al.*, 1988; see also section 3.1).

(iii) *Effects on reproduction and prenatal toxicity*

In a developmental toxicity screening study, 1,1,2-trichloroethane given orally in corn oil at a dose that killed 3/30 pregnant mice caused no developmental toxicity in offspring of survivors (Seidenberg *et al.*, 1986).

(iv) *Genetic and related effects* (Table 4)

The genetic and related effects of 1,1,2-trichloroethane have been reviewed (Infante & Tsongas, 1982).

1,1,2-Trichloroethane did not induce mutation in *Salmonella typhimurium*. It caused chromosome malsegregation in *Aspergillus nidulans* and morphological transformation of BALB/c 3T3 cells. 1,1,2-Trichloroethane bound to DNA *in vitro* (DiRenzo *et al.*, 1982) and to DNA, RNA and protein of various organs following treatment of rodents *in vivo* (Mazzullo *et al.*, 1986). Strong S-phase induction but no unscheduled DNA synthesis was observed in livers of treated mice.

A study in which a negative response was reported in *S. typhimurium* and positive responses in tests for unscheduled DNA synthesis in rat hepatocytes and for transformation in BALB/c 3T3 cells could not be evaluated [details not given] (Milman *et al.*, 1988).

(*b*) *Humans*

(i) *Absorption, distribution, excretion and metabolism*

When ^{38}Cl-1,1,2-trichloroethane was administered by inhalation at a dose of about 5 mg/subject in a single breath to human volunteers [subject weight and gender not stated], about 3% of the administered radiolabel was eliminated in the breath within 1 h. Urinary excretion of ^{38}Cl amounted to < 0.01% of the dose/min (Morgan *et al.*, 1970).

(ii) *Toxic effects*

No data were available to the Working Group.

1,1,2-TRICHLOROETHANE

Table 4. Genetic and related effects of 1,1,2-trichloroethane

Test system	Result		Dose LED/HID	Reference
	Without exogenous metabolic system	With exogenous metabolic system		
SA0, *Salmonella typhimurium* TA100, reverse mutation	–	0	0.0000	Simmon et al. (1977)[a]
SA0, *Salmonella typhimurium* TA100, reverse mutation	–	–	5300.0000	Barber et al. (1981)[a]
SA0, *Salmonella typhimurium* TA100, reverse mutation	–	–	0.0000	Mersch-Sunderman (1989)[b]
SA2, *Salmonella typhimurium* TA102, reverse mutation	–	–	0.0000	Mersch-Sunderman (1989)[c]
SA5, *Salmonella typhimurium* TA1535, reverse mutation	–	–	3900.0000	Rannug et al. (1978)
SA5, *Salmonella typhimurium* TA1535, reverse mutation	–	–	5300.0000	Barber et al. (1981)[a]
SA9, *Salmonella typhimurium* TA98, reverse mutation	–	–	5300.0000	Barber et al. (1981)[a]
SA9, *Salmonella typhimurium* TA98, reverse mutation	–	–	0.0000	Mersch-Sunderman (1989)[b]
SAS, *Salmonella typhimurium* TA97, reverse mutation	–	–	0.0000	Mersch-Sunderman (1989)[b]
ANN, *Aspergillus nidulans*, chromosome malsegregation	+	0	750.0000	Crebelli et al. (1988)
ANG, *Aspergillus nidulans*, genetic crossing over	+	0	750.0000	Crebelli et al. (1988)
TBM, Cell transformation, BALB/c3T3 mouse cells *in vitro*	(+)	0	25.0000	Tu et al. (1985)[a]
???, S-phase synthesis induction, mouse hepatocytes *in vivo*	+	0	0.0000	Mirsalis et al. (1989)
UVM, Unscheduled DNA synthesis, mouse hepatocytes *in vivo*	–	0	1000.0000	Mirsalis et al. (1989)
BVD, Binding to DNA, mouse cells *in vivo*	+	0	0.8000	Mazzullo et al. (1986)
BVD, Binding to DNA, rat cells *in vivo*	+	0	0.8000	Mazzullo et al. (1986)
BVP, Binding to RNA and protein, rat cells *in vivo*	+	0	0.8000	Mazzullo et al. (1986)
BVP, Binding to RNA and protein, mouse cells *in vivo*	+	0	0.8000	Mazzullo et al. (1986)

[a]Closed container
[b]Negative in closed container, standard test, or spot test
[c]Negative in standard test, spot test

(iii) *Effects on reproduction and prenatal toxicity*

No data were available to the Working Group.

(iv) *Genetic and related effects*

No data were available to the Working Group.

3.3 Case reports and epidemiological studies of carcinogenicity to humans

No data were available to the Working Group.

4. Summary of Data Reported and Evaluation

4.1 Exposure data

1,1,2-Trichloroethane is used as an intermediate in the production of vinylidene chloride and, to a lesser extent, as a special-purpose industrial solvent and as a chemical intermediate in other processes. It has been detected in drinking-water as well as in untreated groundwater and surface water in some locations; it may occur mainly as a result of industrial emissions.

4.2 Experimental carcinogenicity data

1,1,2-Trichloroethane was tested for carcinogenicity in a two-year study in male and female B6C3F$_1$ mice and Osborne-Mendel rats by oral administration and in Sprague-Dawley rats by subcutaneous injection. In the studies by oral administration, 1,1,2-trichloroethane produced hepatocellular neoplasms and adrenal phaeochromocytomas in mice of each sex but did not significantly increase the proportion of rats with neoplasms at any site relative to untreated controls. In the study in rats by subcutaneous injection, 1,1,2-trichloroethane did not increase the incidence of neoplasms.

In a screening assay for γ-glutamyltranspeptidase-positive foci in the liver of male Osborne-Mendel rats, 1,1,2-trichloroethane did not increase the number of foci in the liver in the initiation protocol (single injection), but the number was increased in the promotion protocol (repeated injections), with or without initiation by *N*-nitrosodiethylamine.

4.3 Human carcinogenicity data

No data were available to the Working Group.

4.4 Other relevant data

1,1,2-Trichloroethane was not mutagenic to bacteria. In single studies, it induced chromosomal malsegregation in a fungus and transformation in cultured mammalian cells. S-Phase induction, but not unscheduled DNA synthesis, was observed in mice after treatment *in vivo*.

4.5 Evaluation[1]

There is *limited evidence* for the carcinogenicity of 1,1,2-trichloroethane in experimental animals.

No data were available from studies in humans on the carcinogenicity of 1,1,2-trichloroethane.

Overall evaluation

1,1,2-Trichloroethane *is not classifiable as to its carcinogenicity to humans (Group 3)*.

5. References

Aldrich Chemical Co. (1990) *1990-1991 Aldrich Handbook of Fine Chemicals*, Milwaukee, WI, p. 1269

American Conference of Governmental Industrial Hygienists (1989) *TLVs Threshold Limit Values and Biological Exposure Indices for 1989-1990*, Cincinnati, OH, p. 40

American Tokyo Kasei (1988) *TCI American Organic Chemicals 88/89 Catalog*, Portland, OR, p. 1218

Barber, E.D., Donish, W.H. & Mueller, K.R. (1981) A procedure for the quantitative measurement of the mutagenicity of volatile liquids in the Ames *Salmonella*/microsome assay. *Mutat. Res., 90*, 31-48

Bauer, U. (1981) Human exposure to environmental chemicals—Investigations on volatile organic halogenated compounds in water, air, food, and human tissues. III. Communication: results of investigations. *Z. Bakteriol. Mikrobiol. Hyg. Abt. 1 Orig. B, 174*, 200-237

Bonnet, P., Francin, J.-M., Gradiski, D., Raoult, G. & Zissu, D. (1980) Determination of the median lethal concentration of the main chlorinated aliphatic hydrocarbons in the rat (Fr.). *Arch. Mal. prof., 41*, 317-321

---

[1]For definition of the italicized terms, see Preamble, pp. 30-33.

Brodzinsky, R. & Singh, H.B. (1983) *Volatile Organic Chemicals in the Atmosphere: An Assessment of Available Data* (EPA-600/3-83-027a; NTIS PB83-195503), Research Triangle Park, NC, US Environmental Protection Agency

Budavari, S., ed. (1989) *The Merck Index*, 11th ed., Rahway, NJ, Merck & Co., p. 1516

Chemical Information Services Ltd (1988) *Directory of World Chemical Producers—1989/90 Edition*, Oceanside, NY, p. 567

Chemical Information Systems, Inc. (1990) *ISHOW Database*, Baltimore, MD

Cole, R.H., Frederick, R.E., Healy, R.P. & Rolan, R.G. (1984) Preliminary findings of the Priority Pollutant Monitoring Project of the Nationwide Urban Runoff Program. *J. Water Pollut. Control Fed.*, 56, 898-908

Cook, W.A. (1987) *Occupational Exposure Limits—Worldwide*, Cincinnati, OH, American Industrial Hygiene Association, pp. 126, 155, 220

Crebelli, R., Benigni, R., Franekic, J., Conti, G., Conti, L. & Carere, A. (1988) Induction of chromosome malsegregation by halogenated organic solvents in *Aspergillus nidulans*: unspecific or specific mechanism? *Mutat. Res.*, 201, 401-411

DiRenzo, A.B., Gandolfi, A.J. & Sipes, I.G. (1982) Microsomal bioactivation and covalent binding of aliphatic halides to DNA. *Toxicol. Lett.*, 11, 243-252

DiRenzo, A.B., Gandolfi, A.J., Sipes, I.G., Brendel, K. & Byard, J.L. (1984) Effect of O_2 tension on the bioactivation and metabolism of aliphatic halides by primary rat-hepatocyte cultures. *Xenobiotica*, 14, 521-525

Dow Chemical Co. (1990) *Material Safety Data Sheet: 1,1,2-Trichloroethane*, Midland, MI

Easley, D.M., Kleopfer, R.D. & Carasea, A.M. (1981) Gas chromatographic-mass spectrometric determination of volatile organic compounds in fish. *J. Assoc. off. anal. Chem.*, 64, 653-656

Eastman Kodak Co. (1987) *Kodak Laboratory and Research Products* (Catalog No. 53), Rochester, NY, pp. 257, 420

Eller, P.M. (1987) *NIOSH Manual of Analytical Methods*, 3rd ed., Vol. 2, rev. 1 (DHHS (NIOSH) Publ. No. 84-100), Washington DC, US Government Printing Office, pp. 1003-1-1003-6

Gargas, M.L. & Andersen, M.E. (1989) Determining kinetic constants of chlorinated ethane metabolism in the rat from rates of exhalation. *Toxicol. appl. Pharmacol.*, 99, 344-353

Gargas, M.L., Burgess, R.J., Voisard, D.E., Cason, G.H. & Andersen, M.E. (1989) Partition coefficients of low-molecular-weight volatile chemicals in various liquids and tissues. *Toxicol. appl. Pharmacol.*, 98, 87-99

Gupta, K.C., Ulsamer, A.G. & Gammage, R. (1984) Volatile organic compounds in residential air: levels, sources and toxicity. In: *Proceedings of the 77th Annual Meeting of the Air Pollution Control Association, San Francisco, CA*, Pittsburgh, PA, Air Pollution Control Association, Section 84-1.3, pp. 1-9

Hagenmaier, H., Werner, G. & Jäger, W. (1982) Quantitative determination of volatile halogenated hydrocarbons in water samples by capillary gas chromatography and electron capture detection. *Z. Wasser Abwasser Forsch.*, 15, 195-198

Health and Safety Executive (1987) *Occupational Exposure Limits (1985)* (Guidance Note EH 40/87), London, Her Majesty's Stationary Office, p. 21

Henschler, D., Reichert, D. & Metzler, M. (1980) Identification of potential carcinogens in technical grade 1,1,1-trichloroethane. *Int. Arch. occup. environ. Health,* 47, 263-268

Hobara, T., Kobayashi, H., Iwamoto, S. & Sakai, T. (1981) Diminution of 1,1,1- and 1,1,2-trichloroethane in the blood and their excretion by the lungs (Jpn.). *Jpn. J. ind. Health,* 23, 377-382

IARC (1979) *IARC Monographs on the Evaluation of the Carcinogenic Risk of Chemicals to Humans,* Vol. 20, *Some Halogenated Hydrocarbons,* Lyon, pp. 533-543

IARC (1987a) *IARC Monographs on the Evaluation of Carcinogenic Risks to Humans,* Suppl. 7, *Overall Evaluations of Carcinogenicity: An Updating of IARC Monographs Volumes 1 to 43,* Lyon, p. 63

IARC (1987b) *IARC Monographs on the Evaluation of Carcinogenic Risks to Humans,* Suppl. 7, *Overall Evaluations of Carcinogenicity: An Updating of IARC Monographs Volumes 1 to 43,* Lyon, p. 62

IARC (1987c) *IARC Monographs on the Evaluation of Carcinogenic Risks to Humans,* Suppl. 7, *Overall Evaluations of Carcinogenicity: An Updating of IARC Monographs Volumes 1 to 43,* Lyon, pp. 373-376

IARC (1987d) *IARC Monographs on the Evaluation of Carcinogenic Risks to Humans,* Suppl. 7, *Overall Evaluations of Carcinogenicity: An Updating of IARC Monographs Volumes 1 to 43,* Lyon, pp. 376-377

Infante, P.F. & Tsongas, T.A. (1982) Mutagenic and oncogenic effects of chloromethanes, chloroethanes, and halogenated analogues of vinyl chloride. *Environ. Sci. Res.,* 25, 301-327

Ivanetich, K.M. & Van den Honert, L.H. (1981) Chloroethanes: their metabolism by hepatic cytochrome P-450 *in vitro. Carcinogenesis,* 2, 697-702

Jakobson, I., Holmberg, B. & Wahlberg, J.E. (1977) Variations in the blood concentration of 1,1,2-trichloroethane by percutaneous absorption and other routes of administration in the guinea pig. *Acta pharmacol. toxicol.,* 41, 497-506

Kaiser, K.L.E. & Comba, M.E. (1983) Volatile contaminants in the Welland River watershed. *J. Great Lakes Res.,* 9, 274-280

Lodge, J.P., Jr, ed. (1989a) *Methods of Air Sampling and Analysis,* 3rd ed., Chelsea, MI, Lewis Publishers, pp. 678-685

Lodge, J.P., Jr, ed. (1989b) *Methods of Air Sampling and Analysis,* 3rd ed., Chelsea, MI, Lewis Publishers, pp. 78-83

Lodge, J.P., Jr, ed. (1989c) *Methods of Air Sampling and Analysis,* 3rd ed., Chelsea, MI, Lewis Publishers, pp. 171-187

Mazzullo, M., Colacci, A., Grilli, S., Prodi, G. & Arfellini, G. (1986) 1,1,2-Trichloroethane: evidence of genotoxicity from short-term tests. *Jpn. J. Cancer Res. (Gann),* 77, 532-539

Mersch-Sunderman, V. (1989) Examination of the mutagenicity of organic microcontaminations in the environment. II. The mutagenicity of halogenated aliphatic hydrocarbons with the *Salmonella*-microsome test (Ames test) in relation to contamination of ground- and drinking-water (Ger.). *Zbl. Bakt. Hyg. B.,* 187, 230-243

Milman, H.A., Story, D.L., Riccio, E.S., Sivak, A., Tu, A.S., Williams, G.M., Tong, C. & Tyson, C.A. (1988) Rat liver foci and *in vitro* assays to detect initiating and promoting effects of chlorinated ethanes and ethylenes. *Ann. N.Y. Acad. Sci.*, *534*, 521-530

Mirsalis, J.C., Tyson, C.K., Steinmetz, K.L., Loh, E.K., Hamilton, C.M., Bakke, J.P. & Spalding, J.W. (1989) Measurement of unscheduled DNA synthesis and S-phase synthesis in rodent hepatocytes following in vivo treatment: testing of 24 compounds. *Environ. mol. Mutagenesis*, *14*, 155-164

Mitoma, C., Steeger, T., Jackson, S.E., Wheeler, K.P., Rogers, J.H. & Milman, H.A. (1985) Metabolic disposition study of chlorinated hydrocarbons in rats and mice. *Drug chem. Toxicol.*, *8*, 183-194

Morgan, A., Black, A. & Belcher, D.R. (1970) The excretion in breath of some aliphatic halogenated hydrocarbons following administration by inhalation. *Ann. occup. Hyg.*, *13*, 219-233

Morgan, A., Black, A. & Belcher, D.R. (1972) Studies on the absorption of halogenated hydrocarbons and their excretion in breath using ^{38}Cl tracer techniques. *Ann. occup. Hyg.*, *15*, 273-282

National Cancer Institute (1978) *Bioassay of 1,1,2-Trichloroethane for Possible Carcinogenicity. CAS No. 79-00-5* (Technical Report Series No. 74; DHEW (NIH) Publ. No. 78-1324), Washington DC, US Department of Health, Education, and Welfare

National Institute for Occupational Safety and Health (1978) *Chloroethanes: Review of Toxicity. Current Intelligence Bulletin 27* (DHEW (NIOSH) Publ. No. 78-181), Cincinnati, OH, US Department of Health, Education, and Welfare

National Library of Medicine (1989) *Toxic Chemical Release Inventory (TRI) Data Bank: 1,1,2-Trichloroethane*, Bethesda, MD

Norpoth, K., Heger, M., Müller, G., Mohtashamipur, E., Kemena, A. & Witting, C. (1988) Investigations on metabolism and carcinogenicity of 1,1,2-trichloroethane. *J. Cancer Res. clin. Oncol.*, *114*, 158-162

Otson, R., Williams, D.T. & Bothwell, P.D. (1982) Volatile organic chemicals in water at thirty Canadian potable water treatment facilities. *J. Assoc. off. anal. Chem.*, *65*, 1370-1374

Page, G.W. (1981) Comparison of groundwater and surface water for patterns and levels of contamination by toxic substances. *Environ. Sci. Technol.*, *15*, 1475-1481

Pellizzari, E.D. (1982) Analysis for organic vapor emissions near industrial and chemical waste disposal sites. *Environ. Sci. Technol.*, *16*, 781-785

Pfaltz & Bauer (1988) *Organic and Inorganic Chemicals for Research*, Waterbury, CT, p. 378

Pouchert, C.J., ed. (1974) *The Aldrich Library of NMR Spectra*, Vol. 1, Milwaukee, WI, Aldrich Chemical Co., p. 65B

Pouchert, C.J., ed. (1981) *The Aldrich Library of Infrared Spectra*, 3rd ed., Milwaukee, WI, Aldrich Chemical Co., p. 52F

Pouchert, C.J., ed. (1983) *The Aldrich Library of NMR Spectra*, 2nd ed., Vol. 1, Milwaukee, WI, Aldrich Chemical Co., p. 83A

Pouchert, C.J., ed. (1985a) *The Aldrich Library of FT-IR Spectra*, Vol. 1, Milwaukee, WI, Aldrich Chemical Co., p. 85A

Pouchert, C.J., ed. (1985b) *The Aldrich Library of FT-IR Spectra*, Vol. 3, Milwaukee, WI, Aldrich Chemical Co., p. 118B

Rannug, U., Sundvall, A. & Ramel, C. (1978) The mutagenic effect of 1,2-dichloroethane on *Salmonella typhimurium*. I. Activation through conjugation with glutathione *in vitro*. *Chem.-biol. Interact.*, 20, 1-16

Reed, N.R., Reed, W., Beltran, L., Babapour, R. & Hsieh, D.P.H. (1988) *Health Risk Assessment of 1,1,2-Trichloroethane (1,1,2-TCA) in California Drinking Water* (UCD/ET-88/2; PB89-131999), Davis, CA, Department of Environmental Toxicology, University of California-Davis

Riedel-de Haën (1986) *Laboratory Chemicals 1986*, Seelze 1/Hannover, p. 1072

Rosenberg, R., Grahn, O. & Johansson, L. (1975) Toxic effects of aliphatic chlorinated by-products from vinyl chloride production on marine animals. *Water Res.*, 9, 607-612

Sadtler Research Laboratories (1980) *The Sadtler Standard Spectra, 1980 Cumulative Index*, Philadelphia, PA

Sanders, V.M., White, K.L., Jr, Shopp, G.M., Jr & Munson, A.E. (1985) Humoral and cell-mediated immune status of mice exposed to 1,1,2-trichloroethane. *Drug Chem. Toxicol.*, 8, 357-372

Sax, N.I. & Lewis, R.J., Sr (1987) *Hawley's Condensed Chemical Dictionary*, 11th ed., New York, Van Nostrand Reinhold, p. 1176

Seidenberg, J.M., Anderson, D.G. & Becker, R.A. (1986) Validation of an in vivo developmental toxicity screen in the mouse. *Teratog. Carcinog. Mutagenesis*, 6, 361-374

Simmon V.F., Kauhanen, K. & Tardiff, R.G. (1977) Mutagenic activity of chemicals identified in drinking water. In: Scott, D., Bridges, B.A. & Sobels, F.H., eds, *Progress in Genetic Toxicology*, Amsterdam, Elsevier/North-Holland Biomedical Press, pp. 249-258

Singh, H.B., Salas, L.J., Stiles, R. & Shigeishi, H. (1983) *Measurements of Hazardous Organic Chemicals in the Ambient Atmosphere* (EPA-600/3-83-002; NTIS PB83-156935), Research Triangle Park, NC, US Environmental Protection Agency

Sittig, M. (1985) *Handbook of Toxic and Hazardous Chemicals and Carcinogens*, 2nd ed., Park Ridge, NJ, Noyes Publications, pp. 881-883

SKC Inc. (1990) *Comprehensive Catalog and Guide. Air Sampling Products for Worker Monitoring, Chemical Hazard Detection and Industrial Hygiene*, Eighty Four, PA

Story, D.L., Meierhenry, E.F., Tyson, C.A. & Milman, H.A. (1986) Differences in rat liver enzyme-altered foci produced by chlorinated aliphatics and phenobarbital. *Toxicol. ind. Health*, 2, 351-362

Strobel, K. & Grummt, T. (1987) Aliphatic and aromatic halocarbons as potential mutagens in drinking water. III. Halogenated ethanes and ethenes. *Toxicol. environ. Chem.*, 15, 101-128

Takahara, K. (1986) Experimental study on toxicity of trichloroethane. Part 1. Organ distribution of 1,1,1- and 1,1,2-trichloroethanes in exposed mice (Jpn.). *Okayama Igakkai Zasshi*, 98, 1079-1089

Thomas, R., Byrne, M., Gilbert, D., Goyer, M. & Wood, M. (1982) *An Exposure and Risk Assessment for Trichloroethanes* (EPA-440/4-85-018; PB85-220598), Washington DC, Office of Water Regulations and Standards, US Environmental Protection Agency

Tsuruta, H. (1975) Percutaneous absorption of organic solvents. 1. Comparative study of the in vivo percutaneous absorption of chlorinated solvents in mice. *Ind. Health, 13*, 227-236

Tu, A.S., Murray, T.A., Hatch, K.M., Sivak, A. & Milman, H.A. (1985) In vitro transformation of BALB/c-3T3 cells by chlorinated ethanes and ethylenes. *Cancer Lett., 28*, 85-92

United Nations Environment Programme (1990) *International Register of Potentially Toxic Chemicals, Recommendations—Legal Mechanisms*, Geneva

US Environmental Protection Agency (1980) *Ambient Water Quality Criteria for Chlorinated Ethanes* (EPA-440/5-80-029; PB81-117400), Washington DC, Office of Water Regulations and Standards

US Environmental Protection Agency (1986a) Method 8010. Halogenated volatile organics. In: *Test Methods for Evaluating Solid Waste—Physical/Chemical Methods*, 3rd ed. (*EPA No. SW-846*), Washington DC, Office of Solid Waste and Emergency Response

US Environmental Protection Agency (1986b) Method 8240. Gas chromatography/mass spectrometry for volatile organics. In: *Test Methods for Evaluating Solid Waste—Physical/Chemical Methods*, 3rd ed. (EPA No. SW-846), Washington DC, Office of Solid Waste and Emergency Response

US Environmental Protection Agency (1988a) Method TO-14. The determination of volatile organic compounds (VOCs) in ambient air using SUMMA(R) passivated canister sampling and gas chromatographic analysis. In: *Compendium of Methods for the Determination of Toxic Organic Compounds in Ambient Air* (EPA-600/4-89-017; NTIS PB90-116989), Research Triangle Park, NC, Atmospheric Research and Exposure Laboratory, Office of Research and Development

US Environmental Protection Agency (1988b) Methods for organic chemical analysis of municipal and industrial wastewater. Method 601. Purgeable halocarbons. *US Code fed. Regul., Title 40*, Part 136, Appendix A, pp. 267-281

US Environmental Protection Agency (1988c) Methods for organic chemical analysis of municipal and industrial wastewater. Method 624. Purgeables. *US Code fed. Regul., Title 40*, Part 136, Appendix A, pp. 432-446

US Environmental Protection Agency (1988d) Method 502.2. Volatile organic compounds in water by purge and trap capillary column gas chromatography with photoionization and electrolytic conductivity detectors in series. In: *Methods for the Determination of Organic Compounds in Drinking Water* (EPA-600/4-88/039; PB89-220461), Cincinnati, OH, Office of Research and Development, pp. 31-62

US Environmental Protection Agency (1988e) Method 524.2. Measurement of purgeable organic compounds in water by capillary column gas chromatography/mass spectrometry. In: *Methods for the Determination of Organic Compounds in Drinking Water* (EPA-600/4-88/039; PB89-220461), Cincinnati, OH, Office of Research and Development, pp. 285-323

US Environmental Protection Agency (1988f) Methods for organic chemical analysis of municipal and industrial wastewater. Method 1624 Revision B. Volatile organic compounds by isotope dilution GC/MS. *US Code fed. Regul., Title 40*, Part 136, Appendix A, pp. 475-488

US Occupational Safety and Health Administration (1989) Air contaminants—permissible exposure limits (Report No. OSHA 3112). *US Code fed. Regul.*, *Title 29*, Part 1910.1000, pp. 10-73

Van Dyke, R.A. & Gandolfi, A.J. (1975) Characteristics of a microsomal dechlorination system. *Mol. Pharmacol.*, *11*, 809-817

Van Dyke, R.A. & Wineman, C.G. (1971) Enzymatic dechlorination: dechlorination of chloroethanes and propanes *in vitro*. *Biochem. Pharmacol.*, *20*, 463-470

Verschueren, K. (1983) *Handbook of Environmental Data on Organic Chemicals*, 2nd ed., New York, Van Nostrand Reinhold, pp. 1128-1129

Wahlberg, J.E. (1976) Percutaneous toxicity of solvents. A comparative investigation in the guinea pig with benzene, toluene and 1,1,2-trichloroethane. *Ann. occup. Hyg.*, *19*, 115-119

Wahlberg, J.E. & Boman, A. (1979) Comparative percutaneous toxicity of ten industrial solvents in the guinea pig. *Scand. J. Work Environ. Health*, *5*, 345-351

Wallace, L.A., Pellizzari, E., Hartwell, T., Rosenzweig, M., Erickson, M., Sparacino, C. & Zelon, H. (1984) Personal exposure to volatile organic compounds. I. Direct measurements in breathing-zone air, drinking water, food, and exhaled breath. *Environ. Res.*, *35*, 293-319

Weast, R.C., ed. (1989) *CRC Handbook of Chemistry and Physics*, 70th ed., Boca Raton, FL, CRC Press, pp. C-266, D-199

White, K.L., Jr, Sanders, V.M., Barnes, D.W., Shopp, G.M., Jr & Munson, A.E. (1985) Toxicology of 1,1,2-trichloroethane in the mouse. *Drug chem. Toxicol.*, *8*, 333-355

Wiseman, P. (1972) *An Introduction to Industrial Organic Chemistry*, New York, Wiley Interscience, p. 125

Yllner, S. (1971) Metabolism of 1,1,2-trichloroethane-1,2-^{14}C in the mouse. *Acta pharmacol. toxicol.*, *30*, 248-256

Zaki, M.H. (1986) Groundwater contamination with synthetic organic compounds and pesticides in Suffolk County. *Northeast. environ. Sci.*, *5*, 15-22

COBALT AND COBALT COMPOUNDS

COBALT AND COBALT COMPOUNDS

The agents considered herein include (a) metallic cobalt, (b) cobalt alloys (including cobalt-containing medical implants) and (c) cobalt compounds. Organic cobalt-containing agents (e.g., vitamin B_{12}) are not covered comprehensively in this monograph.

1. Chemical and Physical Data

1.1 Synonyms, trade names and molecular formulae

Synonyms, trade names and molecular formulae for cobalt, cobalt alloys and cobalt compounds are presented in Table 1. The cobalt alloys and compounds given in Table 1 are not an exhaustive list, nor are they necessarily the most commercially important cobalt-containing substances; the list indicates the range of cobalt alloys and compounds available.

Table 1. Synonyms (Chemical Abstracts Service names are given in bold type), trade names and atomic or molecular formulae of cobalt and cobalt compounds

Chemical name	Chem. Abstr. Services Reg. No.[a]	Synonyms and trade names	Formulae
Metallic cobalt			
Cobalt	7440-48-4	C.I. 77320; cobalt element; cobalt-59	Co
Cobalt alloys			
Cobalt-chromium alloy[b]	11114-92-4 (91700-55-9)	**Cobalt alloy (nonbase), Co, Cr;** chromium alloy (nonbase), Co, Cr	Co·Cr
Nickel-based cobalt alloy[b]	11068-91-0 (12604-26-1; 12616-60-3; 12616-61-4; 12624-82-7; 12630-37-4; 12636-02-1; 12672-01-4; 12774-12-8; 37323-85-6; 64941-39-5)	**Nickel alloy (base), Ni 47-59, Co 17-20, Cr 13-17, Mo 4.5-5.7, Al 3.7-4.7, Ti 3-4, Fe 0-1, C 0-0.1 (AISI 687)** APK 1; Astroloy; Cabot 700; NiCo18Cr15MoAlTi; Nimonic AP 1; NK17CADT; PM-ATS 380; PWA 1013; R 77; Rene 77; U 700; U 700m; U700PM; Udimet 700	C·Al·Co·Cr·Fe·Mo·Ni·Ti

Table 1 (contd)

Chemical name	Chem. Abstr. Services Reg. No.[a]	Synonyms and trade names	Formulae
Metallic cobalt (contd)			
Cobalt-chromium-nickel-tungsten alloy	12638-07-2 (12618-75-6; 12748-86-6; 37329-48-9; 52827-91-5; 62449-84-7)	**Cobalt alloy (base), Co 48–58, Cr 24–26, Ni 9.5–12, W 7–8, Fe 2, Mn 0–1, Si 0–1, C 0.4–0.6 (ASTM A567-2)** AFNOR K-C25NW; AMS 5382; Co X–40; G–X 55; CoCrNiW 55 25; Haynes Stellite 31; HS 31; 31H114; K-C25NW; MAS 5382; PN 31H114; S–31; Stellite 31; Stellite 31 X 40; Stellite X40; 45VF; X 40	C·Co·Cr·Fe·Mn·Ni·Si·W
Cobalt-chromium-molybdenum alloy[b]	12629-02-6 (8064-15-1; 11068-92-1; 12618-69-8; 55345-18-1; 60382-64-1; 83272-15-5; 85131-98-2; 94076-26-3; 115201-64-4)	**Cobalt alloy (base), Co 56–68, Cr 25–29, Mo 5–6, Ni 1.8–3.8, Fe 0–3, Mn 0–1, Si 0–1, C 0.2–0.3 (AST A567-1)** Akrit CoMo35; AMS 5385D; Celsit 290; F 75; Haynes Stellite 21; HS 21; Protasul-2; Stellite 21; Vinertia; Vitallium; X25CoCrMo62 28 5; Zimaloy	C·Co·Cr·Fe·Mn·Mo·Ni·Si
Cobalt compounds			
Cobalt(II) acetate	71-48-7 (33327-32-1; 68279-06-1; 73005-84-2)	**Acetic acid, cobalt(2+) salt;** bis(acetato)cobalt; cobalt acetate; cobalt(2+) acetate; cobalt diacetate; cobaltous acetate; cobaltous diacetate	$Co(CH_3CO_2)_2$
Cobalt(II) acetate tetrahydrate	6147-53-1	**Bis(acetato)tetraquacobalt**	$Co(CH_3CO_2)_2 \cdot 4H_2O$
Cobalt(III) acetate	917-69-1	**Acetic acid, cobalt(3+) salt;** cobalt(3+) acetate; cobaltic acetate; cobalt triacetate	$Co(CH_3CO_2)_3$
Cobalt(II) carbonate	513-79-1	**Carbonic acid, cobalt(2+) salt (1:1);** C.I. 77353; cobalt carbonate (1:1); cobalt(2+) carbonate; cobalt monocarbonate; cobaltous carbonate	$CoCO_3$

Table 1 (contd)

Chemical name	Chem. Abstr. Services Reg. No.[a]	Synonyms and trade names	Formulae
Cobalt compounds (contd)			
Cobalt(II) carbonate hydroxide (1:1)	12069-68-0	Basic cobalt carbonate; carbonic acid, cobalt complex; cobalt carbonate hydroxide; cobalt, (carbonato)dihydroxydi-; **cobalt, [.mu.-[carbonato-(2-)-0:0′]]dihydroxydi-**	$CoCO_3 \cdot Co(OH)_2$
Cobalt(II) carbonate hydroxide (2:3)	12602-23-2	**Cobalt, bis(carbonato(2-))hexahydroxypenta-;** cobalt, bis(carbonato)hexahydroxypenta-; cobalt carbonate hydroxide; cobalt hydroxide carbonate	$2CoCO_3 \cdot 3Co(OH)_2$
Cobalt(II) carbonate hydroxide (2:3) monohydrate	51839-24-8	Basic cobalt carbonate; carbonic acid, cobalt(2+) salt, basic; **cobalt, bis(carbonato-(2-))hexahydroxypentamonohydrate;** cobaltous carbonate, basic	$2CoCO_3 \cdot 3Co(OH)_2 \cdot H_2O$
Cobalt(II) chloride	7646-79-9 (1332-82-7)	**Cobalt chloride ($CoCl_2$);** cobalt dichloride; cobaltous chloride	$CoCl_2$
Cobalt(II) chloride hexahydrate	7791-13-1	**Cobalt chloride, hexahydrate;** cobalt dichloride hexahydrate; cobaltous chloride hexahydrate	$CoCl_2 \cdot 6H_2O$
Cobalt(II) hydroxide	21041-93-0 (1307-85-3)	Cobalt dihydroxide; **cobalt hydroxide ($Co(OH)_2$);** cobalt(2+) hydroxide; cobaltous hydroxide	$Co(OH)_2$
Cobalt(III) hydroxide	1307-86-4	**Cobalt hydroxide ($Co(OH)_3$);** cobaltic hydroxide; cobalt trihydroxide	$Co(OH)_3$
Cobalt(II) naphthenate	61789-51-3	Cobalt naphthenates; naftolite; naphthenic acid, cobalt salt; **naphthenic acids, cobalt salts**	Unspecified
		Cobalt Nap-All; Naphthex Co; 8SN-Co	
Cobalt(II) nitrate	10141-05-6 (14216-74-1; 19154-72-4)	Cobalt bis(nitrate); cobalt(2+) nitrate; cobaltous nitrate; **nitric acid, cobalt(2+) salt**	$Co(NO_3)_2$

Table 1 (contd)

Chemical name	Chem. Abstr. Services Reg. No.[a]	Synonyms and trade names	Formulae
Cobalt compounds (contd)			
Cobalt(II) nitrate hexahydrate	10026-22-9 (13478-32-5)	Cobalt dinitrate hexahydrate; cobalt nitrate hexahydrate; cobalt(2+) nitrate hexahydrate; cobalt(II) nitrate hydrate; cobaltous nitrate hexahydrate; **nitric acid, cobalt(2+) salt, hexahydrate**	$Co(NO_3)_2 \cdot 6H_2O$
Cobalt(II) molybdenum(VI) oxide	13762-14-6 (12205-99-1; 14566-03-1; 63511-60-4)	Cobalt molybdate; cobalt molybdate(VI); cobalt(2+) molybdate; **cobalt molybdenum oxide ($CoMoO_4$)**; cobaltous molybdate; cobalt monomolybdate; molybdenum cobaltate; molybdenum cobalt oxide; molybdic acid (H_2MoO_4), cobalt(2+) salt (1:1)	$CoMoO_4$
Cobalt(II) oxide	1307-96-6	C.I. 77322; C.I. Pigment Black 13; cobalt black; cobalt monoxide; cobalt monooxide; cobaltous oxide; **cobalt oxide (CoO)**; cobalt(2+) oxide; monocobalt oxide Zaffre	CoO
Cobalt(II,III) oxide	1308-06-1 (12314-25-9; 25729-03-7)	Cobaltic–cobaltous oxide; cobalto–cobaltic oxide; cobalto–cobaltic tetroxide; cobaltosic oxide; **cobalt oxide (Co_3O_4)**; cobalt tetraoxide; tricobalt tetraoxide; tricobalt tetroxide	Co_3O_4
Cobalt(III) oxide	1308-04-9 (12314-25-9; 25729-03-7)	C.I. 77323; cobaltic oxide; **cobalt oxide (Co_2O_3)**; cobalt(3+) oxide; cobalt peroxide; cobalt sesquioxide; cobalt trioxide; dicobalt oxide; dicobalt trioxide	Co_2O_3
Cobalt(III) oxide monohydrate	12016-80-7 (61864-72-0)	**Cobalt hydroxide oxide (Co(OH)O)**; cobalt(III) hydroxide oxide; cobalt oxide hydroxide; cobalt oxyhydroxide	Co(OH)O or $Co_2O_3 \cdot H_2O$

Table 1 (contd)

Chemical name	Chem. Abstr. Services Reg. No.[a]	Synonyms and trade names	Formulae
Cobalt compounds (contd)			
Cobalt(II) sulfate	10124-43-3 (10393-49-4)	Cobalt monosulfate; cobaltous sulfate; cobalt sulfate (1:1); cobalt(2+) sulfate; cobalt sulphate; **sulfuric acid, cobalt(2+) salt (1:1)**	$CoSO_4$
Cobalt(II) sulfide	1317-42-6	Cobalt monosulfide; cobaltous sulfide; cobalt(2+) sulfide	CoS
Dicobalt octacarbonyl	10210-68-1 (12553-61-6; 14525-26-9; 19998-88-0; 24917-04-2; 90043-99-5)	Cobalt, di-.mu.-carbonylhexacarbonyldi-; **cobalt tetracarbonyl dimer**	$[Co(CO)_4]_2$ or $Co_2(CO)_8$
Tetracobalt dodecacarbonyl	17786-31-1 (12083-62-9; 19212-11-4; 19478-05-8; 19495-98-8; 20623-64-7; 28963-39-5)	**Cobalt, tri-.mu.-carbonyl-nonacarbonyltetra-**	$[Co(CO)_3]_4$ or $Co_4(CO)_{12}$

[a]Replaced CAS Registry Numbers are given in parentheses.
[b]Approximately 5000 alloys of cobalt with other metals are listed by the Chemical Abstracts Registry Service, of which cobalt is the base metal for approximately 2000. Chromium is contained in approximately 1400 of these alloys and nickel in approximately 1500. An example of each is listed here.

1.2 Chemical and physical properties of the pure substances

Selected chemical and physical properties of cobalt and cobalt compounds covered in this monograph are presented in Table 2.

Metallic cobalt

Cobalt metal was isolated by the Swedish scientist G. Brandt in 1735; in 1780, T.O. Bergman established cobalt as an element (Donaldson, 1986).

Cobalt exists in two allotropic forms. The hexagonal close-packed form is more stable at temperatures below 417°C, and the face-centred cubic form at

Table 2. Physical properties of cobalt and cobalt compounds[a]

Chemical name	Atomic/molecular weight	Melting-point (°C)	Typical physical description	Solubility
Metallic cobalt				
Cobalt	58.93	1495 (boiling-point, 2870)	Silver-grey, hard, magnetic, ductile, somewhat malleable metal	Practically insoluble in water; Readily soluble in dilute nitric acid; Readily soluble in hydrofluoric acid and readily in sulfuric and hydrochloric acids[b]
Cobalt compounds				
Cobalt(II) acetate (tetrahydrate)	177.03	–	Light-pink crystals	Readily soluble in water
	249.08	Loses four H_2O at 140	Red-violet monoclinic, deliquescent	Soluble in water, dilute acids, pentyl acetate and alcohols
Cobalt(III) acetate	236.07	100 (decomposes)	Dark-green, very hygroscopic powder or green crystals	Soluble in water, acetic acid, ethanol, n-butanol; Aqueous solutions hydrolyse slowly at room temperature, rapidly at 60–70°C
Cobalt(II) carbonate	118.94	Decomposes	Red, trigonal	Practically insoluble in water, ammonium hydroxide, ethanol or methyl acetate; Soluble in acids
Cobalt(II) carbonate hydroxide (2:3)	516.73	Decomposes[c]	Pale-red powder, usually containing some H_2O	Practically insoluble in water; Soluble in dilute acids and ammonium carbonate solution
(monohydrate)	534.74	Decomposes[d]	Violet-red crystals	Insoluble in cold water; Decomposes in hot water; Soluble in acid and ammonium carbonate solution
Cobalt(II) chloride	129.84	724 (in HCl gas) decomposes at 400 on long heating in air	Pale-blue, hygroscopic leaflets; colourless in very thin layers; turns pink on exposure to moist air	Soluble in water (450 g/l at 7°C; 1050 g/l at 96°C), ethanol (544 g/l), acetone (86 g/l), methanol (385 g/l), glycerol and pyridine; Slightly soluble in diethyl ether

Table 2 (contd)

Chemical name	Atomic/molecular weight	Melting-point (°C)	Typical physical description	Solubility
(hexahydrate)	237.93	86; loses four H_2O at 52–56, an additional H_2O by 100 and another H_2O at 110	Pink to red, slightly deliquescent, monoclinic, prismatic; turns blue when heated or when hydrochloric or sulfuric acid is added; slight odour[e]	Soluble in ethanol and in water (767 g/l at 0°C; 1907 g/l at 100°C), acetone, diethyl ether (2.9 g/l) and glycerol
Cobalt(II) hydroxide	92.95	Decomposes	Blue-green or rose-red powder or microscopic crystals	Very slightly soluble in water (0.0032 g/l) Soluble in acid and ammonium salts Insoluble in aqueous hydroxide solutions
Cobalt(III) hydroxide (trihydrate)	219.91	Decomposes; loses H_2O at 100	Black-brown powder	Practically insoluble in water and ethanol Soluble in nitric acid[f], sulfuric acid and hydrochloric acid
Cobalt(II) molybdenum oxide	218.87	–	Grey-green powder	–
Cobalt naphthenate	–[g]	140[h]	Brown, amorphous powder or bluish-red solid[d]	Practically insoluble in water Soluble in ethanol, diethyl ether and oils
Cobalt(II) nitrate	182.96	100–105 (decomposes)	Pale-red powder	Soluble in water
(hexahydrate)	291.03	55–56; loses three H_2O at 55	Red, monoclinic; liquid becomes green and decomposes to the oxide above 74°C	Soluble in water (1338 g/l at 0°C; 2170 g/l at 80°C), ethanol (1000 g/l at 12.5°C), acetone and most organic solvents Slightly soluble in ammonium hydroxide

Table 2 (contd)

Chemical name	Atomic/molecular weight	Melting-point (°C)	Typical physical description	Solubility
Cobalt(II) oxide	74.93	1795±20	Powder or crystals; colour varies from olive-green to red, depending on particle size, but the commercial material is usually dark-grey	Practically insoluble in water, ethanol and ammonium hydroxide. Soluble in acids (hydrochloric, sulfuric, nitric)[f]
Cobalt(II,III) oxide	240.80	895[c]; transition-point to CoO is 900–950	Black or grey crystals	Practically insoluble in water, aqua regia, hydrochloric or nitric acid. Soluble in sulfuric acid and fused sodium hydroxide[d]
Cobalt(III) oxide	165.86	895 (decomposes)	Black-grey crystals	Insoluble in water and ethanol. Soluble in acids
Cobalt(II) sulfate	154.99	735 (decomposes)	Dark-bluish crystals	Soluble in water (362 g/l at 20°C; 830 g/l at 100°C) and methanol (10.4 g/l at 18°C). Insoluble in ammonium hydroxide
(heptahydrate)	281.10	96.8; loses H$_2$O at 41.5, six H$_2$O at 71 and seven H$_2$O at 420	Pink-to-red monoclinic, prismatic	Soluble in water (604 g/l at 3°C; 670 g/l at 70°C), ethanol (25 g/l at 3°C) and methanol (545 g/l at 18°C)
Cobalt(II) sulfide	90.99		Exists in two forms: β-CoS — reddish, silver-white crystals or grey powder;	Practically insoluble in water (0.0038 g/l at 18°C) and soluble in acids
	>1116		α-CoS — black amorphous powder	Soluble in hydrochloric acid

Table 2 (contd)

Chemical name	Atomic/ molecular weight	Melting-point (°C)	Typical physical description	Solubility
Dicobalt octacarbonyl	341.95	Decomposes above 52	Orange crystals or dark-brown microcrystals	Practically insoluble in water Slightly soluble in ethanol Soluble in carbon disulfide and diethyl ether
Tetracobalt dodeca-carbonyl	571.86	–	Black crystals	Slightly soluble in cold water Soluble in benzene

[a]From Weast (1988); Budavari (1989), unless otherwise specified
[b]From Considine (1974)
[c]From CP Chemicals (1989a)
[d]From Sax & Lewis (1987)
[e]From Hall Chemical Co. (undated a)
[f]From Brauer (1965)
[g]The molecular weight of cobalt naphthenate varies, depending on the source of naphthenate and the method of preparation, ranging between 239–409 (6–10.5% cobalt) (US Environmental Protection Agency, 1983)
[h]From Bennett (1974)
[i]From Aldrich Chemical Co. (undated a)

higher temperatures (from 417°C to the melting-point; Considine, 1974). The free energy change is low, however, so that transformation from the face-centred cubic back to the hexagonal close-packed form is slow and may be inhibited by physical form (e.g., grain size or presence of other metals) (Donaldson, 1986).

The main oxidation states of cobalt are Co(2+) and Co(3+). Cobalt is stable to atmospheric oxygen, but when it is heated it is oxidized to the mixed oxide, Co(II,III) oxide (Co_3O_4); at temperatures above 900°C, Co(II) oxide (CoO) is the end-product. Cobalt metal does not combine directly with hydrogen or nitrogen but combines with sulfur, phosphorus and carbon when heated. Cobalt forms a protective layer of sulfide scale when reacted with sulfur at temperatures below 877°C or in an atmosphere of hydrogen sulfide. It forms a mixed oxide-sulfide scale in air containing sulfur dioxide (Donaldson et al., 1986a).

Cobalt also has magnetic properties. Hexagonal cobalt is ferromagnetic. The cubic form is magnetically anisotropic up to about 1000°C and becomes paramagnetic at 1121°C. Single crystals show marked magnetic anisotropy up to about 250°C (Donaldson, 1986).

Cobalt compounds

With the exception of the mixed oxide (Co_3O_4), the major commercial cobalt chemicals are all compounds of cobalt in its stable +2 oxidation state. A few simple salts of cobalt in its +3 oxidation state have been used commercially (e.g., Co_2O_3), and many Co(III) complexes with ligands such as NH_3, CN^-, NO^{2-}, ethylenediaminetetraacetic acid, phthalocyanines and azo dyes have been studied extensively. These electron-donor ligands strongly stabilize Co^{3+} in solution, usually forming octahedral complexes, many of which can be isolated as stable salts. In acid solution, in the absence of such complexing ligands, Co^{2+} is the stable form and Co^{3+} is so unstable that it is reduced rapidly and spontaneously to Co^{2+}, oxidizing water to molecular oxygen. In contrast, in an alkaline solution containing ammonium hydroxide or cyanide, Co^{2+} is readily oxidized by air or hydrogen peroxide to the more stable Co^{3+} complex. The $Co^{2+} \leftrightarrows Co^{3+}$ interconversion is important in many applications of cobalt compounds, including their use as catalysts and as paint driers and in the reactions of vitamin B_{12} (National Research Council, 1977; Donaldson, 1986; Donaldson et al., 1986a,b).

1.3 Technical products and impurities

(a) *Cobalt metal and cobalt alloys*

Cobalt metal is available for industrial use as 'broken' or 'cut' cathodes or electrolytic coarse powder. The cathodes measure 10-25 mm and weigh 20-50 g, with a purity greater than 99.5%. The 'fine', 'extrafine' and 'superfine' cobalt powders manufactured from the cathodes have a submicrometre mean particle size and contain both allotropic crystal forms in varying proportions for different applications. Electrolytic coarse powder has a mean particle size of 4-10 μm (Cobalt Development Institute, 1989). Cobalt is also available as briquets, granules (99.5% cobalt), rondelles, powder (99.995% cobalt or 99.8% cobalt, < 2 μm), ductile strips (95% cobalt, 5% iron), high purity strips (99% cobalt), foil (99.95 or 99.99% cobalt, 0.1-1 mm), rods (99.998% cobalt, 5.0 mm) and wire (> 99.9% cobalt, 0.25-2 mm) (Sax & Lewis, 1987; American Chemical Society, 1988; Aldrich Chemical Co., 1990).

Cobalt alloys can be categorized into six broad types: superalloys (high-temperature alloys), magnetic alloys, hard-metal alloys, high-strength steels, electrodeposited alloys and alloys with special properties (Donaldson, 1986).

Elements used in cobalt alloys are classified in terms of their effect on the transition from the cubic to the hexagonal form. Enlarged-field components, which lower the transition temperature, include aluminium, boron, carbon, copper, iron, manganese, niobium, nickel, tin, titanium and zirconium. Restricted-field components, which raise the transition temperature, include antimony, arsenic, chromium, germanium, iridium, molybdenum, osmium, platinum, rhenium, rhodium, ruthenium, silicon, tantalum and tungsten (Donaldson, 1986).

Cobalt *superalloys*, a term generally applied to immensely strong, hard, wear- and corrosion-resistant alloys, were first introduced in the 1930s. They were developed for use at high temperatures where relatively severe mechanical stressing is encountered and where high surface stability is required. Their superior strength at high temperatures arises from a close-packed face-centred cubic, austentitic lattice system, which can maintain better tensile, rupture and creep properties at elevated temperatures than a body-centred cubic system (Donaldson & Clark, 1985; Donaldson, 1986).

Superalloys are usually either cobalt- or nickel-based. Cobalt-based superalloys typically consist of a cobalt-chromium face-centred cubic solid solution matrix with the following ranges of composition: chromium, 15-29.5%; nickel, $\leq$ 28%; tungsten, $\leq$ 15%; tantalum, $\leq$ 9%; molybdenum, $\leq$ 5.5%; aluminium, $\leq$ 4.3%; titanium, $\leq$ 4%; zirconium, $\leq$ 2.25%; carbon, 0.04-1%; and boron, $\leq$ 0.11%. Small quantities of niobium, yttrium, lanthanum, iron, manganese, silicon and rhenium are present; and the balance is cobalt. Chromium is added to improve resistance to hot corrosion and oxidation. Nickel is added to stabilize the face-centred cubic structure by offsetting the tendency of the refractory metals to initiate transformation to the hexagonal close-packed structure (Donaldson & Clark, 1985).

Nickel-based superalloys were developed from the nickel-chromium alloys that had been used for over 50 years for electrical resistance, which often contain cobalt. They consist of a face-centred cubic, solid solution matrix with the following ranges of composition: chromium, 1.6-28.5%; cobalt, 1.1-22%; tungsten, 0-12.5%; molybdenum, 0-10%; aluminium, 0-6%; titanium, 0-5%; boron, 0-0.62%; carbon, 0.04-0.35%; zirconium, 0-0.13%; small amounts of tantalum, hafnium, iron, manganese, silicon, vanadium, niobium, magnesium and rhenium; and the balance as nickel (Donaldson & Clark, 1985).

Vitallium (CAS No. 12629-02-6), a cobalt-chromium alloy containing 56-68% cobalt with additions of chromium (25-29%), molybdenum (5-6%) and nickel (1.8-3.8%) was developed in 1936 (ASTM A567-1; Planinsek & Newkirk, 1979; Donaldson *et al.*, 1986b; Johnston, 1988; Roskill Information Services, 1989).

Some representative analyses of cobalt-containing alloys are given in Table 3.

Magnetic alloys. Cobalt is the only element capable of increasing the saturation magnetization of iron and is an important constituent of permanent magnets, commercial magnet steel (35% cobalt) and soft-magnet alloys. Representative analyses of some Alnico magnetic alloys (cobalt added to alloys of aluminium, nickel and iron) are given in Table 4. Magnets combining cobalt with rare-earth minerals were developed in 1967. Rare-earth cobalt alloys contain 60-65% cobalt and have the composition RCo_5, where R represents a rare-earth metal (Donaldson, 1986). A samarium-cobalt magnet was commercially available in the early 1970s, and a series of magnets with the composition R_2Co_{17} was marketed in 1980.

In *'hard-metal' alloys* (cemented carbides), cobalt powder is used as a matrix or bonding agent. The most commonly used cemented carbide, tungsten carbide, contains 80-90% by weight of hard metal and 5-10% cobalt, although up to 30% cobalt may be used for certain purposes. The properties of cemented tungsten carbides are sometimes enhanced by addition of the carbides of niobium, tantalum or titanium (Donaldson, 1986).

Cobalt-containing high-strength steels. Although cobalt is not a common alloying element in steel, it can be an important component when high strength is required (Donaldson, 1986). Maraging steels, used in the fabrication of tools and other applications requiring high strength-to-weight ratios, typically contain 8-18% cobalt alloyed with iron, nickel (8-19%), molybdenum (1-14%) and small amounts of aluminium and titanium (Roskill Information Services, 1989).

Cobalt-containing martensitic stainless maraging steels, especially designed for corrosion resistance and high tensile strength, typically contain 5-20% cobalt, 10-15.5% chromium, 0-8.2% nickel, 2-5.5% molybdenum and small amounts of carbon and titanium (Roskill Information Services, 1989).

Table 3. Examples of superalloys containing cobalt (values in weight %)[a]

Trade name	Co	Cr	Ni	Fe	Mo	W	Ta	Nb	Al	Ti	Mn	Si	C	B	Zn
Nimocast alloy 263	20.0	20.0	55.0	0.5	5.8	-	-	-	0.5	2.2	0.5	-	0.06	0.008	0.04
Udimet 500	19.0	18.0	52.0	-	4.2	-	-	-	3.0	3.0	-	-	0.07	0.007	0.05
Hastelloy alloy X	1.5	22.0	47.0	18.5	9.0	0.6	-	-	-	-	0.5	0.5	0.10	-	-
Inconel alloy 617	12.5	22.0	54.0	-	9.0	-	-	-	1.0	-	-	-	0.07	-	-
Haynes alloy 1002	Balance	22.0	16.0	1.5	-	7.0	3.75	-	0.3	0.2	0.7	0.4	0.6	-	0.3
WI-52	63.0	21.0	-	2.0	-	11.0	-	2.0	-	-	0.25	0.25	0.45	-	-
Haynes alloy 188	39.0	22.0	22.0	3.0 max	-	14.0	-	-	-	-	1.25 max	0.4	0.1	-	-
Haynes alloy 556	20.0	22.0	20.0	29.0	3.0	2.5	0.9	0.1	0.3	-	1.5	0.4	0.1	-	-

[a] From Nickel Development Institute (1987)

Table 4. Composition and magnetic properties of Alnico alloys[a]

Composition (%)						Method of manufacture	Coercive force (kA/m)
Co	Ni	Al	Cu	Ti	Nb		
3–5	21–28	11–13	2–4	0–1	–	Cast	36–56
12–14	16–20	9–11	3–6	0–1	–	Cast	40–50
17–20	18–21	8–10	2–4	4–8	–	Cast	60–72
23–25	12–15	7.8–8.5	2–4	0–0.5	–	Field treated	46–52
32–36	14–16	7–8	4	4–6	–	Field treated	110–140
24–25	13–15	7.8–8.5	2–4	–	0–1	Columnar	56–62
32–36	14–16	7–8	4	4–6	0–1	Columnar	110–140

[a]From Donaldson (1986)

The uses and composition of *electrodeposited alloys* and *alloys with special properties* are described below. Typical specifications for one class of special purpose alloys, those used in surgical implants, are given in Table 5.

Table 5. Composition of some cobalt-containing alloys used for surgical implants (%)[a]

Element	Alloy			
	A	B	C	D
Cobalt	Balance	Balance	Balance	Balance
Chromium	27.0–30.0	19.0–21.0	18.0–22.0	26.0–30.0
Molybdenum	5.0–7.0	9.0–10.5	3.0–4.0	5.0–7.0
Nickel	1.0 max	33.0–37.0	15.0–25.0	1.0 max
Iron	0.75 max	1.0 max	4.0–6.0	0.75 max
Carbon	0.35 max	0.025 max	0.05 max	0.35 max
Silicon	1.0 max	0.15 max	0.50 max	1.0 max
Manganese	1.0 max	0.15 max	1.0 max	1.0 max
Nitrogen	NA	NA	NA	0.25 max
Phosphorus	NA	0.015 max	NA	NA
Sulfur	NA	0.010 max	0.010 max	NA
Titanium	NA	1.0 max	0.50–3.50	NA
Tungsten	NA	NA	3.0–4.0	NA

[a]From American Society for Testing and Materials (1984, 1987a,b, 1988)
NA, not applicable

(b) Cobalt compounds

Cobalt(II) acetate is sold by one company as a reddish-pink solution containing 6-9% cobalt and 2% acetic acid (Hall Chemical Co., undated b).

Cobalt(II) acetate tetrahydrate is available at purities up to 100% from several companies as pink to red-violet crystals (BDH Ltd, 1989a; CP Chemicals, 1989b; J.T. Baker, 1989a; Mallinckrodt, 1989a; Hall Chemical Co., undated c). Technical-grade cobalt(II) acetate tetrahydrate, offered by one US company as red crystals, contains a minimum of 23.5% cobalt and small amounts of impurities (iron, 0.005% max; copper, 0.005% max; chlorine, 0.01% max; sulfate ion, 0.05% max; insolubles in acetic acid, 0.03% max; Shepherd Chemical Co., 1987a, 1989a).

Cobalt carbonate is offered by one US company as a reddish-purple powder containing a minimum of 45.5% cobalt and small amounts of impurities (iron, 0.005% max; copper, 0.005% max; lead, 0.005% max; chlorine, 0.01% max; sodium, 0.6% max; insolubles in dilute hydrochloric acid, 0.05% max; cadmium, 0.005% max; sulfate ion, 0.2% max; Shepherd Chemical Co., 1987b, 1989b). Several companies offer cobalt carbonate as a pink powder or red crystals at 90-100% purity (CP Chemicals, 1989c; J.T. Baker, 1989b; Hall Chemical Co., undated d). Basic cobalt carbonate, the primary commercial product, typically contains 45-47% cobalt (Donaldson *et al.*, 1986a).

Cobalt chloride is sold commercially mainly as the hexahydrate or other hydrated form. Cobalt chloride hexahydrate is available from several companies as red crystals in purities up to approximately 100% (BDH Ltd, 1989b; CP Chemicals, 1989c; Mallinckrodt, 1989b; Aldrich Chemical Co., undated b,c; Hall Chemical Co., undated a). Technical-grade cobalt chloride hexahydrate, available from one US company as red crystals, contains a minimum of 24% cobalt and small amounts of impurities (iron, 0.02% max; copper, 0.02% max; sulfate ion, 0.1% max; water insolubles, 0.05% max; Shepherd Chemical Co., 1987c, 1989c). The hexahydrate is also available as a pink-to-red powder at 98-100% purity (J.T. Baker, 1989c) and as a clear reddish aqueous solution containing 14.5% cobalt (Hall Chemical Co., undated e). Cobalt chloride is also available commercially as a clear, purple aqueous solution containing approximately 6% cobalt chloride (Mallinckrodt, 1989c) and as essentially pure (99.999%) hydrated red-violet powder and chunks (Aldrich Chemical Co., undated d).

Anhydrous cobalt chloride is available from two companies as a blue powder at purities up to 97% (BDH Ltd, 1989c; Aldrich Chemical Co., undated e) and from another at a purity of 100% (Hall Chemical Co., undated f).

Cobalt(II) hydroxide is available commercially as a solid containing 62% cobalt and an antioxidant (Donaldson *et al.*, 1986a), as a blue-green, moist press cake (E grade) containing 68% cobalt hydroxide and less than 500 ppm ammonia (Hall Chemical Co., undated g), as a technical grade (95% cobalt hydroxide; Aldrich Chemical Co., 1990) and as a pink powder containing a minimum of 61% cobalt and small amounts of impurities (chlorine, 0.02% max; acetic acid insolubles, 0.2% max; copper, 0.01% max; iron, 0.01% max; manganese, 0.03% max; nickel, 0.3% max; sulfate ion, 0.3% max; Shepherd Chemical Co., 1988a, 1989d).

Cobalt molybdenum oxide is produced by one company in the USA (Chemical Information Services Ltd, 1988).

Commercial grade *cobalt naphthenate* is available as a solution of 65% cobalt naphthenate (6% cobalt) in white spirits (Nuodex, 1986; Hall Chemical Co., undated h). One US company offers 6 and 8% liquid grades; another offers liquid, flake and solid forms (American Chemical Society, 1988). One Canadian company and one US company offer 6% cobalt naphthenate in solution with white spirits and 10.5% flaked cobalt naphthenate (Dussek Campbell Ltd, 1989a,b; Shepherd Chemical Co., 1989e,f).

One US company offers *cobalt nitrate hexahydrate* as a red-brown crystalline powder at 99.999% purity or as red chips in reagent grade or at 99% purity. The reagent grade is 98% pure and contains small amounts of impurities (insolubles, < 0.01%; chloride ion, < 0.002%; copper, < 0.002%; iron, < 0.001%; ammonium, $\leq$ 0.2%; nickel, $\leq$ 0.15%; and sulfate ion, $\leq$ 0.005%) (Aldrich Chemical Co., 1990, undated f,g,h). The hexahydrate is available as pink-to-red crystals at 90-100% purity from three US companies and from one company in the UK (BDH Ltd, 1989d; J.T. Baker, 1989d; Mallinckrodt, 1989d; Hall Chemical Co., undated i). Technical-grade cobalt nitrate hexahydrate is available from one US company as small, red flakes with a slight odour of nitric acid and contains a minimum of 19.8% cobalt, with small amounts of impurities (iron, 0.002% max; copper, 0.005% max; lead, 0.005% max; zinc, 0.05% max; chlorine 0.005% max; sulfate ion, 0.01% max; water insolubles, 0.02% max; Shepherd Chemical Co., 1986a, 1989g). Aqueous cobalt nitrate ($Co(NO_3)_2 \cdot xH_2O$) is available from one US company as a dark-red solution containing approximately 14% cobalt (Hall Chemical Co., undated j).

Cobalt nitrate is also available in 1-2% aqueous nitric acid solution as a laboratory standard containing 1000 ppm cobalt (0.1% w/v; J.T. Baker, 1989e; Aldrich Chemical Co., 1990).

Cobalt(II) oxide is available as a laboratory reagent from one US company as a green, red, grey or black powder at 90-100% purity (70-74% as cobalt), with small amounts of impurities (chloride, 0.02% max; nitrogen compounds as nitrogen, 0.02% max; sulfur compounds as sulfate ion, 0.1% max; iron, 0.1% max; nickel, 0.2% max; insolubles in hydrochloric acid, 0.05%; J.T. Baker, 1989f,g). One company in the UK offers cobalt oxide as a fine, black powder (BDH Ltd, 1989e). Cobalt(II) oxide is also available in ceramic grade (70-71% cobalt), metallurgical grade (76% cobalt) and high-purity powder grade (99.5%; may contain 10 ppm metallic impurities; American Chemical Society, 1988). Cobalt(II) oxide is produced by only a few companies (Chemical Information Services Ltd, 1988) and is not of major commercial importance.

Cobalt(II,III) oxide is available as a black powder at 99.995% purity (Aldrich Chemical Co., undated a), as a black powder with a cobalt content of 72-73% (Aldrich Chemical Co., 1990, undated i) and as a black-grey powder with 71-72% cobalt as cobalt oxide and less than 1% nickel as nickel monoxide (Hall Chemical Co., undated k). Another mixed oxide, containing a ratio of 3:1 cobalt(III) oxide:cobalt(II) oxide, is available at 99.999% purity (Chemical Dynamics Corp., 1989). It is produced by many companies throughout the world.

Cobalt(III) oxide is available in small quantities for laboratory use from one US company as a powder at 99.9996% purity (72.3% as cobalt) with small amounts of impurities (chloride, 80 µg/g; nitrate, 35 µg/g; silicon, 2 µg/g; aluminium, < 1 µg/g; copper, < 0.5 µg/g; iron, 1 µg/g; magnesium, 0.7 µg/g; nickel, 2 µg/g; J.T. Baker, 1989g).

Cobalt sulfide (form unspecified) is sold by one company in the USA (Chemical Information Services Ltd, 1988).

Cobalt sulfate heptahydrate is available from several companies as pink-to-dark-red crystals in purities of 90-100% (BDH Ltd, 1989f; J.T. Baker, 1989h; Mallinckrodt, 1989e; Aldrich Chemical Co., undated j,k; Hall Chemical Co., undated l). Technical-grade cobalt sulfate heptahydrate available from one US company as red-pink crystals contains a minimum of 20.8% cobalt and small amounts of impurities (iron, 0.005% max; copper, 0.002% max; water insolubles, 0.05% max; Shepherd Chemical Co., 1986b, 1989h). The monohydrate is available as pink-to-red crystals with a minimum of 33% cobalt and with small amounts of impurities (iron, 0.007% max; copper, 0.003% max; water insolubles, 0.1% max; Shepherd Chemical Co., 1987d, 1988b), and with a purity of 100% (Hall Chemical Co., undated m).

Cobalt sulfate is also available commercially as a rose-to-dark-red aqueous solution containing approximately 8% cobalt (CP Chemicals, 1989d; Hall Chemical Co., undated n).

2. Production, Use, Occurrence and Analysis

2.1 Production

(a) Cobalt and cobalt alloys

Cobalt, a major constituent of about 70 naturally occurring oxide, sulfide, arsenide and sulfoarsenide minerals, is produced primarily as a by-product of the mining and processing of copper and nickel ores and, to a lesser extent, of silver, zinc, iron, lead and gold ores.

Commercial cobalt production began in Canada in 1905. In 1924, a company in Zaire (then the Belgian Congo) started recovering cobalt during the mining of copper ores, and that country has been the world's largest producer since 1926 (Roskill Information Services, 1989). World mine production of cobalt peaked in the mid-1980s, but the production of refined cobalt metal has been decreasing since the early 1980s because beneficiation and extractive metallurgy are not designed for maximizing the recovery of cobalt (Roskill Information Services, 1989). World mine and metal production of cobalt in 1970-88 is presented in Table 6.

Table 6. World mine and metal production of cobalt, 1970-88 (tonnes)[a]

Year	Mine production	Metal production	Year	Mine production	Metal production
1970	28 985	25 909	1980	37 873	36 720
1971	26 405	27 203	1981	37 363	31 325
1972	30 177	24 645	1982	24 567	19 292
1973	35 746	28 113	1983	37 875	18 084
1974	39 453	30 745	1984	41 075	23 627
1975	37 479	25 275	1985	48 304	26 906
1976	26 024	22 827	1986	48 903	30 673
1977	26 303	25 227	1987	46 382	26 939
1978	32 817	24 780	1988	43 900[b]	25 286[c,d]
1979	36 148	34 317	1989	38 700[b,c]	NA

[a]From Roskill Information Services (1989), unless otherwise specified
[b]From Shedd (1990)
[c]Estimate
[d]From Shedd (1988)
NA, Not available

Between 1983 and 1987, cobalt was mined in amounts greater than 100 tonnes in 16 countries and was refined in 12. The cobalt-producing countries or regions in those years were Albania, Australia, Botswana, Brazil, Canada, China, Cuba, Finland, Morocco, New Caledonia, the Philippines, South Africa, the USSR, Zaire, Zambia and Zimbabwe. The countries that refined cobalt during this period were Belgium, Canada, China, Finland, France, Japan, Norway, South Africa, the USSR, Zaire, Zambia and Zimbabwe (Johnston, 1988; Shedd, 1988).

(i) *Cobalt mining, refining and/or production by country*

Australia: Cobalt is mined but not refined in Australia (Shedd, 1988). In 1986, one company ceased supplying nickel-cobalt sulfides to Japanese refineries and began to supply all of their by-products to a refinery in Finland (Kirk, 1986).

Belgium: Small quantities of partly processed materials containing cobalt have been imported, but information is inadequate to estimate the recovery of cobalt (Kirk, 1986). About one-third of the cobalt exported by Zaire is processed in Belgium, and about half of this production is exported to the USA (Kirk, 1985).

Botswana: One company in Botswana began mining for cobalt in 1973 (Kirk, 1985). The cobalt-containing nickel-copper matte is sent to Norway (74%) and Zimbabwe (26%) for refining (Shedd, 1988); previously, it was refined in the USA (Kirk, 1985).

Brazil: One company began production of electrolytic cobalt in late 1989 at a nickel plant with an initial production capacity of 300 tonnes. It produced a cobalt concentrate which was sent to a Norwegian refinery for processing. Previously, Brazil depended on imports from Canada, Norway, Zaire and Zambia (Kirk, 1987; Shedd, 1988, 1989).

Bulgaria: Bulgaria is known to produce ores that contain cobalt, but information is inadequate to estimate output (Kirk, 1985).

Canada: Cobalt production in Canada began in 1905 (Roskill Information Services, 1989). Three companies currently mine cobalt, and one of these refines it (Shedd, 1988). The intermediate metallurgical product cobalt oxide has been shipped to the UK for further processing, and a nickel-copper cobalt matte has been shipped to Norway (Kirk, 1986, 1987).

China: A primary cobalt deposit mine was equipped in 1986 and has a reported annual output of 45 thousand tonnes of ore (Kirk, 1986). Cobalt mine production in 1987 was estimated to be 270 tonnes (Johnston, 1988). A large deposit of nickel-copper-cobalt was discovered in China in 1988 (Shedd, 1988).

Czechoslovakia: Czechoslovakia is believed to recover cobalt from Cuban nickel-cobalt oxide and oxide sinter (Kirk, 1985; Shedd, 1988).

Finland: In 1986, a company in Finland began processing nickel-cobalt sulfide from Australian nickel oxide production into cobalt and nickel salts (Kirk, 1986). In

1987, a mining and metallurgical cobalt and nickel producing company in Finland suspended production of standard-grade cobalt powder and briquets to focus on producing extra-fine powder and cobalt chemicals. In 1988, the copper-cobalt mine was closed and cobalt concentrates were no longer produced (Kirk, 1987; Shedd, 1988).

Germany: Ores that contain cobalt are produced in Germany, but information is inadequate to estimate output (Kirk, 1985).

Greece: Ores that contain cobalt are produced in Greece, but information is inadequate to estimate output (Roskill Information Services, 1989).

India: A plant projected to open in 1990 can recover approximately 27 tonnes of cobalt per year from a lead-zinc ore mine in India. In addition, recovery of cobalt from lateritic overburden in chromite mines is being studied (Shedd, 1988).

Indonesia: One company in Indonesia produces ores that contain cobalt, but information is inadequate to estimate output (Shedd, 1988).

Japan: Mining of cobalt in Japan ceased in 1986. Two Japanese refiners have received nickel-matte from a Canadian facility in Indonesia and feedstock from Australia and the Philippines (Shedd, 1988).

Morocco: Mining of cobalt was begun in Morocco in the late 1930s (Roskill Information Services, 1989); mining of cobalt as a primary product ceased in 1982, but mining from cobalt-iron-nickel arsenides was resumed in 1988. Beginning in 1988, Morocco agreed to provide China with cobalt concentrate (Shedd, 1988).

New Caledonia: Ores and intermediate metallurgical products have been exported to France, Japan and the USA (Kirk, 1987; Shedd, 1988).

Norway: One company in Norway refines cobalt mostly from nickel-cobalt-copper matte imported from Canada (60%) and Botswana (30%) (Shedd, 1988).

Philippines: Cobalt was recovered as a by-product of nickel mining by a state-owned company in the Philippines until 1986, when the mine was closed. Production of cobalt from the mine peaked at about 1360 tonnes in 1979 (Kirk, 1987; Shedd, 1988).

Poland: Ores that contain cobalt are produced in Poland, but information is inadequate to estimate output (Kirk, 1985).

South Africa: Cobalt is mined and refined in South Africa (Shedd, 1988), and a foreign-owned company produced cobalt as a by-product of platinum mining operations (Kirk, 1987).

Spain: Ores that contain cobalt are produced in Spain, but information is inadequate to estimate output (Kirk, 1985).

Uganda: Construction of a cobalt refinery is planned in conjunction with the rehabilitation of copper mines, which ceased operation in 1979 (Shedd, 1988).

UK: Products of Canadian origin are processed in the UK (Kirk, 1986, 1987).

USA: The USA began mining cobalt in the late 1930s but ceased domestic mine production at the end of 1971. Refining of imported nickel-cobalt matte by the sole US cobalt refinery was discontinued in late 1985. In 1985-88, the USA imported 31% of its cobalt from Zaire, 21% from Zambia, 21% from Canada, 10% from Norway (originating in Canada and Botswana) and 17% from other countries (Shedd, 1990), which include Belgium, Finland, France, Germany, Japan, the Netherlands, South Africa and the UK (Kirk, 1987).

Two companies in the USA produce extra-fine cobalt powder: one is a foreign-owned company that uses imported primary metal; the other is a domestically controlled company that uses cobalt recovered from recycled materials. Seven companies produce cobalt compounds (Shedd, 1990).

USSR: Cobalt is mined and refined in the USSR (Shedd, 1988); in addition, nickel-cobalt sulfide concentrate from Cuba is refined (Kirk, 1985).

Zaire: Cobalt recovery from the mining of copper ores began in 1924, and since 1926 Zaire has been the world's largest producer of cobalt (Roskill Information Services, 1989). Sulfide and oxide concentrates are processed to cobalt metal in the form of cathodes and granules. About one-third of their exports go to Belgium for further processing (Kirk, 1985).

Zambia: Mining of cobalt began in Zambia in the late 1930s (Roskill Information Services, 1989). Cobalt is also mined and refined as a by-product of copper mining (Kirk, 1985; Shedd, 1988).

Zimbabwe: Cobalt is mined and refined in Zimbabwe and is also recovered from nickel-copper matte imported from Botswana (Shedd, 1988).

Mine and metal production of cobalt by country or region with reported outputs for 1984 to 1988 are presented in Tables 7 and 8.

(ii) *Metallurgy*

Cobalt-containing ores vary widely in composition but usually contain less than 1% cobalt. Although each type of ore (arsenide, sulfide or oxide) is processed differently, six general metallurgical processes can be distinguished; depending on the ore's composition, recovery of cobalt may require one or a combination of these techniques. It is important to note that in nearly all cases cobalt is a by-product of the refining of other metals (Roskill Information Services, 1989), especially copper and nickel. Refinery methods therefore are generally not designed to maximize cobalt recovery (Anon., 1990a).

The main sources of cobalt (in decreasing ease of recovery) are ores of copper-cobalt oxides (Zaire) and sulfides (Zaire and Zambia), copper-nickel sulfides (Canada), cobalt-iron-nickel arsenides (Morocco and China) and

Table 7. World mine production of cobalt by country or region, 1984-88[a]

Country	Mine output, metal content (tonnes)				
	1984	1985	1986	1987	1988[b]
Albania	590	590	590	590	590
Australia	938	1 136	1 218	1 200	1 100
Botswana	259	222	162	182	292[c]
Brazil	100	100	150	150	150
Canada	2 330	2 071	2 491	2 495	2 770
Cuba[d]	1 400	1 491	1 500	1 590	2 000
Finland	862	1 094	628	190	182
Morocco	NA	NA	NA	NA	253
New Caledonia[b]	500	677	700	750	800
Philippines	64	913	92	NA	NA
South Africa[b]	682	682	682	727	727
USSR[b]	2 590	2 725	2 815	2 815	2 860
Zaire	25 997	29 226	33 403	29 056[b]	25 425
Zambia	4 625	5 812[b]	5 770[b]	5 950[b]	6 675
Zimbabwe[b]	77	100	76	109	126
Total	41 014	46 838	50 277	45 804	43 950

[a]From Shedd (1988), unless otherwise specified
[b]Estimates
[c]Reported figure
[d]Estimates from reported nickel–cobalt content of granular and powder oxide, oxide sinter and sulfide production
NA, not available

nickel-cobalt oxides (lateritic nickel ore from most other sources) (Planinsek & Newkirk, 1979; Donaldson et al., 1986a; Shedd, 1988).

After crushing and grinding, the first stage of cobalt recovery from ore involves the physical separation of cobalt-containing minerals from other nickel ores and gauge, usually by gravity (arsenide ores) or froth flotation (sulfoarsenide and sulfide ores). Flotation is also used for separating cobalt in oxide and mixed oxide-sulfide ores. Flotation is frequently aided by the addition of xanthates, oils or cyanide to depress cobalt flotation (Donaldson, 1986; Donaldson et al., 1986a); the amount of cobalt in the concentrate is usually enhanced four to eight fold by these operations (Roskill Information Services, 1989).

Cobalt is extracted from ore and concentrated by pyrometallurgical, hydrometallurgical and electrolytic processes alone or in combination. Arsenic-free cobalt concentrates can be mixed with lime and coal and smelted in a reducing

Table 8. World metal production of cobalt by country, 1984-88 (tonnes)[a]

Country	1984	1985	1986	1987	1988[b]
Canada	2 218	2 027	1 994	2 205[b]	2 205
Finland	1 456	2 235	1 350	498	220
France	116	123	100[b]	109[b]	50
Japan	907	1 279	1 340	124	109
Norway	1 193	1 640	1 583	1 603	1 605
South Africa[b]	500	500	500	523	523
USSR[b]	4 725	4 815	5 315	5 315	5 315
Zaire	9 083	10 690	14 513	11 911	10 150
Zambia	3 475	4 365	4 348	4 483	4 995
Zimbabwe	78	92	76	110	126
Total	23 751	27 766	31 119	26 881	25 298

[a]From Shedd (1988)
[b]Estimates

environment to give copper-cobalt alloys. The alloy is further processed to separate copper and cobalt. The most commonly used hydrometallurgical processes involve roasting and leaching of ore concentrates (with acid or alkali solutions), fractional separation of cobalt from other metals in the leachate (by differential sulfide or hydroxide precipitation) and reduction of the cobalt ions to metal (by chemical or electrochemical means) (Donaldson, 1986; Donaldson *et al.*, 1986a; Roskill Information Services, 1989).

The three main processes for leaching cobalt from ores and concentrates are described below.

Acid sulfate leaching can be done by one of four methods: (a) treating oxide ore concentrates with sulfuric acid and reducing agents (SO_2); this is the primary process used in Zaire; (b) water extraction of cobalt sulfate from ores following an oxidizing roast; (c) cobalt sulfate extraction of sulfide ore concentrate following a sulfatizing roast; this method is used in Zaire, Zambia and Finland; or (d) pressure leaching with sulfuric acid, which has recently been introduced in Canada and is useful for arsenic-containing ores. The cobalt is separated from copper, iron, nickel and zinc (when present) by alkalinization and fractional dissolution with sulfide. Cobalt is precipitated as the hydroxide, redissolved and refined by electrolysis or hydrogen reduction to cobalt metal cathode or powder, respectively (Roskill Information Services, 1989).

Acid chloride leaching of ore mattes and recyclable materials is used as an alternative to acid sulfate leaching on oxides, sulfides, arsenides and alloys. This method is usually followed by solvent extraction or ion exchange purification. The soluble chloride complexes are often formed by reaction with chlorine or hydrogen chloride gas or a metal chloride. This method is used in Japan.

Ammoniacal solution leaching gives rise to the hexammine cobalt complex $[Co(NH_3)_6]^{2+}$. This method has been used to treat alloy scrap and laterite or arsenide ores. It is used in Canada for processing lateritic nickel ores. The soluble extract is treated with hydrogen sulfide to produce mixed nickel-cobalt sulfides, which are redissolved in sulfuric acid. Cobalt powder is recovered after the introduction of ammonia and hydrogen under high pressure.

Metallic cobalt can also be recovered directly from purified leachate by electrolysis (electrowinning) after nickel has been removed as the carbonyl. Some cobalt salts can be formed by dissolution of the metal in the corresponding acid. Some refineries utilize cobalt hydroxide to form the oxide and other cobalt compounds directly (Donaldson, 1986; Donaldson *et al.*, 1986a; Roskill Information Services, 1989; Anon., 1990a).

(iii) *Production processes*

Refined *cobalt* is available to the industrial market primarily as broken or cut cathodes (92%) and to a lesser extent as electrolytic coarse powder (3%) and in other forms. The cathode form is further processed to alloys, chemicals and oxide or used in the manufacture of special cobalt powders for cemented carbide by chemical and pyrometallurgical processes. About 2000 tonnes of cobalt cathode are converted to a distinct allotropic mixture, called 'fine powder' or 'extrafine powder', by specialist producers for cemented carbide and diamond polishing. The process involved is a chemical reaction that results in a submicrometre powder with a high proportion of face-centred cubic crystal retained in the mixture. This special material differs from electrolytic coarse powder and from cobalt powders generated during industrial attritive operations, which are predominantly hexagonal crystals (Cobalt Development Institute, 1989).

Cobalt alloys are usually manufactured from broken or cut cathodes by electric arc or by induction melting techniques, although vacuum induction melting is required for some alloys containing metals such as aluminium, titanium, zirconium, boron, yttrium and lanthanum. The resultant master alloy is then remelted and cast into moulds (Donaldson & Clark, 1985; Donaldson, 1986).

An important use of cobalt is in the production of cemented tungsten carbide, also called 'hard metal'. Hard metals are used to tip the edges of drills and cutting tools and for dies, tyre studs and stamping machines (Kipling, 1980). Hard metal is made by a process in which precise weights of tungsten carbide (80-90% by weight) and cobalt metal powder (5-10%) and, in some grades, small amounts of other carbides (titanium, tantalum, niobium and molybdenum) are added and thoroughly mixed in mills. The cobalt thus acts as a matrix; nickel is also used with cobalt as a matrix in some grades. Organic solvents, such as acetone and *n*-hexane, are added for mixing; the mixture is dried, and the organic solvents are evaporated off. The powder is put into frames made of steel or rubber and then pressed into the desired shapes; the pieces are placed on graphite plates and embedded in nitrous aluminium powder; and the pressed material is presintered in hydrogen furnaces at 500-800°C. After presintering, the material has the consistency of chalk, and it is cut, ground, drilled or shaped into the configurations required. The shaped material is finally sintered at temperatures of 1550°C. After sintering, the product approaches the hardness of diamond. Hard-metal products are sand blasted or shot blasted, brazed into holders made of iron using fluoride-based fluxes and then ground with diamond or carborundum wheels. These processes are illustrated in Figure 1 (Kusaka *et al.*, 1986).

The manufacture of some alloys containing cobalt and their further fabrication into engineering parts can be assumed to take place to some extent in almost all industrialized countries. Manufacture specifically of superalloys for aircraft engines is concentrated in the USA, the UK, France, Germany and Japan, but small volumes of manufacture and specialist manufacture occur in several other regions. Use of cobalt in magnetic applications occurs mainly in Japan, but the USA and European countries (particularly Germany, France and the UK) also have large production capacities (Johnston, 1988).

(b) Cobalt compounds

Europe produces 50% of the global amount of cobalt chemicals and 70% of fine cobalt powders (Johnston, 1988). Most cobalt chemicals (75-80%) are produced by six companies in Belgium, Germany, Finland and the USA. A further 6-8% is made by three Japanese companies; minor quantities are made directly from concentrates in France and South Africa; and the balance is shared by a number of small manufacturers serving local markets or specializing in perhaps one group of cobalt products, such as naphthenates for the paint (Sisco *et al.*, 1982) and ink industries.

Most countries—industrialized or not—have a ceramics industry of some kind or size, many of them very ancient, and in each there is some use of cobalt oxide or some manufacture of cobalt pigment. The major world suppliers of cobalt pigments are, however, located in Germany, the USA, the Netherlands and the UK.

Fig. 1. Steps in the manufacture of hard-metal tools[a]

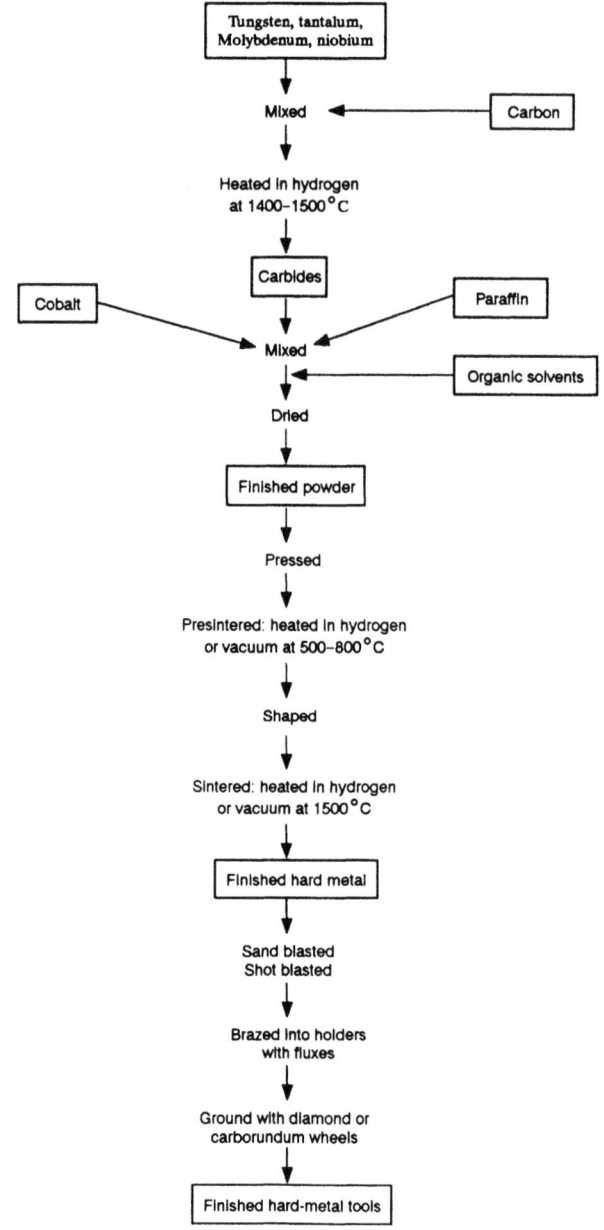

[a]From Kusake *et al.* (1986)

Cobalt(II) acetate is prepared commercially (a) by concentrating solutions of cobalt powder in acetic acid in the presence of oxygen or (b) from cobaltous hydroxide or carbonate and an excess of dilute acetic acid. Preparation of the tetrahydrate involves treatment of cobalt powder in acetic acid solution with hydrogen peroxide (Donaldson *et al.*, 1986a; Budavari, 1989).

Cobalt(III) acetate can be prepared by electrolytic oxidation of cobalt(II) acetate tetrahydrate in glacial acetic acid containing 2% (v/v) water. Another method is oxidation of solutions of cobaltous salts by alkaline persulfates in the presence of acetic acid (Budavari, 1989).

Cobalt(II) chloride can be produced by several processes: (a) from cobalt powder and chlorine, (b) from the acetate and acetyl chloride, (c) by dehydration of the hexahydrate with thionyl chloride and (d) by dissolving cobalt metal, oxide, hydroxide or carbonate in hydrochloric acid (Considine, 1974; Donaldson *et al.*, 1986a; Budavari, 1989). The hexahydrate is prepared by treating an aqueous solution of a cobaltous salt with hydrochloric acid (Budavari, 1989). Solutions of high-purity cobalt chloride and its hexahydrate can be manufactured by dissolving high-purity cobalt metal electrolytically using a dilute hydrochloric acid electrolyte at about 60°C (Donaldson *et al.*, 1986a).

Cobalt(II) carbonate is prepared by heating cobalt sulfate with a solution of sodium bicarbonate. Basic cobalt carbonate (cobalt(II) carbonate hydroxide (2:3) monohydrate) is prepared by adding sodium carbonate to a solution of cobaltous acetate followed by filtration and drying (Sax & Lewis, 1987).

Cobalt(II) hydroxide is prepared commercially as a pink solid by precipitation from a cobalt(II) salt solution with sodium hydroxide. Precipitation at higher temperatures (55-70°C) causes partial oxidation of cobalt(II) to cobalt(III) and yields the pink form, whereas precipitation at lower temperatures yields the blue form. Cobalt(II) hydroxide is prepared *in situ* during the manufacture of secondary batteries: typically, a spongy nickel foam plate is impregnated with an acidic solution of cobalt chloride, nitrate or sulfate, and cobalt(II) hydroxide is precipitated by alkali treatment (Donaldson *et al.*, 1986a).

Cobalt(III) hydroxide can be produced by several methods, e.g., addition of sodium hydroxide to a solution of cobaltic salt, action of chlorine on a suspension of cobaltous hydroxide, or action of sodium hypochlorite ion on a cobaltous salt (Brauer, 1965; Sax & Lewis, 1987).

Cobalt(II) molybdenum(VI) oxide is obtained by raising the pH to 6.4 to coprecipitate the hydroxides of cobalt and molybdenum from mixed solutions of cobalt nitrate and ammonium molybdate. The product is dried at 120°C and calcined at 400°C to give the mixed metal oxide (Donaldson *et al.*, 1986a). This is invariably also mixed with aluminium oxide in commercial manufacture and use.

Cobalt(II) naphthenate is prepared by treating cobalt hydroxide or cobalt acetate with naphthenic acid (Sax & Lewis, 1987), which is recovered as a by-product of petroleum refining. Commercial naphthenic acids used in the production of cobalt naphthenate differ widely in properties and impurities, depending upon the crude oil source and refining processes. All contain 5-25 wt % hydrocarbons, the composition of which corresponds to the petroleum fraction from which the naphthenic acids are derived; and all contain impurities (e.g., phenols, mercaptans and thiophenols) in small quantities (Sisco *et al.*, 1982).

Cobalt(II) nitrate hexahydrate is produced by dissolving cobalt metal, the oxide, hydroxide or carbonate in dilute nitric acid and concentrating the solution (Considine, 1974; Donaldson *et al.*, 1986a).

Cobalt(II) oxide (CoO) containing 78.7% cobalt is usually manufactured by controlled oxidation of the metal at above 900°C, followed by cooling in a protective atmosphere to prevent partial oxidation to cobalt(II,III) oxide (Donaldson *et al.*, 1986a).

Cobalt(II) oxide can also be prepared by additional processing of the white alloy formed during the processing of arsenic-free cobalt-copper ores to remove copper and iron as sulfates and calcining cobalt as the carbonate (Morral, 1979) or by calcination of cobalt carbonate or its oxides at high temperatures in a neutral or slightly reducing atmosphere (Sax & Lewis, 1987).

Another method for preparing cobalt(II) oxide is dissolution of a cobalt salt that is unstable at high temperatures (e.g., cobalt sulfate) in molten sodium sulfate or potassium fluoride. The cobalt salt decomposes, leaving the cobalt(II) oxide, which crystallizes out at high temperatures. The water-soluble salts are then dissolved, leaving cobalt(II) oxide crystals (Wilke, 1964).

Cobalt(II,III) oxide (Co_3O_4) containing 73.44% cobalt can prepared by the controlled oxidation of cobalt metal or cobalt(II) oxide or by thermal decomposition of cobalt(II) salts at temperatures below 900°C. It absorbs oxygen at room temperature but is not transformed to cobalt(III) oxide (Co_2O_3) (Donaldson *et al.*, 1986a).

Pyrohydrolysis of cobalt chloride has also been used to manufacture cobalt(II,III) oxide. The reaction is performed in a spray roaster by heating a fine spray of aqueous solution of cobalt(II) chloride in a countercurrent heating gas stream. The hydrogen chloride gas produced is removed with the exhaust gases, and the cobalt(II,III) oxide falls to the bottom of the furnace (Donaldson *et al.*, 1986a).

Cobalt(III) oxide (Co_2O_3) is derived by heating cobalt compounds (e.g., hydroxides) at low temperature with an excess of air (Sax & Lewis, 1987).

Cobalt(II) sulfate heptahydrate is prepared commercially by dissolving cobalt metal in sulfuric acid (Donaldson *et al.*, 1986a).

α-*Cobalt(II) sulfide* can be precipitated from cobalt nitrate hexahydrate by reaction with hydrogen sulfide and dried for 90 h, the temperature being raised slowly from 100 to 540°C (Brauer, 1965). β-*Cobalt(II) sulfide* can be synthesized by heating fine cobalt powder mixed with fine sulfur powder at 650°C for two to three days. It can also be derived by treating a solution of cobalt chloride with acetic acid, precipitating with hydrogen sulfide and drying for 90 h, the temperature being raised slowly from 100 to 540°C (Brauer, 1965). Cobalt sulfides are normally produced *in situ* as needed, as mixed metal catalysts with molybdenum (Roskill Information Services, 1989).

Dicobalt octacarbonyl is prepared commercially by heating cobalt metal with carbon monoxide at high pressure (200-300 atm) [20.2-30.3 × 10^3 kPa] or by heating a mixture of cobalt(II) acetate with cyclohexane at about 160°C and 300 atm (30.3 × 10^3 kPa) in the presence of a 1:1 mixture of carbon monoxide:hydrogen (Donaldson *et al.*, 1986a). Dicobalt octacarbonyl is frequently prepared *in situ* as needed.

2.2 Use

Cobalt compounds have been used as blue colouring agents in ceramic and glass for thousands of years, although most of the blue colour of ancient glasses and glazes has been found to be due to copper. Cobalt has been found in Egyptian pottery dated at about 2600 BC, in Persian glass beads dating from 2250 BC, in Greek vases and in pottery of Persia and Syria from the Christian era, in Chinese pottery from the Tang (600-900 AD) and Ming (1350-1650 AD) dynasties and in Venetian glass from the early fifteenth century. Leonardo Da Vinci was one of the first artists to use cobalt as a brilliant blue pigment in oil paints. The pigment was probably produced by fusing an ore containing cobalt oxide with potash and silica to produce a glass-like material (a smalt), which was then reduced to the powdered pigment. In the sixteenth century, a blue pigment called zaffre was produced from silver-cobalt-bismuth-nickel-arsenate ores in Saxony (Young, 1960; Donaldson, 1986).

It was not until the twentieth century, however, that cobalt was used for industrial purposes. In 1907, a US scientist, E. Haynes, patented a series of cobalt-chromium alloys known as stellites that were very resistant to corrosion and wear at high temperatures (Kirk, 1985). Cobalt was added to tungsten carbide in 1923 to produce cemented carbides (Anon., 1989) and to permanent magnet alloys known as Alnicos (cobalt added to alloys of aluminium, nickel and iron) in 1933 (Johnston, 1988).

(i) *Cobalt*

Cobalt has many important uses in industry today, and in some major applications there is no suitable replacement. The most important use of metallic cobalt is as an alloying element in superalloys, magnetic and hard-metal alloys, such as stellite and cemented carbides, cobalt-containing high-strength steels, electrodeposited alloys and alloys with special properties. Cobalt salts and oxides are used as pigments in the glass and ceramics industries, as catalysts in the oil and chemical industries, as paint and printing ink driers and as trace metal additives for agricultural and medical uses (Donaldson, 1986).

Most cobalt is used industrially in the form of cobalt metal as an alloying component and in the preparation of cobalt salts. Estimated consumption as primary raw materials, such as cobalt metal, cobalt oxide and cobalt salts, in selected countries in 1979-87, is presented in Table 9. These countries represented approximately 59% of total consumption in the western world in 1979, 71.5% in 1980, 65% in 1981, 65.5% in 1982, 59.4% in 1983, 53% in 1984, 53.6% in 1985 and 62.5% in 1986. Consumption of cobalt in the western world represented approximately 85% of total world consumption from 1983 to 1988 (Roskill Information Services, 1989).

Table 9. Consumption of cobalt in selected countries, 1979-87 (thousand tonnes)[a]

Country	1979	1980	1981	1982	1983	1984	1985	1986	1987[b]
USA	7.9	6.9	5.3	4.3	5.1	5.4	6.1	6.6	6.9
Japan	2.2	1.9	1.5	1.4	1.5	1.8	1.7	1.7	1.8
UK	2.5	2.3	2.00	1.1	1.2	0.91	0.96	1.6	1.33
France	0.95	1.0	0.75	1.5	0.51	0.62	0.48	0.74	NA
Italy	0.23	0.23	0.19	0.23	0.30	0.38	0.36	0.57	0.56
Sweden	0.29	0.39	0.21	0.21	0.17	0.31	0.36	0.36	0.26
Canada	0.12	0.11	0.10	0.09	0.10	0.11	0.16	0.1[c]	NA

[a]From Roskill Information Services (1989)
[b]Preliminary
[c]Estimated

Industrial consumption of cobalt in the western world averaged 4000 tonnes in 1936-46, 7000 in 1947-52, 10 000 in 1953-62, 16 800 in 1963-72, 19 500 in 1973-78, 21 000 in 1979-81 and 17 500 in 1982-84. Recently, less cobalt has been used in alloys and more in chemical applications. Table 10 presents overall estimates of cobalt consumption in western economies by end use.

Table 10. Evolution of cobalt consumption in selected countries (thousand tonnes)[a]

End product	1950	1960	1970	1981	1987
Alloys	2.85	5.95	6.98	8.74	6.83
Hard metals	0.30	0.73	0.78	1.43	2.02
Magnets	2.10	3.77	3.41	2.47	2.15
Ceramics	0.90	1.60	1.55	1.81	2.04
Chemicals	1.35	2.46	2.79	4.56	6.77
Total	7.50	14.50	15.5	19.01	19.81

[a]From Johnston (1988)

(ii) *Cobalt alloys*

Superalloys are used primarily in the manufacture of components for gas turbine and jet engines. Their combined properties of resistance to hot corrosion and high strength at elevated temperatures contribute to their great commercial and strategic importance. They are used in turbine components that operate at temperatures above 540°C, including ducts, cases and liners, as well as the major turbine blade, vane, disc and combustion-can components. Nickel-based superalloys are usually used for gas turbine components such as discs because they are more workable than cobalt-based superalloys; the latter have excellent resistance to thermal shock and hot corrosion and are used for combustor tubes, stator vanes and diaphragms. Superalloys designed to operate for long periods at temperatures above 900°C sacrifice some of their resistance to oxidation and hot corrosion for increased strength. The nickel-based superalloys are more resistant to oxidation than the cobalt-based superalloys because they have a higher aluminium content and form a better aluminium oxide coating on the alloy. The cobalt-based superalloys primarily form a chromium oxide coating which is not as stable, and when they are used in components subject to extremely high operating temperatures, such as turbine blades and nozzle guide vanes, oxidation-resistant protective coatings are required. Two types of coating can be used: intermetallic and overlay coatings. Intermetallic coatings are applied by heat treatment of the surface of the alloy with cement powders containing aluminides or, less often, silicides. Overlay coatings, which are applied by hot vapour deposition methods, are alloys containing aluminium, chromium and yttrium together with nickel, cobalt or iron. Other applications of the superalloys include airframes, chemical reactors, natural gas transmission pipelines, marine equipment and hazardous waste incineration equipment (Donaldson & Clark, 1985; Donaldson, 1986; Kirk, 1987; Cobalt Development Institute, 1989).

Magnetic alloys. Cobalt is used in a wide variety of magnetic applications, including telecommunication systems, magnetic couplings, electromagnets, meters, loudspeakers, permanent magnet motors and repulsion devices. Alnico magnets, invented in the mid-1930s, are used for heavy-duty applications such as automobile anti-skid braking systems. Consumption of Alnicos declined through the 1960s and 1970s due to the introduction in the 1960s of the less powerful but cheaper and smaller ferrite-ceramic combinations of barium and strontium with iron (Kirk, 1985; Donaldson, 1986; Cobalt Development Institute, 1989; Anon., 1990b).

Magnets combining cobalt with rare-earth minerals were developed in 1967 (Johnston, 1988). The first such magnets were samarium-cobalt alloys, but limited supplies of samarium led to the development of competitive neodymium-iron-boron magnets, which became available commercially in 1983. Rare-earth cobalt magnets have remained important because of their power/size advantages in certain applications. In the 1980s, they contributed to the miniaturization of electrical and electronic equipment. They are used as focusing magnets in travelling wave tubes, as magnetic bearings in ultra-high-speed centrifugal separators and inertia wheels, and in actuators, motors, and generators of various sizes, from watches to 100-hp [74.6-kw] motors (Kirk, 1985; Donaldson, 1986; Anon., 1990b).

Magnetic alloys are also used in medicine to provide an external attractive force. For instance, Alnicos have been used to operate a reed switch in implanted heart pacemakers; samarium-cobalt magnets have been used to hold dental plates in mouth reconstruction, to correct funnel chest and to remove magnetic fragments from the posterior portion of the eye. Magnetic cobalt alloys attached to flexible tubes have also been used to remove iron-containing material from the intestinal and bronchial tubes. Platinum-cobalt and samarium-cobalt magnetic alloys are also used as prostheses, to provide a mechanical closing device in situations where muscle function is impaired. They have been used in the treatment of urinary incontinence in women, to close eyelids in patients with facial paralysis and as colostomy closure devices. In addition, rare-earth-cobalt magnets are used in hearing aids (Donaldson *et al.*, 1986b).

Use of cobalt in magnetic alloys in western countries declined from 28% in 1950, 26% in 1960, 22% in 1970 and 13% in 1981 to 10.8% in 1987 (Johnston, 1988).

Hard-metal alloys (cemented carbides) have essential applications in wear-related engineering because of their high strength, corrosion resistance and ability to retain hardness at elevated temperature. 'Fine', 'extrafine' and 'superfine' special cobalt powders are used as the metal matrix or bonding agent in cemented carbides used in cutting, grinding and drilling tools destined for use on hard materials, such as metals and rocks, and in diamond polishing. Annual industrial consumption of these special powders is approximately 2000 tonnes. Applications of cemented carbides include grinding wheels, moulds, seal rings, dies, valves, nozzles, pump liners, wear parts subject to severe shock, hot mill rolls, extrusion and can tooling, cutters and slitters, mining, drilling and tunnelling (Kirk, 1985; Donaldson, 1986; Anon., 1989; Cobalt Development Institute, 1989).

Consumption of cobalt for hard-metal alloys in the western world rose from 4% in 1950 to 10.2% in 1987. The tungsten carbide industry accounted for the majority of use in 1987 and diamond polishing for the rest (Johnston, 1988).

Cobalt-containing high-strength steels (maraging steels) are used in the aerospace industry for the manufacture of helicopter drive shafts, aircraft landing gear components and hinges for swing-wing aircraft. Machine component uses include timing mechanisms in fuel injection pumps, index plates for machine tools, bolts and fasteners, barrels for rapid-firing guns and components for cryogenic applications. They also find use in marine equipment, such as deep-submergence vehicles and foil assemblies on hydrofoil ships. In addition, they are used in the manufacture of tools, especially hot forging and stamping dies, close tolerance plastic moulds and die holders (Roskill Information Services, 1989).

Cobalt-containing martensitic stainless maraging steels have been developed for a variety of applications, including in machine construction, the aerospace industry, the chemical industry and naval engineering (Roskill Information Services, 1989).

Electrodeposited nickel-cobalt alloys have good corrosion resistance in many environments and have been used as protective coatings in the production of mirrors and decorative coatings and for electroforming. Electrodeposited cobalt-tungsten alloys retain their hardness at high temperature and are used to improve the wear resistance of hot forging dies. Electrodeposited cobalt alloys containing iron, nickel, platinum or phosphorus have magnetic properties suitable for use in recording systems and computer applications (Donaldson, 1986).

Alloys with special properties. Some cobalt-containing alloys have special applications as dental material, surgical implants, low expansion alloys and springs. Properties that are suitable for dentistry include ease of casting, resistance to tarnish, compatibility with mouth tissues, high strength and stiffness, and low density. Vitallium, a cobalt-chromium alloy, was used for cast denture bases, complex partial dentures and some types of bridgework. A modified alloy is used to fuse porcelain coatings to crowns *via* a metal bridge.

Cobalt-chromium surgical implant alloys were first used in the 1940s for femoral head cups because of their resistance to corrosion by body fluids; they were subsequently developed for use in bone replacement and bone repair (Donaldson, 1986). The use of metallic implants has played an increasingly important role in orthopaedy: about 500 000 knee, hip and other joint replacements were manufactured in the 1970s (Donaldson *et al.*, 1986b). Total joint arthroplasty using artificial prostheses has become a common surgical technique in the treatment of severely injured or diseased hip joints; other applications include plates, screws and nails. The implantation of each metallic device is associated with the release of metal, either by corrosion, dissolution or wear or some combination of these processes. Although different materials have been used in the fabrication of prostheses, the preferred material for clinically acceptable knee or hip prostheses is the cobalt-chromium-molybdenum alloy (Donaldson *et al.*, 1986b; Cobalt Development Institute, 1989).

A range of iron-nickel-cobalt alloys is used by the electronics industry for sealing metals in glasses (Donaldson, 1986).

A new chemical use of cobalt is in the manufacture of video tapes. Cobalt is used to coat the basic ferric oxide particles to increase coercivity and reconcile opposing properties of erasability and control of stray magnetic effects. Manufacturers of high-quality audio tapes have also applied this development. Thin films containing cobalt phosphate and cobalt-nickel alloy particles are the most important metallic recording materials. The introduction of cobalt-chromium film for perpendicular recording is a potentially very important use of cobalt. Normally, magnetic particles are orientated horizontally on the tape surface; but by getting them to orientate vertically, much closer packing of information is allowed. Magnetic optical recording (using gadolinium-cobalt and terbium-cobalt alloys) and, to a much smaller extent, bubble memory applications also involve cobalt. Another use of cobalt is as an additive in dry electric cells (Donaldson *et al.*, 1988).

(iii) *Cobalt compounds*

Table 11 summarizes the uses of a number of compounds of cobalt. The commercially significant compounds are the oxides, hydroxide, chloride, sulfate, nitrate, phosphate, carbonate, acetate, oxalate and other carboxylic acid derivatives (Donaldson, 1986).

The compounds of cobalt have a variety of end uses. Cobalt oxides and organic compounds are used in paints, ceramics and allied products as decolorizers, dyes, dryers, pigments and oxidizers. Cobalt oxide, used as a ground-coat frit, promotes the adherence of enamel to steel. In the rubber industry, organic cobalt compounds are used to promote the adherence of metal to rubber in steel-belted radial tyres. Cobalt is also used in chemical processes. It is used in the petroleum industry principally as a catalyst for hydrodesulfurization, oxidation, reduction and synthesis of hydrocarbons. The artificial isotope cobalt-60 provides a controllable source of gamma-radiation and is used in physical, chemical and biological research, the treatment of cancer, and in industrial radiography for the investigation of physical strains and imperfections in metals (Kirk, 1985).

Table 11. Industrial uses of cobalt compounds[a]

Compound	Formula	Uses
Acetate(III)	$Co(C_2H_2O_2)_3$	Catalyst
Acetate(II)	$Co(C_2H_2O_2)_2 \cdot 4H_2O$	Driers for lacquers and varnishes, sympathetic inks, catalysts, pigment for oil-cloth, mineral supplement, anodizer, stabilizer for malt beverages
Acetylacetonate	$Co(C_5H_7O_2)_3$	Vapour plating of cobalt
Aluminate	$CoAl_2O_4$	Pigment, catalysts, grain refining
Ammonium sulfate	$CoSO_4(NH_4)_2SO_4 \cdot 6H_2O$	Catalysts, plating solutions
Arsenate	$Co_3(AsO_4)_2 \cdot 8H_2O$	Pigment for paint, glass and porcelain
Bromide	$CoBr_2$	Catalyst, hydrometers
Carbonate	$CoCO_3$	Pigment, ceramics, feed supplements, catalyst
Carbonate (basic)	$2CoCO_3 \cdot Co(OH)_2 \cdot H_2O$	Chemicals
Carbonyl	$Co_2(CO)_8$	Catalyst
Chloride	$CoCl_2 \cdot 6H_2O$	Chemicals, sympathetic inks, hydrometers, plating baths, metal refining, pigment, catalyst
Chromate	$CoCrO_4$	Pigment
Citrate	$Co_3(C_6H_5O_7)_2 \cdot 2H_2O$	Therapeutic agents, vitamin preparations
Dicobalt manganese tetroxide	$MnCo_2O_4$	Catalyst
Dicobalt nickel tetroxide	$NiCo_2O_4$	Catalyst, anode
Dilanthanum tetroxide	La_2CoO_4	Catalyst, anode
2-Ethylhexanoate	$Co(C_8H_{15}O_2)_2$	Paint and varnish drier
Ferrate	$CoFe_2O_4$	Catalyst, pigment
Fluoride(II)	CoF_2	Fluorinating agent
Fluoride(III)	CoF_3	Fluorinating agent
Fluoride	$CoF_2 \cdot 4H_2O$	Catalyst
Fluorosilicate	$CoSiF_6 \cdot 6H_2O$	Ceramics
Formate	$Co(CHO_2)_2 \cdot 2H_2O$	Catalyst
Hydroxide	$Co(OH)_2$	Paints, chemicals, catalysts, printing inks
Iodide	CoI_2	Moisture indicator
Lanthanum trioxide	$LaCoO_3$	Electrode
Linoleate	$Co(C_{18}H_{31}O_2)_2$	Paint and varnish drier
Lithium oxide	$LiCoO_2$	Battery electrode
Manganate	$CoMn_2O_4$	Catalyst, electrocatalyst

Table 11 (contd)

Compound	Formula	Uses
Naphthenate	$Co(C_{11}H_{10}O_2)_2$	Catalyst, paint and varnish drier
Nitrate	$Co(NO_3)_2 \cdot 6H_2O$	Pigments, chemicals, ceramics, feed supplements, catalyst
Oleate	$Co(C_{18}H_{33}O_2)_2$	Paint and varnish drier
Oxalate	CoC_2O_4	Catalysts, cobalt powders
Oxide(II)	CoO	Chemicals, catalysts, pigments
Oxide(II,III)	Co_3O_4	Enamels, semiconductors
Oxides	Mixed metal	Pigments
Phosphate	$Co_3(PO_4)_2 \cdot 8H_2O$	Glazes, enamels, pigments, steel pretreatment
Potassium nitrite	$K_3Co(NO_2)_6 \cdot 1.5H_2O$	Pigment
Resinate	$Co(C_{44}H_{62}O_4)_2$	Paint and varnish drier, catalyst
Sodium oxide	$NaCoO_2$	Battery electrode
Stearate	$Co(C_{18}H_{35}O_2)_2$	Paint and varnish drier, tyre cord adhesives
Succinate	$Co(C_4H_4O_4) \cdot 4H_2O$	Therapeutic agents, vitamin preparations
Sulfamate	$Co(NH_2SO_3) \cdot 3H_2O$	Plating baths
Sulfate	$CoSO_4 \cdot xH_2O$	Chemicals, ceramics, pigments
Sulfide	CoS	Catalysts
Tricobalt tetralanthanum decaoxide	$La_4Co_3O_{10}$	Catalyst
Tungstate	$CoWO_4$	Drier for paints and varnishes

*a*From Donaldson (1986); Donaldson *et al.* (1986a)

Cobalt is an effective catalyst for many organic reactions. Its major use in this way is in hydrotreating catalysts, the active components of which are molybdenum and cobalt sulfides. This type of catalyst is used in the synthesis of fuels (Fischer-Tropsch process). The reactions catalysed by cobalt also include the oxo synthesis, in which olefins and carbon monoxide are combined to form aldehydes. The basic catalyst is cobalt carbonyl ($CO_2(CO)_8$), although other cobalt carbonyls can be used. In both the Fischer-Tropsch and the oxo process, the catalysts are normally generated *in situ* in the reactor. Cobalt catalysts are also used in hydrogenation reactions, such as the hydrogenation of nitriles to amines. Cobalt salts are valuable oxidation catalysts, e.g., for the production of terephthalic acid by the oxidation of *para*-xylene, and the manufacture of phenol by the oxidation of toluene. Cobalt-containing catalysts have also been used for polymerization reactions, e.g., polyethylene production by the Amoco process (Morral, 1979; Donaldson, 1986; Donaldson *et al.*, 1986a; Johnston, 1988; Schrauzer, 1989).

Combinations of the oxides of cobalt and those of aluminium, magnesium, zinc and silicon are constituents of blue and green ceramic glazes and pigments (Donaldson, 1986). Cobalt zinc silicate is used in a blue underglaze paint for porcelain articles; the pigment is specially developed to withstand intense heat (Raffn *et al.*, 1988). Cobalt is also used in the glass industry to impart blue colours and to mask the greenish tinge in glass or porcelain caused by iron impurities (Donaldson, 1986).

Spinels are mixed metal oxides with a special crystal structure, based on magnesium and aluminium oxides ($MgAl_2O_4$). These two metals may be partially replaced in the crystal structures by other metals, such as cobalt(II) and chromium(III). Spinels occur naturally and are also produced synthetically. Some cobalt spinels, such as the cobalt-magnesium-aluminium and cobalt-aluminium oxide spinels, are used as pigments (Donaldson *et al.*, 1986a; Sax & Lewis, 1987).

An important use of cobalt is as a drying agent for paints, varnishes, lacquers and printing inks. In these processes, cobalt oleate, resinate and linoleate have been used, but cobalt naphthenate is the more common ingredient (Buono & Feldman, 1979). Cobalt naphthenate is also added to polyester and silicone resins to promote hardening (Bedello *et al.*, 1984).

Consumption in ceramics was relatively stable from 1950 to 1987, ranging from a low of 9.5% to a high of 12% of the total annual cobalt consumption. Use of cobalt in chemicals in 1987 was almost equal to the amount used in alloys. Consumption in chemicals was 17-18% during 1950-70, 24% in 1981 and 34.2% in 1987; use in chemicals during 1987 represented 42.6% of consumption in Europe and 34.4% in the USA. In 1987, applications were: chemicals, catalysts, paint, ink and rubber additives, 24.9%; unspecified, 3.7%; electronics and magnetic tape, 2.8%; medical and veterinary, 1.5%; and plating and anodizing, 1.3% of total cobalt consumption (see Table 10; Johnston, 1988).

Cobalt(III) acetate has been used as a catalyst in cumene hydroperoxide decomposition (Budavari, 1989). *Cobalt(II) acetate* is used much more commonly, in the manufacture of drying agents for inks and varnishes, as dressings for fabrics, as catalysts and pigments, and in anodizing and agricultural applications. Mixed metal acetates such as cobalt-tin acetate can also be prepared (Donaldson, 1986; Donaldson *et al.*, 1986a; Budavari, 1989). During the 1960s, cobalt(II) acetate, cobalt chloride and cobalt sulfate (see below) were used as foam stabilizers in malt beverages in Canada, Belgium and the USA. In 1964-66, US breweries reportedly added up to 1.5 µg/ml of cobalt in 20-25% of all beer sold (Morral, 1979; Budavari, 1989; Cobalt Development Institute, 1989).

Cobalt(II) carbonate is used in ceramics, as a trace element added to soils and animal feed, as a temperature indicator, as a catalyst and in pigments (Morral, 1979; Sax & Lewis, 1987). Basic cobalt carbonate is often used as a starting material in the manufacture of other chemicals, such as cobalt oxide, cobalt pigments and cobalt salts. It is also used in ceramics and in agriculture (Donaldson, 1986; Donaldson *et al.*, 1986a; Budavari, 1989).

The main use of *cobalt(II) chloride* hexahydrate is as an intermediate in the manufacture of other cobalt salts. It has been used in invisible inks because, when it is heated, the crystal water is liberated and the almost invisible colour changes to dark blue (Suvorov & Cekunova, 1983). Because of its hygroscopic nature, anhydrous cobalt chloride has been used in barometers and as a humidity indicator in hygrometers; the anhydrous form turns from blue to pink when hydrated. Other uses include the absorption of military poison gas and ammonia, as electroplating flux for magnesium refining, as a solid lubricant and dye mordant, in the preparation of catalysts, for painting on glass and porcelain, as a temperature indicator in grinding, as a fertilizer additive, as a trace mineral supplement in animal feed and in magnetic recording materials (Morral, 1979; Donaldson *et al.*, 1986a; Budavari, 1989). The hexahydrate is used to prepare a standard solution of cobalt for analytical purposes (National Library of Medicine, 1989).

Cobalt chloride is also used in the ceramic and glass industries, in pharmaceuticals for the manufacture of vitamin B_{12} and as catalysts for the oxidation in air of toxic waste solutions containing sulfites and antioxidants (Considine, 1974). It was used as a foam stabilizer in malt beverages in the 1960s (see under cobalt acetate above). Cobalt chloride has been used as an adjunct to iron therapy (if cobalt deficiency is suspected) in patients with refractory anaemia to improve haematocrit, haemoglobin and erythrocyte values. Although cobalt stimulates erythropoietin production, it also blocks certain enzymes involved in iron transport and may stimulate erythrocyte production by causing intracellular hypoxia. Therapeutic doses of 20-300 mg per day orally have been used (Goodman & Gilman, 1975; Goodman-Gilman *et al.*, 1985; Berkow, 1987). According to Reynolds (1989), its general therapeutic use is unjustified.

Cobalt(II) hydroxide is used in the manufacture of other cobalt compounds, as a starting material to make driers for paints and printing inks, as a catalyst or starting material for catalysts and in solutions for impregnating electrodes in storage batteries (Morral, 1979; Donaldson *et al.*, 1986a; Budavari, 1989).

Cobalt(III) hydroxide is used as an oxidation catalyst (Sax & Lewis, 1987; Budavari, 1989).

Cobalt molybdenum oxide is used with aluminium oxide as a desulfurization and reforming catalyst in oil refining (Considine, 1974; Donaldson, 1986; Sax & Lewis, 1987).

Cobalt(II) naphthenate is used primarily as a drying agent in paints, inks and varnishes. Additionally, it is used to enhance the adhesion of sulfur-vulcanized rubber to steel and other metals (i.e., in tyres), as a dressing for fabrics, as a catalyst and as an antistatic adhesive (Buono & Feldman, 1979; Donaldson *et al.*, 1986a; Sax & Lewis, 1987).

Cobalt(II) nitrate hexahydrate is used mostly in the preparation of catalysts, in pigments, chemicals, ceramics, feed supplements, battery materials, invisible inks, hair dyes and vitamin B_{12} preparations. It serves as an important source of high-purity cobalt for use in the electronics industry (Considine, 1974; Morral, 1979; Donaldson, 1986; Donaldson *et al.*, 1986a; Budavari, 1989).

Cobalt(II) oxide (CoO) is used as a starting material for the manufacture of other chemicals and catalysts, in pigments such as colour reagents and in ceramics, gas sensors and thermistors (Donaldson *et al.*, 1986a).

Cobalt(II,III) oxide (Co_3O_4) is used in ceramics and enamels as a colorizer and decolorizer, in semiconductors, as a catalyst, in solar collectors, in grinding wheel coolants and as an implant into the oesophagus of cobalt-deficient ruminants (Morral, 1979; Donaldson, 1986; Donaldson *et al.*, 1986a; Sax & Lewis, 1987).

Lilac pigments containing 22-33 wt % *cobalt(III) oxide* (Co_2O_3) and blue-green pigments containing 8-20 wt % cobalt(III) oxide are used in ceramics. A prime enamel has been prepared that contains 0.8 wt % cobalt(III) oxide (Donaldson *et al.*, 1986a). Cobalt(III) oxide monohydrate is used as an oxidation catalyst (Budavari, 1989).

Cobalt(II) sulfate is the preferred source of water-soluble cobalt salts used in the manufacture of other cobalt chemicals and in electroplating, because it has less tendency to deliquesce or dehydrate than the chloride or nitrate. The monohydrate and heptahydrate are used in plating, feed supplements, to make catalysts, magnetic recording materials, anodizing agents and corrosion protection agents (Morral, 1979; Donaldson *et al.*, 1986a; Budavari, 1989). Cobalt sulfate is also used in the manufacture of vitamin B_{12} during the biological fermentation of molasses by *Pseudoneras denitrificans* (Cobalt Development Institute, 1989). Treating cobalt-deficient soil with 100-150 g/acre [247-371 g/ha] of cobalt sulfate prevents cobalt deficiency in ruminant animals (Jones *et al.*, 1977); injection of cobalt sulfate solution through rumenal fistulas and subcutaneous implantation of slow-release cobalt glasses have been used as alternative methods of supplying cobalt (Donaldson *et al.*, 1986b). In the 1960s, cobalt sulfate was used in various countries as a foam stabilizer in beer (see under cobalt acetate above).

Both α- and β-*cobalt sulfides* are used as catalysts for hydrodesulfurization of organic compounds in petroleum refining. The sulfide is generated as needed by passing hydrogen sulfide over mixed cobalt-molybdenum-aluminium oxides in refinery reactors to form catalytic cobalt sulfide *in situ* (Brauer, 1965; Donaldson *et al.*, 1986a; Budavari, 1989).

2.3 Occurrence

(a) *Geological occurrence*

Cobalt is widely distributed throughout the environment. It is thirty-third in abundance among the elements in the earth's crust, accounting for 0.001-0.002%. The largest concentrations of cobalt are found in mafic (igneous rocks rich in magnesium and iron and comparatively low in silica) and ultramafic rocks; the average cobalt content in ultramafic rocks is 270 mg/kg, with a nickel:cobalt ratio of 7. Sedimentary rocks contain varying amounts of cobalt; average values are 4 mg/kg for sandstone, 6 mg/kg for carbonate rocks and 40 mg/kg for clays or shales. Levels of cobalt in metamorphic rock depend on the amount of the element in the original igneous or sedimentary source. Cobalt has also been found in meteorites (Donaldson, 1986; Donaldson *et al.*, 1986b; Weast, 1988; Budavari, 1989).

Cobalt minerals occur in nature as a small percentage of other metal deposits (particularly copper), generally as sulfides, oxides or arsenides, which are the largest mineral sources. Smaltite ($CoAs_2$) has a cobalt content of 25% and is the most important arsenide found in the USA, Canada and Morocco; other arsenides include safflorite ($CoFe)As_2$, skutterudite ($(Co,Fe)As_3$) and the arsenosulfide ($CoAsS$; cobaltite), which contains up to 35% cobalt and is found in Cobalt City, Australia, and in Burma. Carrollite ($(Co,Ni)_2CuS_4$) and linnaeite (Co_3S_4) are sulfides which contain 40-50% cobalt and are found in the African copper belt; siegenite ($(Co,Ni)_3S_4$), which contains 25% cobalt, is found in the mines of Missouri, USA. The supplies of oxides that have the greatest economic importance are heterogenite ($CoO(OH)$) and sphaerocobaltite (containing 50% cobalt) from Katanga, Zaire, and asbolite (obtained from manganese copper) from New Caledonia (Kipling, 1980; Merian, 1985; Donaldson, 1986; Budavari, 1989; Schrauzer, 1989).

(b) *Occupational exposure*

The main route of absorption during occupational exposure to cobalt is *via* the respiratory tract, due to inhalation of dusts, fumes or mists containing cobalt or inhalation of gaseous cobalt carbonyl. Occupational exposures occur during the production of cobalt powder, in hard-metal production, processing and use, and in the use of cobalt-containing pigments and driers. Workers who regenerate spent catalysts may also be exposed to cobalt sulfides.

Occupational exposure to cobalt can be measured by analysis of ambient air levels and by biological monitoring, i.e., analyses of cobalt concentrations in blood or urine (for reviews see Ferioli *et al.*, 1987; Alessio & Dell'Orto, 1988; Angerer, 1989; Angerer *et al.*, 1989). (See also Table 19 and p. 419).

Data on exposure to cobalt measured by air and biological monitoring in various industries and occupations are summarized in Table 12. Where possible, the correlations between the concentrations of cobalt in air and biological body fluids are given. Information available to date on blood and urinary concentrations of cobalt indicates that these tests are suitable for assessing exposure on a group basis. The determination of urinary levels of cobalt seems to offer more advantages than that of blood levels. The biological indicator levels are influenced by the chemical and physical properties of the cobalt compound studied and by the time of sampling. It should be noted that the type of compound, the timing of collection of biological samples (normally at the end of a shift) and the analytical methods differ among the studies.

Using biological indicators, the concentration of cobalt in air was related to that in biological fluids; an exposure to 50 $\mu g/m^3$ cobalt in air was found to be equivalent to a level of 2.5 $\mu g/l$ cobalt in blood and 30 $\mu g/l$ cobalt in urine (Angerer, 1989).

Lehmann *et al.* (1985) took stationary and personal air samples at workplaces during dry grinding (with exhaust facilities) in the mechanical processing of cobalt alloys containing 5-67% cobalt. They found the following airborne concentrations: stationary sampling—total dust, 0.1-0.85 mg/m^3 (median, 0.55 mg/m^3; 13 samples); cobalt in total dust, 0.06-23.3 $\mu g/m^3$ (median, 0.4 $\mu g/m^3$; 13 samples); personal sampling—total dust, 0.42-2.05 mg/m^3 (median, 0.55 mg/m^3; six samples); cobalt in total dust, 0.2-69.1 $\mu g/m^3$ (median, 3.2 $\mu g/m^3$; seven samples).

In dental laboratories, concentrations of cobalt were measured during the preparation and polishing of cobalt-chromium alloys and ranged from 30 to 190 $\mu g/m^3$ (Kempf & Pfeiffer, 1987).

Kusaka *et al.* (1986) carried out extensive personal air monitoring at different stages of hard-metal (cemented carbide) manufacturing and processing; the results, by group of workers, are given in Table 13. A similar study was performed by Lehmann *et al.* (1985), who took stationary and personal air samples during various grinding operations involving hard metal (Table 14). The airborne concentrations of cobalt were mainly below 100 $\mu g/m^3$; higher concentrations were observed mainly during dry and wet grinding operations without ventilation or exhaust facilities. Exposure to cobalt during wet grinding presumably originates not only in the workplace but also from cobalt dissolved in coolants. After one week of use,

Table 12. Occupational exposures to cobalt in various industries and activities[a]

Industry/activity	No. of samples	Sex	Concentration of cobalt in ambient air	Concentration of cobalt in blood and urine	Comments	References
Hard-metal production (two subgroups)	10	M	a. Mean, 0.09 mg/m^3; b. Mean, 0.01 mg/m^3 (personal samples)	Blood: a. Mean, 10.5 μg/l; b. Mean, 0.7 μg/l. Urine: a. Mean, [106] μg/l; b. Mean, [~3] μg/l. Sampling on Friday pm	Significant correlations: air:urine ($r = 0.79$); air:blood ($r = 0.87$); blood:urine ($r = 0.82$)	Alexandersson & Lidums (1979); Alexandersson (1988)
Hard-metal grinding (seven subgroups)	153	—	Up to 61 μg/m^3 (stationary samples)	Median values for all subgroups: serum, 2.1 μg/l; urine, 18 μg/l	Significant correlation: serum (x)/urine (y) $y = 2.69x + 14.68$	Hartung & Schaller (1985)
Hard-metal tool production (11 subgroups)	170 / 5	M / F	Mean, 28–367 μg/m^3 (personal samples)	Blood: mean, 3.3–18.7 μg/l; urine, 10–235 μg/l. Sampling on Wednesday or Thursday at end of shift	Significant correlations (based on mean values): air (x)/urine (y): $y = 0.67x + 0.9$; air (x)/blood (y): $y = 0.004x + 0.23$; urine (x)/blood (y): $y = 0.0065x + 0.23$	Ichikawa et al. (1985)
Hard-metal production (six subgroups)	27	—	Breathable dust range, 0.3–15 mg/m^3 with 4–17% cobalt	Serum: mean, 2.0–18.3 μg/l; urine, 6.4–64.3 μg/g creatinine	Significant correlation: serum/urine, $r = 0.93$	Posma & Dijstelberger (1985)
Hard-metal production	26	M	Range, approx. 0.002–0.1 mg/m^3; median, approx. 0.01 mg/m^3 (personal samples)	Urine: Monday at end of shift (a) up to 36 μg/l; Friday at end of shift (b) up to 63 μg/l	Significant correlations: air (x)/urine (y): (a) $y = 0.29x + 0.83$; (b) $y = 0.70x + 0.80$	Scansetti et al. (1985)
Cobalt powder production	6	—	a. Range, 0.675–10 mg/m^3	Urine: a. mean, 35.1 μg/l; b. mean, 9.6 μg/l; c. mean, 11.7 μg/l. Sampling on Sunday (24 h)	Times of sampling: Monday am for basic exposure level; Friday evening for cumulative exposure level	Pellet et al. (1984)
Presintered tungsten carbide production	15	—	b. Range, 0.120–0.284 mg/m^3			
Hard-metal use	7	—	c. Range, 0.180–0.193 mg/m^3			

Table 12 (contd)

Industry/activity	No. of samples	Sex	Concentration of cobalt in ambient air	Concentration of cobalt in blood and urine	Comments	References
Cobalt powder and cobalt salt production (seven subgroups)	40	M	Mean, 46-1046 µg/m³ (stationary samples)	Blood: mean, 5-48 µg/l; urine: mean, 19-438 µg/l Post-shift sampling	Significant correlations: air/urine; air/blood; blood (x)/urine (y): y = 7.5x + 11.2	Angerer et al. (1985)
Cobalt oxide processing and cobalt salt manufacture	49	M	Median, 0.52 mg/m³; range, 0.1-3.0 mg/m³ (personal samples)	Urine: mean, 0.34 mg/l; range, 0.1-0.9 mg/l	Poor correlation air:urine	Morgan (1983)
Painting porcelain with soluble cobalt salts	46	F	a. Range, 0.07-8.61 mg/m³	Blood: a. Mean, 2.16 µg/l; b. Mean, 0.63 µg/l; Urine: a. Mean, 8.35 µg/ mmol creatinine; b. Mean, 0.13 µg/ mmol creatinine	Significant correlation: blood/urine ($r = 0.88$)	Christensen & Mikkelsen (1986)
Painting porcelain with slightly soluble cobalt salts	15	F	b. Range, 0.05-0.25 mg/m³ (personal samples)			

[a]From Angerer & Heinrich (1988)

Table 13. Airborne concentrations of cobalt at various stages in the manufacture and processing of hard metals[a]

Activity	No. of workers	No. of samples	Concentration of cobalt ($\mu g/m^3$)	
			Mean ± SD	Range
Powder	12	38	688 ± 1075	6–6388
Press				
Rubber	4	19	473 ± 654	48–2905
Machine	25	27	85 ± 95	4–407
Sintering	21	38	28 ± 26	2–145
Shaping	47	129	126 ± 191	6–1155
Grinding				
Wet	131	205	53 ± 106	11–1247
Dry	1	2	1292 ± 179	1113–1471
Electron discharging	5	5	4 ± 1	1–5
Blasting	2	5	3 ± 1	1–4

[a]From Kusaka et al. (1986)

Table 14. Concentrations of cobalt in total dust during hard-metal grinding[a]

Type of grinding/ type of sample	No. of workplaces	No. of companies	Sampling time (h)	Concentration of cobalt ($\mu g/m^3$)	
				Median	Range
Dry grinding with exhaust facilities					
Stationary	16	6	2	3.1	0.1–203.5
Personal	16	5	2	12.3	0.5–223.8
Wet grinding without exhaust facilities					
Stationary	9	5	2	12.8	2.4–90.4
Personal	14	5	2	42.5	7.9–208.0
Wet grinding with exhaust facilities					
Stationary	8	1	2	6.9	1.1–11.8
Personal	7	1	2	13.7	1.3–29.9

[a]From Lehmann et al. (1985)

levels of up to 118 mg/kg were found in the coolant; after four weeks, up to 182 mg/kg were observed (Lehmann et al., 1985). This finding was confirmed by Hartung (1986). Einarsson et al. (1979) studied the dissolution of cobalt in nine commercial cutting fluids one to five days after use in the grinding of hard-metal alloys. After one day, most of the cobalt liberated by grinding was found in solution; this percentage decreased when grinding was continued using the same coolant fluid. Only a small fraction of the cobalt was found as particles in the circulating fluid. The authors concluded that the bulk probably remains in the sediment in the storage tank.

The concentration of cobalt dust was measured in the air of a Danish porcelain factory in 1981. In personal air samples taken for 19 female plate painters, the levels were 0.07-8.61 mg/m3. The cobalt levels in blood and urine were measured in 1982 in 46 female plate underglaze painters exposed to soluble cobalt silicate and in 51 female plate overglaze painters with no exposure to cobalt. The mean levels in the blood of exposed persons (longer than four weeks) were 2.16 µg/l (range, 0.2-24; median, 1.0) compared with 0.24 µg/l in the controls (range, 0.05-0.6; median, 0.2). Mean levels in urine were: 77 µg/l (median, 26; range, 2.2-848) in exposed workers and 0.94 µg/l (median, 0.3; range, 0.05-13.8) in unexposed workers (Mikkelsen et al., 1984; Christensen & Mikkelsen, 1986). In 1984, after conditions in the workplace had been improved, the concentration of cobalt in air had decreased to about 0.05 mg/m3. The mean urinary level of cobalt in 38 of the 46 workers investigated originally who were selected for urine analysis was 2.6 µg/mmol creatinine (range, 0.16-16.1) compared to 4.2 µg/mmol creatinine (range, 0.24-29.1) in 1982. A significant correlation was observed between blood cobalt and creatinine-corrected urinary cobalt levels ($p < 0.001$). In 1982, in a factory using a slightly soluble cobalt silicate, the mean cobalt levels in blood and urine from 15 female plate painters were 0.63 µg/l (median, 0.60; range, 0.37-1.58) and 0.13 µg/mmol creatinine (median, 0.11; range 0.02-0.37), respectively (Christensen & Mikkelsen, 1986; see also Table 12).

(c) *Air*

Levels of cobalt in the ambient air are a function of the extent to which particles of soil are dispersed by the wind. They are higher near factories in which cobalt is used, and atmospheric concentrations of cobalt in remote areas are very low: less than 1 ng/m3 in the Antarctic. In other areas, ambient air concentrations are usually around 1 ng/m3. Levels exceeding 10 ng/m3 have been reported in heavily industrialized cities (Elinder & Friberg, 1986). Combustion of organic materials containing cobalt is reported to be an additional source of emission (Lange, 1983; Angerer & Heinrich, 1988). Coal contains up to 40 mg/kg (average, 1 mg/kg) cobalt (Angerer & Heinrich, 1988), and hard coal contains about 8 mg/kg

(Schrauzer, 1989). Merian (1985) estimated a global annual generation of about 5000 tonnes of cobalt from the burning of coal.

A survey of atmospheric trace elements in the UK in 1977 showed ambient concentrations of cobalt in the range of 0.04-6.5 ng/kg at seven stations sampled. Around 57% of the cobalt content was in a soluble form (Cawse, 1978).

(d) Tobacco smoke

The content of cobalt in cigarettes has been studied by means of neutron activation; different brands of tobacco were found to contain < 0.01-2.3 mg/kg dry weight (Wyttenbach *et al.*, 1976; Iskander, 1986; Iskander *et al.*, 1986). When cigarettes were smoked in a standard smoking machine, 0.5% of the cobalt content of the cigarette was transferred into smoke condensate (Nadkarmi & Ehmann, 1970).

(e) Water and sediments

Uncontaminated samples of fresh water generally contain low concentrations of cobalt, ranging from 0.1 to 10 µg/l (Schrauzer, 1989). Concentrations of 0.1-5 µg/l have been found in drinking-water (Elinder & Friberg, 1986).

Approximately 20 000 tonnes of cobalt are transported annually by rivers to oceans, where they are precipitated (Merian, 1985). A cobalt content of 74 mg/kg has been measured in sediments (Schrauzer, 1989). Natural transport is not significantly affected by mining activities or industrial use. The concentration of cobalt in seawater is normally quite low, at 0.002-0.007 µg/l, the level decreasing with increased depth (Knauer *et al.*, 1982).

(f) Foods and beverages

Human dietary intake of cobalt is highly variable; Table 15 summarizes estimated total intake of cobalt from food in various countries. Most of the cobalt ingested is inorganic: vitamin B_{12}, which occurs almost entirely in food of animal origin, accounts for only a very small fraction. Vegetables contain inorganic cobalt but little or no vitamin B_{12} (Friedrich, 1984; Donaldson *et al.*, 1986b).

Values for the cobalt content of foods vary widely between reports, even among analyses of the same foods, probably owing as much to differences in environmental cobalt levels as to analytical difficulties or inadequate analytical techniques. Green leafy vegetables and fresh cereals are the richest and most variable sources of cobalt (0.2-0.6 µg/g dry mass), while dairy products, refined cereals and sugar contain the least cobalt (0.01-0.03 µg/g dry mass; Donaldson *et al.*, 1986b). Plant products have been estimated to contribute up to 88% of the total cobalt in the Japanese diet (Yamagata *et al.*, 1963). Normal cows' milk contains very little cobalt (average,

Table 15. Total daily intake of cobalt from food *per caput*

Country	Daily intake *per caput* (μg)	Reference
Canada	45–55	Kirkpatrick & Coffin (1974)
Finland	13	Varo & Koivistoinen (1980)
Germany	17	Pfannhauser (1988)
	15	Pfannhauser (1988)
	5–10 (vitamin B_{12} only)	Schormüller (1974)
Hungary	100	Lindner–Szotyori & Gergely (1980)
Italy	9	Pfannhauser (1988)
Japan	19.5	Yamagata *et al.* (1963)
Netherlands	5–7	Pfannhauser (1988)
Spain	25	Barberá & Farré (1986)
UK (vitamin B_{12} only)	7.0	Spring *et al.* (1979)
USA	5–6	Harp & Scoular (1952)
USSR	1.7	Reshetkina (1965)
	31	Nodiya (1972)

about 0.5 μg/l); shelled eggs have been reported to contain 0.03 μg/g (Donaldson *et al.*, 1986b). Varo and Koivistoinen (1980) found concentrations of 30–50 μg/kg dry weight in fish and vegetables; that in meat and dairy products was 10 μg/kg. The daily diet of the 70-kg 'reference man' contains cobalt at 0.01–0.02 mg/kg fresh weight (based on 20–40 μg/day intake) (Donaldson *et al.*, 1986b).

In 15 commercial beers analysed in 1965 using a colorimetric method, the levels of cobalt were well below 0.1 mg/l. When cobalt salts had been added during processing, values of up to 1.1 mg/l were recorded (Elinder & Friberg, 1986).

The cobalt content of five brewed teas averaged 0.2 μg/g (range, 0.16–0.34) and that of seven brewed coffees, 0.75 μg/g (range, 0.42–2.0 μg/g; Horwitz & Van der Linden, 1974).

(g) Soils and plants

In one study, the cobalt content of soils ranged from 1 to 40 mg/kg (Merian, 1985) with an average of 8 mg/kg (Schrauzer, 1989). In general, cobalt tends to be deficient in areas where there is granite, sand or limestone and in volcanic and peaty soils. Good drainage may reduce cobalt content (Kipling, 1980). The solubilities of cobalt compounds are pH-dependent, and cobalt is more mobile in acid soils than in alkaline soils (Schrauzer, 1989).

In industrialized areas, up to 75 mg/kg cobalt have been found in the soil around factories using cobalt powders, and higher concentrations may occur in waste-metal dumps (Kipling, 1980).

The uptake of cobalt by plants is species-dependent: cobalt is hardly detectable in green beans and the level is exceedingly low in radishes (Schrauzer, 1989). Leafy plants, such as lettuce, cabbage and spinach, have a relatively high cobalt content, whereas the content is low in grasses and cereals (Kipling, 1980). It is as yet unknown whether cobalt is essential for plants. In some cases, small amounts of cobalt produce positive growth effects, but these are dose-dependent and may be indirect (Schrauzer, 1989). It has been suggested that the element is necessary for the fixation of nitrogen in vegetables that are relatively rich in cobalt. Cobalt concentrations in pastures vary according to season and the presence of fertilizers (Kipling, 1980).

(h) Human tissues and body fluids

Over the years, there has been a progressive downward adjustment in the reported normal levels of cobalt in human tissues and body fluids as a result of improvements in analytical methodology. Concentrations of cobalt observed in the blood and urine of the general population are summarized in Table 16. The concentrations in body fluids are well below the microgram per litre level; mean concentrations reported in serum range from 0.1 to 0.3 µg/l.

Alexandersson (1988) found that smokers with no occupational exposure had a significantly higher mean cobalt concentration in urine (0.6 µg/l; SD, 0.6) than nonsmokers (0.3 µg/l; SD, 0.1). There was no difference between smokers and nonsmokers in the cobalt levels in blood.

Patients in various stages of renal failure showed a significantly higher serum concentration of cobalt than a control group, but there was no correlation to the degree of renal insufficiency. Haemodialysis did not influence the levels, whereas kidney transplantation reduced them (Lins & Pehrsson, 1984). Values for whole blood were a little higher than serum concentrations but were not well documented (Iyengar & Woittiez, 1988). In urine samples obtained from normal adults, the concentrations of cobalt were reported to be approximately 0.1-2 µg/l (see Table 16). Greatly increased urinary levels have been reported for persons taking multivitamin pills containing cobalt (Reynolds, 1989).

Considerable differences have been found in the levels of cobalt in hair, ranging from 0.4 to 500 µg/kg (Iyengar & Woittiez, 1988).

In autopsy studies, the liver has been shown to contain the highest concentration of cobalt, with individual values ranging from 6 to 151 µg/kg (median, 30 µg/kg) in seven studies. This may be attributed, at least in part, to differences in

Table 16. Concentrations of cobalt in urine, serum and whole blood of persons not exposed occupationally to cobalt

Urine Mean	Urine Range	Serum (or plasma) (μg/l)	Whole blood (μg/l)[a]	Reference
-	-	0.108 ± 0.06	-	Versieck et al. (1978)
-	-	-	0.5 ± 0.1	Alexandersson & Swensson (1979)
0.4 μg/l	0.2–1.2	-	0.5	Alexandersson & Lidums (1979)
-	-	0.195 ± 0.015 (plasma)	-	Kasperek et al. (1981)
0.18 μg/creatinine	-	-	-	Kennedy et al. (1981)
0.38 μg/l	0.1–0.75 μg/l	-	-	Schumacher-Wittkopf & Angerer (1981)
1.3 μg/l	-	-	-	Hartung et al. (1982)
-	-	0.01–1.9	-	Masiak et al. (1982)
-	-	-	0.09 ± 0.02	Ostapczuk et al. (1983)
-	-	0.15 ± 0.07 (plasma)	-	Andersen & Høgetveit (1984)
0.94 μg/l	0.05–13.8 μg/l	-	-	Mikkelsen et al. (1984)
-	4.6 μg/g creatinine	-	-	Posma & Dijstelberger (1985)
0.41 μg/l or 0.28 μg/g creatinine	-	-	-	Scansetti et al. (1985)
2.0 μg/l	-	-	1.9 ± 1.1	Ichikawa et al. (1985)
-	-	0.28	-	Lewis et al. (1985)
0.09 μg/mmol creatinine	0.004–1.21 μg/mmol creatinine	-	0.24 (0.05–0.6)	Christensen & Mikkelsen (1986)
-	-	0.1	-	Hartung (1986)
-	-	0.73 ± 0.10 (plasma)	-	Collecchi et al. (1986)
0.4 μg/l	0.1–2.2 μg/l	-	0.5 (0.1–1.2)	Alexandersson (1988)
0.01 μg/l	-	-	0.2–1.3	Angerer et al. (1989)

[a]Range, or mean ± standard deviation

-, not given

food intake, since this organ stores vitamin B_{12} (Iyengar & Woittiez, 1988). In New Zealand, 96 human liver samples showed a mean concentration of 120 µg/kg wet weight cobalt, with no significant difference between sex, age or regional district (Pickston et al., 1983). Levels of cobalt were lower in liver carcinoma tissue than in normal hepatocytes from the same liver samples (Kostić et al., 1982). The total cobalt content of a 70-kg, unexposed man was estimated to be about 1.5 mg. The total amount of vitamin B_{12} in the body of an adult is about 5 mg, corresponding to 0.25 mg cobalt, of which 50-90% is localized in the liver (Schrauzer, 1989).

Cobalt concentrations in the hearts of patients dying from myocardiopathy associated with the consumption of beer containing cobaltous salts were found to be 10 times higher than in normal cardiac muscle (Sullivan et al., 1968).

(i) *Iatrogenic exposure*

Cobalt is the major constituent (approximately 62%) of porous-coated cobalt-chromium alloys used in surgical implants; therefore, body levels of cobalt (urine, serum) have been used as an index of the wear rate of the prostheses. Table 17 summarizes the results of several investigations on trace metal concentrations in the body fluids of patients with total knee and hip arthroplasty with metal prostheses. Cobalt-containing particles have also been identified by microscopic examination of tissues adjacent to prostheses (Hildebrand et al., 1988; Sunderman et al., 1989).

Certain authors observed significant increases in mean concentrations of cobalt in the serum or urine from patients with various metal implants (especially those with metal-to-metal contact), while others found that the concentrations of this metal were only sporadically elevated. These discordant results may reflect greater rates of release of metals from implants with metal-to-metal *versus* metal-to-polyethylene articular surfaces, as well as differences among the cobalt-containing alloys used (e.g., porous-coated *versus* non-porous surfaces and cemented *versus* cementless implants). Analytical limitations may also play a major role, since the concentrations of cobalt in the serum and urine specimens from control subjects far exceeded the currently accepted ranges. Analytical inaccuracies in previous studies probably resulted from metal contamination during specimen collection, inattention to quality assurance techniques and/or inadequate instrumental sensitivity and specificity (Sunderman et al., 1989).

Raithel et al. (1989) investigated the cobalt content in tissues surrounding hip arthroplasties and in distant muscle samples. From 10 patients with loosening of prostheses, tissue samples were taken from the implanted cup (polyethylene surface to avoid metal-to-metal friction), from the implanted shaft and from the musculus vastus lateralis, and the patients received new hip prostheses. The old cobalt-chromium-molybdenum types (ASTM F 75-74) were replaced after 5-15.5 years

Table 17. Cobalt concentrations in body fluids of patients with total hip or knee arthroplasty[a]

Study	No. of patients	Period of observation	Type of implant	Observations	Concentrations of cobalt		
					Urine	Blood	Synovial fluid
Coleman et al. (1973)	12	3 weeks to 32 months	Hip, cobalt-molybdenum-chromium alloy, cemented, nonporous, with or without polyethylene component (C cast alloy)	Increased cobalt and chromium in blood and urine, only with metal-to-metal contact (no polyethylene)	15-73 µg/l after 1 year	4.5-16 µg/l after 1 year	—
Jones et al. (1975)	4	Not given	Hip, cobalt-chromium-molybdenum alloy, cemented, nonporous, with metal-to-metal contact (C cast alloy)	Increased cobalt in urine and (in one case) in synovial fluid and liver, bone and brain tissues	22-55 µg/l	—	250 µg/l 0.5-3 mg/kg
Miehlke et al. (1981)	30	6 months to 10 years	Knee, cobalt-chromium alloy, cemented, nonporous, with or without polyethylene component	Increased cobalt and chromium in synovial fluid and serum, especially with metal-to-metal contact	—	0.16-79 µg/l	0.36-7200 µg/l
Jorgensen et al. (1983)	10	Not given	Hip, cobalt-chromium-molybdenum alloy, porous-coated or nonporous, cementless	Increased cobalt in urine, especially in patients with porous-coated implants	Porous: mean, 14.2 µg/l; nonporous: mean, 8.4 µg/l	—	—
Black et al. (1983)	15	1 day to 6 months	Hip, cobalt-chromium-nickel alloy, cemented, nonporous, polyethylene cup (cobalt-chromium/UMHWPE THRs[a])	Increase in serum chromium (peak at 15 days), serum nickel (peak at 6 months); normal serum cobalt	—	—	—
Bartolozzi & Black (1985)	14	1 to >30 days	Hip, cobalt-chromium alloy, cemented, nonporous, polyethylene cup	Increase in chromium (serum peak at 10 days, urine peak at 15 days)	Peak, 26.2 ng/mg creatinine	Peak, 39.9 pg/mg protein	—
Pazzaglia et al. (1986)	17	7-15 years	Hip, cobalt-chromium-molybdenum alloy, cemented, nonporous, with or without polyethylene cup	Increased cobalt and chromium in urine and chromium in plasma	0.9-1.05 µg/l	—	—
Jones & Hungerford (1987)	14	1 week to 1 year	Hip, cobalt-chromium alloy, cementless, porous-coated, polyethylene cup (PCA®)	Increased urinary nickel in 2 of 14 patients at 6 months; increased urinary nickel and cobalt in 3 of 4 measured at 1 year	—	—	—

Table 17 (contd)

Study	No. of patients	Period of observation	Type of implant	Observations	Concentrations of cobalt		
					Urine	Blood	Synovial fluid
Braun et al. (1986)	22	5 months to 3 years	Hip, cobalt-chromium-molybdenum alloy, cementless, porous-coated, polyethylene cup, fixed	Increased urinary chromium	-	-	-
Raithel et al. (1989)	15	2 years	Fixed hip, cobalt-chromium-nickel-molybdenum alloy, cemented, nonporous with polyethylene cup	Increased serum cobalt	-	1.8 µg/l	
	10	5-15.5 years (mean, 12.5)	Loose hip, cobalt-chromium-nickel molybdenum alloy, cemented, nonporous with polyethylene cup; old hip replaced	Increased urinary chromium, nickel and cobalt, increased serum nickel	3.8 µg/l	-	
Sunderman et al. (1989)	28	1 day to 2.5 years	Knee or hip, cobalt-chromium alloy (ASTM F-75-82), porous-coated, 10 cemented, 18 cementless with polyethylene	Slight increase in serum and urinary cobalt in knee prostheses. 2 patients, substantially elevated levels (7 weeks and 22 months post-arthroplasty, with loosening of prostheses); serum and urinary chromium levels also elevated in one patient	1 µg/g creatinine (6-120 weeks) 7.7 µg/g creatinine and 5.6 µg/l in the 2 patients	0.15 µg/l (6-120 weeks) 1 and 1.15 µg/l in the 2 patients	

[a]Ultrahigh molecular weight polyethylene (total hip replacements)

(median, 12.5 years). The concentrations of cobalt in the tissues surrounding the shaft ranged from 367 to 6510 µg/kg (median, 868 µg/kg), and those in tissues surrounding the cup, from 98 to 16 293 µg/kg (median, 1080 µg/kg). Muscle tissue contained 24-151 µg/kg (median, 124 µg/kg) cobalt.

Hildebrand *et al.* (1988) also found extremely high concentrations of cobalt, up to three orders of magnitude (140 µg/g dry weight) above the normal values, in connective tissue taken on a Vitallium plate.

(j)   Others

The total concentration of cobalt in cement made in Asia ranged from 8.1 to 14.2 µg/g. The metal existed mainly as insoluble salts; the concentration of water-soluble cobalt was 0.39-0.65 µg/g (Goh *et al.*, 1986). The cobalt content in 42 US cement samples was < 0.5 µg/g (Perone *et al.*, 1974).

The cobalt content of 30 household cleaning products sold in Spain in 1985 ranged from 0.1 to 14 mg/l; the highest levels were found in two bleaches, containing 1.1 and 1.4 mg/l (Vilaplana *et al.*, 1987).

2.4 Regulatory status and guidelines

Occupational exposure limits and guidelines established in different parts of the world are given in Table 18.

Table 18. Occupational exposure limit values for cobalt[a]

Country or region	Year	Concentration (mg/m^3)	Interpretation[b]
Australia	1985	0.1 cobalt, metal fumes and dust	TWA
Belgium	1989	0.05 cobalt, metal dust and fumes (as Co)	TWA
Bulgaria	1985	0.5 cobalt and compounds (as Co); cobalt, metal dust and fumes (as Co)	TWA
Canada	1980	0.1 cobalt as metal dust and fume	TWA
Czechoslovakia	1985	0.05 cobalt and compounds (as Co)	TWA
		0.1 cobalt and compounds (as Co)	max
Denmark	1988	0.1 cobalt carbonyl (as Co); cobalt hydro-carbonyl (as Co)	TWA
		0.05 cobalt in the form of powder, dust and fumes and inorganic compounds (as Co)	TWA
Finland	1987	0.05 cobalt and inorganic compounds (as Co)	TWA

Table 18 (contd)

Country or region	Year	Concentration (mg/m³)	Interpretation[b]
Hungary	1987	0.1 cobalt and compounds (as Co)	TWA
		0.2 cobalt and compounds (as Co)	STEL
Indonesia	1987	0.1 cobalt and compounds (as Co)	TWA
Italy	1987	0.1 cobalt, metal dust and fumes (as Co)	TWA
Mexico	1987	0.1 cobalt, metal dust and fumes (as Co)	TWA
Netherlands	1986	0.1 cobalt, metal dust and fume (as Co)	TWA
Norway	1981	0.05 cobalt and compounds (as Co)	TWA
Poland	1985	0.5 cobalt and compounds (as Co); cobalt, metal dust and fumes (as Co)	TWA
Romania	1985	0.2 cobalt and cobalt oxide and cobalt, metal dust and fumes (as Co)	TWA
		0.5 cobalt and cobalt oxide and cobalt, metal dust and fumes (as Co)	max
Sweden	1988	0.05 cobalt and inorganic compounds (as Co)	TWA
Switzerland	1987	0.1 cobalt dust and compounds (as Co)	TWA
Taiwan	1987	0.1 cobalt, metal dust and fumes (as Co)	TWA
UK	1987	0.1 cobalt and compounds (as Co)	TWA
USA			
ACGIH	1989	0.05 cobalt (as Co) metal dust and fumes	TWA
		0.1 cobalt carbonyl (as Co); cobalt hydrocarbonyl (as Co)	Guide-lines
OSHA	1988	0.1 cobalt (as Co) metal dust and fume	TWA
USSR	1987	0.5 cobalt and compounds (as Co); cobalt, metal dust and fumes (as Co)	max
		0.01 cobalt hydrocarbonyl and decomposition products (as Co)	
Venezuela	1987	0.1 cobalt, metal dust and fumes (as Co)	TWA
Yugoslavia	1985	0.1 cobalt and compounds (as Co); cobalt, metal dust and fumes (as Co)	TWA

[a]From Direktoratet for Arbeidstilsynet (1981); Arbeidsinspectie (1986); Cook (1987); Health and Safety Executive (1987); National Swedish Board of Occupational Health (1987); Arbejdstilsynet (1988); National Institute for Occupational Safety and Health (1988); American Conference of Governmental Industrial Hygienists (ACGIH) (1989); US Occupational Safety and Health Administration (OSHA) (1989); United Nations Environment Programme (1990). Guidelines and standards are generally prepared by scientific bodies and sometimes become official standards, or they are recognized and applied in practice on a voluntary basis as a guide for monitoring the working environment or for technical prevention.

[b]TWA, 8-h time-weighted average; STEL, 10–15-min short-term exposure limit

2.5 Analysis

Typical methods for the analysis of cobalt in air, water, various working materials, food and biological materials are summarized in Table 19.

Table 19. Methods for the analysis of cobalt

Sample matrix	Sample preparation	Assay procedure[a]	Limit of detection	Reference
Urine	Digestion with nitric/sulfuric acid; ion-exchange separation	GF/AAS	0.1 µg/l	Lidums (1979)
	Chelatization, extraction	GF-AAS	0.1 µg/l	Schumacher-Wittkopf & Angerer (1981)
	Dilution with nitric acid	GF-AAS	Not given	Hartung et al. (1983)
	Dilution with nitric acid	GF-AAS	2 µg/l	Pellet et al. (1984)
	Digestion with sulfuric, nitric, perchloric acid; chelation, extraction	F-AAS	1 µg/l	Ichikawa et al. (1985)
	Direct analysis	GF-AAS	6 µg/l	Bouman et al. (1986)
	N,N-Hexamethyleneammonium-hexamethylenedithiocarbamic acid/xylene extraction	GF-AAS (Z)	1 µg/l (0.2 µg/l for 6 ml urine)	Bouman et al. (1986)
	Magnesium nitrate modifier	GF-AAS (Z)	2.6 µg/l	Kimberley et al. (1987)
Blood, urine	Dilution with nitric acid	GF-AAS (Z)	0.1 µg/l	Christensen et al. (1983)
	Protein precipitation; dilution with nitric acid	GF-AAS (Z)	0.1 µg/l	Christensen & Mikkelsen (1986)
Blood	Digestion with nitric, sulfuric acid	GF-AAS	0.1 µg/l	Lidums (1979)
	Dilution and matrix modification	GF-AAS	0.2 µg/l	Delves et al. (1983)
	Freeze-dried, low-temperature ashing; resolved in nitric acid	GF-AAS (Z)	0.8 µg/l	Ichikawa et al. (1985)
Blood, tissues	Digestion with nitric, sulfuric, perchloric acid	ICP	10 µg/kg blood 0.2 µg/kg tissue	National Institute for Occupational Safety and Health (1985)
Blood, serum	Wet digestion with nitric, sulfuric, perchloric acid; chelation, extraction	GF-AAS	0.1 µg/l	Barfoot & Pritchard (1980)
Serum	Dry ashing at 450°C	NAA	Not given	Versieck et al. (1978)

Table 19 (contd)

Sample matrix	Sample preparation	Assay procedure[a]	Limit of detection	Reference
Plasma, urine	Palladium matrix modification	GF-AAS (Z)	0.15 µg/l	Sampson (1988)
Biological materials	Wet digestion with nitric, sulfuric acid	ADPV	1 ng/l in the analyte solution	Ostapczuk et al. (1983)
Air	Digestion with nitric acid	GF-AAS	Not given	Hartung et al. (1983)
	Digestion with nitric, perchloric acid	ICP	1 µg/sample	National Institute for Occupational Safety and Health (1984a)
	Digestion with aqua regia	F-AAS	0.6 µg/sample	National Institute for Occupational Safety and Health (1984b)
	Digestion with hydrochloric, nitric acid	GF-AAS	1 µg/m^3	Ichikawa et al. (1985)
	Digestion with nitric acid	GF-AAS	20 ng/m^3 (sample volume 1.5 m^3)	Kettrup & Angerer (1988)
Seawater	Direct analysis	DPCSV	6 pmol (0.4 ng)	Donat & Bruland (1988)
Water	Chelation with ammonium pyrrolidinedithiocarbamate; preconcentration on activated charcoal	F-AAS, ICP	< 1 µg/l	Berndt et al. (1985)
Food	Dry digestion; triethanolamine electrolyte	Adsorption voltammetry	Not given	Meyer & Neeb (1985)
	Dry digestion; chelation with sodium di(trifluoroethyl)dithiocarbamate	GC	50 ng/sample	Meyer & Neeb (1985)
	Digestion with nitric acid; extraction with cupferron, chloroform	F-AAS	1.4 ng/ml	Barberá et al. (1986)
Milk	Ashing in muffle furnace	GF-AAS (Z)	Not given	Gunshin et al. (1985)

[a]Abbreviations: GF-AAS, graphite furnace-atomic absorption spectrometry; F-AAS, flame atomic absorption spectrometry; Z, background correction for Zeeman effect; ICP, inductively coupled plasma emission spectrometry; NAA, neutron activation analysis; ADPV, adsorption differential pulse voltammetry; DPCSV, differential pulse cathodic stripping voltammetry; GC, gas chromatography

Methods for quantitative analysis include graphite furnace-atomic absorption spectrometry (GF-AAS), inductively coupled plasma emission spectrometry (ICP), neutron activation analysis and electrochemical methods such as differential pulse anodic stripping voltammetry (DPASV). ICP and X-ray fluorescence appear to be too insensitive for the determination of cobalt in environmental and biological matrices; this is also true of the older photometric methods, which also showed lack of specificity.

With NAA, cobalt can be determined at the nanogram per kilogram level. This method offers the advantage that it requires little sample preparation, but its application is restricted to a few highly specialized laboratories. Voltammetry and, in particular, GF-AAS are much more common and permit determination of cobalt at the nanogram per kilogram level. GF-AAS, in comparison to voltammetry, does not usually require complete digestion of the sample, which makes the technique more practicable.

Air samples are collected on cellulose ester membrane filters, wet-digested with nitric and perchloric acids or aqua regia and analysed by AAS or ICP (National Institute for Occupational Safety and Health, 1984a,b; Kettrup & Angerer, 1988). The routine procedures do not permit identification of individual cobalt compounds.

Analysis of cobalt in soil, food, industrial samples and human tissues also requires complete digestion of the matrices. The US Environmental Protection Agency (1983) established standard methods using ICP and GF-AAS for the chemical analysis of water and wastes. An extremely low detection limit of 1.2 ng/l natural water was obtained using cation-exchange liquid chromatography with luminol chemiluminescence (Boyle *et al.*, 1987). A similarly high sensitivity, 0.64 ng/kg, is obtained by photoacoustic spectroscopy after extraction with 2-nitroso-1-naphthol/*meta*-xylene (Kitamori *et al.*, 1986).

Determination of cobalt in whole blood, plasma, serum and urine is used as a biological indicator of exposure to cobalt (Ichikawa *et al.*, 1985; Ferioli *et al.*, 1987; Angerer *et al.*, 1989). Choice of specimen, sampling strategies, specimen collection, transport, storage and contamination control, as well as quality control and quality assurance procedures (Schaller *et al.*, 1987), are of fundamental importance for an adequate monitoring programme. GF-AAS and DPASV are practical and reliable techniques that furnish the requisite sensitivity for measuring cobalt concentrations in biological samples. The detection limits for cobalt determination by GF-AAS analysis with Zeeman background collection are below 0.6 µg/l of body fluids, depending on the type of sample preparation.

Greater sensitivity in DPASV analysis can be achieved by using a dimethylglyoxime-sensitized mercury electrode, which provides detection limits down to 1 ng/l for cobalt in biological media (Ostapczuk et al., 1983, 1984).

Koponen et al. (1982) analysed cobalt-containing airborne dusts from hard-metal manufacturing and grinding processes by AAS and instrumental NAA. The structure of the dusts was studied by scanning electron microscopy with an energy dispersive X-ray. Cobalt was found to exist as separate particles in the dust from the mixing of raw material powders only. In the dusts from the pressing, forming and grinding of hard metal, cobalt appeared mainly in contact with tungsten carbide particles.

3. Biological Data Relevant to the Evaluation of Carcinogenic Risk to Humans

3.1 Carcinogenicity studies in animals

(a) *Inhalation exposure*[1]

Hamster: As part of a larger study, groups of 51 male Syrian golden hamsters (ENG:ELA strain), two months of age, were exposed by inhalation to 0 or 10 mg/m^3 *cobalt[II] oxide* dust (with a mass median diameter of 0.45 μm) for 7 h per day on five days per week for life. Median survival was 16.6 months in treated hamsters compared to 15.3 months in controls. No difference in the incidence of any tumour was observed between the cobalt oxide-treated and untreated hamsters (Wehner et al., 1977). [The Working Group noted the poor survival of the treated and control animals.]

(b) *Intratracheal instillation*

Rat: Groups of 50 male and 50 female Sprague-Dawley rats, ten weeks of age, received intratracheal instillations of 2 or 10 mg/kg bw *cobalt[II] oxide* powder (derived from thermal decomposition of cobalt[II] nitrate; approximately 80% of particles 5-40 μm [purity unspecified]) or 10 mg/kg bw of a *cobalt-aluminium-chromium spinel* (a blue powder [purity unspecified], with the empirical formula Co[II] 0.66, Al 0.7, Cr[III] 0.3, O 3.66, made of a mixture of CoO, Al(OH)$_3$ and Cr$_2$O$_3$ ignited at 1250°C; 80% of particles < 1.5 μm) in saline every two weeks (then

---

[1]The Working Group was aware that an inhalation study of cobalt sulfate heptahydrate was planned in mice and rats (IARC, 1990)

every four weeks from the nineteenth to the thirtieth treatment) for two years (total doses, 78 and 390 mg/kg bw cobalt oxide and 390 mg/kg bw cobalt spinel). Control groups of 50 males and 50 females received instillations of saline only or remained untreated. Animals were allowed to live until natural death or were sacrificed when moribund. No appreciable difference in body weights or survival times was observed between the treated and control groups [exact survival data not given]. Bronchoalveolar proliferation was observed in 0/100 untreated controls, 0/100 saline controls, 51/100 low-dose cobalt oxide-treated rats and 70/100 high-dose cobalt oxide-treated rats, and in 61/100 rats treated with the spinel. [The Working Group noted that the nature of the bronchoalveolar proliferation or possible association with inflammation was not described.] No pulmonary tumour was observed in 100 untreated or 100 saline controls. In the groups treated with the low dose of cobalt oxide, one male and one female developed benign lung tumours; in the groups treated with the high dose of cobalt oxide, one bronchoalveolar carcinoma occurred in a female and three adenocarcinomas and two bronchoalveolar adenomas were observed in males; in the groups receiving the spinel, one squamous-cell carcinoma was observed in males and two squamous-cell carcinomas were observed in females (Steinhoff & Mohr, 1991).

In a smaller experiment by the same authors, groups of 20 female Sprague-Dawley rats, 10 weeks of age, received weekly intratracheal instillations of 10 mg/kg bw *cobalt[II] oxide* for seven weeks and 20 mg/kg bw once every two weeks for 20 treatments (total dose, 470 mg/kg bw), and 20 mg/kg bw benzo[*a*]pyrene following the same dose regimen (total dose, 200 mg/kg bw), with a four-day interval between the two treatments. A further group of 20 females received treatment with benzo[*a*]pyrene alone. Animals were allowed to live their natural lifespan or were sacrificed when moribund [exact survival not stated]. Eight rats treated with cobalt oxide and benzo[*a*]pyrene had squamous-cell carcinomas and one had an adenocarcinoma of the lung. One animal given benzo[*a*]pyrene had a squamous-cell carcinoma of the lung (Steinhoff & Mohr, 1991).

Hamster: In a large experiment to study the effects of particulates on *N*-nitrosodiethylamine (NDEA)-induced respiratory tract carcinogenesis, groups of 25 male and 25 female hamsters [strain unspecified], seven weeks old, were given subcutaneous injections of 0.5 mg NDEA in saline or saline alone once a week for 12 weeks. One week later and once a week thereafter for 30 weeks, 4 mg *cobalt[II,III] oxide* powder (particle size, 0.5-1.0 μm [purity unspecified]) suspended in a gelatin and saline vehicle were administered by intratracheal instillation. Groups of 25 male and 25 female hamsters receiving subcutaneous injections of NDEA or saline and intratracheal instillations of the gelatin-saline vehicle served as controls. At the end of treatment (42 weeks), 39, 43, 33 and 43 animals were still alive in the four groups, respectively. Animals were observed for an additional 43-68 weeks following the last intratracheal instillation. Two of 50 hamsters receiving injections of saline and cobalt oxide by intratracheal instillation developed pulmonary alveolar tumours; 1/50 hamsters receiving injections of saline and gelatin-saline intratracheally developed a tracheal tumour. The incidences of tumours at various sites in hamsters given NDEA with cobalt oxide in gelatin-saline were similar to those in animals receiving NDEA and gelatin-saline alone (Farrell & Davis, 1974).

(c) *Subcutaneous injection*

Rat: In a study designed to monitor cobalt-induced hyperlipidaemia, 20 male Wistar rats, about four weeks of age, received two courses, separated by a nine-day interval, of five daily subcutaneous injections of 40 mg/kg bw *cobalt[II] chloride* [purity unspecified] dissolved in saline, and were observed for 12 months. A control group of 20 males received injections of saline alone. At the end of the observation period, 8/11 surviving treated rats had developed subcutaneous fibrosarcomas (four of which were reported to be distant from the injection site), whereas none of the 19 surviving controls developed a tumour [$p < 0.001$, Fisher's exact test]. Post-mortem examinations were not made on the nine rats that died during the experiment. In a second experiment, 20 male Wistar rats received the same treatment but were observed for eight months. No control group was provided. At the end of this observation period, six of the 16 survivors had subcutaneous fibrosarcomas, including one tumour distant from the site of injection. Four rats that died during the observation period were not autopsied (Shabaan *et al.*, 1977).

Groups of 10 male Sprague-Dawley rats, 10 weeks of age, received subcutaneous injections of saline (two groups) or 2 mg/kg bw *cobalt[II] oxide* [purity unspecified] suspended in saline, five times a week, or subcutaneous injections of 10 mg/kg bw cobalt[II] oxide in saline once a week over a period of two years (total dose, 1000 mg/kg bw). Animals were allowed to live their natural lifespan or were sacrificed when moribund [survival data not given]. Malignant tumours (histiocytomas or sarcomas) developed at the injection site in 0/10, 0/10, 5/10 and 4/10 rats in the four groups, respectively (Steinhoff & Mohr, 1991).

(d) *Subcutaneous implantation*

Rat: Groups of five male and five female Wistar rats, four to six weeks of age, received subcutaneous implants of four pellets (approximately 2 mm in diameter) of either a *cobalt-chromium-molybdenum* (and lesser amounts of nickel) alloy (Vitallium; see p. 374 of this monograph), nickel metal, copper metal, nickel-gallium alloy (60% nickel, 40% gallium) or one of seven other implant materials not known to contain nickel, chromium or cobalt. Animals were observed for up to 27 months [survival of animals receiving cobalt-chromium-molybdenum alloy not given]. Sarcomas (mostly fibrosarcomas and rhabdomyosarcomas) developed around the implants in 5/10 rats that received nickel pellets and in 9/10 rats that received nickel-gallium alloy pellets; no sarcoma developed in rats that received the cobalt-chromium-molybdenum pellets or in any of the other groups (Mitchell et al., 1960).

(e) Intramuscular injection

Mouse: A group of 50 female Swiss mice, two to three months of age, received single intramuscular injections of 10 mg/site of unwashed powdered *cobalt[II] oxide* (particle size, ≤ 5 μm [purity unspecified]) in 10% aqueous penicillin G procaine in each thigh. Within two to six days, 25 mice had died. A further group of 25 females received similar injections of the powdered cobalt oxide that had been washed repeatedly in distilled water; this washed cobalt oxide did not induce acute mortality. The 25 survivors of the first group and the 25 mice from the second group were combined, and 46 were still alive 13 weeks after injection. A control group of 51 female mice similarly received intramuscular injections of penicillin G procaine vehicle (60 000 IU/site) into each thigh; 48 survived 13 weeks after injection. Animals were observed for up to 110 weeks [survival unspecified]. No tumour developed at the injection site in any of the cobalt oxide-treated or control mice. Incidences of tumours at other sites were similar in the treated and control groups (Gilman & Ruckerbauer, 1962).

A group of 30 mice [sex, strain and age unspecified] received intramuscular injections of 0.2 mg cobalt as *cobalt naphthenate* [purity, dosage, schedule, vehicle and duration unspecified] into the right hind limb. Tumours of the muscle in the hind leg developed in eight of the mice (Nowak, 1966). [The Working Group noted the incomplete reporting.]

Rat: A group of 10 male and 10 female hooded rats, two to three months old, received a single intramuscular injection of 28 mg *cobalt metal powder* (spectrographically pure, 400 mesh; 3.5 μm × 3.5 μm to 17 μm × 12 μm with large numbers of long narrow particles of the order of 10 μm × 4 μm) in 0.4 ml fowl serum into the thigh; a control group of ten males and ten females received fowl serum only. Average survival times were 71 weeks in treated males and 61 weeks in treated females; survival of controls was not specified. During the observation period of up to 122 weeks, 4/10 male and 5/10 female treated rats developed sarcomas (mostly rhabdomyosarcomas) at the injection site compared to 0/20 controls. A further group of ten female rats received a single intramuscular injection of 28 mg *cobalt metal powder* in 0.4 ml fowl serum; others received injections of 28 mg zinc powder (five rats) or 28 mg tungsten powder (five rats). Average survival time for cobalt-treated rats was 43 weeks. During the observation period of up to 105 weeks, sarcomas (mostly rhabdomyosarcomas) developed in 8/10 cobalt powder-treated rats; none occurred in the zinc powder- or tungsten powder-treated rats. No other tumour occurred in any of the cobalt-treated or other rats, except for one malignant lymphoma in a zinc-treated rat (Heath, 1954a, 1956).

In a supplementary study, a group of 30 male hooded rats, two to three months of age, received a single intramuscular injection of 28 mg *cobalt metal powder* (spectrographically pure [particle size unspecified]) in 0.4 ml fowl serum into the right thigh; a control group of 15 males received a single injection of fowl serum only. The rats were killed at intervals of one to four weeks after injection or at fortnightly intervals up to 20 weeks after injection, when the first tumour appeared. The author described leukocyte infiltration, muscle fibre necrosis and regeneration and the development of a tumour nodule in one rat (Heath, 1960).

Groups of 10 male and female Wistar rats [sex ratio unspecified], two to three months old, received a single intramuscular injection of 30 mg/site of powdered, reagent-grade *cobalt[II] oxide* (particles ground to ≤ 5 µm and washed repeatedly in distilled water) suspended in 10% aqueous penicillin G procaine or penicillin G procaine (90 000 IU/site) alone into the thigh muscle and were observed for 74 weeks [number of survivors unspecified]. No tumour occurred at the site of injection in the 10 control rats during the study, whereas rhabdomyosarcomas developed at the injection site in 5/10 cobalt oxide-treated rats. Metastases were seen in four of the five tumour-bearing rats. No other neoplasm was noted in control or treated rats (Gilman & Ruckerbauer, 1962).

A group of 30 male and female Wistar rats [sex ratio unspecified], two to three months of age, received simultaneous intramuscular injections of 20 mg/site of powdered *cobalt[II] sulfide* [purity unspecified] (ground to ≤ 5 µm diameter and washed repeatedly in water) suspended in penicillin G procaine into each thigh. A total of 35 sarcomas were observed at the 58 injection sites in the 29 rats that survived 13 weeks after treatment, with a mean latency of 28 weeks. Metastases were noted in 16/29 rats with tumours; no other neoplasm was seen. No control was reported (Gilman, 1962).

Groups of male and female Wistar rats [sex ratio unspecified], two to three months of age, received two simultaneous intramuscular injections (five rats) in each thigh or single injections (19 rats) of *cobalt[II] oxide* (20 mg/site; particle size ≤ 5 µm; washed repeatedly in water) suspended in aqueous procaine G penicillin. No control group was reported. A total of 13 sarcomas (mostly rhabdomyosarcomas) were noted at the 29 injection sites of the 24 rats that survived 13 weeks of treatment (mean latency, 25 weeks). Metastases were noted in 3/12 rats with tumours (Gilman, 1962).

In a series of three experiments, a total of 80 female hooded rats, seven to nine weeks of age, received an intramuscular injection of 28 mg/rat of wear particles, obtained by working in Ringer's solution *in vitro* of artificial hip or knee prostheses made from *cobalt-chromium-molybdenum* alloy (66.5% cobalt, 26.0% chromium, 6.65% molybdenum, 1.12% manganese; particle diameter, down to 0.1 μm [mostly 0.1-1 μm]), in 0.4 ml horse serum and were observed for up to 29 months [survival not specified]. No control group was reported. Sarcomas developed at the injection site in 3/16, 4/14 and 16/50 rats in the three series, respectively. Approximately half of the tumours were rhabdomyosarcomas; the remainder were mostly fibrosarcomas (Heath *et al.*, 1971; Swanson *et al.*, 1973).

(f) *Intramuscular implantation*

Rat: As a follow-up to the studies by Heath and Swanson (see above), groups of female Wistar and hooded rats, weighing 190-310 and 175-220 g, respectively, received intramuscular implants of 28 mg of coarse (100-250 μm diameter; 51 Wistar rats) or fine (0.5-50 μm diameter, 85% 0.5-5 μm; 61 Wistar and 53 hooded rats) particles as a dry powder, obtained by grinding a *cobalt-chromium-molybdenum* alloy (68% cobalt, 28% chromium, 4% molybdenum), and were observed for life. A sham-operated control group of 50 female Wistar rats was available. Survival at two years was 11/51 rats receiving the coarse particles, 7/61 Wistar rats receiving the fine particles, 0/53 hooded rats receiving the fine particles and 5/50 Wistar controls. No tumour was noted at the implantation site of rats treated with either of the alloy particles or in sham-operated control animals (Meachim *et al.*, 1982).

Groups of 15 male and 15 female Sprague-Dawley rats, aged 20-30 days, received intramuscular implants of polished rods (1.6 mm diameter, 8 mm length) of one of three *alloys* (wrought Vitallium: 19-20% chromium, 14-16% tungsten, 9-11% nickel, < 0.15% carbon, < 2% [manganese], < 1% silicon, < 3% iron, balance cobalt; cast Vitallium: 27-30% chromium, 5-7% molybdenum, < 2.5% nickel, < 0.3% carbon, < 1% [manganese], < 1% silicon, < 0.75% iron, balance cobalt; MP$_{35}$N alloy: 19-21% chromium, 33-37% nickel, < 0.025% carbon, < 1% iron, < 0.15% manganese, 9.5-10.5% molybdenum, < 0.15% silicon, 0.65-1% titanium; balance cobalt) and were observed for up to two years [survival unspecified]. Groups of 15 male and 15 female untreated and sham-operated control animals were available. No benign or malignant tumour developed at the implant site in any of the groups receiving metal implants or in either control group. The incidences of malignant tumours at distant sites did not differ significantly among the treated and control groups (Gaechter *et al.*, 1977).

Guinea-pig: A group of 46 female Dunkin-Hartley guinea-pigs, weighing 550-930 g, received intramuscular implants of 28 mg of a powdered *cobalt-chromium-molybdenum* alloy (68% cobalt, 28% chromium, 4% molybdenum; particle diameter, 0.5-50 μm) and were observed for life; 12/46 animals were alive at three years. No control group was reported. No tumour was observed at the implantation site of any guinea-pig; nodular fibroblastic hyperplasia was observed at the implantation site in eight animals (Meachim *et al.*, 1982).

(g) Intra-osseous implantation

Rat: Groups of 10-17 male and 8-15 female Sprague-Dawley rats, 30-43 days of age, received implants of one of seven test materials containing *cobalt alloyed with chromium and nickel, molybdenum, tungsten and/or zirconium*, with traces of other elements (as small rods, 1.6 mm diameter and 4 mm length, powders or porous compacted wire), in the femoral bone and were observed for up to 30 months. Groups of 13 male and 13 female untreated and sham-operated controls were available. Average survival was longer than 22 months. Sarcomas at the implant site were observed in 1/18 rats (males and females given cobalt-based alloy powder containing 41% Co), 3/26 rats (males and females given MP$_{35}$N powder containing 33% Co) and 3/32 rats (males and females given porous compacted wire containing 51% Co). No tumour was observed in two groups of 25 rats given rods containing 69 or 47% cobalt, in two groups of 26 rats given rods containing 0.11 or 33% cobalt, in two groups of 25 and 26 untreated rats, or in a group of 26 sham-treated control rats (Memoli *et al.*, 1986).

(h) Intraperitoneal injection

Mouse: In a screening study based on the enhanced induction of lung tumours, groups of 10 male and 10 female strain A mice, six to eight weeks of age, received intraperitoneal injections of *cobalt[III] acetate* (> 97% pure) in saline three times per week for eight weeks (total doses, 95, 237 and 475 mg/kg bw). After 30 weeks, lung tumours were found in 8/20, 8/20 and 10/17 mice in the respective treatment groups, and in 7/19 saline-treated controls (not significant) (Stoner *et al.*, 1976).

Rat: Groups of 10 male and 10 female Sprague-Dawley rats, 10 weeks of age, received three intraperitoneal injections at two-month intervals of saline or 200 mg/kg bw *cobalt[II] oxide* [purity unspecified] or *cobalt-aluminium-chromium spinel powder* (see above) in saline (total dose, 600 mg/kg bw). Animals were allowed to live their natural lifespan or were sacrificed when moribund [survival not given]. Malignant peritoneal tumours occurred in 1/20 controls (histiocytoma), 14/20 cobalt oxide-treated rats (10 histiocytomas, three sarcomas, one mesothelioma) and 2/20 spinel-treated animals (one histiocytoma, one sarcoma) (Steinhoff & Mohr, 1991).

(i) Intrarenal administration

Rat: Two groups of 20 and 18 female Sprague-Dawley rats, weighing 120-140 g, received a single injection of 5 mg *cobalt[II] sulfide* [reagent grade; purity and particle size unspecified] or 5 mg *metallic cobalt powder* [purity unspecified] suspended in 0.05 ml glycerine into each pole of the right kidney. A group of 16 female rats receiving injections of 0.05 ml glycerine into each pole of the kidney served as controls. After 12 months, all rats were necropsied; no tumour was observed in the kidneys of treated or control rats (Jasmin & Riopelle, 1976). [The Working Group noted the short duration and inadequate reporting of the experiment.]

(j) *Other*

Rat: Two groups of 10 female hooded rats, two to three months of age, received *intrathoracic injections* of 28 mg *cobalt metal powder* (spectrographically pure; particle size, < 400 mesh; 3.5 μm × 3.5 μm to 17 μm × 12 μm, with many long narrow particles of the order of 10 μm × 4 μm) in serum [species unspecified] through the right dome of the diaphragm (first group) or through the fourth left intercostal space (second group) and were observed for up to 28 months. Death occurred within three days of the treatment in 6/10 rats injected through the diaphragm and in 2/10 rats injected through the intercostal space. The remaining rats in the first group (diaphragm) survived 11-28 months and in the second group (intercostal space), 7.5-17.5 months. Of the 12 rats that survived the injection, four developed intrathoracic sarcomas (three of mixed origin, including rhabdomyosarcomatous elements, one rhabdomyosarcoma arising in the intercostal muscles) (Heath & Daniel, 1962).

Rabbit: Twelve male rabbits [strain unspecified], weighing 2-2.5 kg, were given intramuscular, intravenous, intrapleural or intrahepatic injections of *cobalt naphthenate* [purity and dose unspecified]. Within two to six months, tumours developed at the site of injection in eight rabbits, including one pleural mesothelioma, one haemangioendothelioma of the liver, one osteochondroma of the ear and five skeletal muscle tumours (Nowak, 1961). [The Working Group noted the lack of controls, the small number of animals and the incomplete reporting of the experiment.]

A summary of most of these studies is given in Table 20.

3.2 Other relevant data

The metabolism and toxicity of cobalt have been reviewed (Taylor & Marks, 1978; Elinder & Friberg, 1986). Recent interest has centred on the biological monitoring of cobalt, i.e., the determination of cobalt in human biological materials such as blood and urine, and how such data may be used to assess absorption, exposure and possible health risks (Alessio & Dell'Orto, 1988).

Table 20. Summary of animal carcinogenicity studies by form of cobalt

Reference	Species/strain	Sex	Dose schedule	Experimental parameter/observation	Group 0	1	2	3	Comments
Cobalt metal powder									
Heath (1954, 1956)	Rat Hooded	M	i.m., single inj., fowl serum	Dose (mg) Survival (122 weeks) Local sarcoma	0 Not given 0/10	28 4/10			
		F		Dose (mg) Survival (122 weeks) Local sarcoma	0 Not given 0/10	28 5/10	28 8/10		
Heath & Daniel (1962)	Rat Hooded	F	intrathoracic in serum	Dose (mg) Survival (3 days) Thoracic tumour	0	28 12/20 4/12			
Jasmin & Riopelle (1976)	Rat Sprague-Dawley	F	intrarenal	Dose (mg) Survival (12 months) Kidney tumour	0 Not given 0/16	5 0/18			Inadequate
Heath et al. (1971); Swanson et al. (1973)	Rat Hooded	F	i.m., single inj., wear particles from Co/Cr/Mo, in horse serum	Dose (mg) Survival (29 months) Local sarcoma	0 Not given	28 23/80			
Cobalt alloys									
Gaechter et al. (1977)	Rat Sprague-Dawley	M+F	i.m. impl. Co/Cr/W/Ni/C/Mn/Si/Fe (1.6 × 8 mm)	Dose (polished rod) Survival (2 years) Local tumour	0[a] Not given 0/30	0[a] 0/30	1 0/90		No significant difference in distant tumours
Memoli et al. (1986)	Rat Sprague-Dawley	M+F	intraoss. impl., Co/Cr/Ni/Mo/W/Zr	Dose (powder, wire, rod) Survival (30 months) Local sarcoma	0[a] Not given 0/51	0[a] 0/26	1 7/76[b]		
Mitchell et al. (1960)	Rat Wistar	M+F	s.c. impl. Co/Cr/Mo/Ni	Dose (pellets ~ 2-mm diam) Survival (27 months) Local tumour	Not given	0/10			
Meachim et al. (1982)	Rat Wistar and hooded	F	i.m. impl. Co/Cr/Mo fine and coarse particles	Dose (mg) Survival (2 years) Local tumour	0 5/50 0	28 11/51 0	28 7/61 0	28 0/53 0	

COBALT AND COBALT COMPOUNDS

Table 20 (contd)

Reference	Species/strain	Sex	Dose schedule	Experimental parameter/observation	Group 0	1	2	3	Comments
Cobalt alloys (contd)									
Steinhoff & Mohr (1991)	Rat Sprague-Dawley	M+F	3 i.p. inj., Co/Al/Cr spinel powder	Dose (mg/kg bw) Survival (2 years) Local tumour	0 Not given 1/20	200 2/20			
Steinhoff & Mohr (1991)	Rat Sprague-Dawley	M+F	Intratracheal inst. 1 × 2 weeks Co/Al/Cr spinel 2 years	Dose (mg/kg bw) Survival (2 years) Squamous-cell tumour of the lung	0 Not given 0/200	10 3/100			
Meachim et al. (1982)	Guinea-pig	F	i.m. impl. Co/Cr/Mo powder	Dose (mg) Survival (3 years) Local tumour Local fibroblastic hyperplasia		28 12/46 0/46 8/46			
Cobalt(II) oxide									
Gilman & Ruckerbauer (1962)	Mouse Swiss	F	i.m. inj. in each thigh	Dose (mg/site) Survival (13 weeks) Local sarcoma	0 48/51 0/48	10 46/75 0/46	10		
Steinhoff & Mohr (1991)	Rat Sprague-Dawley	M	Intratracheal inst. 1 × 2 weeks 2 years	Dose (mg/kg bw) Survival (2 years) Benign squamous pulmonary tumour Bronchioalveolar adenoma Pulmonary adenocarcinoma Bronchoalveolar adenocarcinoma	0 Not given 0/100 0/100 0/100 0/100	2 1/50 0/50 0/50 0/50	10 0/50 2/50 2/50 1/50		
		F		Dose (mg/kg bw) Survival Bronchoalveolar adenoma Bronchoalveolar carcinoma	0 Not given 0/100 0/100	2 1/50 0/50	10 0/50 1/50		

Table 20 (contd)

Reference	Species/strain	Sex	Dose schedule	Experimental parameter/observation	Group 0	Group 1	Group 2	Group 3	Comments
Cobalt[II] oxide (contd)									
Gilman & Ruckerbauer (1962)	Rat Wistar	M+F	i.m. inj.	Dose (mg/site) Survival (90 days) Local sarcoma	0 10/10 0/10	30 10/10 5/10			
Gilman (1962)	Rat Wistar	M+F	i.m. inj.	Dose (mg/site) Survival (13 weeks) Local sarcoma		20 24/32 13/29 sites			
Steinhoff & Mohr (1991)	Rat Sprague-Dawley	M	s.c. inj. 2 mg/kg bw 5/week or 10 mg/kg bw 1/week for 2 years	Dose (mg/kg bw) Survival (2 years) Local malignant tumour	0 Not given 0/20	2 5/10	10 4/10		
Steinhoff & Mohr (1991)	Rat Sprague-Dawley	M/F	3 i.p. inj. at 2-month intervals	Total dose (mg/kg bw) Survival (2 years) Local malignant tumour	0 Not given 1/20	200 14/20			
Wehner et al. (1977)	Hamster ENG:ELA	M	Inhalation 7 h/day, 5 d/week for life	Dose (mg/m³) Survival (18 months) Reticulum-cell sarcoma Carcinoma Lymphosarcoma Leukaemia Plasma-cell tumour	0 7/51 0/51 0/51 0/51 0/51 1/51	10 9/51 1/51 1/51 0/51 0/51 0/51			No statistical difference
Cobalt[II] sulfide									
Gilman (1962)	Rat Wistar	M+F	i.m. inj.	Dose (mg/site) Survival (13 weeks) Local sarcoma		20 29/30 35/58 sites			
Jasmin & Riopelle (1976)	Rat Sprague-Dawley	F	intrarenal	Dose (mg) Survival (12 months) Kidney tumours	0 Not given 0/16	5 0/20			Inadequate

Table 20 (contd)

Reference	Species/strain	Sex	Dose schedule	Experimental parameter/observation	Group 0	Group 1	Group 2	Group 3	Comments
Cobalt[II] chloride									
Shabaan et al. (1977)	Rat Wistar	M	s.c. inj. 2 x 5 d, 9–d interval	Dose (mg/kg bw) Survival[c] Subcutaneous sarcoma	0 19/20 0/19	40 11/20 8/11	40 16/20 6/16		$p < 0.001$ (Fisher exact test)
Cobalt naphthenate									
Nowak (1966)	Mouse NS	NS	i.m. inj. NS	Dose (mg) Survival Tumour of the striated muscle	0	0.2 8/30			Inadequate
Nowak (1961)	Rabbit	M	i.m. i.v. i. pleural i. hepatic	Dose unspecified	0	5 1 1 1			Inadequate
Cobalt[III] acetate									
Stoner et al. (1976)	Mouse Strain A	M+F	i.p. inj. 3/week, 24 doses	Total dose (mg/kg bw) Survival (30 weeks) Pulmonary tumour	0 19/20 7/19	95 20/20 8/20	237 20/20 8/20	475 17/20 10/17	Not significant

[a]Group 0, untreated; group 1, sham-treated
[b]Powder, 1/18 sarcoma; MP$_{35}$N, 3/26 sarcomas; compacted wire, 3/32 sarcomas
[c]12 months for groups 0 and 1; at 8 months for group 2
NS, not specified

(a) *Experimental systems*

(i) *Absorption, distribution, metabolism and excretion*

Cobalt compounds

The gastrointestinal absorption of radiolabelled cobalt chloride in rats was found to vary between 11 and 34%, depending on the administered dose (0.01-1000 μg/rat). The relative absorption decreased with increasing dose (Taylor, 1962). However, less than 0.5% of cobalt oxide given at an oral dose of 5 mg was absorbed by hamsters (Wehner & Craig, 1972).

The pulmonary absorption of inhaled cobalt(II) oxide (particle size, 1.0-2.5 μm) by hamsters was both rapid and high: about 25% was recovered in the carcass, lung, liver and kidney 24 h after inhalation of 0.8 mg cobalt oxide; essentially all of the cobalt oxide was eliminated by the sixth day after exposure (Wehner & Craig, 1972). Intratracheally instilled cobalt(II) oxide (1.5 μg) was cleared slowly from the rat lung (half-time, 15 days), and only very low concentrations were found in extrapulmonary tissues (Rhoads & Sanders, 1985). After inhalation or instillation of cobalt oxides in dogs and rats, the highest concentrations of cobalt were found in the lungs (Barnes *et al.*, 1976; Rhoads & Sanders, 1985). After rapid initial elimination (half-time, 0.7 days), the half-time of cobalt oxides deposited in the lungs of dogs was 36-86 days (Barnes *et al.*, 1976).

Kreyling *et al.* (1986) exposed beagle dogs by inhalation to radioactive cobalt[II,III] oxide particles of different size (0.3-2.7 μm) and found that small particles were cleared more rapidly from the lungs. Brune *et al.* (1980) exposed rats by inhalation to chromium-cobalt-containing abrasive dust obtained from dental laboratories. The concentration of cobalt in the lung increased with the length of exposure, indicating slow elimination of deposited metal. Histological examination revealed macrophages containing metal particles. The concentration of cobalt was also elevated in liver and kidney, showing that some systemic uptake of cobalt had taken place. Animals given cobalt chloride orally or by injection showed highest concentrations in the liver, with lower concentrations in kidney, pancreas and spleen (Taylor & Marks, 1978; Stenberg, 1983). Relatively high concentrations were also found in myocardium (Stenberg, 1983; Clyne *et al.*, 1988) and in cartilage and bone (Söremark *et al.*, 1979).

The major proportion of parenterally administered cobalt is cleared rapidly from the body, mainly *via* urine: 63% of radioactive cobalt chloride was recovered in the urine of rats within 24 h (Taylor, 1962). After a single intravenous injection of cobalt chloride to rats, about 70 and 7% were recovered in the urine and faeces, respectively, during the first three days (Onkelinx, 1976). Similarly, 73 and 15% of an intravenous dose of cobalt chloride (0.3 mg/kg bw) to rats was eliminated *via* urine and faeces, respectively, within four days (Gregus & Klaassen, 1986). Dogs injected intravenously with 20 µg/kg bw radioactive cobalt sulfate eliminated 40-70% of the label in urine and bile (90% in urine) over a period of 7-13 h (Lee & Wolterink, 1955). In rats, only 2-7% of intravenously injected cobalt chloride was eliminated in the bile (Cikrt & Tichy, 1981; Gregus & Klaassen, 1986).

Autoradiographic examination of pregnant mice injected intravenously with radioactive cobalt chloride revealed high activity in maternal liver, kidney, pancreas and cartilage and in the fetal skeleton and other tissues (Flodh, 1968; Söremark et al., 1979).

Metal alloy implants

In an experiment *in vitro* simulating mechanical stress on four different types of metallic hip prostheses, three of which contained cobalt, more than 1 mg/l cobalt was found in solution, and metal particles with a size down to 0.1 µm were formed as a result of frictional movement (Swanson et al., 1973).

(ii) *Toxic effects*

Cobalt compounds

The oral LD_{50}s for different inorganic cobalt(II) compounds (cobalt fluoride, oxide, phosphate, bromide, chloride, sulfate, nitrate and acetate) in rats ranged from 150 to 500 mg/kg bw anhydrous compound (Speijers et al., 1982). When the amounts were expressed in moles, the variability in toxicity between different compounds ranged from 1.5 to 3 mmol/kg cobalt. Acute effects recorded in the animals included sedation, diarrhoea and decrease in body temperature. All hamsters died after 6-h exposures by inhalation to 100 mg/m^3 cobalt oxide (Wehner & Craig, 1972). Pulmonary haemorrhagia and oedema and death were observed in guinea-pigs exposed by inhalation to cobalt chloride [dose unclear] (Höbel et al., 1972).

Life-time exposure of hamsters to cobalt oxide by inhalation (10 mg/m^3, 7 h per day, five days a week) resulted in emphysema and in hyperplastic and hypertrophic changes in the alveolar epithelium and distal bronchi (Wehner et al., 1977). Exposure of rabbits by inhalation to concentrations of 0.4 or 2 mg/m^3 cobalt chloride for 6 h per day on five days a week for 14-16 weeks produced nodular aggregation of alveolar type II cells, abnormal accumulation of enlarged, vacuolated alveolar macrophages and interstitial inflammation (Johansson et al., 1987).

Daily doses of 2.5-10 mg/kg bw cobalt(II) salts given orally or parenterally caused polycythaemia in rats (Orten & Bucciero, 1948; Hopps et al., 1954; Oskarsson et al., 1981); reduced weight gain was seen as an early sign of general toxicity in some of these studies. Parenteral administration of 10-60 mg/kg bw cobalt chloride caused hyperlipidaemia in rabbits (Caplan & Block, 1963), induction of hepatic haemoxygenase and a decrease in activity of δ-aminolaevulinic synthase and certain cytochrome P450-dependent drug metabolizing enzymes in rats (Maines & Kappas, 1975; Maines *et al.*, 1976; Numazawa *et al.*, 1989).

Myocardial toxicity of cobalt salts has been reported in rats (Grice *et al.*, 1969; Lin & Duffy, 1970; Rona, 1971), guinea-pigs (Mohiuddin *et al.*, 1970; Desselberger & Wegener, 1971), rabbits (Hall & Smith, 1968) and dogs (Sandusky *et al.*, 1981) following long-term dietary (10-100 mg/kg bw) or parenteral (5-30 mg/kg bw) administration. Observed toxic effects included noninflammatory myocardial degeneration, alterations in mitochondria and myofibrils and abnormal electrocardiographic traces.

Metallic cobalt

Intratracheal instillation of metallic cobalt (50 mg/animal; sterile suspension [particle size not given]) caused pulmonary haemorrhage and oedema and death in rats (Harding, 1950).

In miniature swine exposed to 0.1-1 mg/m^3 metallic cobalt particles (0.4-3.6 μm) for 6 h per day on five days per week for three months by inhalation, a progressive decrease in lung compliance was observed. In addition collagenization of alveolar septa in lung biopsies and electrocardiographic changes indicative of cardiomyopathy were observed (Kerfoot *et al.*, 1975).

In contrast to findings with cobalt chloride, exposure of rabbits by inhalation to metallic cobalt dust (0.2-1.3 mg/m^3, 6 h per day, five days per week for four weeks) had no profound effect on alveolar macrophages (Johansson *et al.*, 1980, 1986).

Cobalt released from cobalt metal, alloys or dissolved salts was cytotoxic to chick primary cultures and rodent fibroblast cell lines, inducing cell death, growth inhibition and mitotic abnormalities at concentrations greater than 7.5 μg/ml (Heath, 1954b; Daniel *et al.*, 1963; Bearden, 1976; Bearden & Cooke, 1980; Takahashi & Koshi, 1981).

(iii) *Effects on reproduction and prenatal toxicity*

Reproductive effects: Ingested cobalt chloride (265 mg/kg diet for 98 days, providing an initial dose of 20 mg/kg bw cobalt) induced degenerative and necrotic changes in the seminiferous tubules of rats. Cyanosis and vascular engorgement of the testes were seen on day 35 of treatment, and necrosis, degenerative and necrotic changes in the germinal epithelium and Sertoli cells by day 70. Damaged tubules were present side by side with normal ones. Multinucleated giant cells containing cellular debris were observed in the damaged tubules. Loss of sperm-tail filaments and degeneration of sperm mitochondria were also observed (Corrier *et al.* 1985a; Mollenhauer *et al.*, 1985). The same group of investigators did not find the lesion in sheep treated with 3.0-15.0 mg/kg bw cobalt for 109 days (Corrier *et al.* 1985b).

Intraperitoneal injection of cobalt chloride (1 mg/kg bw cobalt) 16 and 6 days before sacrifice stimulated spermiogenesis and spermatogenesis in the mouse testis (Niebrój, 1967). Intraperitoneal administration of 200 μmol [47.6 mg]/kg bw cobalt chloride for three days to male mice resulted in small but significant decreases in fertility two to three weeks later in an acute study. Similarly, in a chronic study, 100, 200 and 400 mg/l cobalt chloride given in drinking-water *ad libitum* for 7-13 weeks decreased fertility, sperm concentration, sperm mobility and testicular weight in a time-dose-dependent manner (Pedigo *et al.*, 1988). [The Working Group noted that the apparent differences in the results described above may be due to differences in dose and duration of observation.]

Developmental toxicity. Embryonic death was reported following administration to rats of cobalt chloride in the drinking-water either before and during pregnancy (0.05-5 mg/l) or during pregnancy only (0.005-0.05 mg/l) (Nadeenko *et al.*, 1980). In contrast, no developmental toxicity was observed in the offspring of rats given daily doses of 0, 25, 50 or 100 mg/kg bw cobalt chloride by gavage on days 6-15 of gestation, except for a nonsignificant increase in the incidence of stunted fetuses in the groups given 50 and 100 mg/kg (Paternain *et al.* 1988).

Numbers of litters as well as growth and survival of the offspring were reduced in rats that received 12, 24 and 48 mg/kg bw per day cobalt chloride by gavage from day 14 of gestation through day 21 of lactation (Domingo *et al.*, 1985).

Delay in ossification of the skeleton during embryonic and fetal development was observed at gestation day 17 in the offspring of six- to eight-week-old female mice (24-26 g) administered cobalt chloride (0.1 ml of a 5 mM solution [4.8 mg/kg bw]) intravenously on day 8; the effect was not seen when the cobalt was administered on day 3 of pregnancy. There was no change in fetal body weight on day 17 of pregnancy, and no increase in the frequency of resorption or implantation sites compared with controls (Wide, 1984).

In CF-1 mice, cobalt chloride was reported to protect against cleft lip and palate induced by cortisone (Kasirsky *et al.*, 1967).

As reported in an abstract, fetal damage was detected on gestation day 15 in hamsters administered cobalt acetate (40, 60, 80, 100 or 160 mg/kg bw) subcutaneously on day 8 of pregnancy. The resorption rate ranged from 6% at the low dose to 100% at the high dose. Central nervous system defects were reported at the median doses. Similarly, resorptions and central nervous system defects were observed after intraperitoneal injections of 40-70 mg/kg (Gale, 1980). [The Working Group noted that no information on maternal toxicity was reported.]

Studies on the effects of cobalt salts on chick embryos have produced conflicting results, perhaps due to differences in dose and routes of administration. Degeneration of the brain (Ridgway & Karnofsky, 1952), neural tube malformations (Adhikari, 1967), lethality, eye abnormalities and structural defects (Kury & Crosby, 1968; Gilani & Alibai, 1985, abstract) have been reported.

(iv) *Genetic and related effects*

The results of tests for genetic and related effects of cobalt and cobalt compounds, with references, are given in Table 21. Other studies are described in the text.

The genetic toxicology of cobalt and cobalt compounds has been reviewed (Léonard & Lauwerys, 1990). With few exceptions, only soluble cobalt[II] salts have been tested. Only two reports were available on genetic effects of insoluble cobalt sulfide, and no data have been reported on genetic effects of metallic cobalt.

Like other metallic compounds, cobalt compounds are known to be relatively inactive in prokaryotic systems (Rossman, 1981; Swierenga *et al.*, 1987). The precipitation of metal as phosphates in bacterial culture media may contribute to this inactivity (Rossman, 1981; Arlauskas *et al.*, 1985). However, four of 15 cobalt[III] complexes with aromatic ligands were active in a DNA repair assay and were mutagenic to *Salmonella typhimurium* (Schultz *et al.*, 1982). Several other studies of cobalt salts with positive results have been reported in prokaryotes.

Cobalt[II] chloride was inactive in the λ prophage induction assay, and it gave conflicting results in the *Bacillus subtilis rec*$^{+/-}$ growth inhibition assay. In the study with positive results, a preincubation procedure was used. Cobalt[II] chloride was inactive in all but one bacterial mutagenicity test. One study gave positive results in the absence but not in the presence of an exogenous metabolic system.

In bacteria, cobalt[II] chloride was reported to reduce the incidence of spontaneous mutations and to inhibit mutations induced by N-methyl-N'-nitro-N-nitrosoguanidine and Trp-P-1 (Kada & Kanematsu, 1978; Inoue *et al.*, 1981; Mochizuki & Kada, 1982). It was comutagenic with several heteroaromatic compounds (Ogawa *et al.*, 1986, 1987, 1988).

In *Saccharomyces cerevisiae*, cobalt[II] chloride induced gene conversion and mitochondrial but not other types of mutation. Cobalt[II] salts induced chlorophyll mutations, chromosomal aberrations and aneuploidy in plant cells.

In cultured mammalian cells *in vitro*, predominantly positive results were obtained, with induction of DNA-protein cross-linkage, DNA strand breakage and sister chromatid exchange. Chromosomal aberrations were not observed in cultured human cells. [The Working Group noted the low concentrations employed.] Cobalt[II] chloride induced aneuploidy in cultured human lymphocytes. It also induced mutations at the *hprt* locus in Chinese hamster V79 cells, but not, in a single study, at the *tk* locus in mouse lymphoma L5178Y cells.

Cobalt[II] acetate enhanced viral transformation in Syrian hamster embryo cells, and cobalt sulfide induced morphological transformation in Syrian hamster embryo cells; the crystalline form of cobalt sulfide was more active than the amorphous form.

Cobalt[II] chloride administered *in vivo* to Syrian hamsters by intraperitoneal injection induced aneuploidy in bone marrow and testes. In an assay for dominant lethal mutation in mice, reported as an abstract, significant increases in early embryonic losses were observed (Pedigo, 1988).

A mechanism for the genetic effects of soluble Co[II] salts may involve decreased fidelity of DNA polymerase (Sirover & Loeb, 1976). Cobalt[II] chloride caused extensive cleavage of isolated DNA in the presence of hydrogen peroxide; this effect was attributed to the generation of reactive oxygen species at those sites of DNA bound to cobalt ions (Yamamoto *et al.*, 1989).

(*b*) *Humans*

(i) *Absorption, distribution, excretion and metabolism*

The normal concentrations of cobalt in blood and urine from non-occupationally exposed persons are about 0.1-2 µg/l. The levels of cobalt in blood, and particularly in urine, increase in proportion to the level of occupational exposure and can be used for biological monitoring in order to assess individual exposure (Elinder *et al.*, 1988). Increased levels of cobalt have also been found in blood (serum) from uraemic patients (Curtis *et al.*, 1976; Lins & Pehrsson, 1984).

In a patient who died three months after treatment with cobalt[II] chloride (50 mg per day for three months), the myocardial concentration of cobalt was 1.65 mg/kg wet weight, which was 25-80 times higher than that in control samples (0.01-0.06 mg/kg) (Curtis *et al.*, 1976). Increased levels of cobalt were also reported in lung and mediastinal lymph nodes from hard-metal workers with lung disease; concentrations of cobalt were about 100-1000 µg/kg in two lung tissue samples compared to 5 µg/kg wet weight in controls, and 3280 µg/kg in mediastinal lymph nodes compared to > 2 µg/kg in controls (Hillerdal & Hartung, 1983).

The mean urinary excretion within 24 h of radioactive cobalt chloride given orally at 20 µM was estimated to be about 18% (Sorbie *et al.*, 1971). When healthy persons and uraemic patients were given 50 mg cobalt chloride orally, the two healthy volunteers eliminated between 5.7 and 8.3% of the dose *via* the urine within one week; elimination was considerably slower in uraemic patients, confirming the importance of renal clearance (Curtis *et al.*, 1976). High concentrations of radiolabelled cobalt were found in the liver shortly after parenteral administration of cobalt chloride to humans. After eight days, 28-56% and 2-12% of the dose were eliminated *via* the urine and faeces, respectively. A significant component (9-16% of the administered dose) was cleared very slowly, with a biological half-time of about two years (Smith *et al.*, 1972). Similar results, suggesting that a small

Table 21. Summary of studies on genetic and related effects of cobalt

Test system	Result		Dose LED/HID	Reference
	Without exogenous metabolic system	With exogenous metabolic system		
Cobalt(II) salts				
PRB, Prophage induction in *Escherichia coli*	–	0	4.0000	Rossman *et al.* (1984)
BSD, *Bacillus subtilis* rec strains H17/M45, growth inhibition	–	0	325.0000	Nishioka (1975)
BSD, *Bacillus subtilis* rec strain H17, growth inhibition	+	0	325.0000	Kanematsu *et al.* (1980)
BSD, *Bacillus subtilis* rec strain H17, growth inhibition	(+)	0	325.0000	Kanematsu *et al.* (1980)
BSD, *Bacillus subtilis* rec strain H17, growth inhibition	(+)	0	325.0000	Kanematsu *et al.* (1980)
???, *Bacillus subtilis* strain NIG 1125, reverse mutation	–[a]	0	0.0000	Inoue *et al.* (1981)
SAO, *Salmonella typhimurium* TA100, reverse mutation	–	0	130.0000	Tso & Fung (1981)
SAO, *Salmonella typhimurium* TA100, reverse mutation	–	0	0.0000	Arlauskas *et al.* (1985)
SAO, *Salmonella typhimurium* TA100, reverse mutation	–	0	0.0000	Ogawa *et al.* (1986)
SA2, *Salmonella typhimurium* TA102, reverse mutation	–	–	40.0000	Wong (1988)
SA5, *Salmonella typhimurium* TA1535, reverse mutation	–	0	0.0000	Arlauskas *et al.* (1985)
SA5, *Salmonella typhimurium* TA1535, reverse mutation	–	–	40.0000	Wong (1988)
SA7, *Salmonella typhimurium* TA1537, reverse mutation	–	0	0.0000	Arlauskas *et al.* (1985)
SA7, *Salmonella typhimurium* TA1537, reverse mutation	–	0	65000.0000	Ogawa *et al.* (1986)
SA7, *Salmonella typhimurium* TA1537, reverse mutation	+	–	0.0000	Wong (1988)
SA8, *Salmonella typhimurium* TA1538, reverse mutation	–	0	20.0000	Mochizuki & Kada (1982)
SA8, *Salmonella typhimurium* TA1538, reverse mutation	–	0	0.0000	Arlauskas *et al.* (1985)
SA9, *Salmonella typhimurium* TA98, reverse mutation	–[a]	0	20.0000	Mochizuki & Kada (1982)
SA9, *Salmonella typhimurium* TA98, reverse mutation	–	0	0.0000	Arlauskas *et al.* (1985)
SA9, *Salmonella typhimurium* TA98, reverse mutation	–	0	0.0000	Ogawa *et al.* (1986)
SA9, *Salmonella typhimurium* TA98, reverse mutation	+	–	0.0000	Wong (1988)
SAS, *Salmonella typhimurium* TA2637, reverse mutation	–	0	65000.0000	Ogawa *et al.* (1986)
ECW, *Escherichia coli* WP2 *uvrA*, reverse mutation	–[a]	0	0.0000	Arlauskas *et al.* (1985)
EC2, *Escherichia coli* WP2, reverse mutation	–[a]	0	20.0000	Kada & Kanematsu (1978)
SCG, *Saccharomyces cerevisiae* D7, gene conversion	+	0	1300.0000	Fukunaga *et al.* (1982)

Table 21 (contd)

Test system	Result		Dose LED/HID	Reference
	Without exogenous metabolic system	With exogenous metabolic system		
Cobalt(II) salts (contd)				
SCG, *Saccharomyces cerevisiae* D7, gene conversion	(+)	0	0.0000	Singh (1983)
SCG, *Saccharomyces cerevisiae* D7, gene conversion	+	0	1500.0000	Kharab & Singh (1985)
SCF, *Saccharomyces cerevisiae*, petite mutation	+	0	130.0000	Lindegren et al. (1958)
SCF, *Saccharomyces cerevisiae* SBTD-2B, petite mutation	+	0	260.0000	Prazmo et al. (1975)
SCF, *Saccharomyces cerevisiae*, petite mutation	(+)	0	640.0000	Egilsson et al. (1979)
SCF, *Saccharomyces cerevisiae* D7, petite mutation	+	0	750.0000	Kharab & Singh (1987)
SCR, *Saccharomyces cerevisiae* S/M 13–D, erythromycin-resistant mut.	−	0	1300.0000	Putrament et al. (1977)
SCR, *Saccharomyces cerevisiae* D7, *ilv* gene mutation	−	0	1300.0000	Fukunaga et al. (1982)
SCR, *Saccharomyces cerevisiae* D7, *ilv* gene mutation	−	0	0.0000	Singh (1983)
SCR, *Saccharomyces cerevisiae* D7, *ilv* gene mutation	(+)	0	3000.0000	Kharab & Singh (1985)
PLM, *Pisum abyssinicum*, chlorophyll mutation	+	0	0.0000	von Rosen (1964)[b]
ACC, *Allium cepa*, chromosomal aberration	+	0	3.0000	Gori & Zucconi (1957)
???, *Allium cepa*, aneuploidy	+	0	0.0000	Gori & Zucconi (1957)
DIA, DNA strand breaks, Chinese hamster CHO cells	+	0	260.0000	Hamilton-Koch et al. (1986)
DIA, DNA cross-links, Novikoff hepatoma cells	(+)	0	130.0000	Wedrychowski et al. (1986)
G9H, Gene mutation, Chinese hamster V79 cells, *hprt* locus	(+)	0	26.0000	Miyaki et al. (1979)
G9H, Gene mutation, Chinese hamster V79 cells, *hprt* locus	+	0	0.0000	Hartwig et al. (1990)
G5T, Gene mutation. mouse lymphoma L5178Y cells, *tk* locus	−	0	57.0000	Amacher & Paillet (1980)
SIM, Sister chromatid exchange, mouse macrophage P388D1 cell line	+	0	13.0000	Andersen (1983)
T7S, Cell transformation, SA7/Syrian hamster embryo cells	+	0	35.0000	Casto et al. (1979)
T7S, Cell transformation, SA7/Syrian hamster embryo cells	+	0	55.0000	Casto et al. (1979)
DIH, DNA strand breaks, human white blood cells	+	0	6.5000	McLean et al. (1982)
DIH, DNA strand breaks, human diploid fibroblasts	+	0	650.0000	Hamilton-Koch et al. (1986)
DIH, DNA strand breaks, HeLa cells	+	0	0.0000	Hartwig et al. (1990)
SHL, sister chromatid exchanges, human lymphocytes	+	0	1.3000	Andersen (1983)

Table 21 (contd)

Test system	Result		Dose LED/HID	Reference
	Without exogenous metabolic system	With exogenous metabolic system		
Cobalt(II) salts (contd)				
CHF, Chromosomal aberrations, human fibroblasts	–	0	0.0150	Paton & Allison (1972)
CHL, Chromosomal aberrations, human lymphocytes	–	0	0.6000	Voroshilin et al. (1978)
CIH, Chromosomal aberrations, human leukocytes	–	0	0.1500	Paton & Allison (1972)
AIH, Aneuploidy, human lymphocytes	+	0	3.7000	Resende de Souza–Nazareth (1976)
AVA, Aneuploidy, bone marrow and testes of male hamsters	+	0	400.0000	Farah (1983)[c]
Cobalt sulfides				
DIA, DNA strand breaks, Chinese hamster CHO cells	+	0	10.0000	Robison et al. (1982)
TCS, Cell transformation, Syrian hamster embryo cells	+	0	5.0000	Costa et al. (1982)
TCS, Cell transformation, Syrian hamster embryo cells	(+)	0	10.0000	Costa et al. (1982)
Cobalt(III) salts				
BSD, *Bacillus subtilis* rec strain H17, growth inhibition	(+)	0	1375.0000	Kanematsu et al. (1980)

[a]Antimutagenic effect
[b]Or as EDTA chelate
[c]Injected intraperitoneally over nine days

proportion of the cobalt (from either the metal or the oxide) retained after inhalation has a biological half-time in the order of years, were obtained by other investigators (Newton & Rundo, 1970; Hedge et al., 1979).

Measurements of cobalt, chromium and nickel in blood and urine from persons with metallic hip replacements containing a high proportion of these metals have repeatedly shown elevated levels of one or several of them compared to controls or prior to surgery (Coleman et al., 1973; Jones et al., 1975; Hildebrand et al., 1985; Braun et al., 1986; Hildebrand et al., 1988). [The Working Group noted that the analytical accuracy of several of the earlier studies was not confirmed.]

Sunderman et al. (1989) measured the concentrations of chromium, cobalt and nickel in serum and urine samples collected from patients who had undergone bone surgery and had received metallic hip or knee prostheses. Patients were followed for up to two years. The concentration of chromium in serum and urine remained essentially unchanged, whereas the concentration of nickel was markedly increased in both urine and serum collected shortly after the operation (1-14 days). The cobalt concentration, however, displayed a relatively small, slow increase in serum and blood. The highest concentrations were seen after two and 22 months in two patients who had loosening of their prosthesis.

(ii) *Toxic effects*

Pulmonary effects have been regarded as the major occupational problem in relation to cobalt, particularly in the hard-metal industry where cobalt-containing dust is generated. Two types of lung lesions may develop—interstitial fibrosis (so-called 'hard-metal pneumoconiosis') and occupational asthma (Demedts & Ceuppens, 1989). Hard-metal pneumoconiosis is a severe and progressive type of pneumoconiosis which may develop after several years of exposure to cobalt-containing dust at concentrations of 0.1-2 mg/m^3 (for reviews, see Elinder & Friberg, 1986; Sprince et al., 1988). As the dust in the hard-metal industry always contains agents in combination with cobalt (tungsten carbide and sometimes other metals such as titanium and tantalum), it has been questioned whether cobalt is solely responsible for the observed health effects (Brooks, 1981). Diamond polishers exposed to fine dust containing cobalt and diamond had severe lung fibrosis (Demedts et al., 1984).

Symptoms and signs of obstructive lung disease can develop as a result of occupational exposure to cobalt-containing dust during the production of hard metal (Coates & Watson, 1971, 1973; Bech, 1974; Scherrer & Maillard, 1982), but these were also observed in workers in a porcelain factory using cobalt dye (Raffn et al., 1988) and among diamond polishers (Gheysens et al., 1985). This condition, which usually improves after cessation of exposure, is considered to be of allergic origin (Sjögren et al., 1980). Provocation tests with cobalt usually induce a typical asthmatic reaction (Hartmann et al., 1982). Shirakawa et al. (1989) examined eight workers who developed asthma after having worked in a Japanese hard-metal plant. The total number of workers was about 400. The eight asthmatic workers all reacted with a drop in peak expiratory flow rate after an inhalation challenge with cobalt chloride. In four of them, it was possible to identify specific IgE antibodies towards cobalt-conjugated human albumin. This finding supports the hypothesis that cobalt hypersensitivity has a role in hard-metal asthma.

Histopathological findings in lung biopsies from workers with fibrosis (hard-metal pneumoconiosis) and/or obstructive problems (hard-metal asthma) have been published (Coates & Watson, 1971, 1973; Davison et al., 1983; Demedts et al., 1984; Anttila et al., 1986; Cugell et al., 1990). Typical microscopic findings include advanced fibrosis and desquamative interstitial pneumonia of the giant-cell type (Coates & Watson, 1971; Anttila et al., 1986).

Cobalt has an erythropoietic effect and has been used for the treatment of anaemia (Berk et al., 1949; Duckham & Lee, 1976). Berk et al. (1949) gave patients about 100 mg cobalt in the form of cobalt chloride three times a day for several weeks and recorded vomiting and anorexia in some patients, but only mild symptoms in the alimentary tract were seen as side-effects of the treatment in others. Duckham and Lee (1976) used a lower dose of cobalt chloride (25-50 mg cobalt per day) and observed fewer side effects. Polycythaemia has also been reported in heavy drinkers of cobalt-fortified beer (Morin et al., 1971; Alexander, 1972).

Endemic outbreaks of cardiomyopathy with mortality rates of up to 50% were described among heavy consumers (up to 10 l per day) of cobalt-fortified beer (Morin & Daniel, 1967; Kesteloot et al., 1968; Morin et al., 1971; Alexander, 1972). As the daily dose of cobalt ingested by heavy beer drinkers (a few milligrams) was certainly excessive compared to the normal daily intake of cobalt (around 5-50 µg/day), but considerably lower than the doses prescribed to patients with anaemia, it was suggested that the cardiomyopathy had a multicausal origin (Morin & Daniel, 1967; Balazs & Herman, 1976). Three cases of cardiomyopathy, two of which were fatal, were described in workers exposed industrially to cobalt (Barbořík & Dusek, 1972; Kennedy et al., 1981; Alušík et al., 1982).

There are some indications that workers in hard-metal plants have increased morbidity and mortality from cardiovascular disease. Alexandersson and Atterhög (1980) examined workers exposed to cobalt-containing dusts at concentrations of 0.01-0.06 mg/m^3. Symptoms of dyspnoea, 'heavy breathing' and 'tightness in chest' were more prevalent in exposed workers than in controls, but no pulmonary dysfunction was found. In a recent study of 3163 workers exposed to cobalt-containing dusts at concentrations ranging from 0.001 to up to 11 mg/m3 for at least one year, Hogstedt and Alexandersson (1990) found an excess of deaths from ischaemic heart disease (standardized mortality ratio (SMR), 169; 95% confidence interval (CI), 96-275) among workers who had been exposed to 0.02-11 mg/m^3 cobalt for at least 10 years (see also p. 445).

Cobalt may provoke allergic dermatitis (Camarasa, 1967). Of 853 patch-tested workers, about 7% showed allergic reactions to 1% cobalt chloride (Fischer & Rystedt, 1983). Cobalt allergy, which is usually found in people who suffer from other skin allergies and/or eczema (Rystedt & Fischer, 1983), is also seen in other occupational groups, such as offset printers and construction workers handling cobalt-containing cement (Goh *et al.*, 1986).

Cobalt and nickel released from orthopaedic or dental prostheses may precipitate allergic reactions, with local effects and inflammation (Jones *et al.*, 1975; Fernandez *et al.*, 1986; Thomas *et al.*, 1987).

(iii) *Effects on reproduction and prenatal toxicity*

The spontaneous abortion rate appeared to be increased in women who either worked in metal smelting or had spouses working in the metallurgical industry. Exposure to cobalt, arsenic, copper, zinc and sulfur was considered possible in the work setting (Hemminki *et al.*, 1983). [The Working Group noted that the contribution of cobalt, if any, to the increase in abortion rate was not separately identified.]

(iv) *Genetic and related effects*

No data were available to the Working Group.

3.3. Case reports and epidemiological studies of carcinogenicity in humans

(a) *Implanted medical devices*

The first report of development of a sarcoma at the site of a stainless-steel plate prosthesis for a fracture of the humerus was made in 1956 (McDougall, 1956). There have been 17 further reports of single cases of malignant neoplasia at the site of implants of metal-containing fracture plates or joint prostheses. The metal material used was unknown in four cases, stainless-steel in three cases and cobalt-containing alloys in 10 cases. The period between implantation and tumour development ranged from one to 30 years. The tumours described were various types of sarcoma in 14 cases (Delgado, 1958; Castleman & McNeely, 1965; Dube & Fisher, 1972; Arden & Bywaters, 1978; Tayton, 1980; Bagó-Granell *et al.*, 1984; Lee *et al.*, 1984; Penman & Ring, 1984; Swann, 1984; Weber, 1986; Hughes *et al.*, 1987; Ryu *et al.*, 1987; Martin *et al.*, 1988; Ward *et al.*, 1990), one carcinoma (Mazabraud *et al.*, 1989) and lymphoma in two cases (McDonald, 1981; Dodion *et al.*, 1983).

Incident cancers were recorded for a cohort of 1358 persons who received a total hip replacement in New Zealand in the period 1966-73 and were followed up for six months to 17 years (mean, 10.5 years) to the end of 1983 (Gillespie *et al.*, 1988). Total cancer incidence was similar to that expected (164 observed *versus* 179.4 expected on the basis of general population rates; SMR, 91 [95% CI, 78-107]). While the overall cancer risk within 10 years of hip replacement was significantly low (SMR, 74; 95% CI, 61-90, based on 107 observed cases), the risk after 10 or more years was significantly high (SMR, 160; 95% CI, 122-209, based on 57 cases). There was a significant overall increase in the incidence of tumours of the lymphatic and haematopoietic system (21 observed *versus* 12.5 expected; SMR, 168; 95% CI, 106-260). When the five lymphatic and haematopoietic malignancies diagnosed within two years of hip replacement were excluded, this SMR fell to 151 (16 *versus* 10.6 expected [95% CI, 86-245]). There were significant deficits of breast cancer (six observed *versus* 16.6 expected; SMR, 36; 95% CI, 14-82) and of colorectal cancer (21 observed *versus* 33.8 expected; SMR, 62; 95% CI, 39-96). [No specific information on the composition of the hip prostheses was provided.]

(b) *Occupational exposure*

Schulz (1978) reported a cobalt-containing giant-cell tumour of the buccal membrane in a mineral-oil refinery employee five months after a single accidental exposure to dust containing cobalt[II] phthalocyanine.

Saknyn and Shabynina (1970, 1973) examined mortality rates among workers at four nickel plants in the USSR in 1955-67. The workers were exposed to cobalt, but also to various nickel and arsenic compounds. A two- to four-fold increase in the risk for lung cancer was reported. The risks relative to those of inhabitants in the towns in which the plants were located were increased in various parts of the plants, including the cobalt shops (relative risks, 5-13), where there was exposure to cobalt dust but also to nickel sulfates, nickel chlorides and arsenic compounds. A 1.5-3.3-fold increase in stomach cancer risk was also noted. [The observed numbers of deaths were not given, and no allowance was made for potential confounding factors.]

Cuckle *et al.* (1980) studied mortality in 297 men employed in two departments opened in 1937 and 1938 at a nickel refinery in the UK. In one department, a wet treatment plant, nickel sulfate, copper sulfate, 'cobaltic hydrate' and precious metal concentrates were manufactured; in the other, a chemical production department, a range of compounds of nickel, cobalt and selenium were produced. The men had all been first employed in the refinery in or after 1933 and had worked in one or other of the departments for at least 12 months before 1960. They were followed up to 30 June 1980. Overall, there were 105 deaths (SMR, 109 [95% CI, 89-132]). There were 13 deaths from lung cancer (SMR, 131 [95% CI, 70-224]); six of the men who died from lung cancer [SMR, 154; 95% CI, 57-336] had been employed in the precious metal concentration section of the wet treatment plant. When the expected number of lung cancer deaths was estimated from death rates in rural districts of Glamorganshire (where the refinery was located), rather than in the population of England and Wales as a whole, the SMR was [172; 95% CI, 92-295]. Excess mortality from lung cancer occurred mainly less than 20 years from first employment in the refinery (SMR, 178 [95% CI, 65-387]) and among men who had been employed for six or more years (SMR, 138 [95% CI, 55-283]). Among the 1173 workers employed in the whole refinery in or after 1930 (International Committee on Nickel Carcinogenesis in Man, 1990), those first employed in 1930-39 had a SMR for lung cancer of 154 (95% CI, 97-233), those first employed in 1940-49 a SMR of 130 (95% CI, 71-218) and those first employed after 1950 a SMR of 77 (95% CI, 33-152). Cuckle *et al.* (1980) did not attribute any increase in risk in this cohort to exposure to cobalt.

Mur *et al.* (1987) followed up 1143 workers with at least one year of employment betwen 1950 and 1980 in an electrochemical plant producing cobalt and sodium in France. Altogether, 24.9% of the cohort were migrants (mainly North Africans and Italians). Vital status was established for 99.5% of the French-born workers and for 81.3% of the migrants. A total of 213 deaths occurring before 1981 was identified; cause of death was determined for 80% by interview with attending physicians and from hospital records. After adjustment for unknown causes of death (assuming that the distribution by cause of death was similar to that of cases with known cause of death), a SMR of 90 (95% CI, 44-159, based on nine cases) was observed for the total cohort for cancer of the lung, using mortality rates for France as a reference. For workers employed only in cobalt production, the SMR for lung cancer was 466 (95% CI, 146-1064, based on four observed cases). [The migrants may have had different rates of lung cancer from French-born workers, but the proportion of migrant workers in the different departments of the plants was not reported.] A case-control analysis was performed of lung cancer cases and controls, matched for year of birth, year of death and smoking habits. [The quality and manner of collection of information on smoking is unclear.] An odds ratio of 4.0 [95% CI, 1.6-9.9; calculated by the Working Group using an unmatched analysis] was associated with ever having worked in cobalt production. Workers in cobalt production were also exposed to unknown levels of forms of nickel and arsenic. [It is not known whether workers in other areas also had such exposure. No analysis based on latency or duration of exposure was presented.]

Hogstedt and Alexandersson (1990) reported on 3163 male Swedish workers with at least one year of exposure to cobalt-containing hard-metal dust at one of three hard-metal manufacturing plants in 1940-82 who were followed up during the period 1951-82. There were four categories of exposure (with estimated ambient air concentrations prior to 1970): occasionally present in rooms where hard metal was handled (less than 2 µg/m^3 Co); continuously present in rooms where hard metal was handled, but own work not involving hard metal (1-5 µg/m³ Co); manufacturing hard-metal objects (10-30 µg/m³ Co); and exposed to cobalt in powder form when manufacturing hard-metal objects (60-11 000 µg/m³ Co). No specific information was given on exposure to other substances in this cohort, but the workers were exposed to a number of substances that are used in the production of hard metal, such as tungsten carbide. There were 292 deaths among persons under 80 years of age during the study period; the SMRs relative to that of the male Swedish population were 96 (95% CI, 85-108) for mortality from all causes and 105 (95% CI, 82-132) for all incident tumours (73 cases). There were 17 cases of lung cancer *versus* 12.7 expected (SMR, 134; 95% CI, 77-213). With more than 10 years of exposure time and more than 20 years since first exposure, there were seven cases of lung cancer *versus* 2.5 expected (SMR, 278; 95% CI, 111-572); there were three cases of cancer of the lung *versus* 1.3 expected in the two lower exposure groups, and four cases of lung cancer *versus* 1.2 expected in the two higher exposure groups. A survey carried out at the end of the 1970s among hard-metal workers in Sweden showed that their smoking habits were not different from those of the male Swedish population (Alexandersson, 1979).

4. Summary of Data Reported and Evaluation

4.1 Exposure data

Cobalt is widely distributed in the environment; it is the thirty-third most abundant element in the earth's crust. Cobalt is obtained primarily as a by-product of the mining and processing of copper and nickel ores and is a constituent of about 70 naturally occurring oxide, sulfide, arsenide and sulfoarsenide minerals. Cobalt is extracted from ore and concentrated by pyrometallurgical, hydrometallurgical and electrolytic processes alone or in combination. Refined metallic cobalt is available to the industrial market as cathodes and to a lesser extent as powders; oxides and other compounds are also available.

Cobalt compounds have been used as pigments in glass and ceramics in many countries for thousands of years. Since the beginning of the twentieth century, the major uses of cobalt have been in the production of metal alloys, such as superalloys and magnetic alloys, as well as high-strength steels and hard-metal cemented carbides. At the end of the 1980s, about one-third of the cobalt used was in the production of cobalt chemicals, which are used primarily as catalysts and pigments.

The main route of occupational exposure is *via* the respiratory tract by inhalation of dusts, fumes and mists containing cobalt. Exposures have been measured in hard-metal production, processing and use and in porcelain painting. Occupational exposure to cobalt is regulated in many countries.

Cobalt occurs in vegetables *via* uptake from soil, and vegetables account for the major part of human dietary intake of cobalt. Animal-derived foods, particularly liver, contain cobalt in the form of vitamin B_{12}. Cobalt is also found in air, water and tobacco smoke. Human tissues and fluids normally contain low levels of cobalt, which may be increased as a result of occupational exposures. Cobalt concentrations in tissue, serum and urine can be increased in patients with implants made of cobalt-containing alloys. Cobalt-containing particles have been detected in tissues immediately adjacent to such prostheses.

4.2 Experimental carcinogenicity data

Cobalt metal powder was tested in two experiments in rats by intramuscular injection and in one experiment by intrathoracic injection, producing sarcomas at the injection site.

A finely powdered *cobalt-chromium-molybdenum alloy* was tested in rats by intramuscular injection, producing sarcomas at the injection site. In two other experiments in rats, coarsely or finely ground cobalt-chromium-molybdenum alloy implanted in muscle or pellets of cobalt-chromium-molybdenum alloy implanted subcutaneously did not induce sarcomas. Implantation in the rat femur of three different *cobalt-containing alloys*, in the form of powder, rod or compacted wire, resulted in a few local sarcomas. In another experiment, intramuscular implantation of polished rods consisting of three different cobalt-containing alloys did not produce local sarcomas. In an experiment in guinea-pigs, intramuscular implantation of a *cobalt-chromium-molybdenum alloy* powder did not produce local tumours.

Intraperitoneal injection of a *cobalt-chromium-aluminium spinel* in rats produced a few local malignant tumours, and intratracheal instillation of this spinel in rats was associated with the occurrence of a few pulmonary squamous-cell carcinomas.

In two experiments in rats, intramuscular injection of *cobalt[II] oxide* powder produced sarcomas at the injection site. In an experiment in mice, intramuscular injection of cobalt oxide powder did not produce local tumours. Intratracheal instillation of cobalt oxide powder in rats was associated with a few benign and malignant pulmonary tumours. In a study limited by poor survival, hamsters administered a cobalt oxide dust by inhalation showed no increase in the incidence of pulmonary tumours. In two experiments in rats by subcutaneous and intraperitoneal injection, cobalt oxide powder produced local malignant tumours.

Cobalt[II] sulfide powder was tested in one study in rats by intramuscular injection, producing a high incidence of local sarcomas.

Cobalt[II] chloride was tested in one study in rats by repeated subcutaneous injection, producing many local and a few distant subcutaneous sarcomas.

Cobalt[II,III] oxide was tested in one experiment in hamsters to determine the effects of various particulates on carcinogenesis induced by *N*-nitrosodiethylamine. Intratracheal instillation of cobalt[II,III] oxide did not increase the incidence of pulmonary tumours over that in appropriate control groups.

Studies in mice and rabbits with *cobalt naphthenate* could not be evaluated.

In a screening test for lung adenomas by intraperitoneal injection, *cobalt[III] acetate* did not increase the incidence of lung tumours in strain A mice.

Interpretation of the available evidence for the carcinogenicity of cobalt in experimental animals was difficult because many of the reports failed to include sufficient details on results of statistical analyses, on survival and on control groups. Further, statistical analyses could not be performed by the Working Group in the absence of specific information on survival and on whether the neoplasms were fatal. Nevertheless, weight was given in the evaluation to the consistent occurrence of tumours at the site of administration and to the histological types of tumours observed.

4.3 Human carcinogenicity data

A number of single cases of malignant tumours, mostly sarcomas, have been reported at the site of orthopaedic implants containing cobalt. In one cohort study of people with a hip prosthesis, there was a significant increase in the incidence of lymphatic and haematopoietic malignancies, and significant deficits of breast and colorectal cancers. Overall cancer incidence was significantly lower than expected in the first 10 years after surgery, but significantly higher than expected after 10 or more years. No data were provided on the composition of the prostheses in this study.

Four cohort studies on the association between industrial exposure to cobalt and death from cancer were reviewed, two of which provided information for the evaluation. In a French electrochemical plant, there was a significant increase in the risk for lung cancer among workers in cobalt production, who were also exposed to nickel and arsenic, but not among workers in other departments of the factory. In a study in Sweden of hard-metal workers with documented exposure to cobalt-containing dusts, a significant increase in lung cancer risk was seen in people exposed for more than 10 years whose exposure had begun more than 20 years previously.

Interpretation of the available evidence on the possible association between occupational exposure to cobalt and cancer in humans is made difficult by the fact that in three of the four studies there was concurrent exposure to other potentially carcinogenic substances, including forms of nickel and arsenic. In the Swedish study, there was concurrent exposure to other components of hard-metal dust.

4.4 Other relevant data

Occupational exposure to cobalt-containing dusts can cause fibrotic changes in the lung and can precipitate asthma. Cardiotoxic effects have been reported in exposed humans; in particular, cardiomyopathy can occur after prolonged oral intake.

Cobalt[II] chloride reduced fertility in male mice.

Cobalt[II] compounds had weak or no genetic effect in bacteria; some cobalt[III] complexes with heterocyclic ligands were active.

In single studies with an extensive range of eukaryotes, including animal and human cells *in vitro*, cobalt[II] compounds induced DNA damage, mutation, sister chromatid exchange and aneuploidy. Gene conversion and mutation in eukaryotes and DNA damage in human cells were observed in several studies. There was some evidence that these compounds can also induce aneuploidy in hamsters *in vivo*. In single studies, cobalt[II] sulfide induced DNA damage and transformation in cultured mammalian cells.

4.5 Evaluation[1]

There is *inadequate evidence* for the carcinogenicity of cobalt and cobalt compounds in humans.

There is *sufficient evidence* for the carcinogenicity of cobalt metal powder in experimental animals.

There is *limited evidence* for the carcinogenicity of metal alloys containing cobalt, chromium and molybdenum in experimental animals.

There is *sufficient evidence* for the carcinogenicity of cobalt[II] oxide in experimental animals.

There is *limited evidence* for the carcinogenicity of cobalt[II] sulfide in experimental animals.

There is *limited evidence* for the carcinogenicity of cobalt[II] chloride in experimental animals.

There is *inadequate evidence* for the carcinogenicity of cobalt-aluminium-chromium spinel, cobalt[II,III] oxide, cobalt naphthenate and cobalt[III] acetate in experimental animals.

---

[1]For definition of the italicized terms, see Preamble, pp. 30-33.

Overall evaluation

Cobalt and cobalt compounds *are possibly carcinogenic to humans (Group 2B)*.

5. References

Adhikari, S. (1967) Effects of cobalt chloride on chick embryos. *Anat. Anz. Bd*, *120*, 75-83

Aldrich Chemical Co. (1990) *1989-1990 Aldrich Catalog/Handbook of Fine Chemicals*, Milwaukee, WI, pp. 339-342

Aldrich Chemical Co. (undated a) *Material Safety Data Sheet 20311-4: Cobalt(II,III) Oxide, 99.995%*, Milwaukee, WI

Aldrich Chemical Co. (undated b) *Material Safety Data Sheet 20218-5: Cobalt(II) Chloride Hexahydrate*, Milwaukee, WI

Aldrich Chemical Co. (undated c) *Material Safety Data Sheet 25559-9: Cobalt(II) Chloride Hexahydrate, 98%, ACS Reagent*, Milwaukee, WI

Aldrich Chemical Co. (undated d) *Material Safety Data Sheet 20308-4: Cobalt(II) Chloride Hydrate, 99.999%*, Milwaukee, WI

Aldrich Chemical Co. (undated e) *Material Safety Data Sheet 23269-6: Cobalt(II) Chloride, 97%*, Milwaukee, WI

Aldrich Chemical Co. (undated f) *Material Safety Data Sheet 20310-6: Cobalt(II) Nitrate Hydrate, 99.999%*, Milwaukee, WI

Aldrich Chemical Co. (undated g) *Material Safety Data Sheet 23037-5: Cobalt(II) Nitrate Hexahydrate, 99%*, Milwaukee, WI

Aldrich Chemical Co. (undated h) *Material Safety Data Sheet 23926-7: Cobalt(II) Nitrate Hexahydrate, 98%, ACS Reagent*, Milwaukee, WI

Aldrich Chemical Co. (undated i) *Material Safety Data Sheet 22164-3: Cobalt(II,III) Oxide*, Milwaukee, WI

Aldrich Chemical Co. (undated j) *Material Safety Data Sheet 22959-8: Cobalt(II) Sulfate Hydrate, 99.999%*, Milwaukee, WI

Aldrich Chemical Co. (undated k) *Material Safety Data Sheet 23038-3: Cobalt(II) Sulfate Hydrate*, Milwaukee, WI

Alessio, L. & Dell'Orto, A. (1988) Biological monitoring of cobalt. In: Clarkson, T.W., Friberg, L., Nordberg, G.F. & Sager, P.R., eds, *Biological Monitoring of Toxic Metals*, New York, Plenum Press, pp. 407-416

Alexander, C.S. (1972) Cobalt-beer cardiomyopathy. A clinical and pathologic study of twenty-eight cases. *Am. J. Med.*, *53*, 395-417

Alexandersson, R. (1988) Blood and urinary concentrations as estimators of cobalt exposure. *Arch. environ. Health*, *43*, 299-303

Alexandersson, R. & Atterhög, J.-H. (1980) Studies on effects of exposure to cobalt. VII. Heart effects of exposure to cobalt in the Swedish hard-metal industry (Swed.). *Arbete Hälsa*, *9*, 1-21

Alexandersson, R. & Lidums, V. (1979) Studies on the effects of exposure to cobalt. VII. Cobalt concentrations in blood and urine as exposure indicators (Swed.). *Arbete Hälsa*, *8*, 2-23

Alexandersson, R. & Swensson, Å. (1979) Studies on the pulmonary reaction of workers exposed to cobalt in the tungsten carbide industry. *Arch. hig. Rada. Toksikol.*, *30* (Suppl.), 355-361

Alušík, S., Černhorsky, J. & Barbořík, M. (1982) Cobalt cardiomyopathy (Czech.). *Vnitrní. Lék.*, *28*, 493-497

Amacher, D.E. & Paillet, S.C. (1980) Induction of trifluorothymidine-resistant mutants by metal ions in L5178Y/TK$^{+/-}$ cells. *Mutat. Res.*, *78*, 279-288

American Chemical Society (1988) *Chemcyclopedia 1989*, Washington DC, pp. 184, 265

American Conference of Governmental Industrial Hygienists (1989) *Threshold Limit Values and Biological Exposure Indices for 1989-1990*, Cincinnati, OH, p. 18

American Society for Testing and Materials (1984) *Standard Specification for Wrought Cobalt-Nickel-Chromium-Molybdenum Alloy for Surgical Implant Applications* (ASTM F 562-84), Philadelphia, pp. 1-4

American Society for Testing and Materials (1987a) *Standard Specification for Cast Cobalt-Chromium-Molybdenum Alloy for Surgical Implant Applications* (ASTM F 75-87), Philadelphia, pp. 1-2

American Society for Testing and Materials (1987b) *Standard Specification for Cobalt-Chromium-Molybdenum Alloy for Surgical Implants* (ASTM F 799-87), Philadelphia, pp. 1-3

American Society for Testing and Materials (1988) *Standard Specification for Wrought Cobalt-Nickel-Chromium-Molybdenum-Tungsten-Iron Alloy for Surgical Implant Applications* (ASTM F 563-88), Philadelphia, pp. 1-3

Andersen, O. (1983) Effects of coal combustion products and metal compounds on sister chromatid exchange (SCE) in a macrophagelike cell line. *Environ. Health Perspectives*, *47*, 239-253

Andersen, I. & Høgetveit, A.C. (1984) Analysis of cobalt in plasma by electrothermal atomic absorption spectrometry. *Fresenius Z. anal. Chem.*, *318*, 41-44

Angerer, J. (1989) Cobalt. In: Henschler, D. & Lehnert, G., eds, *Biologische Arbeitsstoff-Toleranzwerte (BAT-Werte), Arbeitsmedizinisch-toxikologische Bergründungen* [Biological Occupational Tolerance Value, Occupational Medical-toxicological Basis], Weinheim, VCH-Verlag, pp. 1-13

Angerer, J. & Heinrich, R. (1988) Cobalt. In: Seiler, H.G. & Sigel, H., *Handbook on Toxicity of Inorganic Compounds*, New York, Marcel Dekker, pp. 251-264

Angerer, J., Heinrich, R., Szadkowski, D. & Lehnert, G. (1985) Occupational exposure to cobalt powder and salts—biological monitoring and health effects. In: Lekkas, T.D., ed., *Proceedings of an International Conference on Heavy Metals in the Environment, Athens, September 1985*, Vol. 2, Luxembourg, Commission of the European Communities, pp. 11-13

Angerer, J., Heinrich-Ramm, R. & Lehnert, G. (1989) Occupational exposure to cobalt and nickel: biological monitoring. *Int. J. environ. anal. Chem.*, *35*, 81-88

Anon. (1989) The role of cobalt in cemented carbides. *Cobalt News*, 89, 2-3

Anon. (1990a) The extraction of cobalt from its ores. *Cobalt News*, 90, 8-10

Anon. (1990b) Mutual attraction—magnets and the Cobalt Development Institute. *Cobalt News*, 90, 2-5

Anttila, S., Sutinen, S., Paananen, M., Kreus, K.-E., Sivonen, S.J., Grekula, A. & Alapieti, T. (1986) Hard metal lung disease: a clinical, histological, ultrastructural and X-ray microanalytical study. *Eur. J. respir. Dis.*, 69, 83-94

Arbeidsinspectie (Labour Inspection) (1986) *De Nationale MAC-Lijst 1986* [National MAC-List 1986] (P 145), Voorburg, Ministry of Social Affairs and Work Environment, p. 15

Arbejdstilsynet (Labour Inspection) (1988) *Graensevaerdier for Stoffer og Materialer* [Limit Values for Substances and Materials] (At-anvisning No. 3.1.0.2), Copenhagen, p. 14

Arden, G.P. & Bywaters, E.G.L. (1978) Tissue reaction. In: Arden, G.B. & Ansel, B.M., eds, *Surgical Management of Juvenile Chronic Polyarthritis*, London, Academic Press, pp. 253-275

Arlauskas, A., Baker, R.S.U., Bonin, A.M., Tandon, R.K., Crisp, P.T. & Ellis, J. (1985) Mutagenicity of metal ions in bacteria. *Environ. Res.*, 36, 379-388

Bagó-Granell, J., Aguirre-Canyadell, M., Nardi, J. & Tallada, N. (1984) Malignant fibrous histiocytoma of bone at the site of a total hip arthroplasty. A case report. *J. Bone Joint Surg.*, 66B, 38-40

Balazs, T. & Herman, E.H. (1976) Toxic cardiomyopathies. *Ann. clin. Lab. Sci.*, 6, 467-476

Barberá, R. & Farré, R. (1986) Cobalt content of foods and diets in a Spanish population. *Nahrung*, 30, 565-567

Barberá, R., Irles, J. & Farré, R. (1986) Elimination of iron interference and use of APDC [ammonium pyrrolidine dithiocarbamate] and NaDDC [sodium diethyldithiocarbamate] as chelating agents in the determination of cobalt in foods by atomic absorption spectrometry. *Atomic Spectr.*, 7, 151-154

Barbořík, M. & Dusek, J. (1972) Cardiomyopathy accompanying industrial cobalt exposure. *Br. Heart J.*, 34, 113-116

Barfoot, R.A. & Pritchard, J.G. (1980) Determination of cobalt in blood. *Analyst*, 105, 551-557

Barnes, J.E., Kanapilly, G.M. & Newton, G.J. (1976) Cobalt-60 oxide aerosols: methods of production and short-term retention and distribution kinetics in the beagle dog. *Health Phys.*, 30, 391-398

Bartolozzi, A. & Black, J. (1985) Chromium concentrations in serum, blood clot and urine from patients following total hip arthroplasty. *Biomaterials*, 6, 2-8

BDH Ltd (1989a) *Health and Safety Information 00492: Cobalt(II) Acetate*, Poole

BDH Ltd (1989b) *Health and Safety Information 00496: Cobalt(II) Chloride, 6-Hydrate*, Poole

BDH Ltd (1989c) *Health and Safety Information 00495: Cobalt(II) Chloride Anhydrous*, Poole

BDH Ltd (1989d) *Health and Safety Information 00497: Cobalt(II) Nitrate, 6-Hydrate*, Poole

BDH Ltd (1989e) *Health and Safety Information 00501: Cobalt(II) Oxide*, Poole

BDH Ltd (1989f) *Health and Safety Information 00561: Cobalt(II) Sulphate, 7-Hydrate*, Poole

Bearden, L.J. (1976) The toxicity of two prosthetic metals (cobalt and nickel) to cultured fibroblasts (Abstract). *Diss. Abstr. int. B*, *37*, 1785-B

Bearden, L.J. & Cooke, F.W. (1980) Growth inhibition of cultured fibroblasts by cobalt and nickel. *J. biomed. Materials Res.*, *14*, 289-309

Bech, A.O. (1974) Hard metal disease and tool room grinding. *J. Soc. occup. Med.*, *24*, 11-16

Bedello, P.G., Goitre, M., Alovisi, V. & Cane, D. (1984) Contact dermatitis caused by cobalt naphthenate. *Contact Derm.*, *11*, 247-264

Bennett, M. (1974) *Concise Chemical and Technical Dictionary*, New York, Chemical Publisher, p. 262

Berk, L., Burchenal, J.H. & Castle, W.B. (1949) Erythropoietic effect of cobalt in patients with or without anemia. *New Engl. J. Med.*, *240*, 754-761

Berkow, R., ed. (1987) *The Merck Manual of Diagnosis and Therapy*, 15th ed., Rahway, NJ, Merck, Sharp & Dohme Research Laboratories, pp. 899, 949

Berndt, H., Harms, U. & Sonneborn, M. (1985) Multielement trace preconcentration from water on activated carbon for sample pretreatment for atomic spectroscopy (Flame-AAS, ICP/OES) (Ger.). *Fresenius Z. anal. Chem.*, *322*, 329-333

Black, J., Maitin, E.C., Gelman, H. & Morris, D.M. (1983) Serum concentrations of chromium, cobalt and nickel after total hip replacement: a six month study. *Biomaterials*, *4*, 160-164

Bouman, A.A., Platenkamp, A.J. & Posma, F.D. (1986) Determination of cobalt in urine with flameless atomic absorption spectroscopy. Comparison of direct analysis using Zeeman background correction and indirect analysis using extraction in organic solution. *Ann. clin. Biochem.*, *23*, 346-350

Boyle, E.A., Handy, B. & van Geen, A. (1987) Cobalt determination in natural waters using cation-exchange liquid chromatography with luminol chemiluminescence detection. *Anal. Chem.*, *59*, 1499-1503

Brauer, G., ed. (1965) *Handbook of Preparative Inorganic Chemistry*, 2nd ed., London, Academic Press, pp. 1519-1525

Braun, E., Schmitt, D., Nabet, F., Legras, B., Coudane, H. & Molé, D. (1986) Urinary concentration of cobalt and chromium in patients with a total uncemented hip prosthesis (Fr.). *Int. Orthopaed. (SICOT)*, *10*, 277-282

Brooks, S.M. (1981) Lung disorders resulting from the inhalation of metals. *Clin. Chest Med.*, *2*, 235-254

Brune, D., Kjaerheim, A., Paulsen, G. & Beltesbrekke, H. (1980) Pulmonary deposition following inhalation of chromium-cobalt grinding dust in rats and distribution in other tissues. *Scand. J. dent. Res.*, *88*, 543-551

Budavari, S., ed. (1989) *The Merck Index*, 11th ed., Rahway, NJ, Merck & Co., pp. 379-382

Buono, F.J. & Feldman, M.L. (1979) Driers and metallic soaps. In: Mark, H.F., Othmer, D.F., Overberger, C.G., Seaborg, G.T. & Grayson, M., eds, *Kirk-Othmer Encyclopedia of Chemical Technology*, Vol. 8, 3rd ed., New York, John Wiley & Sons, pp. 34-49

Camarasa, J.M.G. (1967) Cobalt contact dermatitis. *Acta dermatol. venereol.*, *47*, 287-292

Caplan, R.M. & Block, W.D. (1963) Experimental production of hyperlipemia in rabbits by cobaltous chloride. *J. invest. Dermatol.*, *40*, 199-203

Castleman, B. & McNeely, B.U. (1965) Case records of the Massachusetts General Hospital. Case 38-1965. Presentation of case. *New Engl. J. Med.*, 273, 494-504

Casto, B.C., Meyers, J. & DiPaolo, J.A. (1979) Enhancement of viral transformation for evaluation of the carcinogenic or mutagenic potential of inorganic metal salts. *Cancer Res.*, 39, 193-198

Cawse, P.A. (1978) *A Survey of Atmospheric Trace Elements in the UK. Results for 1977* (AERE-R9164), Harwell, Atomic Energy Research Establishment, Environmental and Medical Sciences Division

Chemical Dynamics Corp. (1989) *The 1989-90 Chemalog. Catalog/Handbook of Biochemicals, Organic Chemicals and Inorganic Chemicals*, South Plainfield, NJ, p. 165

Chemical Information Services Ltd (1988) *Directory of World Chemical Producers 1989/90 Edition*, Oceanside, NY

Christensen, J.M. & Mikkelsen, S. (1986) Cobalt concentration in whole blood and urine from pottery plate painters exposed to cobalt paint. In: Lakkas, T.D., ed., *Proceedings of an International Conference, Heavy Metals in the Environment, Athens, September 1985*, Vol. 2, Luxembourg, Commission of the European Communities, pp. 86-88

Christensen, J.M., Mikkelsen, S. & Skov, A. (1983) A direct determination of cobalt in blood and urine by Zeeman atomic absorption spectrophotometry. In: Brown, S.S. & Savory, J., eds, *Chemical Toxicology and Clinical Chemistry of Metals*, London, Academic Press, pp. 65-68

Cikrt, M. & Tichý, M. (1981) Biliary excretion of cobalt in rats. *J. Hyg. Epidemiol. Microbiol. Immunol.*, 25, 364-368

Clyne, N., Lins, L.-E., Pehrsson, S.K., Lundberg, Å. & Werner, J. (1988) Distribution of cobalt in myocardium, skeletal muscle and serum in exposed and unexposed rats. *Trace Elem. Med.*, 5, 52-54

Coates, E.O., Jr & Watson, J.H.L. (1971) Diffuse interstitial lung disease in tungsten carbide workers. *Ann. intern. Med.*, 75, 709-716

Coates, E.O., Jr & Watson, J.H.L. (1973) Pathology of the lung in tungsten carbide workers using light and electron microscopy. *J. occup. Med.*, 15, 280-286

Cobalt Development Institute (1989) *Cobalt and Its Compounds*, Slough

Coleman, R.F., Herrington, J. & Scales, J.T. (1973) Concentration of wear products in hair, blood and urine after total hip replacement. *Br. med. J.*, i, 527-529

Collecchi, P., Esposito, M., Brera, S., Mora, E., Mazzucotelli, A. & Uddone, M. (1986) The distribution of arsenic and cobalt in patients with laryngeal carcinoma. *J. appl. Toxicol.*, 6, 287-289

Considine, D.M., ed. (1974) *Chemical and Process Technology Encyclopedia*, New York, McGraw Hill, pp. 302-307

Cook, W.A. (1987) *Occupational Exposure Limits—Worldwide*, Washington DC, American Industrial Hygiene Association, pp. 119, 133-134, 175

Corrier, D.E., Mollenhauer, H.H., Clark, D.E., Hare, M.F. & Elissalde, M.H. (1985a) Testicular degeneration and necrosis induced by dietary cobalt. *Vet. Pathol.*, 22, 610-616

Corrier, D.E., Rowe, L.D., Clark, D.E. & Hare, M.F. (1985b) Tolerance and effect of chronic dietary cobalt on sheep. *Vet. hum. Toxicol.*, 28, 216-219

Costa, M., Heck, J.D. & Robison, S.H. (1982) Selective phagocytosis of crystalline metal sulfide particles and DNA strand breaks as a mechanism for the induction of cellular transformation. *Cancer Res.*, *42*, 2757-2763

CP Chemicals (1989a) *Material Safety Data Sheet 4500: Cobalt Carbonate, Basic*, Fort Lee, NJ

CP Chemicals (1989b) *Material Safety Data Sheet 4745: Cobalt Acetate*, Fort Lee, NJ

CP Chemicals (1989c) *Material Safety Data Sheet 4746: Cobalt Chloride*, Fort Lee, NJ

CP Chemicals (1989d) *Material Safety Data Sheet 4580: Cobalt Sulfate Solution*, Fort Lee, NJ

Cuckle, H., Doll, R. & Morgan, L.G. (1980) Mortality study of men working with soluble nickel compounds. In: Brown, S.S & Sunderman, F.W., Jr, eds, *Nickel Toxicology*, London, Academic Press, pp. 11-14

Cugell, D.W., Morgan, W.K.C., Perkins, D.G. & Rubin, A. (1990) The respiratory effects of cobalt. *Arch. intern. Med.*, *150*, 177-183

Curtis, J.R., Goode, G.C., Herrington, J. & Urdaneta, L.E. (1976) Possible cobalt toxicity in maintenance hemodialysis patients after treatment with cobaltous chloride: a study of blood and tissue cobalt concentrations in normal subjects and patients with terminal renal failure. *Clin. Nephrol.*, *5*, 61-65

Daniel, M., Dingle, J.T., Webb, M. & Heath, J.C. (1963) The biological action of cobalt and other metals. I. The effects of cobalt on the morphology and metabolism of rat fibroblasts in vitro. *Br. J. exp. Pathol.*, *44*, 163-176

Davison, A.G., Haslam, P.L., Corrin, B., Coutts, I.I., Dewar, A., Riding, W.D., Studdy, P.R. & Newman-Taylor, A.J. (1983) Interstitial lung disease and asthma in hard-metal workers: bronchoalveolar lavage, ultrastructural, and analytical findings and results of bronchial provocation tests. *Thorax*, *38*, 119-128

Delgado, E.R. (1958) Sarcoma following a surgically treated fractured tibia. A case report. *Clin. Orthopaed.*, *12*, 315-318

Delves, H.T., Mensikov, R. & Hinks, L. (1983) Direct determination of cobalt in whole-blood by electrothermal atomization and atomic absorption spectroscopy. In: Braetter, P. & Schramel, P., eds, *Trace Elements—Analytical Chemistry in Medicine and Biology*, Vol. 2, Berlin, Walter de Gruyter & Co., pp. 1123-1127

Demedts, M. & Ceuppens, J.L. (1989) Respiratory diseases from hard metal or cobalt exposure—solving an enigma. *Chest*, *95*, 2-3

Demedts, M., Gheysens, B., Nagels, J., Verbeken, E., Lauweryns, J., van den Eeckhout, A., Lahaye, D. & Gyselen, A. (1984) Cobalt lung in diamond polishers. *Am. Rev. respir. Dis.*, *130*, 130-135

Desselberger, U. & Wegener, H.-H. (1971) Experimental investigations on alcohol, cobalt, and combined alcohol-cobalt poisoning in guinea-pigs (Ger.). *Beitr. Pathol.*, *142*, 150-176

Direktoratet for Arbeidstilsynet (Directorate for Labour Inspection) (1981) *Administrative Normer for Forurensning i Arbeidsatmosfaere 1981* [Administrative Norms for Pollution in Work Atmosphere 1981] (No. 361), Oslo, p. 15

Dodion, P., Putz, P., Amiri-Lamraski, M.H., Efira, A., de Martelaere, E. & Heimann, R. (1983) Immunoblastic lymphoma at the site of an infected vitallium bone plate. *Histopathology*, *6*, 807-813

Domingo, J.L., Paternain, J.L., Llobet, J.M. & Corbella, J. (1985) Effects of cobalt on postnatal development and late gestation in rats upon oral administration. *Rev. esp. Fisiol.*, *41*, 293-298

Donaldson, J.D. (1986) Cobalt and cobalt compounds. In: Gerhartz, W., Yamamoto, Y.S., Campbell, F.T., Pfefferkorn, R. & Rounsaville, J.F., eds, *Ullmann's Encyclopedia of Industrial Chemistry*, 5th ed., Weinheim, VCH-Verlag, pp. 281-313

Donaldson, J.D. & Clark, S.J. (1985) *Cobalt in Superalloys*, Slough, Cobalt Development Institute

Donaldson, J.D., Clark, S.J. & Grimes, S.M. (1986a) *Cobalt in Chemicals*, Slough, Cobalt Development Institute

Donaldson, J.D., Clark, S.J. & Grimes, S.M. (1986b) *Cobalt in Medicine, Agriculture and the Environment*, Slough, Cobalt Development Institute

Donaldson, J.D., Clark, S.J. & Grimes, S.M. (1988) *Cobalt in Electronic Technology*, Slough, Cobalt Development Institute

Donat, J.R. & Bruland, K.W. (1988) Direct determination of dissolved cobalt and nickel in seawater by differential pulse cathodic stripping voltammetry preceded by adsorptive collection of cyclohexane-1,2-dione dioxime complexes. *Anal. Chem.*, *60*, 240-244

Dube, V.E. & Fisher, D.E. (1972) Hemangioendothelioma of the leg following metallic fixation of the tibia. *Cancer*, *30*, 1260-1266

Duckham, J.M. & Lee, H.A. (1976) The treatment of refractory anaemia of chronic renal failure with cobalt chloride. *Q.J. Med. new Ser.*, *178*, 277-294

Dussek Campbell Ltd (1989a) *Material Data Safety Sheet B00221: 6% Cobalt Naphthenate*, Belleville, Québec

Dussek Campbell Ltd (1989b) *Material Data Safety Sheet B00226: 10.5% Cobalt Naphthenate Flaked*, Belleville, Québec

Egilsson, V., Evans, I.H. & Wilkie, D. (1979) Toxic and mutagenic effects of carcinogens on the mitochondria of *Saccharomyces cerevisiae*. *Mol. gen. Genet.*, *174*, 39-46

Einarsson, Ö., Eriksson, E., Lindstedt, G. & Wahlberg, J.E. (1979) Dissolution of cobalt from hard metal alloys by cutting fluids. *Contact Derm.*, *5*, 129-132

Elinder, C.-G. & Friberg, L. (1986) Cobalt. In: Friberg, L., Nordberg, G.F. & Vouk, V.B., eds, *Handbook on the Toxicology of Metals*, 2nd ed., Amsterdam, Elsevier, pp. 211-232

Elinder, C.-G., Gerhardsson, L. & Oberdörster, G. (1988) Biological monitoring of toxic metals—overview. In: Clarkson, T.W., Friberg, L., Nordberg, G.F. & Sager, P.R., eds, *Biological Monitoring of Toxic Metals*, New York, Plenum Press, pp. 1-71

Farah, S.B. (1983) The in vivo effect of cobalt chloride on chromosomes. *Rev. Brasil. Genet. VI*, *3*, 433-442

Farrell, R.L. & Davis, G.W. (1974) The effects of particulates on respiratory carcinogenesis by diethylnitrosamine. In: Karbo, E. & Paris, J.R., eds, *Experimental Lung Cancer: Carcinogenesis and Bioassays*, New York, Springer, pp. 219-233

Ferioli, A., Roi, R. & Alessio, L. (1987) Biological indicators for the assessment of human exposure to industrial chemicals. In: Alessio, L., Berlin, A., Boni, M. & Roi, R., eds, *CEC-Industrial Health and Safety* (EUR 11135 EN), Luxembourg, Commission of the European Communities, pp. 48-61

Fernandez, J.P., Veron, C., Hildebrand, H.F. & Martin, P. (1986) Nickel allergy to dental prostheses (Short communication). *Contact Derm.*, *14*, 312

Fischer, T. & Rystedt, I. (1983) Cobalt allergy in hard metal workers. *Contact Derm.*, *9*, 115-121

Flodh, H. (1968) Autoradiographic studies on distribution of radiocobalt chloride in pregnant mice. *Acta radiol. ther. phys. biol.*, *7*, 121-128

Friedrich, W. (1984) Vitamins. In: Mark, H.F., Othmer, D.F., Overberger, C.G., Seaborg, G.T. & Grayson, M., eds, *Kirk-Othmer Encyclopedia of Chemical Technology*, Vol. 24, 3rd ed., New York, John Wiley & Sons, pp. 158-185

Fukunaga, M., Kurachi, Y. & Mizuguchi, Y. (1982) Action of some metal ions on yeast chromosomes. *Chem. pharm. Bull.*, *30*, 3017-3019

Gaechter, A., Alroy, J., Andersson, G.B.J., Galante, J., Rostoker, W. & Schajowicz, F. (1977) Metal carcinogenesis. A study of the carcinogenic activity of solid metal alloys in rats. *J. Bone Joint Surg.*, *59A*, 622-624

Gale, T.F. (1980) Does cobalt damage the hamster embryo? A preliminary report (Abstract). *Anat. Rec.*, *196*, 232A

Gheysens, B., Auwerx, J., Van den Eeckhout, A. & Demedts, M. (1985) Cobalt-induced bronchial asthma in diamond polishers. *Chest*, *88*, 740-744

Gilani, S.H. & Alibai, Y. (1985) The effects of heavy metals on the chick embryo development (Abstract). *Anat. Rec.*, *211*, 68A-69A

Gillespie, W.J., Frampton, C.M.A., Henderson, R.J. & Ryan, P.M. (1988) The incidence of cancer following total hip replacement. *J. Bone Joint Surg.*, *70B*, 539-542

Gilman, J.P.W. (1962) Metal carcinogenesis. II. A study on the carcinogenic activity of cobalt, copper, iron, and nickel compounds. *Cancer Res.*, *22*, 158-162

Gilman, J.P.W. & Ruckerbauer, G.M. (1962) Metal carcinogenesis. I. Observations on the carcinogenicity of a refinery dust, cobalt oxide, and colloidal thorium dioxide. *Cancer Res.*, *22*, 152-157

Goh, C.L., Kwok, S.F. & Gan, S.L. (1986) Cobalt and nickel content of Asian cements. *Contact Derm.*, *15*, 169-172

Goodman, L.S. & Gilman, A., eds (1975) *The Pharmacological Basis of Therapeutics*, 5th ed., New York, MacMillan, p. 905

Goodman-Gilman, A., Goodman, L.S., Rall, T.W. & Mured, F., eds (1985) *Goodman and Gilman's The Pharmacological Basis of Therapeutics*, 7th ed., New York, MacMillan, p. 1319

Gori, C. & Zucconi, L. (1957) Cytological activity induced by a group of inorganic compounds in *Allium cepa* (Ital.). *Caryologia*, *10*, 29-45

Gregus, Z. & Klaassen, C.D. (1986) Disposition of metals in rats: a comparative study of fecal, urinary and biliary excretion and tissue distribution of eighteen metals. *Toxicol. appl. Pharmacol.*, *85*, 24-38

Grice, H.C., Goodman, T., Munro, I.C., Wiberg, G.S. & Morrison, A.B. (1969) Myocardial toxicity of cobalt in the rat. *Ann. N.Y. Acad. Sci.*, *156*, 189-194

Gunshin, H., Yoshikawa, M., Doudou, T. & Kato, N. (1985) Trace elements in human milk, cow's milk, and infant formula. *Agric. Biol. Chem.*, *49*, 21-26

Hall Chemical Co. (undated a) *Material Safety Data Sheet HCC-85-CO-05: Cobalt Chloride Hexahydrate*, Wickliffe, OH

Hall Chemical Co. (undated b) *Material Safety Data Sheet HCC-85-CO-02: Rev 1: Cobalt Acetate Solution*, Wickliffe, OH

Hall Chemical Co. (undated c) *Material Safety Data Sheet HCC-85-CO-01 Rev. 1: Cobalt Acetate Tetrahydrate*, Wickliffe, OH

Hall Chemical Co. (undated d) *Material Safety Data Sheet HCC-85-CO-06: Rev. 2: Cobalt Carbonate*, Wickliffe, OH

Hall Chemical Co. (undated e) *Material Safety Data Sheet HCC-85-CO-08: Cobalt Chloride Solution*, Wickliffe, OH

Hall Chemical Co. (undated f) *Material Safety Data Sheet HCC-85-CO-07: Cobalt Chloride Anhydrous*, Wickliffe, OH

Hall Chemical Co. (undated g) *Material Safety Data Sheet HCC-85-CO-25: Cobalt Hydroxide Press Cake (E Grade)*, Wickliffe, OH

Hall Chemical Co. (undated h) *Material Safety Data Sheet HCC-85-CO-06 Rev. 1: Cobalt Naphthenate*, Wickliffe, OH

Hall Chemical Co. (undated i) *Material Safety Data Sheet HCC-85-CO-11: Cobalt Nitrate Hexahydrate*, Wickliffe, OH

Hall Chemical Co. (undated j) *Material Safety Data Sheet HCC-85-CO-12: Cobalt Nitrate Solution*, Wickliffe, OH

Hall Chemical Co. (undated k) *Material Safety Data Sheet HCC-85-CO-14D: Cobalt Oxide*, Wickliffe, OH

Hall Chemical Co. (undated l) *Material Safety Data Sheet HCC-85-CO-16: Cobalt Sulfate Heptahydrate*, Wickliffe, OH

Hall Chemical Co. (undated m) *Material Safety Data Sheet HCC-85-CO-17: Cobalt Sulfate Monohydrate*, Wickliffe, OH

Hall Chemical Co. (undated n) *Material Safety Data Sheet HCC-85-CO-18: Cobalt Sulfate Solution*, Wickliffe, OH

Hall, J.L. & Smith, E.B. (1968) Cobalt heart disease. An electron microscopic and histochemical study in the rabbit. *Arch. Pathol.*, 86, 403-412

Hamilton-Koch, W., Snyder, R.D. & LaVelle, J.M. (1986) Metal-induced DNA damage and repair in human diploid fibroblasts and Chinese hamster ovary cells. *Chem.-biol. Interactions*, 59, 17-28

Harding, H.E. (1950) Notes on the toxicology of cobalt metal. *Br. J. ind. Med.*, 7, 76-78

Harp, M.J. & Scoular, F.I. (1952) Cobalt metabolism of young college women on self-selected diets. *J. Nutr.*, 47, 67-72

Hartmann, A., Wüthrich, B. & Bolognini, G. (1982) Occupational lung diseases in the production and processing of hard metals. An allergic event (Ger.). *Schweiz. med. Wochenschr.*, 112, 1137-1141

Hartung, M. (1986) *Lungenfibrosen bei Hartmetallschleifern—Bedeutung der Cobalteinwirkung* [Lung Fibrosis in Hard-metal Grinding—Significance of Cobalt Activity] (Publication Series of Main Associations of Industrial Societies), Bonn, Köllen-Druck & Verlag

Hartung, M. & Schaller, K.-H. (1985) Occupational medical significance of cobalt exposure in hard-metal grinding (Ger.). In: Bolt, H.M., Piekarski, C. & Rutenfranz, J., eds, *Aktuelle arbeitsmedizinische Probleme in der Schwerindustrie. Theorie und Praxis biologischer Toleranzwerte für Arbeitsstoffe (BAT-Werte). Bedeutung neuer Technologien für die arbeitsmedizinische Praxis. Arbeitsmedizinisches Kolloquium der gewerblichen Berufsgenossenschaften* [Actual Occupational Medical Problems in Heavy Industry. Theory and Practice of Biological Tolerance Values for Industrial Substances. Significance of New Technologies for Occupational and Medical Practice. Occupational Medical Colloquium of Industrial Societies], Stuttgart, Gentner Verlag, pp. 55-63

Hartung, M., Schaller, K.-H. & Brand, E. (1982) On the question of the pathogenic importance of cobalt for hard metal fibrosis of the lung. *Int. Arch. occup. environ. Health*, *50*, 53-57

Hartung, M., Schaller, K.-H., Kentner, M., Weltle, D. & Valentin, H. (1983) Studies on exposure to cobalt in different branches of industry (Ger.). *Arbeitsmed. Sozialmed. Präventivmed.*, *4*, 73-75

Hartwig, A., Kasten, U., Boakye-Dankwa, K., Schlopegrell, R. & Beyersmann, D. (1990) Uptake and genotoxicity of micromolar concentrations of cobalt chloride in mammalian cells. *Toxicol. environ. Chem.*, *28*, 205-215

Health and Safety Executive (1987) *Occupational Exposure Limits* (Guidance Note EH 40/87), London, Her Majesty's Stationery Office, p. 11

Heath, J.C. (1954a) Cobalt as a carcinogen. *Nature*, *173*, 822-823

Heath, J.C. (1954b) The effect of cobalt on mitosis in tissue culture. *Exp. Cell Res.*, *6*, 311-320

Heath, J.C. (1956) The production of malignant tumours by cobalt in the rat. *Br. J. Cancer*, *10*, 668-673

Heath, J.C. (1960) The histogenesis of malignant tumours induced by cobalt in the rat. *Br. J. Cancer*, *14*, 478-482

Heath, J.C. & Daniel, M.R. (1962) The production of malignant tumours by cobalt in the rat: intrathoracic tumours. *Br. J. Cancer*, *16*, 473-478

Heath, J.C., Freeman, M.A.R. & Swanson, S.A.V. (1971) Carcinogenic properties of wear particles from prostheses made in cobalt-chromium alloy. *Lancet*, *i*, 564-566

Hedge, A.G., Thakker, D.M. & Bhat, I.S. (1979) Long-term clearance of inhaled ^{60}Co. *Health Phys.*, *36*, 732-734

Hemminki, K., Niemi, M.-L., Kyyronen, P., Koskinen, K. & Vainio, H. (1983) Spontaneous abortion as risk indicator in metal exposure. In: Clarkson, T.W., Nordberg, G.F. & Sager, P.R., eds, *Reproductive and Developmental Toxicity of Metals*, New York, Plenum Press, pp. 369-380

Hildebrand, H.F., Roumazeille, B., Decoulx, J., Herlant-Peers, M.C., Ostapczuk, P., Stoeppler, M. & Mercier, J.M. (1985) Biological consequences of long-term exposure to orthopedic implants. In: Brown, S.S. & Sunderman, F.W., Jr, eds, *Progress in Nickel Toxicology*, Oxford, Blackwell Scientific Publications, pp. 169-172

Hildebrand, H.F., Ostapczuk, P., Mercier, J.F., Stoeppler, M., Roumazeille, B. & Decoulx, J. (1988) Orthopaedic implants and corrosion products. Ultrastructural and analytical studies of 65 patients. In: Hildebrand, H.F. & Champy, M., eds, *Biocompatibility of Co-Cr-Ni Alloys* (NATO-ASI Series A, Vol. 158), London, Plenum Publishing, pp. 133-153

Hillerdal, G. & Hartung, M. (1983) On cobalt in tissues from hard metal workers. *Int. Arch. occup. environ. Health*, 53, 89-90

Höbel, M., Maroske, D., Wegener, K. & Eichler, O. (1972) Toxic effects of $CoCl_2$-, Co[Co-EDTA]- and Na_2[Co-EDTA]-containing aerosols on the rat and the distribution of (Co-EDTA)$^{-2}$ in guinea-pig organs (Ger.). *Arch. int. Pharmacodyn.*, 198, 213-222

Hogstedt, C. & Alexandersson, R. (1990) Mortality among hard-metal workers (Swed.). *Arbete Hälsa*, 21, 1-26

Hopps, H.C., Stanley, A.J. & Shideler, A.M. (1954) Polycythemia induced by cobalt. III. Histologic studies with evaluation of toxicity of cobaltous chloride. *Am. J. clin. Pathol.*, 24, 1374-1380

Horwitz, C. & Van der Linden, S.E. (1974) Cadmium and cobalt in tea and coffee and their relationship to cardiovascular disease. *S.A. med. J.*, 48, 230-233

Hughes, A.W., Sherlock, D.A., Hamblen, D.L. & Reid, R. (1987) Sarcoma at the site of a single hip screw. A case report. *J. Bone Joint Surg.*, 69B, 470-472

IARC (1990) *Information Bulletin on the Survey of Chemicals Being Tested for Carcinogenicity*, No. 14, Lyon, p. 206

Ichikawa, Y., Kusaka, Y. & Goto, S. (1985) Biological monitoring of cobalt exposure, based on cobalt concentrations in blood and urine. *Int. Arch. occup. environ. Health*, 55, 269-276

Inoue, T., Ohta, Y., Sadaie, Y. & Kada, T. (1981) Effect of cobaltous chloride on spontaneous mutation induction in a *Bacillus subtilis* mutator strain. *Mutat. Res.*, 91, 41-45

International Committee on Nickel Carcinogenesis in Man (1990) Report. *Scand. J. Work Environ. Health*, 16, 1-82

Iskander, F.Y. (1986) Egyptian and foreign cigarettes. II. Determination of trace elements in tobacco, ash and wrapping paper. *J. radioanal. nucl. Chem.*, 97, 107-112

Iskander, F.Y., Bauer, T.L. & Klein, D.E. (1986) Determination of 28 elements in American cigarette tobacco by neutron-activation analysis. *Analyst*, 111, 107-109

Iyengar, V. & Woittiez, J. (1988) Trace elements in human clinical specimens: evaluation of literature data to identify reference values. *Clin. Chem.*, 34, 474-481

Jasmin, G. & Riopelle, J.L. (1976) Renal carcinomas and erythrocytosis in rats following intrarenal injection of nickel subsulfide. *Lab. Invest.*, 35, 71-78

Johansson, A., Lundborg, M., Hellström, P.-Å., Camner, P., Keyser, T.R., Kirton, S.E. & Natusch, D.F.S. (1980) Effect of iron, cobalt, and chromium dust on rabbit alveolar macrophages: a comparison with the effects of nickel dust. *Environ. Res.*, 21, 165-176

Johansson, A., Lundborg, M., Wiernik, A., Jarstrand, C. & Camner, P. (1986) Rabbit alveolar macrophages after long-term inhalation of soluble cobalt. *Environ. Res.*, 41, 488-496

Johansson, A., Robertson, B. & Camner, P. (1987) Nodular accumulation of type II cells and inflammatory lesions caused by inhalation of low cobalt concentrations. *Environ. Res.*, 43, 227-243

Johnston, J.M. (1988) *Cobalt 87. A Market Research Study of Cobalt in 1987*, Slough, Cobalt Development Institute

Jones, L.C. & Hungerford, D.S. (1987) Urinary metal ion levels in patients implanted with porous coated total hip prosthesis. *Trans. orthopaed. Res. Soc.*, 32, 317

Jones, D.A., Lucas, H.K., O'Driscoll, M., Price, C.H.G. & Wibberley, B. (1975) Cobalt toxicity after McKee hip arthroplasty. *J. Bone Joint Surg.*, 57B, 289-296

Jones, L.M., Booth, N.H. & McDonald, L.E., eds (1977) *Veterinary Pharmacology and Therapeutics*, 4th ed., Ames, IA, The Iowa State University Press, p. 800

Jorgensen, T.J., Munno, F., Mitchell, T.G. & Hungerford, D. (1983) Urinary cobalt levels in patients with porous Austin-Moore prostheses. *Clin. Orthopaed. rel. Res.*, 176, 124-126

J.T. Baker (1989a) *Material Safety Data Sheet, C4895-02: Cobalt Acetate, 4-Hydrate*, Phillipsburg, NJ

J.T. Baker (1989b) *Material Safety Data Sheet, C4917: Cobalt Carbonate*, Phillipsburg, NJ

J.T. Baker (1989c) *Material Safety Data Sheet, C4928-05: Cobalt Chloride, 6-Hydrate*, Phillipsburg, NJ

J.T. Baker (1989d) *Material Safety Data Sheet, C4939-02: Cobalt Nitrate, 6-Hydrate*, Phillipsburg, NJ

J.T. Baker (1989e) *Material Safety Data Sheet, C4884-02: Cobalt, 1000 ppm (0.100% w/v)*, Phillipsburg, NJ

J.T. Baker (1989f) *Material Safety Data Sheet, C4961-06: Cobalt Oxide*, Phillipsburg, NJ

J.T. Baker (1989g) *Laboratory Reagents and Chromatography Products*, Phillipsburg, NJ, pp. 44-45

J.T. Baker (1989h) *Material Safety Data Sheet, C4972-02: Cobalt Sulfate, 7-Hydrate*, Phillipsburg, NJ

Kada, T. & Kanematsu, N. (1978) Reduction of N-methyl-N'-nitro-N-nitrosoguanidine-induced mutations by cobalt chloride in *Escherichia coli*. *Proc. Jpn. Acad.*, 54B, 234-237

Kanematsu, N., Hara, M. & Kada, T. (1980) Rec assay and mutagenicity studies on metal compounds. *Mutat. Res.*, 77, 109-116

Kasirsky, G., Gautieri, R.F. & Mann, D.E., Jr (1967) Inhibition of cortisone-induced cleft palate in mice by cobaltous chloride. *J. pharm. Sci.*, 56, 1330-1332

Kasperek, K., Kiem, J., Iyengar, G.V. & Feinendegen, L.E. (1981) Concentration differences between serum and plasma of the elements cobalt, iron, mercury, rubidium, selenium and zinc determined by neutron activation analysis. *Sci. total Environ.*, 17, 133-143

Kempf, E. & Pfeiffer, W. (1987) Health hazards through dust in dental laboratories (Ger.). *Arbeitsmed. Sozialmed. Präventivmed.*, 22, 13-18

Kennedy, A., Dornan, J.D. & King, R. (1981) Fatal myocardial disease associated with industrial exposure to cobalt. *Lancet*, i, 412-414

Kerfoot, E.J., Fredrick, W.G. & Domeier, E. (1975) Cobalt metal inhalation studies on miniature swine. *Am. ind. Hyg. Assoc. J.*, 36, 17-25

Kesteloot, H., Roelandt, J., Willems, J., Claes, J.H. & Joossens, J.V. (1968) An enquiry into the role of cobalt in the heart disease of chronic beer drinkers. *Circulation*, *37*, 854-864

Kettrup, A. & Angerer, J. (1988) *Luftanalysen. Analytische Methoden zur Prüfung gesundheitsschädlicher Arbeitsstoffe* [Air Analysis. Analytical Methods for Investigation of Noxious Industrial Compounds], Vol. 1, Weinheim, VCH-Verlag

Kharab, P. & Singh, I. (1985) Genotoxic effects of potassium dichromate, sodium arsenite, cobalt chloride and lead nitrate in diploid yeast. *Mutat. Res.*, *155*, 117-120

Kharab, P. & Singh, I. (1987) Induction of respiratory deficiency in yeast by salts of chromium, arsenic, cobalt and lead. *Indian J. exp. Biol.*, *25*, 141-142

Kimberley, M.M., Bailey, G.G. & Paschal, D.C. (1987) Determination of urinary cobalt using matrix modification and graphite furnace atomic absorption spectrometry with Zeeman-effect background correction. *Analyst*, *112*, 287-290

Kipling, M.D. (1980) Cobalt. In: Waldron, H.A., ed., *Metals in the Environment*, London, Academic Press, pp. 133-153

Kirk, W.S. (1985) Cobalt. In: *Mineral Facts and Problems, 1985 Edition* (Preprint from Bulletin 675), Washington DC, Bureau of Mines, US Department of the Interior, pp. 1-8

Kirk, W.S. (1986) Cobalt. In: *Preprint from the 1986 Bureau of Mines Minerals Yearbook*, Washington DC, Bureau of Mines, US Department of the Interior, pp. 1-8

Kirk, W.S. (1987) Cobalt. In: *Preprint from the 1987 Bureau of Mines Minerals Yearbook*, Washington DC, Bureau of Mines, US Department of the Interior, pp. 1-8

Kirkpatrick, D.C. & Coffin, D.E. (1974) The trace metal content of representative Canadian diets in 1970 and 1971. *J. Inst. Can. Sci. technol. aliment.*, *7*, 56-58

Kitamori, T., Suzuki, K., Sawada, T., Gohshi, Y. & Motojima, K. (1986) Determination of sub-part-per-trillion amounts of cobalt by extraction and photoacoustic spectroscopy. *Anal. Chem.*, *58*, 2275-2278

Knauer, G.A., Martin, J.H. & Gordon, R.M. (1982) Cobalt in north-east Pacific waters. *Nature*, *297*, 49-51

Koponen, M., Gustafsson, T. & Kalliomäki, P.-L. (1982) Cobalt in hard metal manufacturing dusts. *Am. ind. Hyg. Assoc. J.*, *43*, 645-651

Kostić, K., Drašković, R., Dordević, M. & Stanković, S. (1982) Distribution of zinc, iron and cobalt in selected samples of cirrhotic and cancerous liver tissues (Slav.). *Radiol. Jugosl.*, *16*, 217-220

Kreyling, W.G., Ferron, G.A. & Haider, B. (1986) Metabolic fate of inhaled Co aerosols in beagle dogs. *Health Phys.*, *51*, 773-795

Kury, G. & Crosby, R.J. (1968) Studies on the development of chicken embryos exposed to cobaltous chloride. *Toxicol. appl. Pharmacol.*, *13*, 199-206

Kusaka, Y., Yokoyama, K., Sera, Y., Yamamoto, S., Sone, S., Kyono, H., Shisakawa, T. & Goto, S. (1986) Respiratory diseases in hard metal workers: an occupational hygiene study in a factory. *Br. J. ind. Med.*, *43*, 474-485

Lange, M. (1983) Emission sources and emission standards for carcinogenic substances (Ger.). *Staub-Reinhalt. Luft*, *43*, 309-317

Lee, C.-C. & Wolterink, L.F. (1955) Urinary excretion, tubular reabsorption and biliary excretion of cobalt 60 in dogs. *Am. J. Physiol.*, *183*, 167-172

Lee, Y.-S., Pho, R.W.H. & Nather, A. (1984) Malignant fibrous histiocytoma at site of metal implant. *Cancer, 54,* 2286-2289

Lehmann, E., Fröhlich, N. & Minkwitz, R. (1985) Exposure to beryllium and cobalt through grinding alloys and hard metals (Ger.). *Zbl. Arbeitsmed., 35,* 106-113

Léonard, A. & Lauwerys, R. (1990) Mutagenicity, carcinogenicity and teratogenicity of cobalt metal and cobalt compounds. *Mutat. Res., 239,* 17-27

Lewis, S.A., O'Haver, T.C. & Harnly, J.M. (1985) Determination of metals at the microgram-per-liter level in blood serum by simultaneous multielement atomic absorption spectrometry with graphite furnace atomization. *Anal. Chem., 57,* 2-5

Lidums, V.V. (1979) Determination of cobalt in blood and urine by electrothermal atomic absorption spectrometry. *Atomic Absorp. Newsl., 18,* 71-72

Lin, J.H. & Duffy, J.L. (1970) Cobalt-induced myocardial lesions in rats. *Lab. Invest., 23,* 158-162

Lindegren, C.C., Nagai, S. & Nagai, H. (1958) Induction of respiratory deficiency in yeast by manganese, copper, cobalt and nickel. *Nature, 182,* 446-448

Lindner-Szotyori, L. & Gergely, A. (1980) On the supply of some essential trace elements to the Hungarian population (Ger.). *Nahrung, 24,* 829-837

Lins, L.E. & Pehrsson, S.K. (1984) Cobalt in serum and urine related to renal function. *Trace Elements Med., 1,* 172-174

Maines, M.D. & Kappas, A. (1975) Cobalt stimulation of heme degradation in the liver. *J. biol. Chem., 250,* 4171-4177

Maines, M.D., Janoušek, V., Tomio, J.M. & Kappas, A. (1976) Cobalt inhibition of synthesis and induction of δ-aminolevulinate synthase in liver. *Proc. natl Acad. Sci. USA, 73,* 1499-1503

Mallinckrodt (1989a) *Material Safety Data Sheet: Cobalt Acetate,* St Louis, MO

Mallinckrodt (1989b) *Material Safety Data Sheet: Cobalt Chloride,* St Louis, MO

Mallinckrodt (1989c) *Material Safety Data Sheet: Cobalt Chloride CS,* St Louis, MO

Mallinckrodt (1989d) *Material Safety Data Sheet: Cobalt Nitrate,* St Louis, MO

Mallinckrodt (1989e) *Material Safety Data Sheet: Cobalt Sulfate,* St Louis, MO

Martin, A., Bauer, T.W., Manley, M.T. & Marks, K.E. (1988) Osteosarcoma at the site of total hip replacement. *J. Bone Joint Surg., 70A,* 1561-1567

Masiak, M., Owczarek, H., Skowron, S. & Zmijewska, W. (1982) Serum levels of certain trace elements (Ag, Co, Cr) in healthy subjects. II. *Acta physiol. pol., 33,* 65-73

Mazabraud, A., Florent, J. & Laurent, M. (1989) A case of epidermoid carcinoma developing in contact with a hip prosthesis (Fr.). *Bull. Cancer, 76,* 573-581

McDonald, I. (1981) Malignant lymphoma associated with internal fixation of a fractured tibia. *Cancer, 48,* 1009-1011

McDougall, A. (1956) Malignant tumour at site of bone plating. *J. Bone Joint Surg., 38B,* 709-713

McLean, J.R., McWilliams, R.S., Kaplan, J.G. & Birnboim, H.C. (1982) Rapid detection of DNA strand breaks in human peripheral blood cells and animal organs following treatment with physical and chemical agents. In: Bora, K.C., Douglas, G.R. & Nestmann, E.R., eds, *Progress in Mutation Research*, Vol. 3, Amsterdam, Elsevier Biomedical Press, pp. 137-141

Meachim, G., Pedley, R.B. & Williams, D.F. (1982) A study of sarcogenicity associated with Co-Cr-Mo particles implanted in animal muscle. *J. biomed. Mat. Res.*, 16, 407-416

Memoli, V.A., Urban, R.M., Alroy, J. & Galante, J.O. (1986) Malignant neoplasms associated with orthopedic implant materials in rats. *J. orthopaed. Res.*, 4, 346-355

Merian, E. (1985) Introduction on environmental chemistry and global cycles of chromium, nickel, cobalt, beryllium, arsenic, cadmium and selenium, and their derivatives. In: Merian, E., Frei, R.W., Härdi, W. & Schlatter, C., eds, *Carcinogenic and Mutagenic Metal Compounds*, New York, Gordon and Breach, pp. 3-32

Meyer, A. & Neeb, R. (1985) Determination of cobalt and nickel in some biological matrices—comparison of chelate gas chromatography and adsorption voltammetry (Ger.). *Fresenius Z. anal. Chem.*, 321, 235-241

Miehlke, R., Henke, G. & Ehrenbrink, H. (1981) Cobalt and chromium concentrations analysed by activation in synovial fluid and in blood after implantation of knee prostheses (Ger.). *Z. Orthopaed.*, 119, 767-768

Mikkelsen, S., Raffn, E., Altman, D., Groth, S. & Christensen, J.M. (1984) *Helbred og Kobolt. En Tvaersnitsundersøgelse af Plattemalere* [Health and Cobalt. A Cross-sectional Study of Plate Painters], Copenhagen, Arbejdsmitjøfondet

Mitchell, D.F., Shankwalker, G.B. & Shazer, S. (1960) Determining the tumorigenicity of dental materials. *J. dent. Res.*, 39, 1023-1028

Miyaki, M., Akamatsu, N., Ono, T. & Koyama, H. (1979) Mutagenicity of metal cations in cultured cells from Chinese hamster. *Mutat. Res.*, 68, 259-263

Mochizuki, H. & Kada, T. (1982) Antimutagenic action of cobaltous chloride on Trp-P-1-induced mutations in *Salmonella typhimurium* TA98 and TA1538. *Mutat. Res.*, 95, 145-157

Mohiuddin, S.M., Taskar, P.K., Rheault, M., Roy, P.-E., Chenard, J. & Morin, Y. (1970) Experimental cobalt cardiomyopathy. *Am. Heart J.*, 80, 532-543

Mollenhauer, H.H., Corrier, D.E., Clark, D.E., Hare, M.F. & Elissalde, M.H. (1985) Effects of dietary cobalt on testicular structure. *Virchows. Arch.* (Cell Pathol.), 49, 241-248

Morgan, L.G. (1983) A study into the health and mortality of men exposed to cobalt and oxides. *J. Soc. occup. Med.*, 33, 181-186

Morin, Y. & Daniel, P. (1967) Quebec beer-drinkers' cardiomyopathy: etiological considerations. *Can. med. Assoc. J.*, 97, 926-928

Morin, Y., Têtu, A. & Mercier, G. (1971) Cobalt cardiomyopathy: clinical aspects. *Br. Heart J.*, 33 (Suppl.), 175-178

Morral, F.R. (1979) Cobalt compounds. In: Mark, H.F., Othmer, D.F., Overberger, C.G., Seaborg, G.T. & Grayson, M., eds, *Kirk-Othmer Encyclopedia of Chemical Technology*, Vol. 6, 3rd ed., New York, NY, John Wiley & Sons, pp. 495-510

Mur, J.M., Moulin, J.J., Charruyer-Seinerra, M.P. & Lafitte, J. (1987) A cohort mortality study among cobalt and sodium workers in an electrochemical plant. *Am. J. ind. Med.*, *11*, 75-81

Nadeenko, V.G., Lenchenko, V.G., Saichenko, S.P., Arkhipenko, T.A. & Radovskaya, T.L. (1980) Embryotoxic action of cobalt administered per os (Russ.). *Gig. Sanit.*, *2*, 6-8

Nadkarni, R.A. & Ehmann, W.D. (1970) Further analyses of University of Kentucky reference and alkaloid series cigarettes by instrumental neutron activation analysis. *Radiochem. radioanal. Lett.*, *4*, 325-335

National Institute for Occupational Safety and Health (1984a) *Manual of Analytical Methods. Elements*, 3rd ed., Cincinnati, OH, pp. 7300-1—7300-5

National Institute for Occupational Safety and Health (1984b) *Manual of Analytical Methods. Cobalt and Compounds*, 3rd ed., Cincinnati, OH, pp. 7027-1—7027-3

National Institute for Occupational Safety and Health (1985) *Manual of Analytical Methods. Elements in Blood or Tissue*, 3rd ed., Cincinnati, OH, p. 8005-1

National Institute for Occupational Safety and Health (1988) NIOSH recommendations for occupational safety and health standards 1988. *Morb. Mort. wkly Rep.*, *37*, 4-7, 8-9

National Library of Medicine (1989) *Hazardous Substances Data Bank: Cobalt Chloride*, Bethesda, MD

National Research Council (1977) *Drinking Water and Health*, Vol. 1, Washington DC, pp. 138-140, 206-213, 216-221, 246-250, 302-304, 308

National Swedish Board of Occupational Safety and Health (1987) *Hygieniska Gränsvärden* [Hygienic Limit Values] (Ordinance 1987:12), Solna, p. 28

Newton, D. & Rundo, J. (1970) The long-term retention of inhaled cobalt-60. *Health Phys.*, *21*, 377-384

Nickel Development Institute (1987) *Nickel Base Alloys*, Toronto, Ontario

Niebrój, T.K. (1967) Influence of cobalt on the histophysiology of mouse testis. *Endokrynol. pol.*, *18*, 1-13

Nishioka, H. (1975) Mutagenic activities of metal compounds in bacteria. *Mutat. Res.*, *31*, 185-189

Nodiya, P.I. (1972) A study of the cobalt and nickel balance in students of an occupational technical school (Russ.). *Gig. Sanit.*, *5*, 108-109

Nowak, H.F. (1961) The pathogenesis of neoplasia in the rabbit under the influence of polyester resin additions (Pol.). *Akad. Medycz. Jul. Marchl. Bialymstoku*, *7*, 323-348

Nowak, H.F. (1966) Neoplasia in mouse skeletal muscles under the influence of polyester resin activator. *Arch. immunol. ther. exp.*, *14*, 774-778

Numazawa, S., Oguro, T., Yoshida, T. & Kuroiwa, Y. (1989) Comparative studies on the inducing effects of cobalt chloride and co-protoporphyrin on hepatic ornithine decarboxylase and heme oxygenase in rats. *J. Pharmacobio-dyn.*, *12*, 50-59

Nuodex (1986) *Material Safety Data Sheet 1013507: Chemical Naphthenate Solution*, Piscataway, NJ

Ogawa, H.I., Sakata, K., Inouye, T., Jyosui, S., Niyitani, Y., Kakimoto, K., Morishita, M., Tsuruta, S. & Kato, Y. (1986) Combined mutagenicity of cobalt(II) salt and heteroaromatic compounds in *Salmonella typhimurium*. *Mutat. Res.*, *172*, 97-104

Ogawa, H.I., Sakata, K., Liu, S.-Y., Mino, H., Tsuruta, S. & Kato, Y. (1987) Cobalt(II) salt-quinoline compound interaction: combined mutagenic activity in *Salmonella typhimurium* and strength of coordinate bond in the mixtures. *Jpn. J. Genet.*, 62, 485-491

Ogawa, H.I., Liu, S.-Y., Sakata, K., Niyitani, Y., Tsuruta, S. & Kato, Y. (1988) Inverse correlation between combined mutagenicity in *Salmonella typhimurium* and strength of coordinate bond in mixtures of cobalt (II) and 4-substituted pyridines. *Mutat. Res.*, 204, 117-121

Onkelinx, C. (1976) Compartment analysis of cobalt(II) metabolism in rats of various ages. *Toxicol. appl. Pharmacol.*, 38, 425-438

Orten, J.M. & Bucciero, M.C. (1948) The effect of cysteine, histidine, and methionine on the production of polycythemia by cobalt. *J. biol. Chem.*, 176, 961-968

Oskarsson, A., Reid, M.C. & Sunderman, F.W., Jr (1981) Effects of cobalt chloride, nickel chloride, and nickel subsulfide upon erythropoiesis in rats. *Ann. clin. Lab. Sci.*, 11, 165-172

Ostapczuk, P., Valenta, P., Stoeppler, M. & Nürnberg, H.W. (1983) Voltammetric determination of nickel and cobalt in body fluids and other biological materials. In: Brown, S.S. & Savory, J., eds, *Chemical Toxicology and Clinical Chemistry of Metals*, London, Academic Press, pp. 61-64

Ostapczuk, P., Goedde, M., Stoeppler, M. & Nürnberg, H.W. (1984) Control and routine determination of Zn, Cd, Pb, Cu, Ni and Co with differential pulse voltammetry in materials from the German environmental specimen bank (Ger.). *Fresenius Z. anal. Chem.*, 317, 252-256

Paternain, J.L., Domingo, J.L. & Corbella, J. (1988) Developmental toxicity of cobalt in the rat. *J. Toxicol. environ. Health*, 24, 193-200

Paton, G.R. & Allison, A.C. (1972) Chromosome damage in human cell cultures induced by metal salts. *Mutat. Res.*, 16, 332-336

Pazzaglia, U.E., Minoia, C., Gualtieri, G., Gualtieri, I., Riccardi, C. & Ceciliani, L. (1986) Metal ions in body fluids after arthroplasty. *Acta orthop. scand.*, 57, 415-418

Pedigo, N.G. (1988) Effects of acute and chronic administration of cobaltous chloride on male reproductive function in mice (Abstract). *Diss. Abstr. int. B*, 48, 2279-B

Pedigo, N.G., George, W.J. & Anderson, M.B. (1988) The effect of acute and chronic exposure to cobalt on male reproduction in mice. *Reprod. Toxicol.*, 2, 45-53

Pellet, F., Perdrix, A., Vincent, M. & Mallion, J.-M. (1984) Biological levels of urinary cobalt (Fr.). *Arch. Mal. prof.*, 45, 81-85

Penman, H.G. & Ring, P.A. (1984) Osteosarcoma in association with total hip replacement. *J. Bone Joint Surg.*, 66B, 632-634

Perone, V.B., Moffitt, A.E., Jr, Possick, P.A., Key, M.M., Danzinger, S.J. & Gellin, G.A. (1974) The chromium, cobalt and nickel contents of American cement and their relationship to cement dermatitis. *Am. ind. Hyg. Assoc. J.*, 35, 301-306

Pfannhauser, W. (1988) *Essentielle Spurenelemente in der Nahrung* [Essential Trace Elements in the Environment], Berlin, Springer-Verlag, pp. 67-79

Pickston, L., Lewin, J.F., Drysdale, J.M., Smith, J.M. & Bruce, J. (1983) Determination of potentially toxic metals in human livers in New Zealand. *J. anal. Toxicol.*, 7, 2-6

Planinsek, F. & Newkirk, J.B. (1979) Cobalt and cobalt alloys. In: Mark, H.F., Othmer, D.F., Overberger, C.G., Seaborg, G.T. & Grayson, M., eds, *Kirk-Othmer Encyclopedia of Chemical Technology*, 3rd ed., Vol. 6, New York, John Wiley & Sons, pp. 481-494

Posma, F.D. & Dijstelberger, S.K. (1985) Serum and urinary cobalt levels as indicators of cobalt exposure in hard metal workers. In: Lekkas, T.D., ed., *Proceedings of an International Conference, Heavy Metals in the Environment, Athens, September 1985*, Luxembourg, Commission of the European Communities, pp. 89-91

Prazmo, W., Balbin, E., Baranowska, H., Ejchart, A. & Putrament, A. (1975) Manganese mutagenesis in yeast. II. Condition of induction and characteristics of mitochondrial respiratory deficient *Saccharomyces cerevisiae* mutants induced with manganese and cobalt. *Genet. Res. Camb.*, 26, 21-29

Putrament, A., Baranowska, H., Ejchart, A. & Jachymczyk, W. (1977) Manganese mutagenesis in yeast. VI. Mn^{2+} uptake, mitDNA replication and E^R induction. Comparison with other divalent cations. *Mol. gen. Genet.*, 151, 69-76

Raffn, E., Mikkelsen, S., Altman, D.G., Christensen, J.M. & Groth, S. (1988) Health effects due to occupational exposure to cobalt blue dye among plate painters in a porcelain factory in Denmark. *Scand. J. Work Environ. Health*, 14, 378-384

Raithel, H.J., Hennig, F. & Schaller, K.H. (1989) Nickel-, chromium- and cobalt-burden of human body after joint endoprosthesis (Ger.). In: *Frankfurter Implantat-Kongress 'Neuere Biomaterialien für die Endoprothetik'* [First Frankfurt Implant Congress on New Biomaterials for Endoprosthesis], *30-31 October 1989, Frankfurt*, pp. 1-8

Resende de Souza-Nazareth, H. (1976) Effect of cobalt chloride on disjunction. Preliminary results (Port.). *Cien. Cult.*, 28, 1472-1475

Reshetkina, L.P. (1965) On the content of zinc, iron, copper and cobalt in some products entering the children's diet in Prikarpatie (Russ.). *Vopr. Pitan.*, 24, 68-72

Reynolds, J.E.F., ed. (1989) *Martindale, The Extra Pharmacopoeia*, London, The Pharmaceutical Press, pp. 1260-1261, 1559

Rhoads, K. & Sanders, C.L. (1985) Lung clearance, translocation, and acute toxicity of arsenic, beryllium, cadmium, cobalt, lead, selenium, vanadium, and ytterbium oxides following deposition in rat lung. *Environ. Res.*, 36, 359-378

Ridgway, L.P. & Karnofsky, D.A. (1952) The effects of metals on the chick embryo: toxicity and production of abnormalities in development. *Ann. N.Y. Acad. Sci.*, 55, 203-215

Robison, S.H., Cantoni, O. & Costa, M. (1982) Strand breakage and decreased molecular weight of DNA induced by specific metal compounds. *Carcinogenesis*, 3, 657-662

Rona, G. (1971) Experimental aspects of cobalt cardiomyopathy. *Br. Heart J.*, 33 (Suppl.), 171-174

von Rosen, G. (1964) Mutations induced by the action of metal ions in Pisum. II. Further investigations on the mutagenic action of metal ions and comparison with the activity of ionizing radiation. *Hereditas*, 51, 89-134

Roskill Information Services Ltd (1989) *The Economics of Cobalt*, 6th ed., London, pp. i-iii, 1-12, 19, 81-82, 120-130, 141-156, 202-212

Rossman, T.G. (1981) Effect of metals on mutagenesis and DNA repair. *Environ. Health Perspect.*, 40, 189-195

Rossman, T.G., Molina, M. & Meyer, L.W. (1984) The genetic toxicology of metal compounds: I. Induction of λ prophage in *E. coli* WP2$_s$(λ). *Environ. Mutagenesis*, 6, 59-69

Rystedt, I. & Fischer, T. (1983) Relationship between nickel and cobalt sensitization in hard metal workers. *Contact Derm.*, 9, 195-200

Ryu, R.K.N., Bovill, E.G., Jr, Skinner, H.B. & Murray, W.R. (1987) Soft tissue sarcoma associated with aluminum oxide ceramic total hip arthroplasty. A case report. *Clin. Orthopaed. rel. Res.*, 216, 207-212

Saknyn, A.V. & Shabynina, N.K. (1970) Some statistical data on the carcinogenous hazards for workers engaged in the production of nickel from oxidized ores (Russ.). *Gig. Tr. prof. Zabol.*, 14, 10-13

Saknyn, A.V. & Shabynina, N.K. (1973) Epidemiology of malignant new growths at nickel smelters (Russ.). *Gig. Tr. prof. Zabol.*, 17, 25-29

Sampson, B. (1988) Determination of cobalt in plasma and urine by electrothermal atomisation atomic absorption spectrometry using palladium matrix modification. *J. anal. atomic Spectrom.*, 3, 465-469

Sandusky, G.E., Crawford, M.P. & Roberts, E.D. (1981) Experimental cobalt cardiomyopathy in the dog: a model for cardiomyopathy in dogs and man. *Toxicol. appl. Pharmacol.*, 60, 263-278

Sax, N.I. & Lewis, R.J. (1987) *Hawley's Condensed Chemical Dictionary*, 11th ed., New York, Van Nostrand Reinhold, pp. 291-296

Scansetti, G., Lamon, S., Talarico, S., Botta, G.C., Spinelli, P., Sulotto, F. & Fantoni, F. (1985) Urinary cobalt as a measure of exposure in the hard metal industry. *Int. Arch. occup. environ. Health*, 57, 19-26

Schaller, K.H., Angerer, J., Lehnert, G., Valentin, H. & Weltle, D. (1987) External quality control programmes in the toxicological analysis of biological material in the field of occupational medicine—experiences from three round-robins in the Federal Republic of Germany. *Fresenius Z. anal. Chem.*, 326, 643-646

Scherrer, M. & Maillard, J.-M. (1982) Hard-metal pneumopathy (Ger.). *Schweiz. med. Wochenschr.*, 112, 198-207

Schormüller, J. (1974) *Lehrbuch der Lebensmittelchemie* [Textbook of Food Chemistry], Berlin, Springer-Verlag, p. 118

Schrauzer, G.N. (1989) Cobalt. In: Merian, E., ed., *Metals and Their Compounds in the Environment. Occurrence, Analysis, and Biological Relevance*, Weinheim, VCH-Verlag, pp. 2-8-1—2-8-11

Schultz, P.N. Warren, G., Kosso, C. & Rogers, S. (1982) Mutagenicity of a series of hexacoordinate cobalt(III) compounds. *Mutat. Res.*, 102, 393-400

Schulz, G. (1978) Giant-cell tumours after exposure to a dust containing cobaltous phthalocyanine (Merx-catalyst) (Ger.). *Staub-Reinhalt. Luft*, 38, 480-481

Schumacher-Wittkopf, E. & Angerer, J. (1981) A practical method for the determination of cobalt in urine (Ger.). *Int. Arch. occup. environ. Health*, 49, 77-81

Shabaan, A.A., Marks, V., Lancaster, M.C. & Dufeu, G.N. (1977) Fibrosarcomas induced by cobalt chloride ($CoCl_2$) in rats. *Lab. Anim.*, 11, 43-46

Shedd, K.B. (1988) Cobalt. In: *Minerals Yearbook 1988*, Washington DC, Bureau of Mines, US Department of the Interior, pp. 1-10

Shedd, K.B. (1989) Cobalt in October 1989. In: *Mineral Industry Surveys: Cobalt Monthly*, Washington DC, Bureau of Mines, Department of the Interior, p. 1

Shedd, K.B. (1990) Cobalt. In: *Mineral Commodity Summaries 1990*, Washington DC, Bureau of Mines, Department of the Interior, pp. 48-49

Shepherd Chemical Co. (1986a) *Technical Data Sheet: Cobalt Nitrate, Technical*, Cincinnati, OH

Shepherd Chemical Co. (1986b) *Technical Data Sheet: Cobalt Sulfate, Technical*, Cincinnati, OH

Shepherd Chemical Co. (1987a) *Technical Data Sheet: Cobalt Acetate, Technical*, Cincinnati, OH

Shepherd Chemical Co. (1987b) *Technical Data Sheet: Cobalt Carbonate, Technical*, Cincinnati, OH

Shepherd Chemical Co. (1987c) *Technical Data Sheet: Cobalt Chloride, Technical*, Cincinnati, OH

Shepherd Chemical Co. (1987d) *Technical Data Sheet: Cobalt Sulfate, Monohydrate*, Cincinnati, OH

Shepherd Chemical Co. (1988a) *Technical Data Sheet: Cobalt Hydroxide, Technical*, Cincinnati, OH

Shepherd Chemical Co. (1988b) *Material Safety Data Sheet: Cobalt Sulfate, Monohydrate*, Cincinnati, OH

Shepherd Chemical Co. (1989a) *Material Safety Data Sheet: Cobalt Acetate, Tetrahydrate*, Cincinnati, OH

Shepherd Chemical Co. (1989b) *Material Safety Data Sheet: Cobalt Carbonate*, Cincinnati, OH

Shepherd Chemical Co. (1989c) *Material Safety Data Sheet: Cobaltous Chloride*, Cincinnati, OH

Shepherd Chemical Co. (1989d) *Material Safety Data Sheet: Cobalt Hydroxide*, Cincinnati, OH

Shepherd Chemical Co. (1989e) *Material Safety Data Sheet: Cobalt Naphthenate*, Cincinnati, OH

Shepherd Chemical Co. (1989f) *Material Safety Data Sheet: Cobalt Naphthenate Mixture*, Cincinnati, OH

Shepherd Chemical Co. (1989g) *Material Safety Data Sheet: Cobalt Nitrate, Hexahydrate*, Cincinnati, OH

Shepherd Chemical Co. (1989h) *Material Safety Data Sheet: Cobalt Sulfate, Heptahydrate*, Cincinnati, OH

Shirakawa, T., Kusaka, Y., Fujimura, N., Goto, S., Kato, M., Heki, S. & Morimoto, K. (1989) Occupational asthma from cobalt sensitivity in workers exposed to hard metal dust. *Chest*, 95, 29-37

Singh, I. (1983) Induction of reverse mutation and mitotic gene conversion by some metal compounds in *Saccharomyces cerevisiae*. *Mutat. Res.*, 117, 149-152

Sirover, M.A. & Loeb, L.A. (1976) Metal activation of DNA synthesis. *Biochem. biophys. Res. Commun.*, *70*, 812-817

Sisco, W.E., Bastian, W.E. & Weierich, E.G. (1982) Naphthenic acids. In: Mark, H.F., Othmer, D.F., Overberger, C.G., Seaborg, G.T. & Grayson, M., eds, *Kirk-Othmer Encyclopedia of Chemical Technology*, Vol. 15, 3rd ed., New York, John Wiley & Sons, pp. 749-753

Sjögren, I., Hillerdal, G., Andersson, A. & Zetterström, O. (1980) Hard metal lung disease: importance of cobalt in coolants. *Thorax*, *35*, 653-659

Smith, T., Edmonds, C.J. & Barnaby, C.F. (1972) Absorption and retention of cobalt in man by whole-body counting. *Health Phys.*, *22*, 359-367

Sorbie, J., Olatunbosun, D., Corbett, W.E.N., Valberg, L.S., Ludwig, J. & Jones, C. (1971) Cobalt excretion test for the assessment of body iron stores. *Can. med. Assoc. J.*, *104*, 777-782

Söremark, R., Diab, M. & Arvidson, K. (1979) Autoradiographic study of distribution patterns of metals which occur as corrosion products from dental restorations. *Scand. J. dent. Res.*, *87*, 450-458

Speijers, G.J.A., Krajnc, E.I., Berkvens, J.M. & van Logten, M.J. (1982) Acute oral toxicity of inorganic cobalt compounds in rat. *Food chem. Toxicol.*, *20*, 311-314

Sprince, N.L., Oliver, L.C., Eisen, E.A., Greene, R.A. & Chamberlin, R.I. (1988) Cobalt exposure and lung disease in tungsten carbide production. *Am. Rev. respir. Dis.*, *138*, 1220-1226

Spring, J.A., Robertson, J. & Buss, D.H. (1979) Trace nutrients. 3. Magnesium, copper, zinc, vitamin B_6, vitamin B_{12} and folic acid in the British household food supply. *Br. J. Nutr.*, *41*, 487-493

Steinhoff, D. & Mohr, U. (1991) On the question of a carcinogenic action of cobalt-containing compounds. *Exp. Pathol.* (in press)

Stenberg, T. (1983) The distribution in mice of radioactive cobalt administered by two different methods. *Acta odontol. scand.*, *41*, 143-148

Stoner, G.D., Shimkin, M.B., Troxell, M.C., Thompson, T.L. & Terry, L.S. (1976) Test for carcinogenicity of metallic compounds by the pulmonary tumor response in strain A mice. *Cancer Res.*, *36*, 1744-1747

Sullivan, J., Parker, M. & Carson, S.B. (1968) Tissue cobalt content in 'beer drinkers' myocardiopathy'. *J. Lab. clin. Med.*, *71*, 893-896

Sunderman, F.W., Jr, Hopfer, S.M., Swift, T., Rezuke, W.N., Ziebka, L., Highman, P., Edwards, B., Folcik, M. & Gossling, H.R. (1989) Cobalt, chromium, and nickel concentrations in body fluids of patients with porous-coated knee or hip prostheses. *J. orthoped. Res.*, *7*, 307-315

Suvorov, I.M. & Cekunova, M.P. (1983) Cobalt, alloys and compounds. In: Parmeggiani, L. ed., *Encyclopedia of Occupational Health and Safety*, 3rd (rev.) ed., Geneva, International Labour Office, pp. 493-495

Swann, M. (1984) Malignant soft-tissue tumour at the site of a total hip replacement. *J. Bone Joint Surg.*, *66B*, 629-631

Swanson, S.A.V., Freeman, M.A.R. & Heath, J.C. (1973) Laboratory tests on total joint replacement prostheses. *J. Bone Joint Surg.*, *55B*, 759-773

Swierenga, S.H.H., Gilman, J.P.W. & McLean, J.R. (1987) Cancer risk from inorganics. *Cancer Metastasis Rev.*, *6*, 113-154

Takahashi, H. & Koshi, K. (1981) Solubility and cell toxicity of cobalt, zinc and lead. *Ind. Health*, *19*, 47-59

Taylor, D.M. (1962) The absorption of cobalt from the gastro-intestinal tract of the rat. *Phys. Med. Biol.*, *6*, 445-451

Taylor, A. & Marks, V. (1978) Cobalt: a review. *J. hum. Nutr.*, *32*, 165-177

Tayton, K.J.J. (1980) Ewing's sarcoma at the site of a metal plate. *Cancer*, *45*, 413-415

Thomas, R.H.M., Rademarker, M., Goddard, N.J. & Munro, D.D. (1987) Severe eczema of the hands due to an orthopaedic plate made of Vitallium. *Br. med. J.*, *294*, 106-107

Tso, W.-W. & Fung, W.-P. (1981) Mutagenicity of metallic cations. *Toxicol. Lett.*, *8*, 195-200

United Nations Environment Programme (1990) *International Register of Potentially Toxic Chemicals, Recommendations—Legal Mechanisms*, Geneva

US Environmental Protection Agency (1983) *Methods for Chemical Analysis of Water and Wastes* (EPA 600/4/79-020), Cincinnati, OH, Environmental Monitoring and Support Laboratory

US Occupational Safety and Health Administration (1989) Air contaminants. *US Code fed. Regul.*, *Title 29*, Part 1910.1000

Varo, P. & Koivistoinen, P. (1980) Mineral element composition of Finnish foods. XII. General discussion and nutritional evaluation. *Acta agric. scand.*, Suppl. 22, 165-171

Versieck, J., Hoste, J., Barbier, F., Steyaert, H., De Rudder, J. & Michels, H. (1978) Determination of chromium and cobalt in human serum by neutron activation analysis. *Clin. Chem.*, *24*, 303-308

Vilaplana, J., Grimalt, F., Romaguera, C. & Mascaro, J.M. (1987) Cobalt content of household cleaning products. *Contact Derm.*, *16*, 139-141

Voroshilin, S.I., Plotko, E.G., Fink, T.V. & Nikiforova, V.J. (1978) Cytogenetic effect of inorganic compounds of tungsten, zinc, cadmium, and cobalt on animal and human somatic cells (Russ.). *Tsitol. Genet.*, *12*, 241-243

Ward, J.J., Thornbury, D.D., Lemons, J.E. & Dunham, W.K. (1990) Metal-induced sarcoma: a case report and literature review. *Clin. Orthopaed. rel. Res.*, *252*, 299-306

Weast, R.C., ed. (1988) *CRC Handbook of Chemistry and Physics*, 70th ed., Boca Raton, FL, CRC Press, pp. B-13—B-14, B-86—B-88

Weber, P.C. (1986) Epithelioid sarcoma in association with total knee replacement. A case report. *J. Bone Joint Surg.*, *68B*, 824-826

Wedrychowski, A., Schmidt, W.N. & Hnilica, L.S. (1986) DNA-protein crosslinking by heavy metals in Novikoff hepatoma. *Arch. Biochem. Biophys.*, *251*, 397-402

Wehner, A.P. & Craig, D.K. (1972) Toxicology of inhaled NiO and CoO in Syrian golden hamsters. *Am. ind. Hyg. Assoc. J.*, *33*, 146-155

Wehner, A.P., Busch, R.H., Olson, R.J. & Craig, D.K. (1977) Chronic inhalation of cobalt oxide and cigarette smoke by hamsters. *Am. ind. Hyg. Assoc. J.*, *38*, 338-346

Wide, M. (1984) Effect of short-term exposure to five industrial metals on the embryonic and fetal development of the mouse. *Environ. Res.*, *33*, 47-53

Wilke, K.T. (1964) Preparation of crystalline metal oxides from molten solutions (Ger.). *Z. anorg. allgem. Chem.*, *330*, 164-169

Wong, P.K. (1988) Mutagenicity of heavy metals. *Bull. environ. Contam. Toxicol.*, *40*, 597-603

Wyttenbach, A., Bajo, S. & Haekkinen, A. (1976) Determination of 16 elements in tobacco by neutron activation analysis. *Beitr. Tabakforsch.*, *8*, 247-249

Yamagata, N., Kurioka, W. & Shimizu, T. (1963) Balance of cobalt in Japanese people and diet. *J. Radiat. Res.*, *4*, 8-15

Yamamoto, K., Inoue, S., Yamazaki, A., Yoshinaga, T. & Kawanishi, S. (1989) Site-specific DNA damage induced by cobalt(II) ion and hydrogen peroxide: role of singlet oxygen. *Chem. Res. Toxicol.*, *2*, 234-239

Young, R.S., ed. (1960) *Cobalt. Its Chemistry, Metallurgy, and Uses*, New York, Reinhold, pp. 1-10

SUMMARY OF FINAL EVALUATIONS

Agent	Degree of evidence of carcinogenicity[a,b]		Overall evaluation of carcinogenicity to humans[b]
	Human	Animal	
Bromodichloromethane	I	S	2B
Bromoethane	ND	L	3
Bromoform	I	L	3
Chlorinated drinking-water	I	I	3
Chlorodibromomethane	I	L	3
Chloroethane	ND	L	3
Cobalt and cobalt compounds	I		2B
Cobalt metal powder		S	
Cobalt-aluminium-chromium spinel		I	
Cobalt-chromium-molybdenum alloys		L	
Cobalt[II] oxide		S	
Cobalt[II] sulfide		L	
Cobalt[II] chloride		L	
Cobalt[II,III] oxide		I	
Cobalt naphthenate		I	
Cobalt[III] acetate		I	
Halogenated acetonitriles			
Bromochloroacetonitrile	ND	I	3
Chloroacetonitrile	ND	I	3
Dibromoacetonitrile	ND	I	3
Dichloroacetonitrile	ND	I	3
Trichloroacetonitrile	ND	I	3
Hypochlorite salts	ND	I	3
Sodium chlorite	ND	I	3
1,1,2-Trichloroethane	ND	L	3

[a] I, inadequate evidence; S, sufficient evidence; L, limited evidence; ND, no data
[b] For definitions of degrees of evidence and groupings of evaluations, see Preamble, pp. 30-33.

APPENDIX 1

SUMMARY TABLES OF GENETIC AND RELATED EFFECTS

APPENDIX 1

Summary table of genetic and related effects of chlorinated drinking-water

Nonmammalian systems												Mammalian systems																			
Prokary-otes		Lower eukaryotes				Plants				Insects				In vitro												In vivo					
														Animal cells						Human cells						Animals				Humans	
D	G	D	R	G	C	A	D	G	C	R	G	C	A	D	G	S	M	C	A	T	I	D	G	S	M	C	A	T	I	D	G	S	M	C	DL	A	D	S	M	C	A
Surface water, chlorinated, not concentrated																																									
−				$-^1$																	−																				
Surface water, chlorinated, concentrated																																									
+																$-^1$	$+^1$	$+^1$	$+^1$																						
Ground and spring water, chlorinated, concentrated																																									
+																$+^1$		$-^1$																							
Surface water, chlorinated and either chlorine dioxide or ozone treated, concentrated																																									
+																																									

A, aneuploidy; C, chromosomal aberrations; D, DNA damage; DL, dominant lethal mutation; G, gene mutation; I, inhibition of intercellular communication; M, micronuclei; R, mitotic recombination and gene conversion; S, sister chromatid exchange; T, cell transformation

In completing the tables, the following symbols indicate the consensus of the Working Group with regard to the results for each endpoint:

+ considered to be positive for the specific endpoint and level of biological complexity
$+^1$ considered to be positive, but only one valid study was available to the Working Group
− considered to be negative
$-^1$ considered to be negative, but only one valid study was available to the Working Group
? considered to be equivocal or inconclusive (e.g., there were contradictory results from different laboratories; there were confounding exposures; the results were equivocal)

Summary table of genetic and related effects of sodium chlorite

Nonmammalian systems													Mammalian systems																				
Prokaryotes	Lower eukaryotes				Plants				Insects				In vitro													In vivo							
													Animal cells							Human cells						Animals					Humans			
D	G	A	D	R	G	A	D	G	C	R	G	C	A	D	G	S	M	C	A	T	I	D	G	S	M	C	A	T	I	D	G	S	M	C	DL	A	D	S	M	C	A
$+^1$																	$+^1$														?	$-^1$			$-^1$	$-^1$					

$-^1$ sperm morphology in B6C3F₁ mice

A, aneuploidy; C, chromosomal aberrations; D, DNA damage; DL, dominant lethal mutation; G, gene mutation; I, inhibition of intercellular communication; M, micronuclei; R, mitotic recombination and gene conversion; S, sister chromatid exchange; T, cell transformation

In completing the tables, the following symbols indicate the consensus of the Working Group with regard to the results for each endpoint:

+   considered to be positive for the specific endpoint and level of biological complexity
+¹   considered to be positive, but only one valid study was available to the Working Group
−   considered to be negative
−¹   considered to be negative, but only one valid study was available to the Working Group; sperm morphology in B6C3F₁ mice
?   considered to be equivocal or inconclusive (e.g. there were contradictory results from different laboratories; the results were equivocal)

APPENDIX 1

Summary table of genetic and related effects of sodium hypochlorite

Nonmammalian systems				Mammalian systems			
Prokaryotes	Lower eukaryotes	Plants	Insects	In vitro		In vivo	
				Animal cells	Human cells	Animals	Humans

Prokaryotes	Lower eukaryotes			Plants			Insects				Animal cells								Human cells								Animals							Humans				
D	G	D	R	A	D	G	C	R	G	C	A	D	G	S	M	C	A	T	I	D	G	S	M	C	A	T	I	D	G	S	M	C	DL	A	D	S	M	C	A
+¹	+														+¹								+¹			-¹					-, -¹				-, -¹				

+¹ micronuclei in newt larvae
+¹ sperm morphology in B6C3F₁ mice

A, aneuploidy; C, chromosomal aberrations; D, DNA damage; DL, dominant lethal mutation; G, gene mutation; I, inhibition of intercellular communication; M, micronuclei; R, mitotic recombination and gene conversion; S, sister chromatid exchange; T, cell transformation

In completing the tables, the following symbols indicate the consensus of the Working Group with regard to the results for each endpoint:

+ considered to be positive for the specific endpoint and level of biological complexity
+¹ considered to be positive, but only one valid study was available to the Working Group; micronuclei in newt larvae
− considered to be negative
−¹ considered to be negative, but only one valid study was available to the Working Group; sperm morphology in B6C3F₁ mice
? considered to be equivocal or inconclusive (e.g., there were contradictory results from different laboratories; there were confounding exposures; the results were equivocal)

Summary table of genetic and related effects of bromodichloromethane

Nonmammalian systems															Mammalian systems																				
Prokaryotes	Lower eukaryotes					Plants				Insects					In vitro																In vivo				
															Animal cells								Human cells								Animals					Humans					
D	D	G	R	G	A	A	D	G	C	R	G	C	A		D	G	S	M	C	A	T	I	D	G	S	M	C	A	T	I	D	G	S	M	C	DL	A	D	S	M	C	A
+															+¹		-		+						+¹								+¹	-¹								

A, aneuploidy; C, chromosomal aberrations; D, DNA damage; DL, dominant lethal mutation; G, gene mutation; I, inhibition of intercellular communication; M, micronuclei; R, mitotic recombination and gene conversion; S, sister chromatid exchange; T, cell transformation

In completing the tables, the following symbols indicate the consensus of the Working Group with regard to the results for each endpoint:

+ considered to be positive for the specific endpoint and level of biological complexity
+¹ considered to be positive, but only one valid study was available to the Working Group
− considered to be negative
−¹ considered to be negative, but only one valid study was available to the Working Group
? considered to be equivocal or inconclusive (e.g., there were contradictory results from different laboratories; there were confounding exposures; the results were equivocal)

APPENDIX 1

Summary table of genetic and related effects of bromoform

Nonmammalian systems																	Mammalian systems																			
Prokaryotes		Lower eukaryotes					Plants				Insects			In vitro														In vivo							
														Animal cells							Human cells							Animals							Humans
D	G	D	R	G	A	D	D	G	C	R	G	C	A	D	G	S	M	C	A	T	D	G	S	M	C	A	T	D	G	S	M	C	DL	A	D	S	M	C	A
	?										+¹			+¹	?		+							+¹				-¹		+	?		-¹						

+¹ mitotic arrest in plants

A, aneuploidy; C, chromosomal aberrations; D, DNA damage; DL, dominant lethal mutation; G, gene mutation; I, inhibition of intercellular communication; M, micronuclei; R, mitotic recombination and gene conversion; S, sister chromatid exchange; T, cell transformation

In completing the tables, the following symbols indicate the consensus of the Working Group with regard to the results for each endpoint:

+ considered to be positive for the specific endpoint and level of biological complexity
+¹ considered to be positive, but only one valid study was available to the Working Group
– considered to be negative
–¹ considered to be negative, but only one valid study was available to the Working Group
? considered to be equivocal or inconclusive (e.g. there were contradictory results from different laboratories; there were confounding exposures; the results were equivocal)

Summary table of genetic and related effects of chlorodibromomethane

Nonmammalian systems																	Mammalian systems																				
Prokaryotes		Lower eukaryotes				Plants					Insects						In vitro																In vivo						
																	Animal cells								Human cells								Animals							Humans				
D	G	D	R	G	A	D	A	D	G	C	C	R	G	C	A	D	G	S	M	C	A	T	I	D	G	S	M	C	A	T	I	D	G	S	M	C	DL	A	D	S	M	C	A
+				+¹	-¹															+¹						+¹								+¹		-¹							

A, aneuploidy; C, chromosomal aberrations; D, DNA damage; DL, dominant lethal mutation; G, gene mutation; I, inhibition of intercellular communication; M, micronuclei; R, mitotic recombination and gene conversion; S, sister chromatid exchange; T, cell transformation

In completing the tables, the following symbols indicate the consensus of the Working Group with regard to the results for each endpoint:

+   considered to be positive for the specific endpoint and level of biological complexity
+¹ considered to be positive, but only one valid study was available to the Working Group
−   considered to be negative
−¹ considered to be negative, but only one valid study was available to the Working Group
?   considered to be equivocal or inconclusive (e.g., there were contradictory results from different laboratories; there were confounding exposures; the results were equivocal)

APPENDIX 1

Summary table of genetic and related effects of halogenated acetonitriles

	Nonmammalian systems														Mammalian systems																									
	Prokaryotes		Lower eukaryotes				Plants				Insects				In vitro														In vivo											
															Animal cells							Human cells							Animals							Humans				
	D	G	R	A	D	G	A	D	G	C	R	G	C	A	D	G	S	M	C	A	T	D	G	S	M	C	A	T	D	G	S	M	C	DL	A	D	S	M	C	A
Bromochloroacetonitrile	+																+¹	-¹																						
Chloroacetonitrile		-¹																				+¹																		
Dibromoacetonitrile	-¹										-¹						+¹	-¹																						
Dichloroacetonitrile	+	-¹										+¹					+¹	-¹				+¹																		
Trichloroacetonitrile	-¹																+¹	-¹				+¹																		

-¹ sperm morphology in B6C3F₁ mice

A, aneuploidy; C, chromosomal aberrations; D, DNA damage; DL, dominant lethal mutation; G, gene mutation; I, inhibition of intercellular communication; M, micronuclei; R, mitotic recombination and gene conversion; S, sister chromatid exchange; T, cell transformation

In completing the tables, the following symbols indicate the consensus of the Working Group with regard to the results for each endpoint:

+   considered to be positive for the specific endpoint and level of biological complexity
+¹   considered to be positive, but only one valid study was available to the Working Group
−   considered to be negative
−¹   considered to be negative, but only one valid study was available to the Working Group; sperm morphology in mice
?   considered to be equivocal or inconclusive (e.g., there were contradictory results from different laboratories; there were confounding exposures; the results were equivocal)

Summary table of genetic and related effects of bromoethane

Nonmammalian systems													Mammalian systems																			
Prokary-otes	Lower eukaryotes			Plants			Insects					*In vitro*													*In vivo*							
												Animal cells							Human cells						Animals						Humans	
D	G	D	R	G	A	D	G	C	G	R	G	C	A	D	G	S	M	C	A	T	I	D	G	S	M	C	A	T	I	D	G	S	M	C	DL	A	D	S	M	C	A
+											$-^1$				$+^1$			$-^1$																							

A, aneuploidy; C, chromosomal aberrations; D, DNA damage; DL, dominant lethal mutation; G, gene mutation; I, inhibition of intercellular communication; M, micronuclei; R, mitotic recombination and gene conversion; S, sister chromatid exchange; T, cell transformation

In completing the tables, the following symbols indicate the consensus of the Working Group with regard to the results for each endpoint:

+ considered to be positive for the specific endpoint and level of biological complexity
$+^1$ considered to be positive, but only one valid study was available to the Working Group
− considered to be negative
$-^1$ considered to be negative, but only one valid study was available to the Working Group
? considered to be equivocal or inconclusive (e.g., there were contradictory results from different laboratories; there were confounding exposures; the results were equivocal)

APPENDIX 1

Summary table of genetic and related effects of chloroethane

Nonmammalian systems											Mammalian systems																		
Prokary-otes	Lower eukaryotes			Plants			Insects				In vitro													In vivo					
											Animal cells							Human cells						Animals				Humans	
D	G	D	R	G	A	D	G	C	R	G	C	A	D	G	S	M	C	A	T	I	D	G	S	M	C	A	T	I	D	G	S	M	C	DL	A	D	S	M	C	A
+⁻¹																					⁻¹																			

A, aneuploidy; C, chromosomal aberrations; D, DNA damage; DL, dominant lethal mutation; G, gene mutation; I, inhibition of intercellular communication; M, micronuclei; R, mitotic recombination and gene conversion; S, sister chromatid exchange; T, cell transformation

In completing the table, the following symbols indicate the consensus of the Working Group with regard to the results for each endpoint:

+   considered to be positive for the specific endpoint and level of biological complexity
+¹ considered to be positive, but only one valid study was available to the Working Group
–   considered to be negative
–¹ considered to be negative, but only one valid study was available to the Working Group
?   considered to be equivocal or inconclusive (e.g., there were contradictory results from different laboratories; there were confounding exposures; the results were equivocal)

Summary table of genetic and related effects of 1,1,2-trichloroethane

Nonmammalian systems													Mammalian systems																			
Prokaryotes	Lower eukaryotes				Plants				Insects				In vitro															In vivo				
													Animal cells									Human cells						Animals				Humans
D	G	D	R	G	A	D	G	C	R	G	C	A	D	G	S	M	C	A	T	I	D	G	S	M	C	A	T	D	G	S	M	C	DL	A	D	S	M	C	A
-					+¹											+¹														?*									

*,+¹ DNA binding, S-phase induction; –¹ unscheduled DNA synthesis

A, aneuploidy; C, chromosomal aberrations; D, DNA damage; DL, dominant lethal mutation; G, gene mutation; I, inhibition of intercellular communication; M, micronuclei; R, mitotic recombination and gene conversion; S, sister chromatid exchange; T, cell transformation

In completing the tables, the following symbols indicate the consensus of the Working Group with regard to the results for each endpoint:

+ considered to be positive for the specific endpoint and level of biological complexity
+¹ considered to be positive, but only one valid study was available to the Working Group
− considered to be negative
−¹ considered to be negative, but only one valid study was available to the Working Group
? considered to be equivocal or inconclusive (e.g., there were contradictory results from different laboratories; there were confounding exposures; the results were equivocal)

APPENDIX 1

Summary table of genetic and related effects of cobalt and cobalt compounds

	Nonmammalian systems															Mammalian systems																							
	Prokaryotes		Lower eukaryotes				Plants				Insects					In vitro																In vivo						
																Animal cells								Human cells								Animals							Humans				
	D	G	D	R	G	A	A	D	G	C	R	G	C	A	D	G	S	M	C	A	T	I	D	G	S	M	C	A	T	I	D	G	S	M	C	DL	A	D	S	M	C	A
Cobalt(II) chloride	?	−		+	+										+¹	+¹	+¹						+	+¹	+¹				+¹													+¹
Cobalt sulfate									+¹																			−														
Cobalt nitrate								+¹													+¹																					
Cobalt acetate																				+¹																						
Cobalt molybdenate																				+¹																						
Cobalt sulfide															+¹					+¹																						

A, aneuploidy; C, chromosomal aberrations; D, DNA damage; DL, dominant lethal mutation; G, gene mutation; I, inhibition of intercellular communication; M, micronuclei; R, mitotic recombination and gene conversion; S, sister chromatid exchange; T, cell transformation

In completing the tables, the following symbols indicate the consensus of the Working Group with regard to the results for each endpoint:

+ considered to be positive for the specific endpoint and level of biological complexity

+¹ considered to be positive, but only one valid study was available to the Working Group

− considered to be negative

−¹ considered to be negative, but only one valid study was available to the Working Group; sperm morphology in mice

? considered to be equivocal or inconclusive (e.g., there were contradictory results from different laboratories; there were confounding exposures; the results were equivocal)

APPENDIX 2

ACTIVITY PROFILES
FOR GENETIC AND RELATED EFFECTS

APPENDIX 2

ACTIVITY PROFILES
FOR GENETIC AND RELATED EFFECTS

Methods

The x-axis of the activity profile (Waters *et al.*, 1987, 1988) represents the bioassays in phylogenetic sequence by endpoint, and the values on the y-axis represent the logarithmically transformed lowest effective doses (LED) and highest ineffective doses (HID) tested. The term 'dose', as used in this report, does not take into consideration length of treatment or exposure and may therefore be considered synonymous with concentration. In practice, the concentrations used in all the in-vitro tests were converted to µg/ml, and those for in-vivo tests were expressed as mg/kg bw. Because dose units are plotted on a log scale, differences in molecular weights of compounds do not, in most cases, greatly influence comparisons of their activity profiles. Conventions for dose conversions are given below.

Profile-line height (the magnitude of each bar) is a function of the LED or HID, which is associated with the characteristics of each individual test system – such as population size, cell-cycle kinetics and metabolic competence. Thus, the detection limit of each test system is different, and, across a given activity profile, responses will vary substantially. No attempt is made to adjust or relate responses in one test system to those of another.

Line heights are derived as follows: for negative test results, the highest dose tested without appreciable toxicity is defined as the HID. If there was evidence of extreme toxicity, the next highest dose is used. A single dose tested with a negative result is considered to be equivalent to the HID. Similarly, for positive results, the LED is recorded. If the original data were analysed statistically by the author, the dose recorded is that at which the response was significant ($p < 0.05$). If the available data were not analysed statistically, the dose required to produce an effect is estimated as follows: when a dose-related positive response is observed with two or more doses, the lower of the doses is taken as the LED; a single dose resulting in a positive response is considered to be equivalent to the LED.

In order to accommodate both the wide range of doses encountered and positive and negative responses on a continuous scale, doses are transformed logarithmically, so that effective (LED) and ineffective (HID) doses are represented by positive and negative numbers, respectively. The response, or logarithmic dose unit (LDU$_{ij}$), for a given test system i and chemical j is represented by the expressions

$$LDU_{ij} = -\log_{10}(\text{dose}), \text{ for HID values; LDU} \leq 0$$
and (1)
$$LDU_{ij} = -\log_{10}(\text{dose} \times 10^{-5}), \text{ for LED values; LDU} \geq 0.$$

These simple relationships define a dose range of 0 to –5 logarithmic units for ineffective doses (1–100 000 µg/ml or mg/kg bw) and 0 to +8 logarithmic units for effective doses (100 000–0.001 µg/ml or mg/kg bw). A scale illustrating the LDU values is shown in Figure 1. Negative responses at doses less than 1 µg/ml (mg/kg bw) are set equal to 1. Effectively, an LED value $\geq$100 000 or an HID value $\leq$1 produces an LDU = 0; no quantitative information is gained from such extreme values. The dotted lines at the levels of log dose units 1 and –1 define a 'zone of uncertainty' in which positive results are reported at such high doses (between 10 000 and 100 000 µg/ml or mg/kg bw) or negative results are reported at such low dose levels (1 to 10 µg/ml or mg/kg bw) as to call into question the adequacy of the test.

Fig. 1. Scale of log dose units used on the y-axis of activity profiles

Positive (µg/ml or mg/kg bw)		Log dose units	
0.001		8	—
0.01		7	–
0.1		6	–
1.0		5	–
10		4	–
100		3	–
1000		2	–
10 000		1	–
100 000	1	0	—
	10	–1	–
	100	–2	–
	1000	–3	–
	10 000	–4	–
	100 000	–5	—
	Negative (µg/ml or mg/kg bw)		

LED and HID are expressed as µg/ml or mg/kg bw.

APPENDIX 2

In practice, an activity profile is computer generated. A data entry programme is used to store abstracted data from published reports. A sequential file (in ASCII) is created for each compound, and a record within that file consists of the name and Chemical Abstracts Service number of the compound, a three-letter code for the test system (see below), the qualitative test result (with and without an exogenous metabolic system), dose (LED or HID), citation number and additional source information. An abbreviated citation for each publication is stored in a segment of a record accessing both the test data file and the citation file. During processing of the data file, an average of the logarithmic values of the data subset is calculated, and the length of the profile line represents this average value. All dose values are plotted for each profile line, regardless of whether results are positive or negative. Results obtained in the absence of an exogenous metabolic system are indicated by a bar (–), and results obtained in the presence of an exogenous metabolic system are indicated by an upward-directed arrow (↑). When all results for a given assay are either positive or negative, the mean of the LDU values is plotted as a solid line; when conflicting data are reported for the same assay (i.e., both positive and negative results), the majority data are shown by a solid line and the minority data by a dashed line (drawn to the extreme conflicting response). In the few cases in which the numbers of positive and negative results are equal, the solid line is drawn in the positive direction and the maximal negative response is indicated with a dashed line.

Profile lines are identified by three-letter code words representing the commonly used tests. Code words for most of the test systems in current use in genetic toxicology were defined for the US Environmental Protection Agency's GENE-TOX Program (Waters, 1979; Waters & Auletta, 1981). For IARC Monographs Supplement 6, Volume 44 and subsequent volumes, including this publication, codes were redefined in a manner that should facilitate inclusion of additional tests. Naming conventions are described below.

Data listings are presented in the text and include endpoint and test codes, a short test code definition, results [either with (M) or without (NM) an exogenous activation system], the associated LED or HID value and a short citation. Test codes are organized phylogenetically and by endpoint from left to right across each activity profile and from top to bottom of the corresponding data listing. Endpoints are defined as follows: A, aneuploidy; C, chromosomal aberrations; D, DNA damage; F, assays of body fluids; G, gene mutation; H, host-mediated assays; I, inhibition of intercellular communication; M, micronuclei; P, sperm morphology; R, mitotic recombination or gene conversion; S, sister chromatid exchange; and T, cell transformation.

Dose conversions for activity profiles

Doses are converted to µg/ml for in-vitro tests and to mg/kg bw per day for in-vivo experiments.

1. In-vitro test systems

   (a) Weight/volume converts directly to µg/ml.

   (b) Molar (M) concentration × molecular weight = mg/ml = 10^3 µg/ml; mM concentration × molecular weight = µg/ml.

   (c) Soluble solids expressed as % concentration are assumed to be in units of mass per volume (i.e., 1% = 0.01 g/ml = 10 000 µg/ml; also, 1 ppm = 1 µg/ml).

   (d) Liquids and gases expressed as % concentration are assumed to be given in units of volume per volume. Liquids are converted to weight per volume using the density (D) of the solution (D = g/ml). Gases are converted from volume to mass using the ideal gas law, PV = nRT. For exposure at 20–37°C at standard atmospheric pressure, 1% (v/v) = 0.4 µg/ml × molecular weight of the gas. Also, 1 ppm (v/v) = 4×10^{-5} µg/ml × molecular weight.

   (e) In microbial plate tests, it is usual for the doses to be reported as weight/plate, whereas concentrations are required to enter data on the activity profile chart. While remaining cognisant of the errors involved in the process, it is assumed that a 2-ml volume of top agar is delivered to each plate and that the test substance remains in solution within it; concentrations are derived from the reported weight/plate values by dividing by this arbitrary volume. For spot tests, a 1-ml volume is used in the calculation.

   (f) Conversion of particulate concentrations given in µg/cm² are based on the area (A) of the dish and the volume of medium per dish; i.e., for a 100-mm dish: A = πR^2 = $\pi \times (5 \text{ cm})^2$ = 78.5 cm². If the volume of medium is 10 ml, then 78.5 cm² = 10 ml and 1 cm² = 0.13 ml.

2. In-vitro systems using in-vivo activation

   For the body fluid–urine (BF–) test, the concentration used is the dose (in mg/kg bw) of the compound administered to test animals or patients.

3. In-vivo test systems

   (a) Doses are converted to mg/kg bw per day of exposure, assuming 100% absorption. Standard values are used for each sex and species of rodent, including body weight and average intake per day, as reported by Gold *et al.* (1984). For example, in a test using male mice fed 50 ppm of the agent in the diet, the standard food intake per day is 12% of body weight, and the conversion is dose = 50 ppm × 12% = 6 mg/kg bw per day.

Standard values used for humans are: weight – males, 70 kg; females, 55 kg; surface area, 1.7 m^2; inhalation rate, 20 l/min for light work, 30 l/min for mild exercise.

(b) When reported, the dose at the target site is used. For example, doses given in studies of lymphocytes of humans exposed *in vivo* are the measured blood concentrations in µg/ml.

Codes for test systems

For specific nonmammalian test systems, the first two letters of the three-symbol code word define the test organism (e.g., SA- for *Salmonella typhimurium*, EC- for *Escherichia coli*). If the species is not known, the convention used is -S-. The third symbol may be used to define the tester strain (e.g., SA8 for *S. typhimurium* TA1538, ECW for *E. coli* WP2*uvr*A). When strain designation is not indicated, the third letter is used to define the specific genetic endpoint under investigation (e.g., —D for differential toxicity, —F for forward mutation, —G for gene conversion or genetic crossing-over, —N for aneuploidy, —R for reverse mutation, —U for unscheduled DNA synthesis). The third letter may also be used to define the general endpoint under investigation when a more complete definition is not possible or relevant (e.g., —M for mutation, —C for chromosomal aberration).

For mammalian test systems, the first letter of the three-letter code word defines the genetic endpoint under investigation: A— for aneuploidy, B— for binding, C— for chromosomal aberration, D— for DNA strand breaks, G— for gene mutation, I— for inhibition of intercellular communication, M— for micronucleus formation, R— for DNA repair, S— for sister chromatid exchange, T— for cell transformation and U— for unscheduled DNA synthesis.

For animal (i.e., non-human) test systems *in vitro*, when the cell type is not specified, the code letters -IA are used. For such assays *in vivo*, when the animal species is not specified, the code letters -VA are used. Commonly used animal species are identified by the third letter (e.g., —C for Chinese hamster, —M for mouse, —R for rat, —S for Syrian hamster).

For test systems using human cells *in vitro*, when the cell type is not specified, the code letters -IH are used. For assays on humans *in vivo*, when the cell type is not specified, the code letters -VH are used. Otherwise, the second letter specifies the cell type under investigation (e.g., -BH for bone marrow, -LH for lymphocytes).

Some other specific coding conventions used for mammalian systems are as follows: BF- for body fluids, HM- for host-mediated, —L for leucocytes or lymphocytes *in vitro* (-AL, animals; -HL, humans), -L- for leucocytes *in vivo* (-LA, animals; -LH, humans), —T for transformed cells.

Note that these are examples of major conventions used to define the assay code words. The alphabetized listing of codes must be examined to confirm a specific code word. As might be expected from the limitation to three symbols, some codes do not fit the naming conventions precisely. In a few cases, test systems are defined by first-letter code words, for example: MST, mouse spot test; SLP, mouse specific locus test, postspermatogonia; SLO, mouse specific locus test, other stages; DLM, dominant lethal test in mice; DLR, dominant lethal test in rats; MHT, mouse heritable translocation test.

The genetic activity profiles and listings that follow were prepared in collaboration with Environmental Health Research and Testing Inc. (EHRT) under contract to the US Environmental Protection Agency; EHRT also determined the doses used. The references cited in each genetic activity profile listing can be found in the list of references in the appropriate monograph.

References

Garrett, N.E., Stack, H.F., Gross, M.R. & Waters, M.D. (1984) An analysis of the spectra of genetic activity produced by known or suspected human carcinogens. *Mutat. Res., 134*, 89–111

Gold, L.S., Sawyer, C.B., Magaw, R., Backman, G.M., de Veciana, M., Levinson, R., Hooper, N.K., Havender, W.R., Bernstein, L., Peto, R., Pike, M.C. & Ames, B.N. (1984) A carcinogenic potency database of the standardized results of animal bioassays. *Environ. Health Perspect., 58*, 9–319

Waters, M.D. (1979) *The GENE-TOX program*. In: Hsie, A.W., O'Neill, J.P. & McElheny, V.K., eds, *Mammalian Cell Mutagenesis: The Maturation of Test Systems* (Banbury Report 2), Cold Spring Harbor, NY, CHS Press, pp. 449–467

Waters, M.D. & Auletta, A. (1981) The GENE-TOX program: genetic activity evaluation. *J. chem. Inf. comput. Sci., 21*, 35–38

Waters, M.D., Stack, H.F., Brady, A.L., Lohman, P.H.M., Haroun, L. & Vainio, H. (1987) Appendix 1: Activity profiles for genetic and related tests. In: *IARC Monographs on the Evaluation of the Carcinogenic Risk of Chemicals to Humans*, Suppl. 6, *Genetic and Related Effects: An Update of Selected* IARC Monographs *from Volumes 1 to 42*, Lyon, IARC, pp. 687–696

Waters, M.D., Stack, H.F., Brady, A.L., Lohman, P.H.M., Haroun, L. & Vainio, H. (1988) Use of computerized data listings and activity profiles of genetic and related effects in the review of 195 compounds. *Mutat. Res., 205*, 295–312

APPENDIX 2

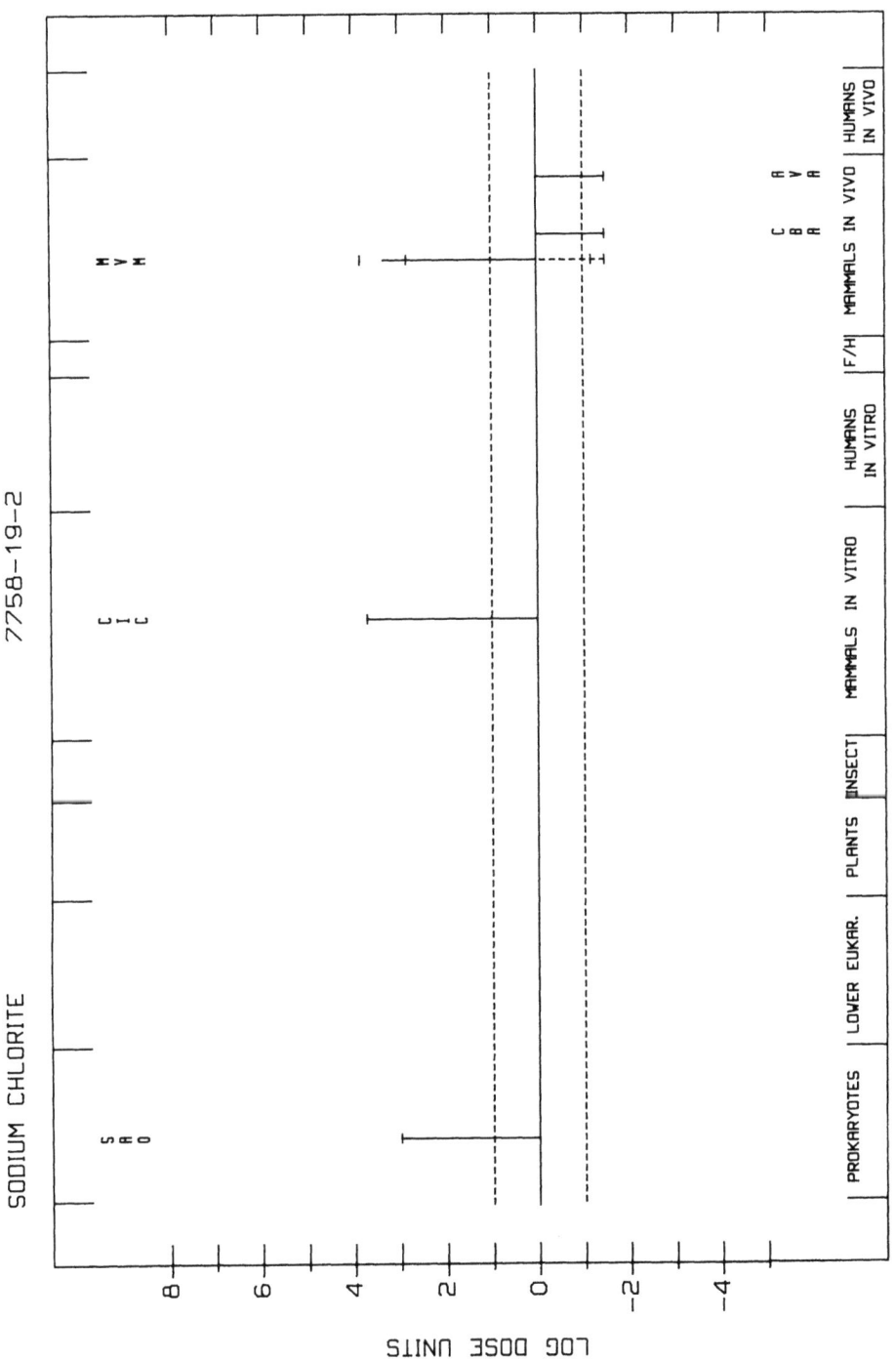

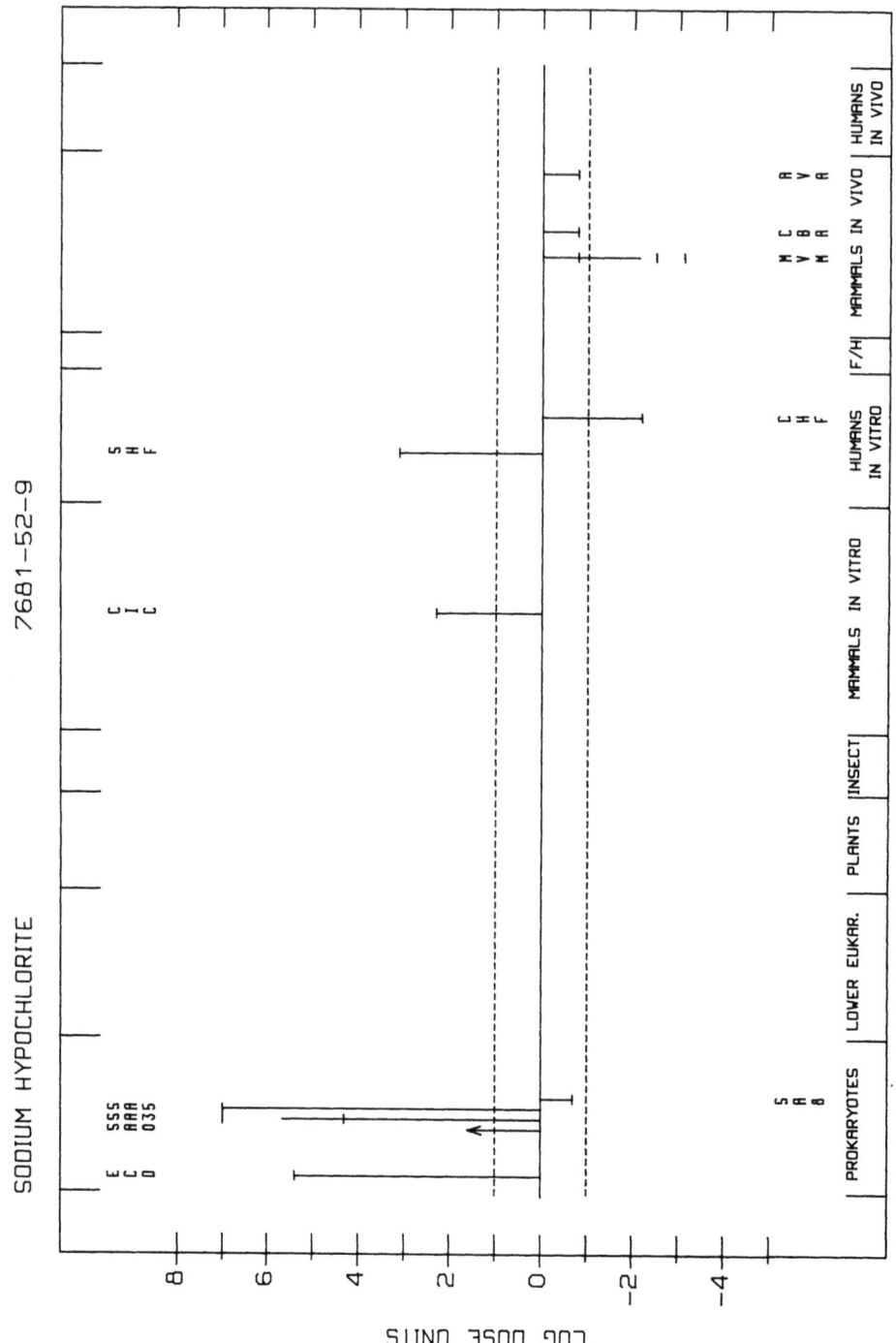

APPENDIX 2

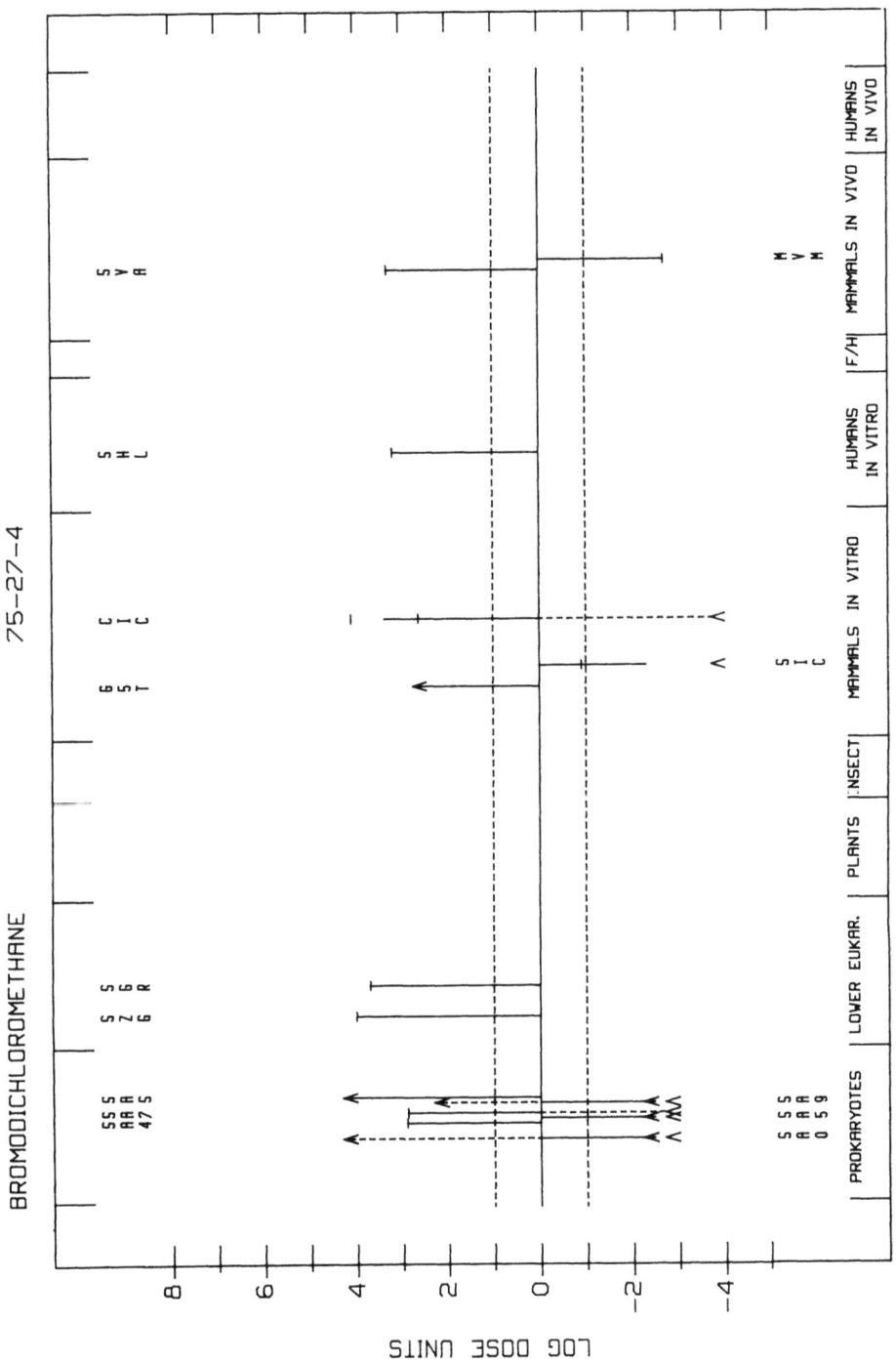

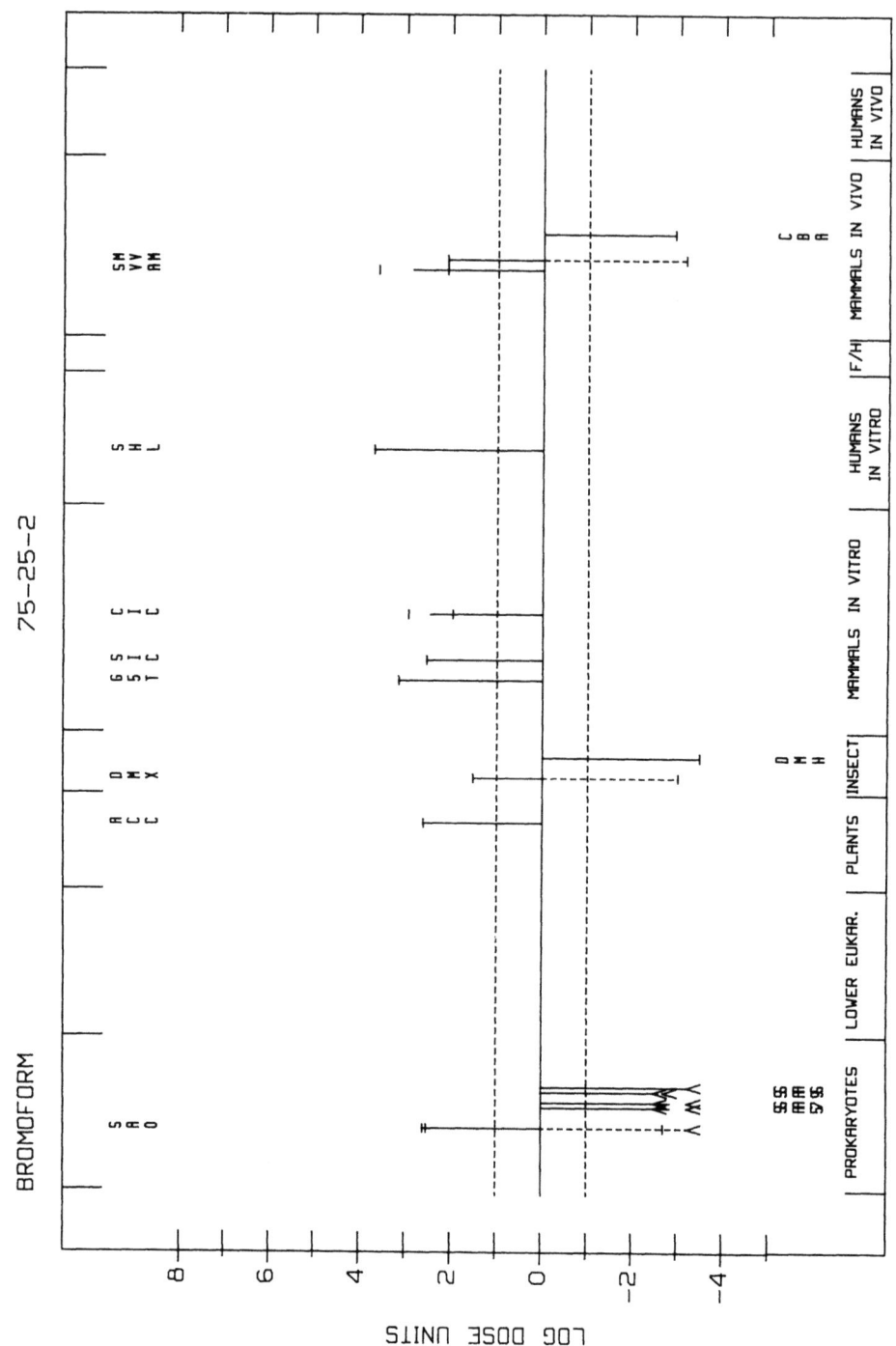

APPENDIX 2

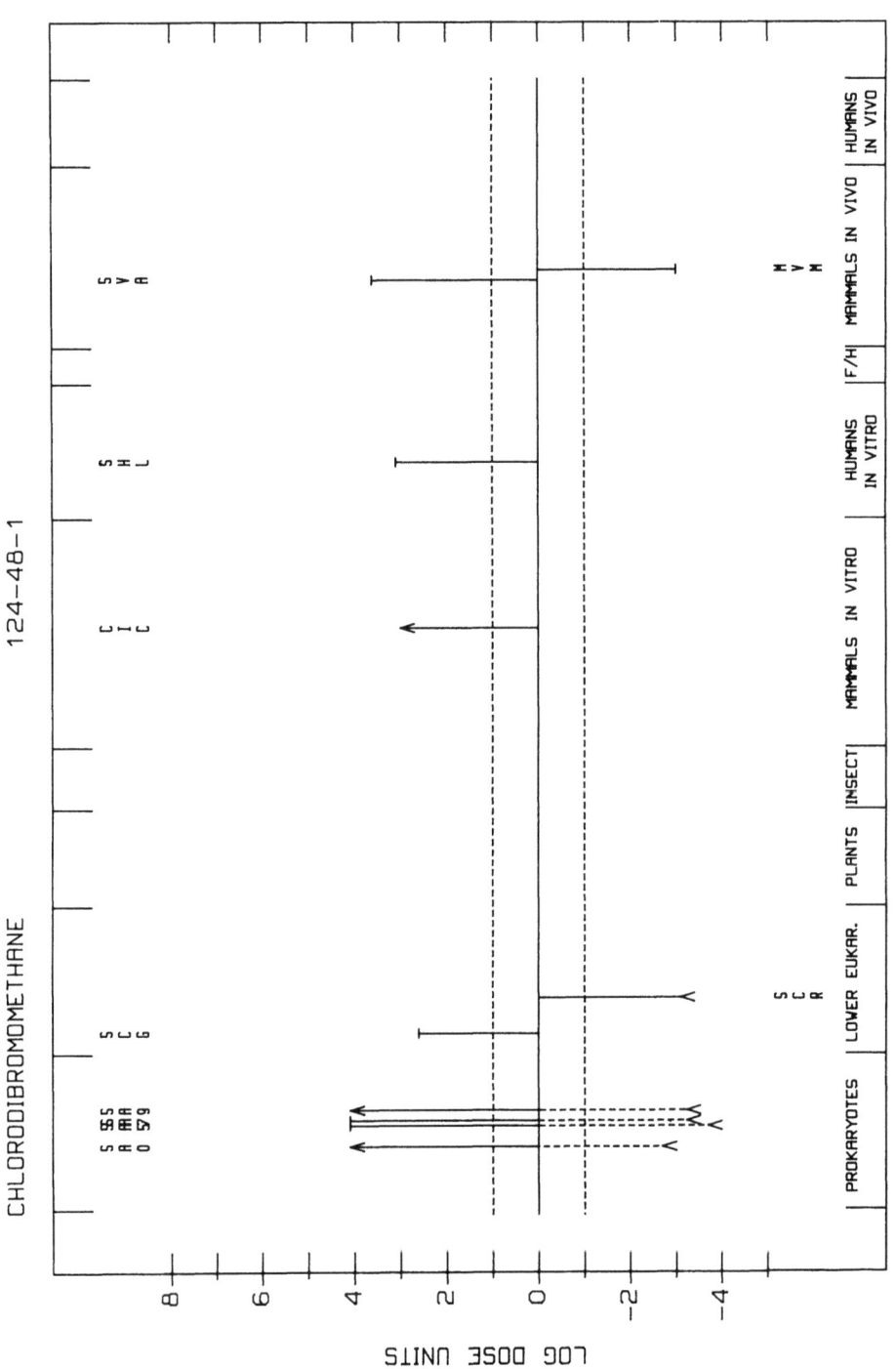

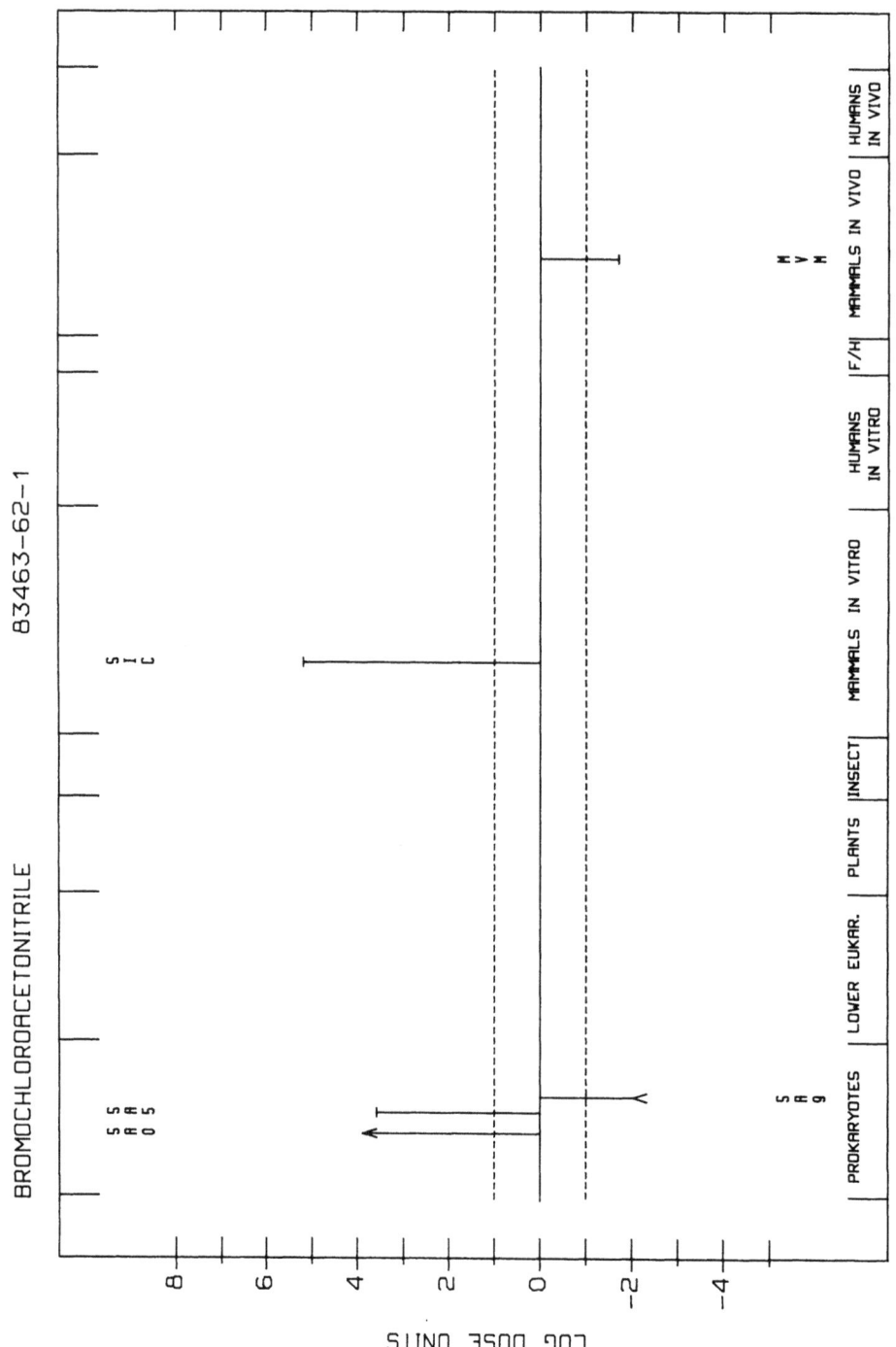

APPENDIX 2

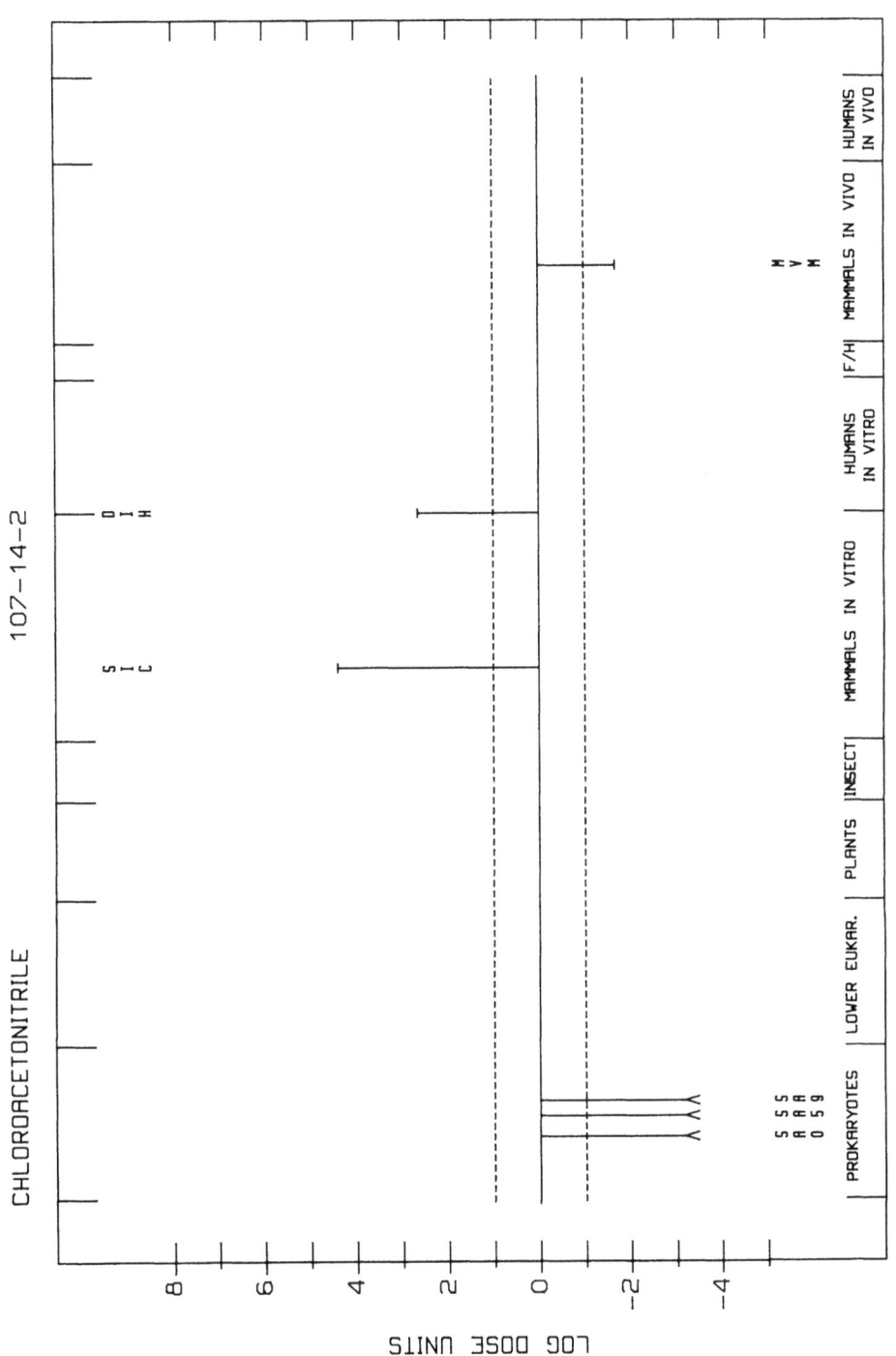

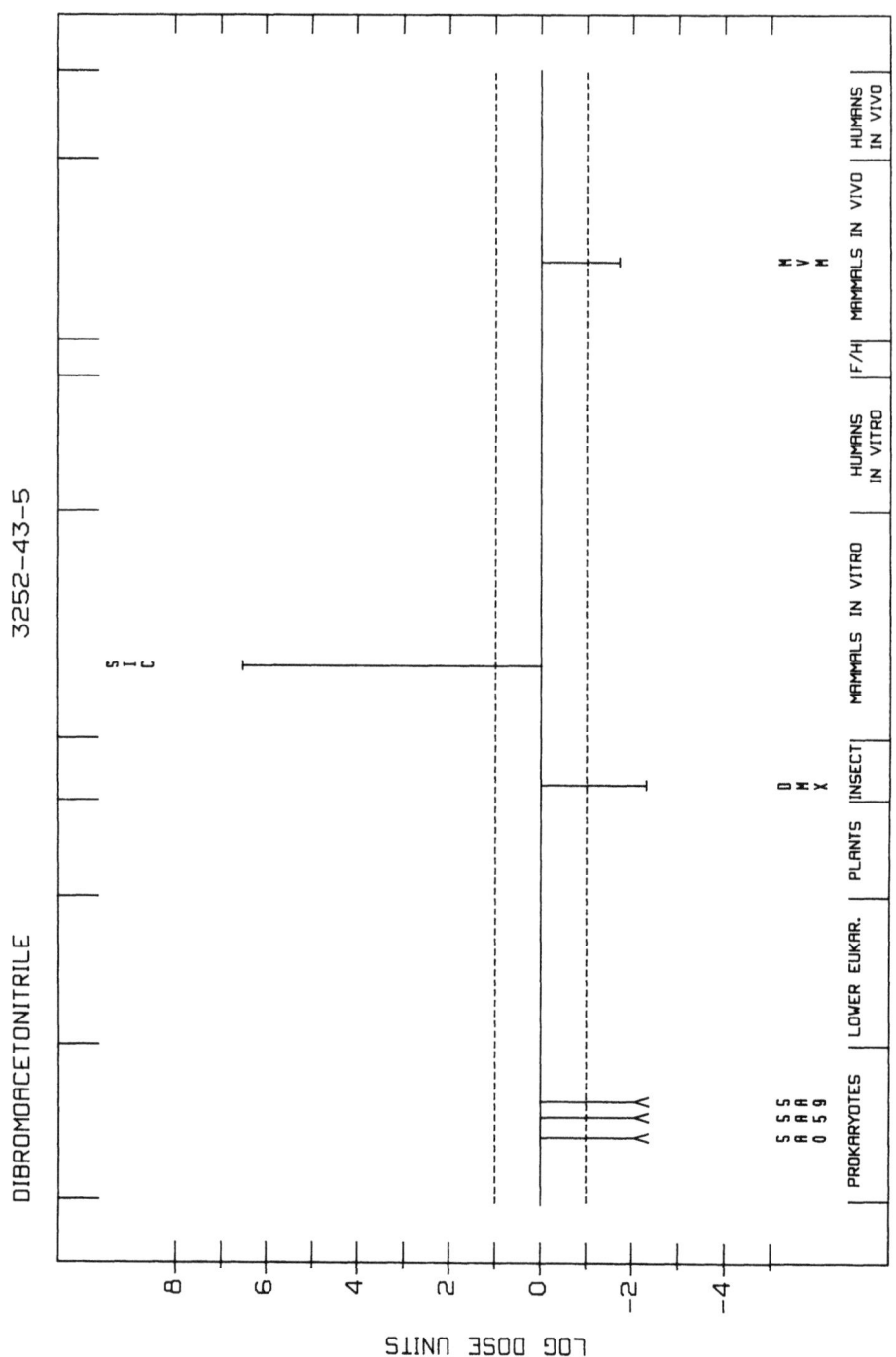

APPENDIX 2

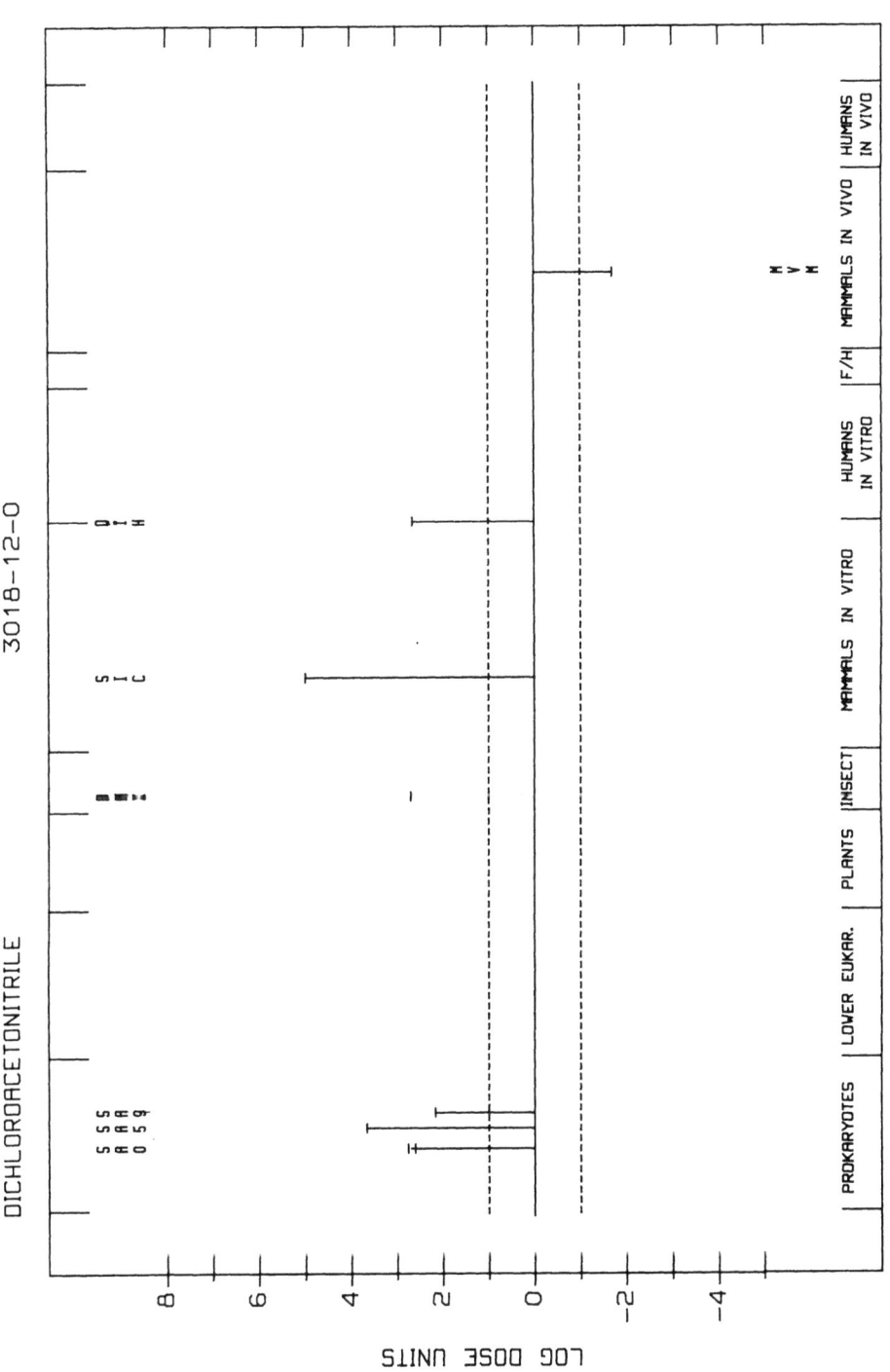

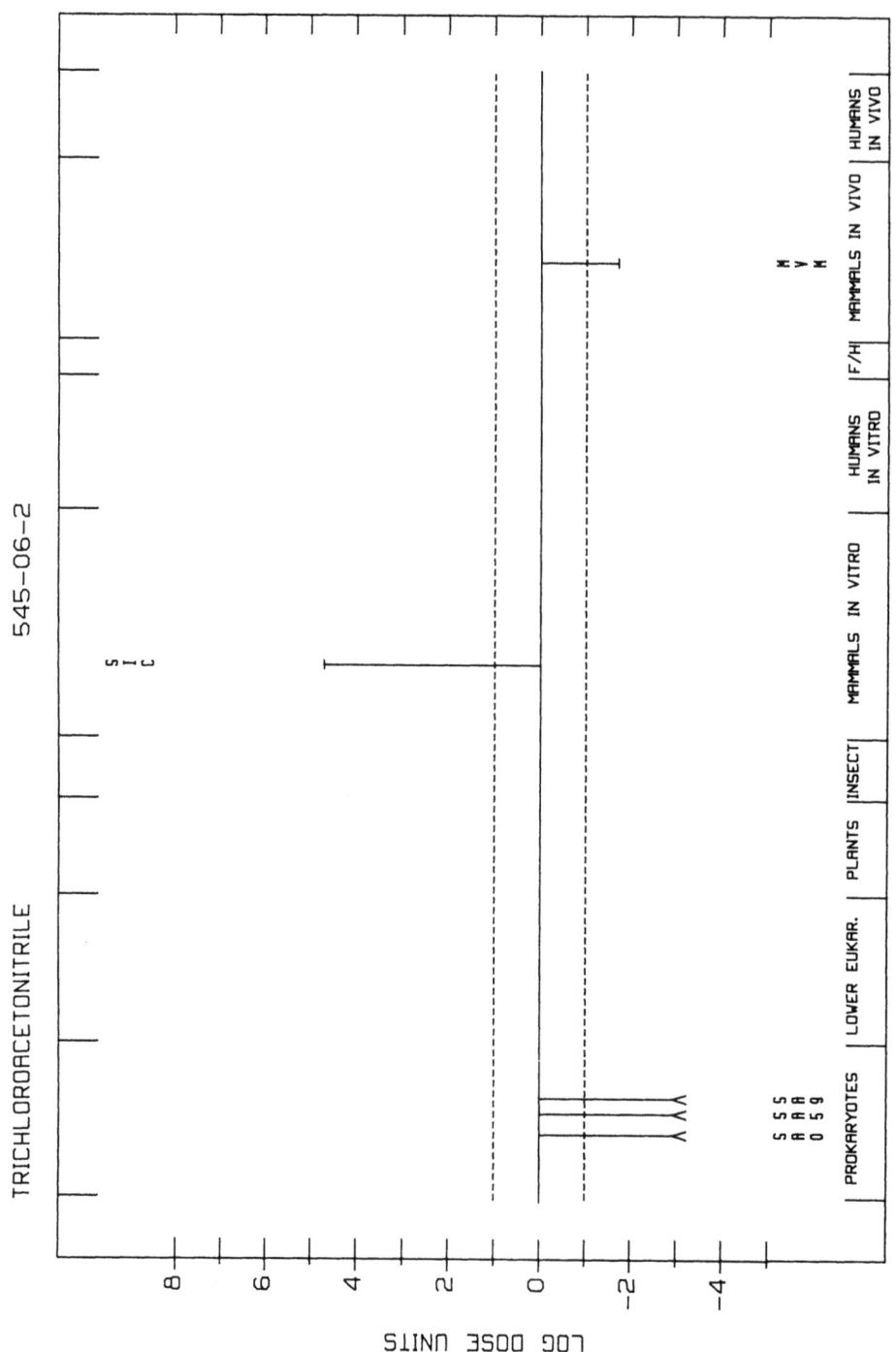

APPENDIX 2

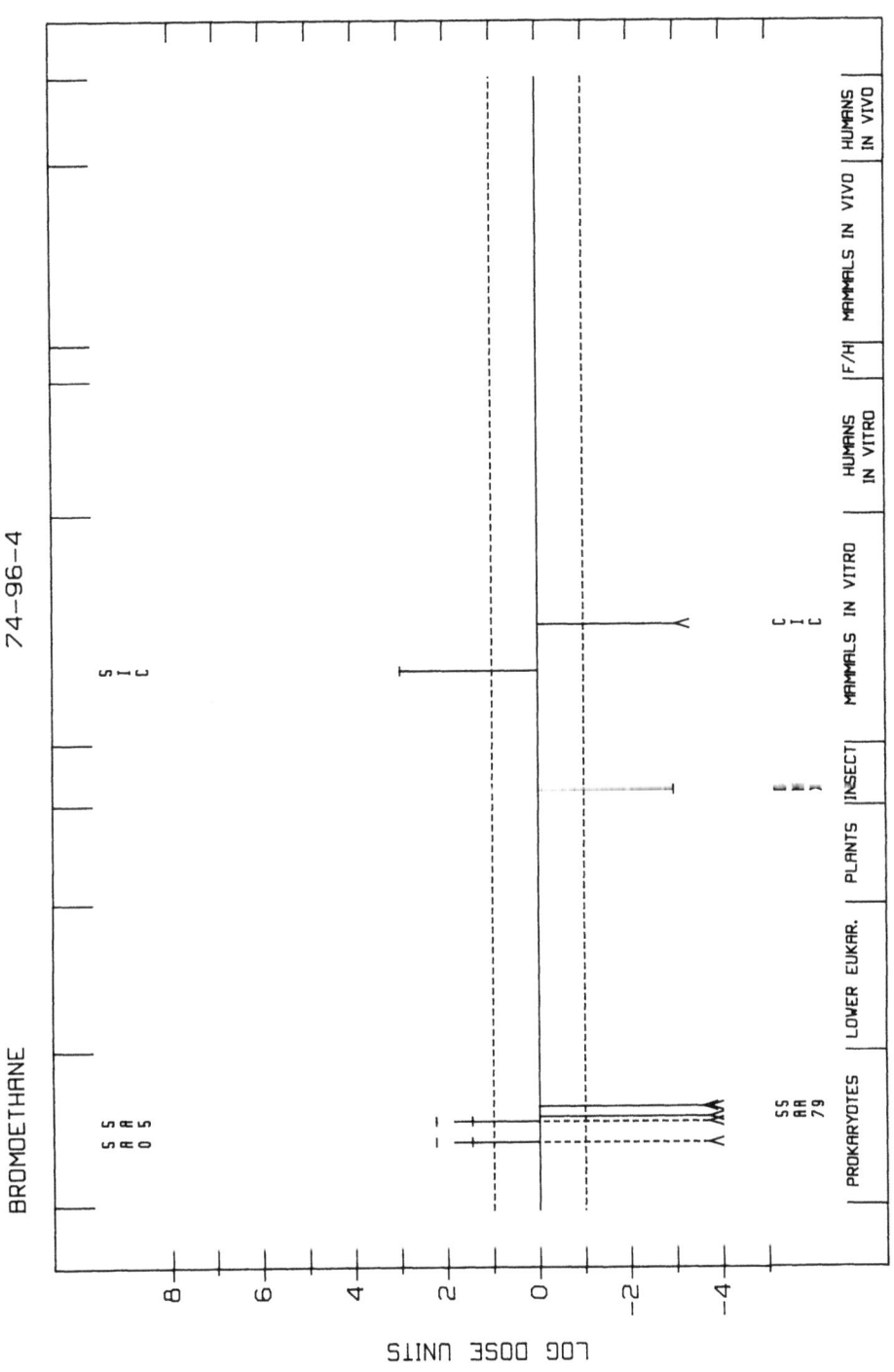

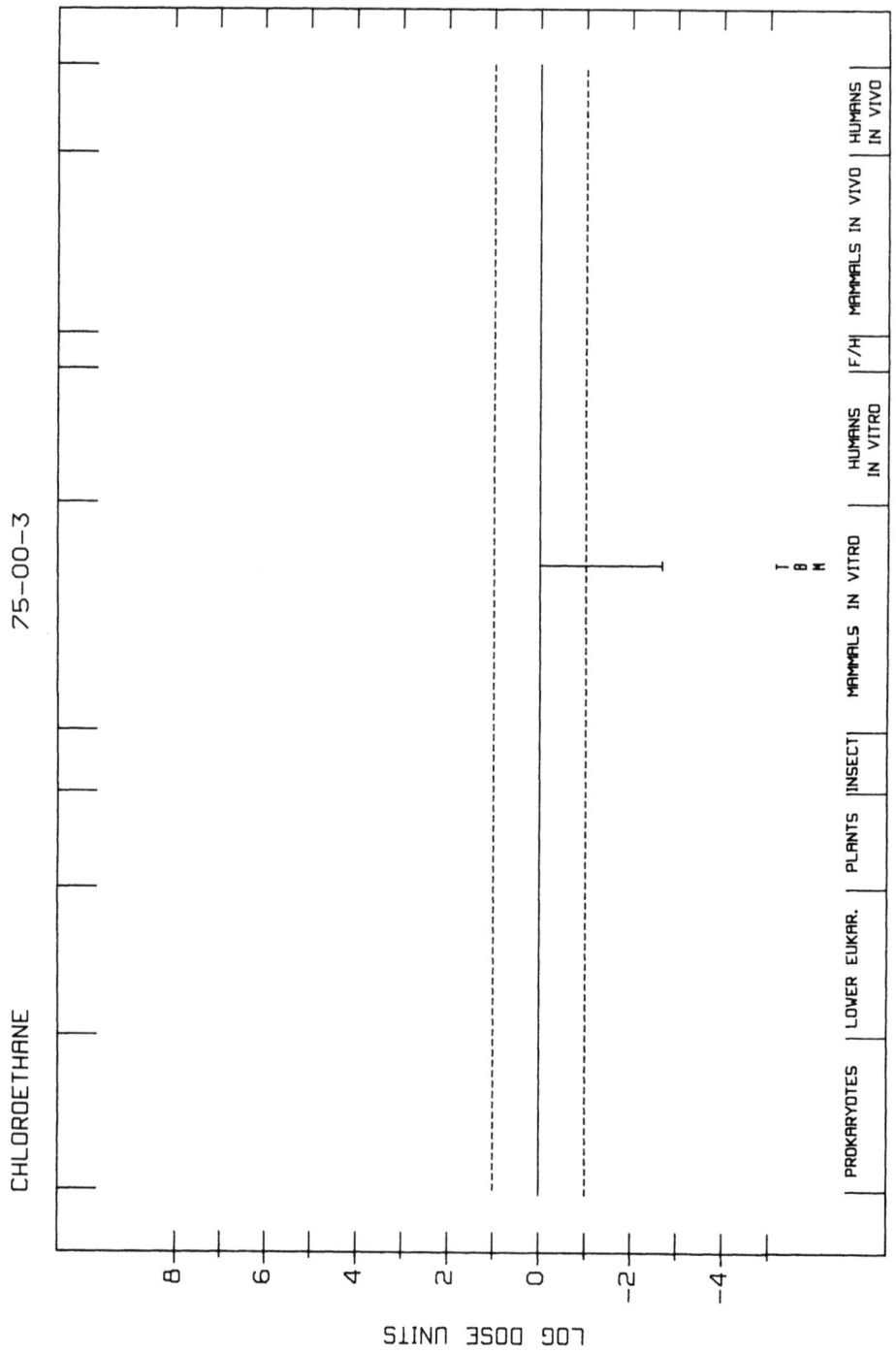

APPENDIX 2

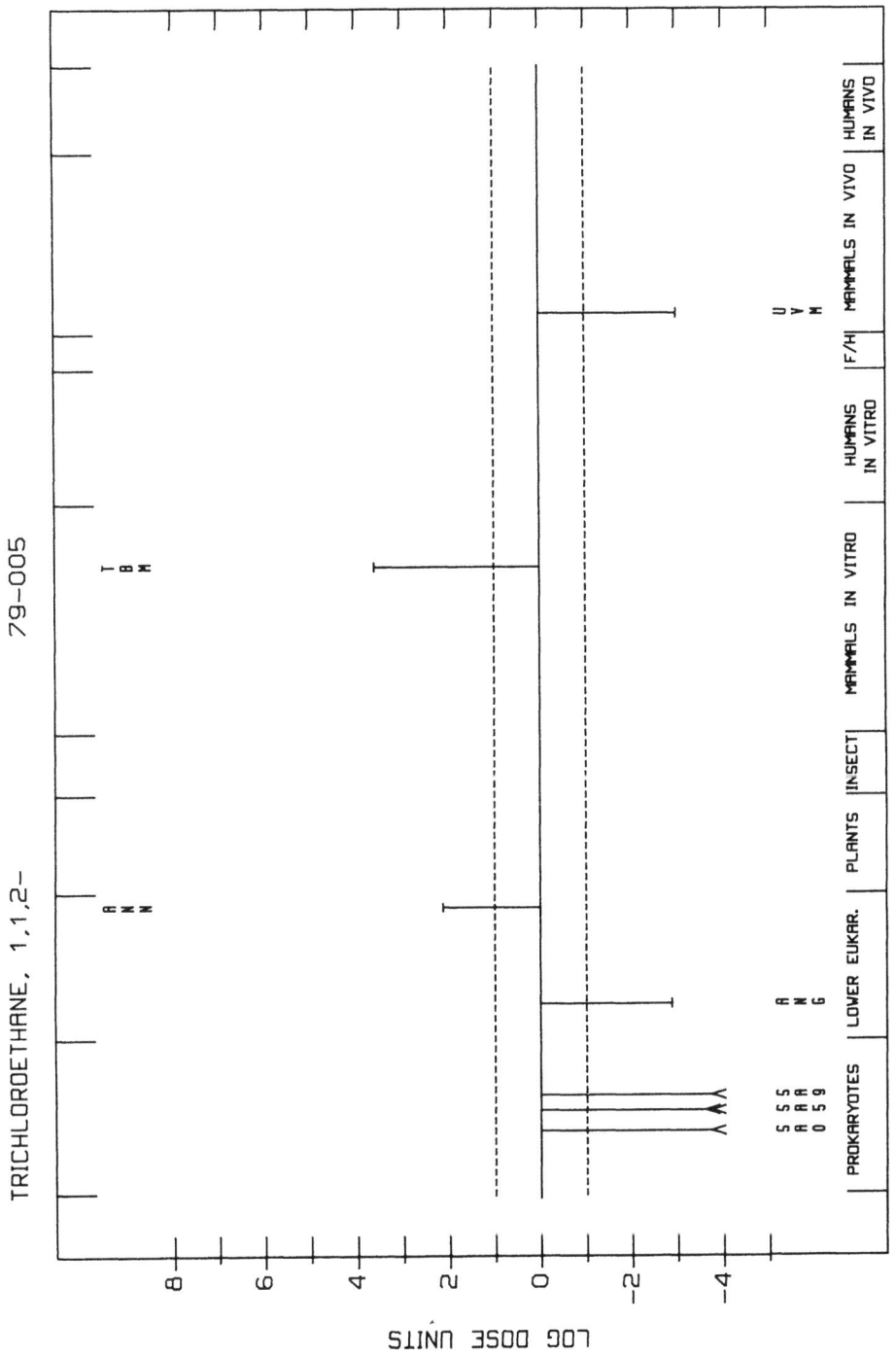

510 IARC MONOGRAPHS VOLUME 52

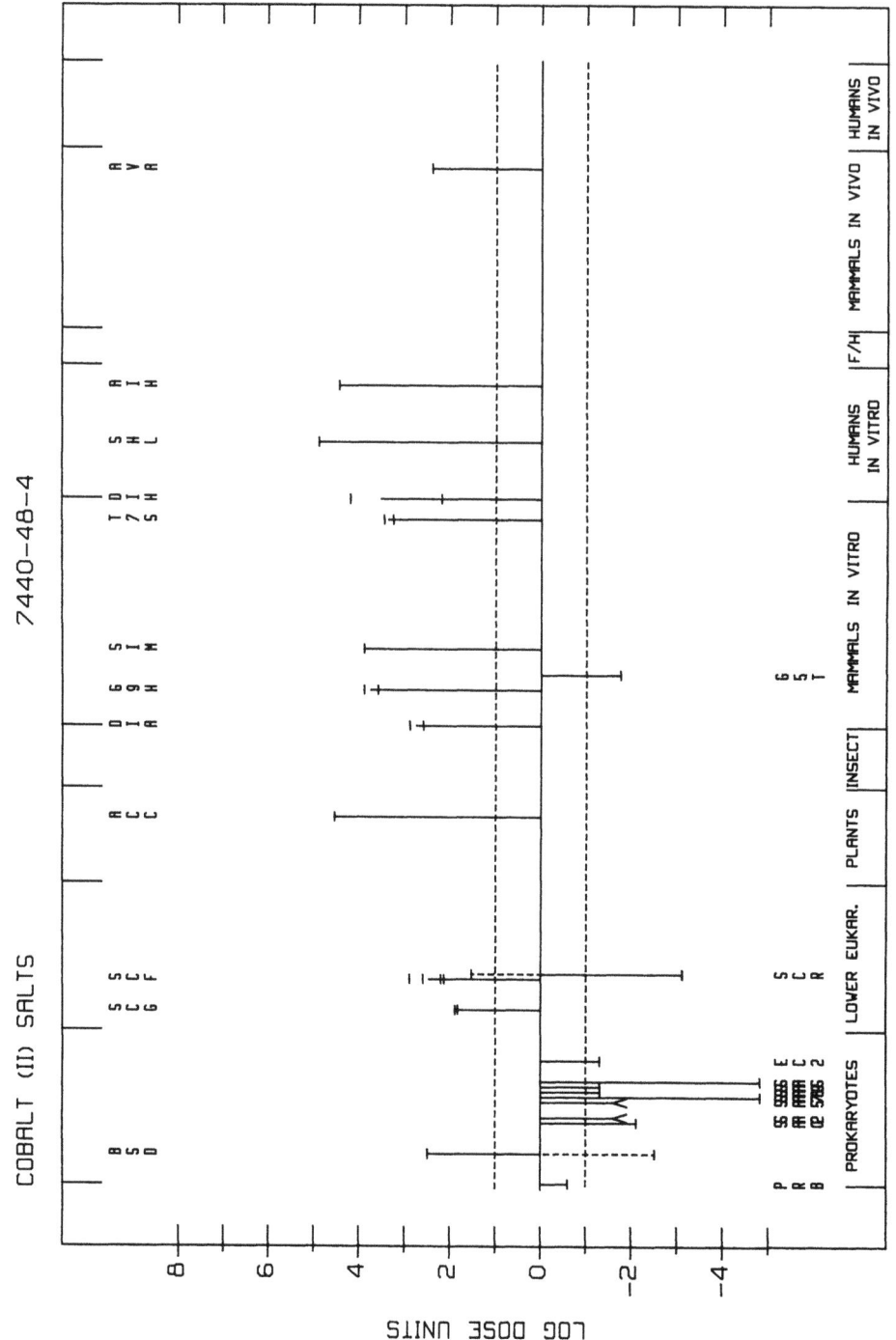

APPENDIX 2

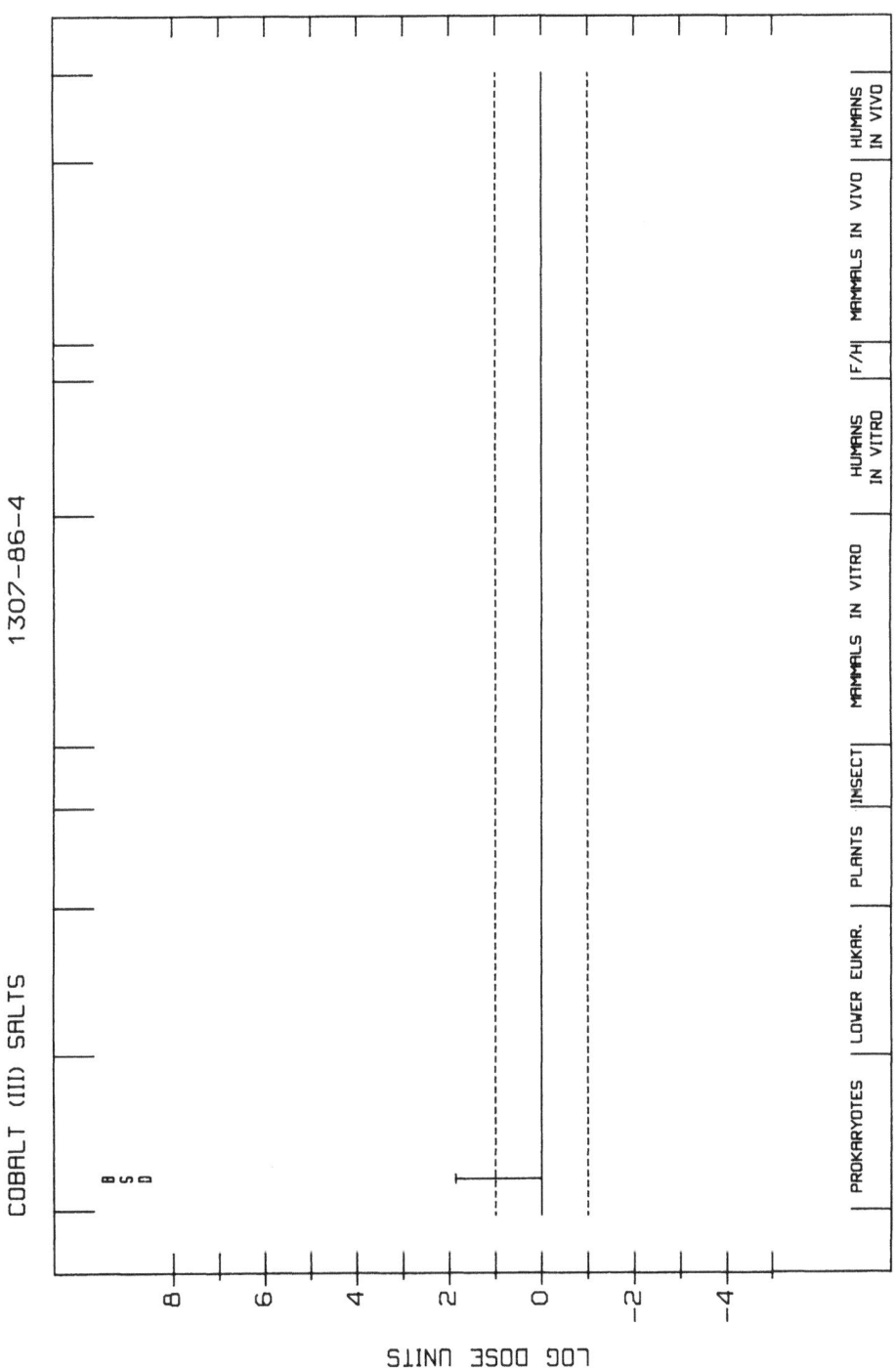

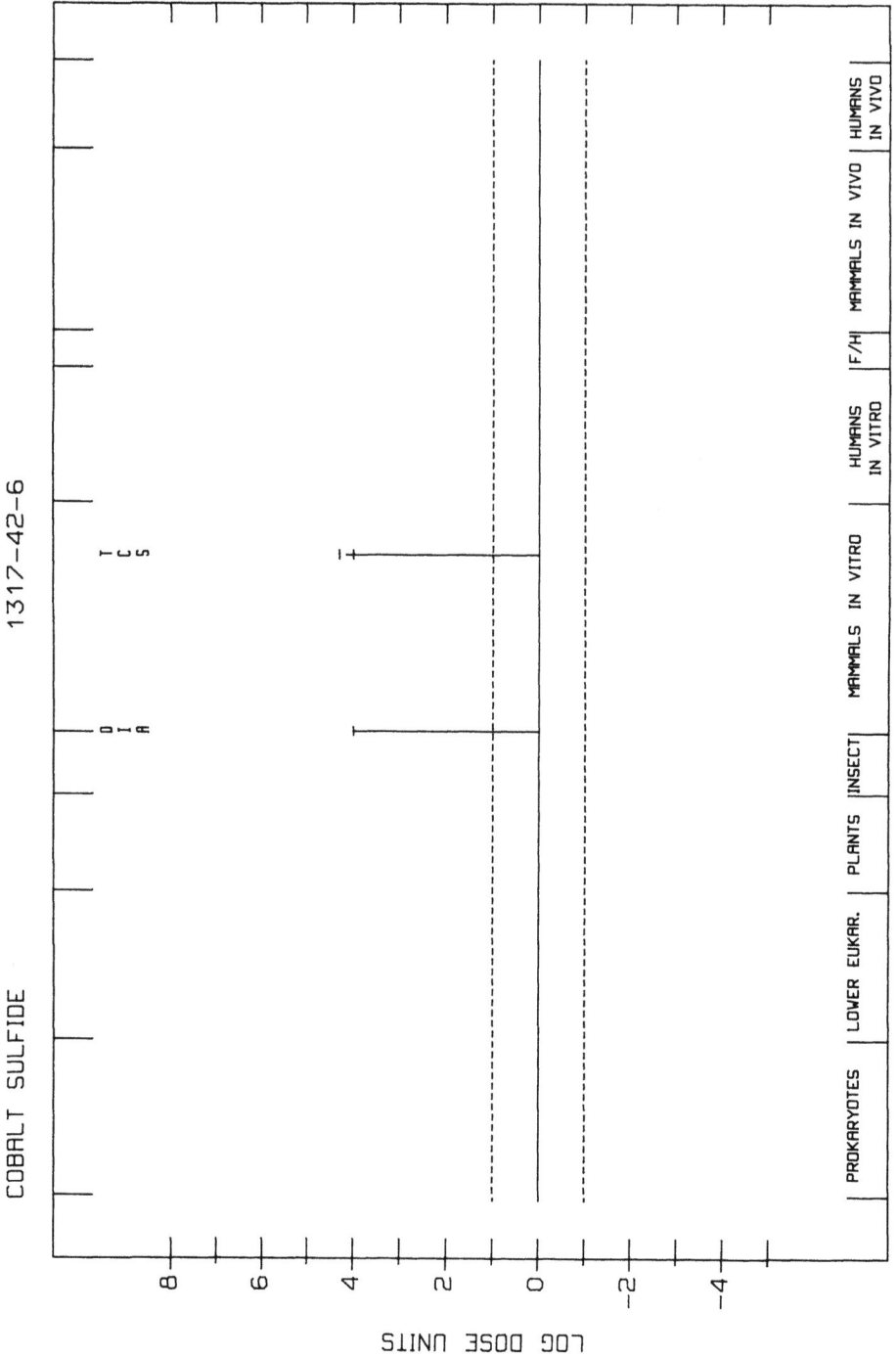

SUPPLEMENTARY CORRIGENDA TO VOLUMES 1–51

Volume 37

p. 90	(*a*) para 1	*Delete from* Axéll *to end.*
p. 117		*Delete reference* Axéll (1976).
p. 131	Schuman *et al.* (1982)	*Delete* (Abstract No. 1172), p. 207

Volume 41

p. 300	para 3	*Add before closing square bracket:* The data reported are probably from the same study reported by Jukes and Shaffer, 1960.

Volume 49

p. 497	*At end of comment to Newhouse* et al. *(1985) (UK), add* (90% CI)

Volume 51

p. 120	*In comments to Phillips & Snowdon (1985), add* for trend *after* p *three times*

CUMULATIVE CROSS INDEX TO *IARC MONOGRAPHS ON THE EVALUATION OF CARCINOGENIC RISKS TO HUMANS*

The volume, page and year are given. References to corrigenda are given in parentheses.

A

A-α-C	*40*, 245 (1986); *Suppl. 7*, 56 (1987)
Acetaldehyde	*36*, 101 (1985) (*corr. 42*, 263); *Suppl. 7*, 77 (1987)
Acetaldehyde formylmethylhydrazone (*see* Gyromitrin)	
Acetamide	*7*, 197 (1974); *Suppl. 7*, 389 (1987)
Acetaminophen (*see* Paracetamol)	
Acridine orange	*16*, 145 (1978); *Suppl. 7*, 56 (1987)
Acriflavinium chloride	*13*, 31 (1977); *Suppl. 7*, 56 (1987)
Acrolein	*19*, 479 (1979); *36*, 133 (1985); *Suppl. 7*, 78*(1987);
Acrylamide	*39*, 41 (1986); *Suppl. 7*, 56 (1987)
Acrylic acid	*19*, 47 (1979); *Suppl. 7*, 56 (1987)
Acrylic fibres	*19*, 86 (1979); *Suppl. 7*, 56 (1987)
Acrylonitrile	*19*, 73 (1979); *Suppl. 7*, 79 (1987)
Acrylonitrile-butadiene-styrene copolymers	*19*, 91 (1979); *Suppl. 7*, 56 (1987)
Actinolite (*see* Asbestos)	
Actinomycins	*10*, 29 (1976) (*corr. 42*, 255); *Suppl. 7*, 80 (1987)
Adriamycin	*10*, 43 (1976); *Suppl. 7*, 82 (1987)
AF-2	*31*, 47 (1983); *Suppl. 7*, 56 (1987)
Aflatoxins	*1*, 145 (1972) (*corr. 42*, 251); *10*, 51 (1976); *Suppl. 7*, 83 (1987)
Aflatoxin B_1 (*see* Aflatoxins)	
Aflatoxin B_2 (*see* Aflatoxins)	
Aflatoxin G_1 (*see* Aflatoxins)	
Aflatoxin G_2 (*see* Aflatoxins)	
Aflatoxin M_1 (*see* Aflatoxins)	
Agaritine	*31*, 63 (1983); *Suppl. 7*, 56 (1987)

Alcohol drinking	*44*
Aldrin	5, 25 (1974); *Suppl. 7*, 88 (1987)
Allyl chloride	36, 39 (1985); *Suppl. 7*, 56 (1987)
Allyl isothiocyanate	36, 55 (1985); *Suppl. 7*, 56 (1987)
Allyl isovalerate	36, 69 (1985); *Suppl. 7*, 56 (1987)
Aluminium production	34, 37 (1984); *Suppl. 7*, 89 (1987)
Amaranth	8, 41 (1975); *Suppl. 7*, 56 (1987)
5-Aminoacenaphthene	16, 243 (1978); *Suppl. 7*, 56 (1987)
2-Aminoanthraquinone	27, 191 (1982); *Suppl. 7*, 56 (1987)
para-Aminoazobenzene	8, 53 (1975); *Suppl. 7*, 390 (1987)
ortho-Aminoazotoluene	8, 61 (1975) (*corr.* 42, 254);. *Suppl. 7*, 56 (1987)
para-Aminobenzoic acid	16, 249 (1978); *Suppl. 7*, 56 (1987)
4-Aminobiphenyl	*1*, 74 (1972) (*corr.* 42, 251); *Suppl. 7*, 91 (1987)
2-Amino-3,4-dimethylimidazo[4,5-*f*]quinoline (*see* MeIQ)	
2-Amino-3,8-dimethylimidazo[4,5-*f*]quinoxaline (*see* MeIQx)	
3-Amino-1,4-dimethyl-5*H*-pyrido[4,3-*b*]indole (*see* Trp-P-1)	
2-Aminodipyrido[1,2-*a*:3',2'-*d*]imidazole (*see* Glu-P-2)	
1-Amino-2-methylanthraquinone	27, 199 (1982); *Suppl. 7*, 57 (1987)
2-Amino-3-methylimidazo[4,5-*f*]quinoline (*see* IQ)	
2-Amino-6-methyldipyrido[1,2-*a*:3',2'-*d*]-imidazole (*see* Glu-P-1)	
2-Amino-3-methyl-9*H*-pyrido[2,3-*b*]indole (*see* MeA-α-C)	
3-Amino-1-methyl-5*H*-pyrido[4,3-*b*]indole (*see* Trp-P-2)	
2-Amino-5-(5-nitro-2-furyl)-1,3,4-thiadiazole	7, 143 (1974); *Suppl. 7*, 57 (1987)
4-Amino-2-nitrophenol	16, 43 (1978); *Suppl.7*, 57 (1987)
2-Amino-5-nitrothiazole	31, 71 (1983); *Suppl. 7*, 57 (1987)
2-Amino-9*H*-pyrido[2,3-*b*]indole [*see* A-α-C]	
11-Aminoundecanoic acid	39, 239 (1986); *Suppl. 7*, 57 (1987)
Amitrole	7, 31 (1974); 41, 293 (1986) (*corr.* 52, 513; *Suppl. 7*, 92 (1987)
Ammonium potassium selenide (*see* Selenium and selenium compounds)	
Amorphous silica (*see also* Silica)	*Suppl. 7*, 341 (1987)
Amosite (*see* Asbestos)	
Ampicillin	50, 153 (1990)
Anabolic steroids (*see* Androgenic (anabolic) steroids)	
Anaesthetics, volatile	*11*, 285 (1976); *Suppl. 7*, 93 (1987)
Analgesic mixtures containing phenacetin (*see also* Phenacetin)	*Suppl. 7*, 310 (1987)
Androgenic (anabolic) steroids	*Suppl. 7*, 96 (1987)
Angelicin and some synthetic derivatives (*see also* Angelicins)	40, 291 (1986)
Angelicin plus ultraviolet radiation (*see also* Angelicin and some synthetic derivatives)	*Suppl. 7*, 57 (1987)
Angelicins	*Suppl. 7*, 57 (1987)

Aniline	4, 27 (1974) (corr. 42, 252); 27, 39 (1982); Suppl. 7, 99 (1987)
ortho-Anisidine	27, 63 (1982); Suppl. 7, 57 (1987)
para-Anisidine	27, 65 (1982); Suppl. 7, 57 (1987)
Anthanthrene	32, 95 (1983); Suppl. 7, 57 (1987)
Anthophyllite (see Asbestos)	
Anthracene	32, 105 (1983); Suppl. 7, 57 (1987)
Anthranilic acid	16, 265 (1978); Suppl. 7, 57 (1987)
Antimony trioxide	47, 291 (1989)
Antimony trisulfide	47, 291 (1989)
ANTU (see 1-Naphthylthiourea)	
Apholate	9, 31 (1975); Suppl. 7, 57 (1987)
Aramite®	5, 39 (1974); Suppl. 7, 57 (1987)
Areca nut (see Betel quid)	
Arsanilic acid (see Arsenic and arsenic compounds)	
Arsenic and arsenic compounds	1, 41 (1972); 2, 48 (1973); 23, 39 (1980); Suppl. 7, 100 (1987)
Arsenic pentoxide (see Arsenic and arsenic compounds)	
Arsenic sulphide (see Arsenic and arsenic compounds)	
Arsenic trioxide (see Arsenic and arsenic compounds)	
Arsine (see Arsenic and arsenic compounds)	
Asbestos	2, 17 (1973) (corr. 42, 252); 14 (1977) (corr. 42, 256); Suppl. 7, 106 (1987) (corr. 45, 283)
Attapulgite	42, 159 (1987); Suppl. 7, 117 (1987)
Auramine (technical-grade)	1, 69 (1972) (corr. 42, 251); Suppl. 7, 118 (1987)
Auramine, manufacture of (see also Auramine, technical-grade)	Suppl. 7, 118 (1987)
Aurothioglucose	13, 39 (1977); Suppl. 7, 57 (1987)
Azacitidine	26, 37 (1981); Suppl. 7, 57 (1987); 50, 47 (1990)
5-Azacytidine (see Azacitidine)	
Azaserine	10, 73 (1976) (corr. 42, 255); Suppl. 7, 57 (1987)
Azathioprine	26, 47 (1981); Suppl. 7, 119 (1987)
Aziridine	9, 37 (1975); Suppl. 7, 58 (1987)
2-(1-Aziridinyl)ethanol	9, 47 (1975); Suppl. 7, 58 (1987)
Aziridyl benzoquinone	9, 51 (1975); Suppl. 7, 58 (1987)
Azobenzene	8, 75 (1975); Suppl. 7, 58 (1987)

B

Barium chromate (see Chromium and chromium compounds)
Basic chromic sulphate (see Chromium and chromium compounds)
BCNU (see Bischloroethyl nitrosourea)

Benz[a]acridine	32, 123 (1983); Suppl. 7, 58 (1987)
Benz[c]acridine	3, 241 (1973); 32, 129 (1983); Suppl. 7, 58 (1987)
Benzal chloride (see also α-Chlorinated toluenes)	29, 65 (1982); Suppl. 7, 148 (1987)
Benz[a]anthracene	3, 45 (1973); 32, 135 (1983); Suppl. 7, 58 (1987)
Benzene	7, 203 (1974) (corr. 42, 254); 29, 93, 391 (1982); Suppl. 7, 120 (1987)
Benzidine	1, 80 (1972); 29, 149, 391 (1982); Suppl. 7, 123 (1987)
Benzidine-based dyes	Suppl. 7, 125 (1987)
Benzo[b]fluoranthene	3, 69 (1973); 32, 147 (1983); Suppl. 7, 58 (1987)
Benzo[j]fluoranthene	3, 82 (1973); 32, 155 (1983); Suppl. 7, 58 (1987)
Benzo[k]fluoranthene	32, 163 (1983); Suppl. 7, 58 (1987)
Benzo[ghi]fluoranthene	32, 171 (1983); Suppl. 7, 58 (1987)
Benzo[a]fluorene	32, 177 (1983); Suppl. 7, 58 (1987)
Benzo[b]fluorene	32, 183 (1983); Suppl. 7, 58 (1987)
Benzo[c]fluorene	32, 189 (1983); Suppl. 7, 58 (1987)
Benzo[ghi]perylene	32, 195 (1983); Suppl. 7, 58 (1987)
Benzo[c]phenanthrene	32, 205 (1983); Suppl. 7, 58 (1987)
Benzo[a]pyrene	3, 91 (1973); 32, 211 (1983); Suppl. 7, 58 (1987)
Benzo[e]pyrene	3, 137 (1973); 32, 225 (1983); Suppl. 7, 58 (1987)
para-Benzoquinone dioxime	29, 185 (1982); Suppl. 7, 58 (1987)
Benzotrichloride (see also α-Chlorinated toluenes)	29, 73 (1982); Suppl. 7, 148 (1987)
Benzoyl chloride	29, 83 (1982) (corr. 42, 261); Suppl. 7, 126 (1987)
Benzoyl peroxide	36, 267 (1985); Suppl. 7, 58 (1987)
Benzyl acetate	40, 109 (1986); Suppl. 7, 58 (1987)
Benzyl chloride (see also α-Chlorinated toluenes)	11, 217 (1976) (corr. 42, 256); 29, 49 (1982); Suppl. 7, 148 (1987)
Benzyl violet 4B	16, 153 (1978); Suppl. 7, 58 (1987)
Bertrandite (see Beryllium and beryllium compounds)	
Beryllium and beryllium compounds	1, 17 (1972); 23, 143 (1980) (corr. 42, 260); Suppl. 7, 127 (1987)

Beryllium acetate (see Beryllium and beryllium compounds)
Beryllium acetate, basic (see Beryllium and beryllium compounds)
Beryllium–aluminium alloy (see Beryllium and beryllium compounds)
Beryllium carbonate (see Beryllium and beryllium compounds)
Beryllium chloride (see Beryllium and beryllium compounds)
Beryllium–copper alloy (see Beryllium and beryllium compounds)

Beryllium–copper–cobalt alloy (*see* Beryllium and beryllium compounds)
Beryllium fluoride (*see* Beryllium and beryllium compounds)
Beryllium hydroxide (*see* Beryllium and beryllium compounds)
Beryllium–nickel alloy (*see* Beryllium and beryllium compounds)
Beryllium oxide (*see* Beryllium and beryllium compounds)
Beryllium phosphate (*see* Beryllium and beryllium compounds)
Beryllium silicate (*see* Beryllium and beryllium compounds)
Beryllium sulphate (*see* Beryllium and beryllium compounds)
Beryl ore (*see* Beryllium and beryllium compounds)

Betel quid	37, 141 (1985); *Suppl. 7*, 128 (1987)
Betel-quid chewing (*see* Betel quid)	
BHA (*see* Butylated hydroxyanisole)	
BHT (*see* Butylated hydroxytoluene)	
Bis(1-aziridinyl)morpholinophosphine sulphide	9, 55 (1975); *Suppl. 7*, 58 (1987)
Bis(2-chloroethyl)ether	9, 117 (1975); *Suppl. 7*, 58 (1987)
N,N-Bis(2-chloroethyl)-2-naphthylamine	4, 119 (1974) (*corr.* 42, 253); *Suppl. 7*, 130 (1987)
Bischloroethyl nitrosourea (*see also* Chloroethyl nitrosoureas)	26, 79 (1981); *Suppl. 7*, 150 (1987)
1,2-Bis(chloromethoxy)ethane	15, 31 (1977); *Suppl. 7*, 58 (1987)
1,4-Bis(chloromethoxymethyl)benzene	15, 37 (1977); *Suppl. 7*, 58 (1987)
Bis(chloromethyl)ether	4, 231 (1974) (*corr.* 42, 253); *Suppl. 7*, 131 (1987)
Bis(2-chloro-1-methylethyl)ether	41, 149 (1986); *Suppl. 7*, 59 (1987)
Bis(2,3-epoxycyclopentyl)ether	47, 231 (1989)
Bisphenol A diglycidyl ether (*see* Glycidyl ethers)	
Bitumens	35, 39 (1985); *Suppl. 7*, 133 (1987)
Bleomycins	26, 97 (1981); *Suppl. 7*, 134 (1987)
Blue VRS	16, 163 (1978); *Suppl. 7*, 59 (1987)
Boot and shoe manufacture and repair	25, 249 (1981); *Suppl. 7*, 232 (1987)
Bracken fern	40, 47 (1986); *Suppl. 7*, 135 (1987)
Brilliant Blue FCF	16, 171 (1978) (*corr.* 42, 257); *Suppl. 7*, 59 (1987)
Bromochloroacetonitrile (*see* Halogenated acetonitriles)	
Bromodichloromethane	52, 179 (1991)
Bromoethane	52, 299 (1991)
Bromoform	52, 213 (1991)
1,3-Butadiene	39, 155 (1986) (*corr.* 42, 264); *Suppl. 7*, 136 (1987)
1,4-Butanediol dimethanesulphonate	4, 247 (1974); *Suppl. 7*, 137 (1987)
n-Butyl acrylate	39, 67 (1986); *Suppl. 7*, 59 (1987)
Butylated hydroxyanisole	40, 123 (1986); *Suppl. 7*, 59 (1987)
Butylated hydroxytoluene	40, 161 (1986); *Suppl. 7*, 59 (1987)
Butyl benzyl phthalate	29, 193 (1982) (*corr.* 42, 261); *Suppl. 7*, 59 (1987)
β-Butyrolactone	11, 225 (1976); *Suppl. 7*, 59 (1987)

γ-Butyrolactone   *11*, 231 (1976); *Suppl. 7*, 59 (1987)

C

Cabinet-making (*see* Furniture and cabinet-making)
Cadmium acetate (*see* Cadmium and cadmium compounds)
Cadmium and cadmium compounds   *2*, 74 (1973); *11*, 39 (1976)
                                                       (*corr. 42*, 255);
                                                       Suppl. 7, 139 (1987)

Cadmium chloride (*see* Cadmium and cadmium compounds)
Cadmium oxide (*see* Cadmium and cadmium compounds)
Cadmium sulphate (*see* Cadmium and cadmium compounds)
Cadmium sulphide (*see* Cadmium and cadmium compounds)
Caffeine   *51*, 291 (1991)
Calcium arsenate (*see* Arsenic and arsenic compounds)
Calcium chromate (*see* Chromium and chromium compounds)
Calcium cyclamate (*see* Cyclamates)
Calcium saccharin (*see* Saccharin)
Cantharidin   *10*, 79 (1976); *Suppl. 7*, 59 (1987)
Caprolactam   *19*, 115 (1979) (*corr. 42*, 258);
                                                       39, 247 (1986) (*corr. 42*, 264);
                                                       Suppl. 7, 390 (1987)
Captan   *30*, 295 (1983); *Suppl. 7*, 59 (1987)
Carbaryl   *12*, 37 (1976); *Suppl. 7*, 59 (1987)
Carbazole   *32*, 239 (1983); *Suppl. 7*, 59 (1987)
3-Carbethoxypsoralen   *40*, 317 (1986); *Suppl. 7*, 59 (1987)
Carbon blacks   *3*, 22 (1973); *33*, 35 (1984); *Suppl.*
                                                       7, 142 (1987)
Carbon tetrachloride   *1*, 53 (1972); *20*, 371 (1979);
                                                       Suppl. 7, 143 (1987)
Carmoisine   *8*, 83 (1975); *Suppl. 7*, 59 (1987)
Carpentry and joinery   *25*, 139 (1981); *Suppl. 7*, 378 (1987)
Carrageenan   *10*, 181 (1976) (*corr. 42*, 255); *31*,
                                                       79 (1983); *Suppl. 7*, 59 (1987)
Catechol   *15*, 155 (1977); *Suppl. 7*, 59 (1987)
CCNU (*see* 1-(2-Chloroethyl)-3-cyclohexyl-1-nitrosourea)
Ceramic fibres (*see* Man-made mineral fibres)
Chemotherapy, combined, including alkylating agents
    (*see* MOPP and other combined chemotherapy including
    alkylating agents)
Chlorambucil   *9*, 125 (1975); *26*, 115 (1981);
                                                       Suppl. 7, 144 (1987)
Chloramphenicol   *10*, 85 (1976); *Suppl. 7*, 145 (1987);
                                                       50, 169 (1990)
Chlorendic acid   *48*, 45 (1990)

Chlordane (*see also* Chlordane/Heptachlor)	20, 45 (1979) (*corr.* 42, 258)
Chlordane/Heptachlor	Suppl. 7, 146 (1987)
Chlordecone	20, 67 (1979); Suppl. 7, 59 (1987)
Chlordimeform	30, 61 (1983); Suppl. 7, 59 (1987)
Chlorinated dibenzodioxins (other than TCDD)	15, 41 (1977); Suppl. 7, 59 (1987)
Chlorinated drinking-water	52, 45 (1991)
Chlorinated paraffins	48, 55 (1990)
α-Chlorinated toluenes	Suppl. 7, 148 (1987)
Chlormadinone acetate (*see also* Progestins; Combined oral contraceptives)	6, 149 (1974); 21, 365 (1979)
Chlornaphazine (*see* N,N-Bis(2-chloroethyl)-2-naphthylamine)	
Chloroacetonitrile (*see* Halogenated acetonitriles)	
Chlorobenzilate	5, 75 (1974); 30, 73 (1983); Suppl. 7, 60 (1987)
Chlorodibromomethane	52, 243 (1991)
Chlorodifluoromethane	41, 237 (1986) (*corr.* 51, 483); Suppl. 7, 149 (1987)
Chloroethane	52, 315 (1991)
1-(2-Chloroethyl)-3-cyclohexyl-1-nitrosourea (*see also* Chloroethyl nitrosoureas)	26, 137 (1981) (*corr.* 42, 260); Suppl. 7, 150 (1987)
1-(2-Chloroethyl)-3-(4-methylcyclohexyl)-1-nitrosourea (*see also* Chloroethyl nitrosoureas)	Suppl. 7, 150 (1987)
Chloroethyl nitrosoureas	Suppl. 7, 150 (1987)
Chlorofluoromethane	41, 229 (1986); Suppl. 7, 60 (1987)
Chloroform	1, 61 (1972); 20, 401 (1979); Suppl. 7, 152 (1987)
Chloromethyl methyl ether (technical-grade) (*see also* Bis(chloromethyl)ether)	4, 239 (1974)
(4-Chloro-2-methylphenoxy)acetic acid (*see* MCPA)	
Chlorophenols	Suppl. 7, 154 (1987)
Chlorophenols (occupational exposures to)	41, 319 (1986)
Chlorophenoxy herbicides	Suppl. 7, 156 (1987)
Chlorophenoxy herbicides (occupational exposures to)	41, 357 (1986)
4-Chloro-*ortho*-phenylenediamine	27, 81 (1982); Suppl. 7, 60 (1987)
4-Chloro-*meta*-phenylenediamine	27, 82 (1982); Suppl. 7, 60 (1987)
Chloroprene	19, 131 (1979); Suppl. 7, 160 (1987)
Chloropropham	12, 55 (1976); Suppl. 7, 60 (1987)
Chloroquine	13, 47 (1977); Suppl. 7, 60 (1987)
Chlorothalonil	30, 319 (1983); Suppl. 7, 60 (1987)
para-Chloro-*ortho*-toluidine and its strong acid salts (*see also* Chlordimeform)	16, 277 (1978); 30, 65 (1983); Suppl. 7, 60 (1987); 48, 123 (1990)
Chlorotrianisene (*see also* Nonsteroidal oestrogens)	21, 139 (1979)
2-Chloro-1,1,1-trifluoroethane	41, 253 (1986); Suppl. 7, 60 (1987)
Chlorozotocin	50, 65 (1990)
Cholesterol	10, 99 (1976); 31, 95 (1983); Suppl. 7, 161 (1987)

Chromic acetate (*see* Chromium and chromium compounds)
Chromic chloride (*see* Chromium and chromium compounds)
Chromic oxide (*see* Chromium and chromium compounds)
Chromic phosphate (*see* Chromium and chromium compounds)
Chromite ore (*see* Chromium and chromium compounds)
Chromium and chromium compounds *2*, 100 (1973); *23*, 205 (1980); *Suppl. 7*, 165 (1987); *49*, 49 (1990) (*corr. 51*, 483)

Chromium carbonyl (*see* Chromium and chromium compounds)
Chromium potassium sulphate (*see* Chromium and chromium compounds)
Chromium sulphate (*see* Chromium and chromium compounds)
Chromium trioxide (*see* Chromium and chromium compounds)
Chrysazin (*see* Dantron)
Chrysene *3*, 159 (1973); *32*, 247 (1983); *Suppl. 7*, 60 (1987)

Chrysoidine *8*, 91 (1975); *Suppl. 7*, 169 (1987)
Chrysotile (*see* Asbestos)
Ciclosporin *50*, 77 (1990)
CI Disperse Yellow 3 *8*, 97 (1975); *Suppl. 7*, 60 (1987)
Cimetidine *50*, 235 (1990)
Cinnamyl anthranilate *16*, 287 (1978); *31*, 133 (1983); *Suppl. 7*, 60 (1987)
Cisplatin *26*, 151 (1981); *Suppl. 7*, 170 (1987)
Citrinin *40*, 67 (1986); *Suppl. 7*, 60 (1987)
Citrus Red No. 2 *8*, 101 (1975) (*corr. 42*, 254); *Suppl. 7*, 60 (1987)

Clofibrate *24*, 39 (1980); *Suppl. 7*, 171 (1987)
Clomiphene citrate *21*, 551 (1979); *Suppl. 7*, 172 (1987)
Coal gasification *34*, 65 (1984); *Suppl. 7*, 173 (1987)
Coal-tar pitches (*see also* Coal-tars) *Suppl. 7*, 174 (1987)
Coal-tars *35*, 83 (1985); *Suppl. 7*, 175 (1987)
Cobalt[III] acetate (*see* Cobalt and cobalt compounds)
Cobalt-aluminium-chromium spinel (*see* Cobalt and cobalt compounds)
Cobalt and cobalt compounds *52*, 363 (1991)
Cobalt[II] chloride (*see* Cobalt and cobalt compounds)
Cobalt–chromium alloy (*see* Chromium and chromium compounds)
Cobalt-chromium-molybdenum alloys (*see* Cobalt and cobalt compounds)
Cobalt metal powder (*see* Cobalt and cobalt compounds)
Cobalt naphthenate (*see* Cobalt and cobalt compounds)
Cobalt[II] oxide (*see* Cobalt and cobalt compounds)
Cobalt[II,III] oxide (*see* Cobalt and cobalt compounds)
Cobalt[II] sulfide (*see* Cobalt and cobalt compounds)
Coffee *51*, 41 (1991) (*corr. 52*, 513)

Coke production	*34*, 101 (1984); *Suppl. 7*, 176 (1987)
Combined oral contraceptives (*see also* Oestrogens, progestins and combinations)	*Suppl. 7*, 297 (1987)
Conjugated oestrogens (*see also* Steroidal oestrogens)	*21*, 147 (1979)
Contraceptives, oral (*see* Combined oral contraceptives; Sequential oral contraceptives)	
Copper 8-hydroxyquinoline	*15*, 103 (1977); *Suppl. 7*, 61 (1987)
Coronene	*32*, 263 (1983); *Suppl. 7*, 61 (1987)
Coumarin	*10*, 113 (1976); *Suppl. 7*, 61 (1987)
Creosotes (*see also* Coal-tars)	*Suppl. 7*, 177 (1987)
meta-Cresidine	*27*, 91 (1982); *Suppl. 7*, 61 (1987)
para-Cresidine	*27*, 92 (1982); *Suppl. 7*, 61 (1987)
Crocidolite (*see* Asbestos)	
Crude oil	*45*, 119 (1989)
Crystalline silica (*see also* Silica)	*Suppl. 7*, 341 (1987)
Cycasin	*1*, 157 (1972) (*corr. 42*, 251); *10*, 121 (1976); *Suppl. 7*, 61 (1987)
Cyclamates	*22*, 55 (1980); *Suppl. 7*, 178 (1987)
Cyclamic acid (*see* Cyclamates)	
Cyclochlorotine	*10*, 139 (1976); *Suppl. 7*, 61 (1987)
Cyclohexanone	*47*, 157 (1989)
Cyclohexylamine (*see* Cyclamates)	
Cyclopenta[*cd*]pyrene	*32*, 269 (1983); *Suppl. 7*, 61 (1987)
Cyclopropane (*see* Anaesthetics, volatile)	
Cyclophosphamide	*9*, 135 (1975); *26*, 165 (1981); *Suppl. 7*, 182 (1987)

D

2,4-D (*see also* Chlorophenoxy herbicides; Chlorophenoxy herbicides, occupational exposures to)	*15*, 111 (1977)
Dacarbazine	*26*, 203 (1981); *Suppl. 7*, 184 (1987)
Dantron	*50*, 265 (1990)
D & C Red No. 9	*8*, 107 (1975); *Suppl. 7*, 61 (1987)
Dapsone	*24*, 59 (1980); *Suppl. 7*, 185 (1987)
Daunomycin	*10*, 145 (1976); *Suppl. 7*, 61 (1987)
DDD (*see* DDT)	
DDE (*see* DDT)	
DDT	*5*, 83 (1974) (*corr. 42*, 253); *Suppl. 7*, 186 (1987)
Decabromodiphenyl oxide	*48*, 73 (1990)
Diacetylaminoazotoluene	*8*, 113 (1975); *Suppl. 7*, 61 (1987)
N,N'-Diacetylbenzidine	*16*, 293 (1978); *Suppl. 7*, 61 (1987)
Diallate	*12*, 69 (1976); *30*, 235 (1983); *Suppl. 7*, 61 (1987)

2,4-Diaminoanisole	16, 51 (1978); 27, 103 (1982); Suppl. 7, 61 (1987)
4,4'-Diaminodiphenyl ether	16, 301 (1978); 29, 203 (1982); Suppl. 7, 61 (1987)
1,2-Diamino-4-nitrobenzene	16, 63 (1978); Suppl. 7, 61 (1987)
1,4-Diamino-2-nitrobenzene	16, 73 (1978); Suppl. 7, 61 (1987)
2,6-Diamino-3-(phenylazo)pyridine (see Phenazopyridine hydrochloride)	
2,4-Diaminotoluene (see also Toluene diisocyanates)	16, 83 (1978); Suppl. 7, 61 (1987)
2,5-Diaminotoluene (see also Toluene diisocyanates)	16, 97 (1978); Suppl. 7, 61 (1987)
ortho-Dianisidine (see 3,3'-Dimethoxybenzidine)	
Diazepam	13, 57 (1977); Suppl. 7, 189 (1987)
Diazomethane	7, 223 (1974); Suppl. 7, 61 (1987)
Dibenz[a,h]acridine	3, 247 (1973); 32, 277 (1983); Suppl. 7, 61 (1987)
Dibenz[a,j]acridine	3, 254 (1973); 32, 283 (1983); Suppl. 7, 61 (1987)
Dibenz[a,c]anthracene	32, 289 (1983) (corr. 42, 262); Suppl. 7, 61 (1987)
Dibenz[a,h]anthracene	3, 178 (1973) (corr. 43, 261); 32, 299 (1983); Suppl. 7, 61 (1987)
Dibenz[a,j]anthracene	32, 309 (1983); Suppl. 7, 61 (1987)
7H-Dibenzo[c,g]carbazole	3, 260 (1973); 32, 315 (1983); Suppl. 7, 61 (1987)
Dibenzodioxins, chlorinated (other than TCDD) (see Chlorinated dibenzodioxins (other than TCDD))	
Dibenzo[a,e]fluoranthene	32, 321 (1983); Suppl. 7, 61 (1987)
Dibenzo[h,rst]pentaphene	3, 197 (1973); Suppl. 7, 62 (1987)
Dibenzo[a,e]pyrene	3, 201 (1973); 32, 327 (1983); Suppl. 7, 62 (1987)
Dibenzo[a,h]pyrene	3, 207 (1973); 32, 331 (1983); Suppl. 7, 62 (1987)
Dibenzo[a,i]pyrene	3, 215 (1973); 32, 337 (1983); Suppl. 7, 62 (1987)
Dibenzo[a,l]pyrene	3, 224 (1973); 32, 343 (1983); Suppl. 7, 62 (1987)
Dibromoacetonitrile (see Halogenated acetonitriles)	
1,2-Dibromo-3-chloropropane	15, 139 (1977); 20, 83 (1979); Suppl. 7, 191 (1987)
Dichloroacetonitrile (see Halogenated acetonitriles)	
Dichloroacetylene	39, 369 (1986); Suppl. 7, 62 (1987)
ortho-Dichlorobenzene	7, 231 (1974); 29, 213 (1982); Suppl. 7, 192 (1987)
para-Dichlorobenzene	7, 231 (1974); 29, 215 (1982); Suppl. 7, 192 (1987)

3,3′-Dichlorobenzidine	*4*, 49 (1974); *29*, 239 (1982); *Suppl. 7*, 193 (1987)
trans-1,4-Dichlorobutene	*15*, 149 (1977); *Suppl. 7*, 62 (1987)
3,3′-Dichloro-4,4′-diaminodiphenyl ether	*16*, 309 (1978); *Suppl. 7*, 62 (1987)
1,2-Dichloroethane	*20*, 429 (1979); *Suppl. 7*, 62 (1987)
Dichloromethane	*20*, 449 (1979); *41*, 43 (1986); *Suppl. 7*, 194 (1987)
2,4-Dichlorophenol (*see* Chlorophenols; Chlorophenols, occupational exposures to)	
(2,4-Dichlorophenoxy)acetic acid (*see* 2,4-D)	
2,6-Dichloro-*para*-phenylenediamine	*39*, 325 (1986); *Suppl. 7*, 62 (1987)
1,2-Dichloropropane	*41*, 131 (1986); *Suppl. 7*, 62 (1987)
1,3-Dichloropropene (technical-grade)	*41*, 113 (1986); *Suppl. 7*, 195 (1987)
Dichlorvos	*20*, 97 (1979); *Suppl. 7*, 62 (1987)
Dicofol	*30*, 87 (1983); *Suppl. 7*, 62 (1987)
Dicyclohexylamine (*see* Cyclamates)	
Dieldrin	*5*, 125 (1974); *Suppl. 7*, 196 (1987)
Dienoestrol (*see also* Nonsteroidal oestrogens)	*21*, 161 (1979)
Diepoxybutane	*11*, 115 (1976) (*corr. 42*, 255); *Suppl. 7*, 62 (1987)
Diesel and gasoline engine exhausts	*46*, 41 (1989)
Diesel fuels	*45*, 219 (1989) (*corr. 47*, 505)
Diethyl ether (*see* Anaesthetics, volatile)	
Di(2-ethylhexyl)adipate	*29*, 257 (1982); *Suppl. 7*, 62 (1987)
Di(2-ethylhexyl)phthalate	*29*, 269 (1982) (*corr. 42*, 261); *Suppl. 7*, 62 (1987)
1,2-Diethylhydrazine	*4*, 153 (1974); *Suppl. 7*, 62 (1987)
Diethylstilboestrol	*6*, 55 (1974); *21*, 173 (1979) (*corr. 42*, 259); *Suppl. 7*, 273 (1987)
Diethylstilboestrol dipropionate (*see* Diethylstilboestrol)	
Diethyl sulphate	*4*, 277 (1974); *Suppl. 7*, 198 (1987)
Diglycidyl resorcinol ether	*11*, 125 (1976); *36*, 181 (1985); *Suppl. 7*, 62 (1987)
Dihydrosafrole	*1*, 170 (1972); *10*, 233 (1976); *Suppl. 7*, 62 (1987)
1,8-Dihydroxyanthraquinone (*see* Dantron)	
Dihydroxybenzenes (*see* Catechol; Hydroquinone; Resorcinol)	
Dihydroxymethylfuratrizine	*24*, 77 (1980); *Suppl. 7*, 62 (1987)
Dimethisterone (*see also* Progestins; Sequential oral contraceptives)	*6*, 167 (1974); *21*, 377 (1979)
Dimethoxane	*15*, 177 (1977); *Suppl. 7*, 62 (1987)
3,3′-Dimethoxybenzidine	*4*, 41 (1974); *Suppl. 7*, 198 (1987)
3,3′-Dimethoxybenzidine-4,4′-diisocyanate	*39*, 279 (1986); *Suppl. 7*, 62 (1987)
para-Dimethylaminoazobenzene	*8*, 125 (1975); *Suppl. 7*, 62 (1987)
para-Dimethylaminoazobenzenediazo sodium sulphonate	*8*, 147 (1975); *Suppl. 7*, 62 (1987)

trans-2-[(Dimethylamino)methylimino]-5-[2-(5-nitro-2-furyl)-vinyl]-1,3,4-oxadiazole	7, 147 (1974) (*corr.* 42, 253); *Suppl.* 7, 62 (1987)
4,4'-Dimethylangelicin plus ultraviolet radiation (*see also* Angelicin and some synthetic derivatives)	*Suppl.* 7, 57 (1987)
4,5'-Dimethylangelicin plus ultraviolet radiation (*see also* Angelicin and some synthetic derivatives)	*Suppl.* 7, 57 (1987)
Dimethylarsinic acid (*see* Arsenic and arsenic compounds)	
3,3'-Dimethylbenzidine	*1*, 87 (1972); *Suppl.* 7, 62 (1987)
Dimethylcarbamoyl chloride	*12*, 77 (1976); *Suppl.* 7, 199 (1987)
Dimethylformamide	*47*, 171 (1989)
1,1-Dimethylhydrazine	*4*, 137 (1974); *Suppl.* 7, 62 (1987)
1,2-Dimethylhydrazine	*4*, 145 (1974) (*corr.* 42, 253); *Suppl.* 7, 62 (1987)
Dimethyl hydrogen phosphite	*48*, 85 (1990)
1,4-Dimethylphenanthrene	*32*, 349 (1983); *Suppl.* 7, 62 (1987)
Dimethyl sulphate	*4*, 271 (1974); *Suppl.* 7, 200 (1987)
3,7-Dinitrofluoranthene	*46*, 189 (1989)
3,9-Dinitrofluoranthene	*46*, 195 (1989)
1,3-Dinitropyrene	*46*, 201 (1989)
1,6-Dinitropyrene	*46*, 215 (1989)
1,8-Dinitropyrene	*33*, 171 (1984); *Suppl.* 7, 63 (1987); *46*, 231 (1989)
Dinitrosopentamethylenetetramine	*11*, 241 (1976); *Suppl.* 7, 63 (1987)
1,4-Dioxane	*11*, 247 (1976); *Suppl.* 7, 201 (1987)
2,4'-Diphenyldiamine	*16*, 313 (1978); *Suppl.* 7, 63 (1987)
Direct Black 38 (*see also* Benzidine-based dyes)	*29*, 295 (1982) (*corr.* 42, 261)
Direct Blue 6 (*see also* Benzidine-based dyes)	*29*, 311 (1982)
Direct Brown 95 (*see also* Benzidine-based dyes)	*29*, 321 (1982)
Disperse Blue 1	*48*, 139 (1990)
Disperse Yellow 3	*48*, 149 (1990)
Disulfiram	*12*, 85 (1976); *Suppl.* 7, 63 (1987)
Dithranol	*13*, 75 (1977); *Suppl.* 7, 63 (1987)
Divinyl ether (*see* Anaesthetics, volatile)	
Dulcin	*12*, 97 (1976); *Suppl.* 7, 63 (1987)

E

Endrin	5, 157 (1974); *Suppl.* 7, 63 (1987)
Enflurane (*see* Anaesthetics, volatile)	
Eosin	*15*, 183 (1977); *Suppl.* 7, 63 (1987)
Epichlorohydrin	*11*, 131 (1976) (*corr.* 42, 256); *Suppl.* 7, 202 (1987)
1,2-Epoxybutane	*47*, 217 (1989)
1-Epoxyethyl-3,4-epoxycyclohexane	*11*, 141 (1976); *Suppl.* 7, 63 (1987)
3,4-Epoxy-6-methylcyclohexylmethyl-3,4-epoxy-6-methyl-cyclohexane carboxylate	*11*, 147 (1976); *Suppl.* 7, 63 (1987)

cis-9,10-Epoxystearic acid	11, 153 (1976); Suppl. 7, 63 (1987)
Erionite	42, 225 (1987); Suppl. 7, 203 (1987)
Ethinyloestradiol (see also Steroidal oestrogens)	6, 77 (1974); 21, 233 (1979)
Ethionamide	13, 83 (1977); Suppl. 7, 63 (1987)
Ethyl acrylate	19, 57 (1979); 39, 81 (1986); Suppl. 7, 63 (1987)
Ethylene	19, 157 (1979); Suppl. 7, 63 (1987)
Ethylene dibromide	15, 195 (1977); Suppl. 7, 204 (1987)
Ethylene oxide	11, 157 (1976); 36, 189 (1985) (corr. 42, 263); Suppl. 7, 205 (1987)
Ethylene sulphide	11, 257 (1976); Suppl. 7, 63 (1987)
Ethylene thiourea	7, 45 (1974); Suppl. 7, 207 (1987)
Ethyl methanesulphonate	7, 245 (1974); Suppl. 7, 63 (1987)
N-Ethyl-N-nitrosourea	1, 135 (1972); 17, 191 (1978); Suppl. 7, 63 (1987)
Ethyl selenac (see also Selenium and selenium compounds)	12, 107 (1976); Suppl. 7, 63 (1987)
Ethyl tellurac	12, 115 (1976); Suppl. 7, 63 (1987)
Ethynodiol diacetate (see also Progestins; Combined oral contraceptives)	6, 173 (1974); 21, 387 (1979)
Eugenol	36, 75 (1985); Suppl. 7, 63 (1987)
Evans blue	8, 151 (1975); Suppl. 7, 63 (1987)

F

Fast Green FCF	16, 187 (1978); Suppl. 7, 63 (1987)
Ferbam	12, 121 (1976) (corr. 42, 256); Suppl. 7, 63 (1987)
Ferric oxide	1, 29 (1972); Suppl. 7, 216 (1987)
Ferrochromium (see Chromium and chromium compounds)	
Fluometuron	30, 245 (1983); Suppl. 7, 63 (1987)
Fluoranthene	32, 355 (1983); Suppl. 7, 63 (1987)
Fluorene	32, 365 (1983); Suppl. 7, 63 (1987)
Fluorides (inorganic, used in drinking-water)	27, 237 (1982); Suppl. 7, 208 (1987)
5-Fluorouracil	26, 217 (1981); Suppl. 7, 210 (1987)
Fluorspar (see Fluorides)	
Fluosilicic acid (see Fluorides)	
Fluroxene (see Anaesthetics, volatile)	
Formaldehyde	29, 345 (1982); Suppl. 7, 211 (1987)
2-(2-Formylhydrazino)-4-(5-nitro-2-furyl)thiazole	7, 151 (1974) (corr. 42, 253); Suppl. 7, 63 (1987)
Frusemide (see Furosemide)	
Fuel oils (heating oils)	45, 239 (1989) (corr. 47, 505)
Furazolidone	31, 141 (1983); Suppl. 7, 63 (1987)
Furniture and cabinet-making	25, 99 (1981); Suppl. 7, 380 (1987)
Furosemide	50, 277 (1990)

2-(2-Furyl)-3-(5-nitro-2-furyl)acrylamide (*see* AF-2)
Fusarenon-X   *11*, 169 (1976); *31*, 153 (1983);
                                                           Suppl. 7, 64 (1987)

G

Gasoline   *45*, 159 (1989) (*corr. 47*, 505)
Gasoline engine exhaust (*see* Diesel and gasoline engine exhausts)
Glass fibres (*see* Man-made mineral fibres)
Glasswool (*see* Man-made mineral fibres)
Glass filaments (*see* Man-made mineral fibres)
Glu-P-1   *40*, 223 (1986); *Suppl. 7*, 64 (1987)
Glu-P-2   *40*, 235 (1986); *Suppl. 7*, 64 (1987)
L-Glutamic acid, 5-[2-(4-hydroxymethyl)phenylhydrazide]
  (*see* Agaratine)
Glycidaldehyde   *11*, 175 (1976); *Suppl. 7*, 64 (1987)
Glycidyl ethers   *47*, 237 (1989)
Glycidyl oleate   11, 183 (1976); *Suppl. 7*, 64 (1987)
Glycidyl stearate   *11*, 187 (1976); *Suppl. 7*, 64 (1987)
Griseofulvin   *10*, 153 (1976); *Suppl. 7*, 391 (1987)
Guinea Green B   *16*, 199 (1978); *Suppl. 7*, 64 (1987)
Gyromitrin   *31*, 163 (1983); *Suppl. 7*, 391 (1987)

H

Haematite   *1*, 29 (1972); *Suppl. 7*, 216 (1987)
Haematite and ferric oxide   *Suppl. 7*, 216 (1987)
Haematite mining, underground, with exposure to radon   *1*, 29 (1972); *Suppl. 7*, 216 (1987)
Hair dyes, epidemiology of   *16*, 29 (1978); *27*, 307 (1982)
Halogenated acetonitriles   *52*, 269 (1991)
Halothane (*see* Anaesthetics, volatile)
α-HCH (*see* Hexachlorocyclohexanes)
β-HCH (*see* Hexachlorocyclohexanes)
γ-HCH (*see* Hexachlorocyclohexanes)
Heating oils (*see* Fuel oils)
Heptachlor (*see also* Chlordane/Heptachlor)   *5*, 173 (1974); *20*, 129 (1979)
Hexachlorobenzene   *20*, 155 (1979); *Suppl. 7*, 219 (1987)
Hexachlorobutadiene   *20*, 179 (1979); *Suppl. 7*, 64 (1987)
Hexachlorocyclohexanes   *5*, 47 (1974); *20*, 195 (1979)
                                                           (*corr. 42*, 258); *Suppl. 7*, 220 (1987)
Hexachlorocyclohexane, technical-grade (*see* Hexachloro-
  cyclohexanes)
Hexachloroethane   *20*, 467 (1979); *Suppl. 7*, 64 (1987)
Hexachlorophene   *20*, 241 (1979); *Suppl. 7*, 64 (1987)

Hexamethylphosphoramide	*15*, 211 (1977); *Suppl. 7*, 64 (1987)
Hexoestrol (*see* Nonsteroidal oestrogens)	
Hycanthone mesylate	*13*, 91 (1977); *Suppl. 7*, 64 (1987)
Hydralazine	*24*, 85 (1980); *Suppl. 7*, 222 (1987)
Hydrazine	*4*, 127 (1974); *Suppl. 7*, 223 (1987)
Hydrochlorothiazide	*50*, 293 (1990)
Hydrogen peroxide	*36*, 285 (1985); *Suppl. 7*, 64 (1987)
Hydroquinone	*15*, 155 (1977); *Suppl. 7*, 64 (1987)
4-Hydroxyazobenzene	*8*, 157 (1975); *Suppl. 7*, 64 (1987)
17α-Hydroxyprogesterone caproate (*see also* Progestins)	*21*, 399 (1979) (*corr. 42*, 259)
8-Hydroxyquinoline	*13*, 101 (1977); *Suppl. 7*, 64 (1987)
8-Hydroxysenkirkine	*10*, 265 (1976); *Suppl. 7*, 64 (1987)
Hypochlorite salts	*52*, 159 (1991)

I

Indeno[1,2,3-*cd*]pyrene	*3*, 229 (1973); *32*, 373 (1983); *Suppl. 7*, 64 (1987)
IQ	*40*, 261 (1986); *Suppl. 7*, 64 (1987)
Iron and steel founding	*34*, 133 (1984); *Suppl. 7*, 224 (1987)
Iron-dextran complex	*2*, 161 (1973); *Suppl. 7*, 226 (1987)
Iron-dextrin complex	*2*, 161 (1973) (*corr. 42*, 252); *Suppl. 7*, 64 (1987)
Iron oxide (*see* Ferric oxide)	
Iron oxide, saccharated (*see* Saccharated iron oxide)	
Iron sorbitol citric acid complex	*2*, 161 (1973); *Suppl. 7*, 64 (1987)
Isatidine	*10*, 269 (1976); *Suppl. 7*, 65 (1987)
Isoflurane (*see* Anaesthetics, volatile)	
Isoniazid (*see* Isonicotinic acid hydrazide)	
Isonicotinic acid hydrazide	*4*, 159 (1974); *Suppl. 7*, 227 (1987)
Isophosphamide	*26*, 237 (1981); *Suppl. 7*, 65 (1987)
Isopropyl alcohol	*15*, 223 (1977); *Suppl. 7*, 229 (1987)
Isopropyl alcohol manufacture (strong-acid process) (*see also* Isopropyl alcohol)	*Suppl. 7*, 229 (1987)
Isopropyl oils	*15*, 223 (1977); *Suppl. 7*, 229 (1987)
Isosafrole	*1*, 169 (1972); *10*, 232 (1976); *Suppl. 7*, 65 (1987)

J

Jacobine	*10*, 275 (1976); *Suppl. 7*, 65 (1987)
Jet fuel	*45*, 203 (1989)
Joinery (*see* Carpentry and joinery)	

K

Kaempferol	*31*, 171 (1983); *Suppl. 7*, 65 (1987)

Kepone (*see* Chlordecone)

L

Lasiocarpine	*10*, 281 (1976); *Suppl. 7*, 65 (1987)
Lauroyl peroxide	*36*, 315 (1985); Suppl. 7, 65 (1987)
Lead acetate (*see* Lead and lead compounds)	
Lead and lead compounds	*1*, 40 (1972) (*corr. 42*, 251); *2*, 52, 150 (1973); *12*, 131 (1976); *23*, 40, 208, 209, 325 (1980); *Suppl. 7*, 230 (1987)
Lead arsenate (*see* Arsenic and arsenic compounds)	
Lead carbonate (*see* Lead and lead compounds)	
Lead chloride (*see* Lead and lead compounds)	
Lead chromate (*see* Chromium and chromium compounds)	
Lead chromate oxide (*see* Chromium and chromium compounds)	
Lead naphthenate (*see* Lead and lead compounds)	
Lead nitrate (*see* Lead and lead compounds)	
Lead oxide (*see* Lead and lead compounds)	
Lead phosphate (*see* Lead and lead compounds)	
Lead subacetate (*see* Lead and lead compounds)	
Lead tetroxide (*see* Lead and lead compounds)	
Leather goods manufacture	*25*, 279 (1981); *Suppl. 7*, 235 (1987)
Leather industries	*25*, 199 (1981); *Suppl. 7*, 232 (1987)
Leather tanning and processing	*25*, 201 (1981); *Suppl. 7*, 236 (1987)
Ledate (*see also* Lead and lead compounds)	*12*, 131 (1976)
Light Green SF	*16*, 209 (1978); *Suppl. 7*, 65 (1987)
Lindane (*see* Hexachlorocyclohexanes)	
The lumber and sawmill industries (including logging)	*25*, 49 (1981); *Suppl. 7*, 383 (1987)
Luteoskyrin	*10*, 163 (1976); *Suppl. 7*, 65 (1987)
Lynoestrenol (*see also* Progestins; Combined oral contraceptives)	*21*, 407 (1979)

M

Magenta	*4*, 57 (1974) (*corr. 42*, 252); *Suppl. 7*, 238 (1987)
Magenta, manufacture of (*see also* Magenta)	*Suppl. 7*, 238 (1987)
Malathion	*30*, 103 (1983); *Suppl. 7*, 65 (1987)
Maleic hydrazide	*4*, 173 (1974) (*corr. 42*, 253); *Suppl. 7*, 65 (1987)
Malonaldehyde	*36*, 163 (1985); *Suppl. 7*, 65 (1987)
Maneb	*12*, 137 (1976); *Suppl. 7*, 65 (1987)
Man-made mineral fibres	*43*, 39 (1988)
Mannomustine	*9*, 157 (1975); *Suppl. 7*, 65 (1987)
Mate	*51*, 273 (1991)

MCPA (*see also* Chlorophenoxy herbicides; Chlorophenoxy herbicides, occupational exposures to)	*30*, 255 (1983)
MeA-α-C	*40*, 253 (1986); *Suppl. 7*, 65 (1987)
Medphalan	*9*, 168 (1975); *Suppl. 7*, 65 (1987)
Medroxyprogesterone acetate	*6*, 157 (1974); *21*, 417 (1979) (*corr. 42*, 259); *Suppl. 7*, 289 (1987)
Megestrol acetate (*see* also Progestins; Combined oral contraceptives)	
MeIQ	*40*, 275 (1986); *Suppl. 7*, 65 (1987)
MeIQx	*40*, 283 (1986); *Suppl. 7*, 65 (1987)
Melamine	*39*, 333 (1986); *Suppl. 7*, 65 (1987)
Melphalan	*9*, 167 (1975); *Suppl. 7*, 239 (1987)
6-Mercaptopurine	*26*, 249 (1981); *Suppl. 7*, 240 (1987)
Merphalan	*9*, 169 (1975); *Suppl. 7*, 65 (1987)
Mestranol (*see also* Steroidal oestrogens)	*6*, 87 (1974); *21*, 257 (1979) (*corr. 42*, 259)
Methanearsonic acid, disodium salt (*see* Arsenic and arsenic compounds)	
Methanearsonic acid, monosodium salt (*see* Arsenic and arsenic compounds	
Methotrexate	*26*, 267 (1981); *Suppl. 7*, 241 (1987)
Methoxsalen (*see* 8-Methoxypsoralen)	
Methoxychlor	*5*, 193 (1974); *20*, 259 (1979); *Suppl. 7*, 66 (1987)
Methoxyflurane (*see* Anaesthetics, volatile)	
5-Methoxypsoralen	*40*, 327 (1986); *Suppl. 7*, 242 (1987)
8-Methoxypsoralen (*see also* 8-Methoxypsoralen plus ultraviolet radiation)	*24*, 101 (1980)
8-Methoxypsoralen plus ultraviolet radiation	*Suppl. 7*, 243 (1987)
Methyl acrylate	*19*, 52 (1979); *39*, 99 (1986); *Suppl. 7*, 66 (1987)
5-Methylangelicin plus ultraviolet radiation (*see also* Angelicin and some synthetic derivatives)	*Suppl. 7*, 57 (1987)
2-Methylaziridine	*9*, 61 (1975); *Suppl. 7*, 66 (1987)
Methylazoxymethanol acetate	*1*, 164 (1972); *10*, 131 (1976); *Suppl. 7*, 66 (1987)
Methyl bromide	*41*, 187 (1986) (*corr. 45*, 283); *Suppl. 7*, 245 (1987)
Methyl carbamate	*12*, 151 (1976); *Suppl. 7*, 66 (1987)
Methyl-CCNU [*see* 1-(2-Chloroethyl)-3-(4-methylcyclohexyl)-1-nitrosourea]	
Methyl chloride	*41*, 161 (1986); *Suppl. 7*, 246 (1987)
1-, 2-, 3-, 4-, 5- and 6-Methylchrysenes	*32*, 379 (1983); *Suppl. 7*, 66 (1987)
N-Methyl-*N*,4-dinitrosoaniline	*1*, 141 (1972); *Suppl. 7*, 66 (1987)
4,4′-Methylene bis(2-chloroaniline)	*4*, 65 (1974) (*corr. 42*, 252); *Suppl. 7*, 246 (1987)

4,4'-Methylene bis(N,N-dimethyl)benzenamine	27, 119 (1982); Suppl. 7, 66 (1987)
4,4'-Methylene bis(2-methylaniline)	4, 73 (1974); Suppl. 7, 248 (1987)
4,4'-Methylenedianiline	4, 79 (1974) (corr. 42, 252); 39, 347 (1986); Suppl. 7, 66 (1987)
4,4'-Methylenediphenyl diisocyanate	19, 314 (1979); Suppl. 7, 66 (1987)
2-Methylfluoranthene	32, 399 (1983); Suppl. 7, 66 (1987)
3-Methylfluoranthene	32, 399 (1983); Suppl. 7, 66 (1987)
Methylglyoxal	51, 443 (1991)
Methyl iodide	15, 245 (1977); 41, 213 (1986); Suppl. 7, 66 (1987)
Methyl methacrylate	19, 187 (1979); Suppl. 7, 66 (1987)
Methyl methanesulphonate	7, 253 (1974); Suppl. 7, 66 (1987)
2-Methyl-1-nitroanthraquinone	27, 205 (1982); Suppl. 7, 66 (1987)
N-Methyl-N'-nitro-N-nitrosoguanidine	4, 183 (1974); Suppl. 7, 248 (1987)
3-Methylnitrosaminopropionaldehyde (see 3-(N-Nitrosomethylamino)propionaldehyde)	
3-Methylnitrosaminopropionitrile (see 3-(N-Nitrosomethylamino)propionitrile)	
4-(Methylnitrosamino)-4-(3-pyridyl)-1-butanal (see 4-(N-Nitrosomethylamino)-4-(3-pyridyl)-1-butanal)	
4-(Methylnitrosamino)-1-(3-pyridyl)-1-butanone (see 4-(N-Nitrosomethylamino)-1-(3-pyridyl)-1-butanone)	
N-Methyl-N-nitrosourea	1, 125 (1972); 17, 227 (1978); Suppl. 7, 66 (1987)
N-Methyl-N-nitrosourethane	4, 211 (1974); Suppl. 7, 66 (1987)
Methyl parathion	30, 131 (1983); Suppl. 7, 392 (1987)
1-Methylphenanthrene	32, 405 (1983); Suppl. 7, 66 (1987)
7-Methylpyrido[3,4-c]psoralen	40, 349 (1986); Suppl. 7, 71 (1987)
Methyl red	8, 161 (1975); Suppl. 7, 66 (1987)
Methyl selenac (see also Selenium and selenium compounds)	12, 161 (1976); Suppl. 7, 66 (1987)
Methylthiouracil	7, 53 (1974); Suppl. 7, 66 (1987)
Metronidazole	13, 113 (1977); Suppl. 7, 250 (1987)
Mineral oils	3, 30 (1973); 33, 87 (1984) (corr. 42, 262); Suppl. 7, 252 (1987)
Mirex	5, 203 (1974); 20, 283 (1979) (corr. 42, 258); Suppl. 7, 66 (1987)
Mitomycin C	10, 171 (1976); Suppl. 7, 67 (1987)
MNNG (see N-Methyl-N'-nitro-N-nitrosoguanidine)	
MOCA (see 4,4'-Methylene bis(2-chloroaniline))	
Modacrylic fibres	19, 86 (1979); Suppl. 7, 67 (1987)
Monocrotaline	10, 291 (1976); Suppl. 7, 67 (1987)
Monuron	12, 167 (1976); Suppl. 7, 67 (1987)
MOPP and other combined chemotherapy including alkylating agents	Suppl. 7, 254 (1987)
Morpholine	47, 199 (1989)

5-(Morpholinomethyl)-3-[(5-nitrofurfurylidene)amino]-2-oxazolidinone	*7*, 161 (1974); *Suppl. 7*, 67 (1987)
Mustard gas	*9*, 181 (1975) (*corr. 42*, 254); *Suppl. 7*, 259 (1987)
Myleran (*see* 1,4-Butanediol dimethanesulphonate)	

N

Nafenopin	*24*, 125 (1980); *Suppl. 7*, 67 (1987)
1,5-Naphthalenediamine	*27*, 127 (1982); *Suppl. 7*, 67 (1987)
1,5-Naphthalene diisocyanate	*19*, 311 (1979); *Suppl. 7*, 67 (1987)
1-Naphthylamine	*4*, 87 (1974) (*corr. 42*, 253); *Suppl. 7*, 260 (1987)
2-Naphthylamine	*4*, 97 (1974); *Suppl. 7*, 261 (1987)
1-Naphthylthiourea	*30*, 347 (1983); *Suppl. 7*, 263 (1987)
Nickel acetate (*see* Nickel and nickel compounds)	
Nickel ammonium sulphate (*see* Nickel and nickel compounds)	
Nickel and nickel compounds	*2*, 126 (1973) (*corr. 42*, 252); *11*, 75 (1976); *Suppl. 7*, 264 (1987) (*corr. 45*, 283); *49*, 257 (1990)
Nickel carbonate (*see* Nickel and nickel compounds)	
Nickel carbonyl (*see* Nickel and nickel compounds)	
Nickel chloride (*see* Nickel and nickel compounds)	
Nickel-gallium alloy (*see* Nickel and nickel compounds)	
Nickel hydroxide (*see* Nickel and nickel compounds)	
Nickelocene (*see* Nickel and nickel compounds)	
Nickel oxide (*see* Nickel and nickel compounds)	
Nickel subsulphide (*see* Nickel and nickel compounds)	
Nickel sulphate (*see* Nickel and nickel compounds)	
Niridazole	*13*, 123 (1977); *Suppl. 7*, 67 (1987)
Nithiazide	*31*, 179 (1983); *Suppl. 7*, 67 (1987)
Nitrilotriacetic acid and its salts	*48*, 181 (1990)
5-Nitroacenaphthene	*16*, 319 (1978); *Suppl. 7*, 67 (1987)
5-Nitro-*ortho*-anisidine	*27*, 133 (1982); *Suppl. 7*, 67 (1987)
9-Nitroanthracene	*33*, 179 (1984); *Suppl. 7*, 67 (1987)
7-Nitrobenz[*a*]anthracene	*46*, 247 (1989)
6-Nitrobenzo[*a*]pyrene	*33*, 187 (1984); *Suppl. 7*, 67 (1987); *46*, 255 (1989)
4-Nitrobiphenyl	*4*, 113 (1974); *Suppl. 7*, 67 (1987)
6-Nitrochrysene	*33*, 195 (1984); *Suppl. 7*, 67 (1987); *46*, 267 (1989)
Nitrofen (technical-grade)	*30*, 271 (1983); *Suppl. 7*, 67 (1987)
3-Nitrofluoranthene	*33*, 201 (1984); *Suppl. 7*, 67 (1987)
2-Nitrofluorene	*46*, 277 (1989)
Nitrofural	*7*, 171 (1974); *Suppl. 7*, 67 (1987); *50*, 195 (1990)

5-Nitro-2-furaldehyde semicarbazone (see Nitrofural)
Nitrofurantoin 50, 211 (1990)
Nitrofurazone (see Nitrofural)
1-[(5-Nitrofurfurylidene)amino]-2-imidazolidinone 7, 181 (1974); Suppl. 7, 67 (1987)
N-[4-(5-Nitro-2-furyl)-2-thiazolyl]acetamide 1, 181 (1972); 7, 185 (1974); Suppl. 7, 67 (1987)

Nitrogen mustard 9, 193 (1975); Suppl. 7, 269 (1987)
Nitrogen mustard N-oxide 9, 209 (1975); Suppl. 7, 67 (1987)
1-Nitronaphthalene 46, 291 (1989)
2-Nitronaphthalene 46, 303 (1989)
3-Nitroperylene 46, 313 (1989)
2-Nitropropane 29, 331 (1982); Suppl. 7, 67 (1987)
1-Nitropyrene 33, 209 (1984); Suppl. 7, 67 (1987); 46, 321 (1989)

2-Nitropyrene 46, 359 (1989)
4-Nitropyrene 46, 367 (1989)
N-Nitrosatable drugs 24, 297 (1980) (corr. 42, 260)
N-Nitrosatable pesticides 30, 359 (1983)
N'-Nitrosoanabasine 37, 225 (1985); Suppl. 7, 67 (1987)
N'-Nitrosoanatabine 37, 233 (1985); Suppl. 7, 67 (1987)
N-Nitrosodi-n-butylamine 4, 197 (1974); 17, 51 (1978); Suppl. 7, 67 (1987)

N-Nitrosodiethanolamine 17, 77 (1978); Suppl. 7, 67 (1987)
N-Nitrosodiethylamine 1, 107 (1972) (corr. 42, 251); 17, 83 (1978) (corr. 42, 257); Suppl. 7, 67 (1987)

N-Nitrosodimethylamine 1, 95 (1972); 17, 125 (1978) (corr. 42, 257); Suppl. 7, 67 (1987)

N-Nitrosodiphenylamine 27, 213 (1982); Suppl. 7, 67 (1987)
para-Nitrosodiphenylamine 27, 227 (1982) (corr. 42, 261); Suppl. 7, 68 (1987)

N-Nitrosodi-n-propylamine 17, 177 (1978); Suppl. 7, 68 (1987)
N-Nitroso-N-ethylurea (see N-Ethyl-N-nitrosourea)
N-Nitrosofolic acid 17, 217 (1978); Suppl. 7, 68 (1987)
N-Nitrosoguvacine 37, 263 (1985); Suppl. 7, 68 (1987)
N-Nitrosoguvacoline 37, 263 (1985); Suppl. 7, 68 (1987)
N-Nitrosohydroxyproline 17, 304 (1978); Suppl. 7, 68 (1987)
3-(N-Nitrosomethylamino)propionaldehyde 37, 263 (1985); Suppl. 7, 68 (1987)
3-(N-Nitrosomethylamino)propionitrile 37, 263 (1985); Suppl. 7, 68 (1987)
4-(N-Nitrosomethylamino)-4-(3-pyridyl)-1-butanal 37, 205 (1985); Suppl. 7, 68 (1987)
4-(N-Nitrosomethylamino)-1-(3-pyridyl)-1-butanone 37, 209 (1985); Suppl. 7, 68 (1987)
N-Nitrosomethylethylamine 17, 221 (1978); Suppl. 7, 68 (1987)
N-Nitroso-N-methylurea (see N-Methyl-N-nitrosourea)
N-Nitroso-N-methylurethane (see N-Methyl-N-methylurethane)
N-Nitrosomethylvinylamine 17, 257 (1978); Suppl. 7, 68 (1987)

N-Nitrosomorpholine	17, 263 (1978); Suppl. 7, 68 (1987)
N'-Nitrosonornicotine	17, 281 (1978); 37, 241 (1985); Suppl. 7, 68 (1987)
N-Nitrosopiperidine	17, 287 (1978); Suppl. 7, 68 (1987)
N-Nitrosoproline	17, 303 (1978); Suppl. 7, 68 (1987)
N-Nitrosopyrrolidine	17, 313 (1978); Suppl. 7, 68 (1987)
N-Nitrososarcosine	17, 327 (1978); Suppl. 7, 68 (1987)
Nitrosoureas, chloroethyl (see Chloroethyl nitrosoureas)	
5-Nitro-ortho-toluidine	48, 169 (1990)
Nitrous oxide (see Anaesthetics, volatile)	
Nitrovin	31, 185 (1983); Suppl. 7, 68 (1987)
NNA (see 4-(N-Nitrosomethylamino)-4-(3-pyridyl)-1-butanal)	
NNK (see 4-(N-Nitrosomethylamino)-1-(3-pyridyl)-1-butanone)	
Nonsteroidal oestrogens (see also Oestrogens, progestins and combinations)	Suppl. 7, 272 (1987)
Norethisterone (see also Progestins; Combined oral contraceptives)	6, 179 (1974); 21, 461 (1979)
Norethynodrel (see also Progestins; Combined oral contraceptives	6, 191 (1974); 21, 461 (1979) (corr. 42, 259)
Norgestrel (see also Progestins, Combined oral contraceptives)	6, 201 (1974); 21, 479 (1979)
Nylon 6	19, 120 (1979); Suppl. 7, 68 (1987)

O

Ochratoxin A	10, 191 (1976); 31, 191 (1983) (corr. 42, 262); Suppl. 7, 271 (1987)
Oestradiol-17β (see also Steroidal oestrogens)	6, 99 (1974); 21, 279 (1979)
Oestradiol 3-benzoate (see Oestradiol-17β)	
Oestradiol dipropionate (see Oestradiol-17β)	
Oestradiol mustard	9, 217 (1975)
Oestradiol-17β-valerate (see Oestradiol-17β)	
Oestriol (see also Steroidal oestrogens)	6, 117 (1974); 21, 327 (1979)
Oestrogen-progestin combinations (see Oestrogens, progestins and combinations)	
Oestrogen-progestin replacement therapy (see also Oestrogens, progestins and combinations)	Suppl. 7, 308 (1987)
Oestrogen replacement therapy (see also Oestrogens, progestins and combinations)	Suppl. 7, 280 (1987)
Oestrogens (see Oestrogens, progestins and combinations)	
Oestrogens, conjugated (see Conjugated oestrogens)	
Oestrogens, nonsteroidal (see Nonsteroidal oestrogens)	
Oestrogens, progestins and combinations	6 (1974); 21 (1979); Suppl. 7, 272 (1987)
Oestrogens, steroidal (see Steroidal oestrogens)	
Oestrone (see also Steroidal oestrogens)	6, 123 (1974); 21, 343 (1979) (corr. 42, 259)

Oestrone benzoate (*see* Oestrone)
Oil Orange SS   8, 165 (1975); *Suppl. 7*, 69 (1987)
Oral contraceptives, combined (*see* Combined oral contraceptives)
Oral contraceptives, investigational (*see* Combined oral
   contraceptives)
Oral contraceptives, sequential (*see* Sequential oral contraceptives)
Orange I   8, 173 (1975); *Suppl. 7*, 69 (1987)
Orange G   8, 181 (1975); *Suppl. 7*, 69 (1987)
Organolead compounds (*see also* Lead and lead compounds)   *Suppl. 7*, 230 (1987)
Oxazepam   13, 58 (1977); *Suppl. 7*, 69 (1987)
Oxymetholone (*see also* Androgenic (anabolic) steroids)   *13*, 131 (1977)
Oxyphenbutazone   *13*, 185 (1977); *Suppl. 7*, 69 (1987)

P

Paint manufacture and painting (occupational exposures in) 47, 329 (1989)
Panfuran S (*see also* Dihydroxymethylfuratrizine)   24, 77 (1980); *Suppl. 7*, 69 (1987)
Paper manufacture (*see* Pulp and paper manufacture)
Paracetamol   *50*, 307 (1990)
Parasorbic acid   *10*, 199 (1976) (*corr. 42*, 255);
                                                            Suppl. 7, 69 (1987)
Parathion   30, 153 (1983); *Suppl. 7*, 69 (1987)
Patulin   *10*, 205 (1976); *40*, 83 (1986);
                                                            Suppl. 7, 69 (1987)
Penicillic acid   *10*, 211 (1976); *Suppl. 7*, 69 (1987)
Pentachloroethane   *41*, 99 (1986); *Suppl. 7*, 69 (1987)
Pentachloronitrobenzene (*see* Quintozene)
Pentachlorophenol (*see also* Chlorophenols; Chlorophenols, 20, 303 (1979)
   occupational exposures to)
Perylene   32, 411 (1983); *Suppl. 7*, 69 (1987)
Petasitenine   *31*, 207 (1983); *Suppl. 7*, 69 (1987)
Petasites japonicus (*see* Pyrrolizidine alkaloids)
Petroleum refining (occupational exposures in)   45, 39 (1989)
Some petroleum solvents   47, 43 (1989)
Phenacetin   *13*, 141 (1977); *24*, 135 (1980);
                                                            Suppl. 7, 310 (1987)
Phenanthrene   32, 419 (1983); *Suppl. 7*, 69 (1987)
Phenazopyridine hydrochloride   8, 117 (1975); *24*, 163 (1980)
                                                            (*corr. 42*, 260); *Suppl. 7*, 312 (1987)
Phenelzine sulphate   24, 175 (1980); *Suppl. 7*, 312 (1987)
Phenicarbazide   *12*, 177 (1976); *Suppl. 7*, 70 (1987)
Phenobarbital   *13*, 157 (1977); *Suppl. 7*, 313 (1987)
Phenol   47, 263 (1989) (*corr. 50*, 385)
Phenoxyacetic acid herbicides (*see* Chlorophenoxy herbicides)
Phenoxybenzamine hydrochloride   9, 223 (1975); *24*, 185 (1980);
                                                            Suppl. 7, 70 (1987)

Phenylbutazone	*13*, 183 (1977); *Suppl. 7*, 316 (1987)
meta-Phenylenediamine	*16*, 111 (1978); *Suppl. 7*, 70 (1987)
para-Phenylenediamine	*16*, 125 (1978); *Suppl. 7*, 70 (1987)
Phenyl glycidyl ether (*see* Glycidyl ethers)	
N-Phenyl-2-naphthylamine	*16*, 325 (1978) (*corr. 42*, 257); *Suppl. 7*, 318 (1987)
ortho-Phenylphenol	*30*, 329 (1983); *Suppl. 7*, 70 (1987)
Phenytoin	*13*, 201 (1977); *Suppl. 7*, 319 (1987)
Piperazine oestrone sulphate (*see* Conjugated oestrogens)	
Piperonyl butoxide	*30*, 183 (1983); *Suppl. 7*, 70 (1987)
Pitches, coal-tar (*see* Coal-tar pitches)	
Polyacrylic acid	*19*, 62 (1979); *Suppl. 7*, 70 (1987)
Polybrominated biphenyls	*18*, 107 (1978); *41*, 261 (1986); *Suppl. 7*, 321 (1987)
Polychlorinated biphenyls	*7*, 261 (1974); *18*, 43 (1978) (*corr. 42*, 258); *Suppl. 7*, 322 (1987)
Polychlorinated camphenes (*see* Toxaphene)	
Polychloroprene	*19*, 141 (1979); *Suppl. 7*, 70 (1987)
Polyethylene	*19*, 164 (1979); *Suppl. 7*, 70 (1987)
Polymethylene polyphenyl isocyanate	*19*, 314 (1979); *Suppl. 7*, 70 (1987)
Polymethyl methacrylate	*19*, 195 (1979); *Suppl. 7*, 70 (1987)
Polyoestradiol phosphate (*see* Oestradiol-17β)	
Polypropylene	*19*, 218 (1979); *Suppl. 7*, 70 (1987)
Polystyrene	*19*, 245 (1979); *Suppl. 7*, 70 (1987)
Polytetrafluoroethylene	*19*, 288 (1979); *Suppl. 7*, 70 (1987)
Polyurethane foams	*19*, 320 (1979); *Suppl. 7*, 70 (1987)
Polyvinyl acetate	*19*, 346 (1979); *Suppl. 7*, 70 (1987)
Polyvinyl alcohol	*19*, 351 (1979); *Suppl. 7*, 70 (1987)
Polyvinyl chloride	*7*, 306 (1974); *19*, 402 (1979); *Suppl. 7*, 70 (1987)
Polyvinyl pyrrolidone	*19*, 463 (1979); *Suppl. 7*, 70 (1987)
Ponceau MX	*8*, 189 (1975); *Suppl. 7*, 70 (1987)
Ponceau 3R	*8*, 199 (1975); *Suppl. 7*, 70 (1987)
Ponceau SX	*8*, 207 (1975); *Suppl. 7*, 70 (1987)
Potassium arsenate (*see* Arsenic and arsenic compounds)	
Potassium arsenite (*see* Arsenic and arsenic compounds)	
Potassium bis(2-hydroxyethyl)dithiocarbamate	*12*, 183 (1976); *Suppl. 7*, 70 (1987)
Potassium bromate	*40*, 207 (1986); *Suppl. 7*, 70 (1987)
Potassium chromate (*see* Chromium and chromium compounds)	
Potassium dichromate (*see* Chromium and chromium compounds)	
Prednimustine	*50*, 115 (1990)
Prednisone	*26*, 293 (1981); *Suppl. 7*, 326 (1987)
Procarbazine hydrochloride	*26*, 311 (1981); *Suppl. 7*, 327 (1987)
Proflavine salts	*24*, 195 (1980); *Suppl. 7*, 70 (1987)

Progesterone (*see also* Progestins; Combined oral contraceptives)	6, 135 (1974); 21, 491 (1979) (corr. 42, 259)
Progestins (*see also* Oestrogens, progestins and combinations)	Suppl. 7, 289 (1987)
Pronetalol hydrochloride	13, 227 (1977) (corr. 42, 256); Suppl. 7, 70 (1987)
1,3-Propane sultone	4, 253 (1974) (corr. 42, 253); Suppl. 7, 70 (1987)
Propham	12, 189 (1976); Suppl. 7, 70 (1987)
β-Propiolactone	4, 259 (1974) (corr. 42, 253); Suppl. 7, 70 (1987)
n-Propyl carbamate	12, 201 (1976); Suppl. 7, 70 (1987)
Propylene	19, 213 (1979); Suppl. 7, 71 (1987)
Propylene oxide	11, 191 (1976); 36, 227 (1985) (corr. 42, 263); Suppl. 7, 328 (1987)
Propylthiouracil	7, 67 (1974); Suppl. 7, 329 (1987)
Ptaquiloside (*see also* Bracken fern)	40, 55 (1986); Suppl. 7, 71 (1987)
Pulp and paper manufacture	25, 157 (1981); Suppl. 7, 385 (1987)
Pyrene	32, 431 (1983); Suppl. 7, 71 (1987)
Pyrido[3,4-*c*]psoralen	40, 349 (1986); Suppl. 7, 71 (1987)
Pyrimethamine	13, 233 (1977); Suppl. 7, 71 (1987)
Pyrrolizidine alkaloids (*see* Hydroxysenkirkine; Isatidine; Jacobine; Lasiocarpine; Monocrotaline; Retrorsine; Riddelliine; Seneciphylline; Senkirkine)	

Q

Quercetin (*see also* Bracken fern)	31, 213 (1983); Suppl. 7, 71 (1987)
para-Quinone	15, 255 (1977); Suppl. 7, 71 (1987)
Quintozene	5, 211 (1974); Suppl. 7, 71 (1987)

R

Radon	43, 173 (1988) (corr. 45, 283)
Reserpine	10, 217 (1976); 24, 211 (1980) (corr. 42, 260); Suppl. 7, 330 (1987)
Resorcinol	15, 155 (1977); Suppl. 7, 71 (1987)
Retrorsine	10, 303 (1976); Suppl. 7, 71 (1987)
Rhodamine B	16, 221 (1978); Suppl. 7, 71 (1987)
Rhodamine 6G	16, 233 (1978); Suppl. 7, 71 (1987)
Riddelliine	10, 313 (1976); Suppl. 7, 71 (1987)
Rifampicin	24, 243 (1980); Suppl. 7, 71 (1987)
Rockwool (*see* Man-made mineral fibres)	
The rubber industry	28 (1982) (corr. 42, 261); Suppl. 7, 332 (1987)
Rugulosin	40, 99 (1986); Suppl. 7, 71 (1987)

S

Saccharated iron oxide	2, 161 (1973); *Suppl. 7*, 71 (1987)
Saccharin	22, 111 (1980) (*corr. 42*, 259); *Suppl. 7*, 334 (1987)
Safrole	*1*, 169 (1972); *10*, 231 (1976); *Suppl. 7*, 71 (1987)
The sawmill industry (including logging) (*see* The lumber and sawmill industry (including logging))	
Scarlet Red	8, 217 (1975); *Suppl. 7*, 71 (1987)
Selenium and selenium compounds	9, 245 (1975) (*corr. 42*, 255); *Suppl. 7*, 71 (1987)
Selenium dioxide (*see* Selenium and selenium compounds)	
Selenium oxide (*see* Selenium and selenium compounds)	
Semicarbazide hydrochloride	12, 209 (1976) (*corr. 42*, 256); *Suppl. 7*, 71 (1987)
Senecio jacobaea L. (*see* Pyrrolizidine alkaloids)	
Senecio longilobus (*see* Pyrrolizidine alkaloids)	
Seneciphylline	10, 319, 335 (1976); *Suppl. 7*, 71 (1987)
Senkirkine	10, 327 (1976); *31*, 231 (1983); *Suppl. 7*, 71 (1987)
Sepiolite	42, 175 (1987); *Suppl. 7*, 71 (1987)
Sequential oral contraceptives (*see also* Oestrogens, progestins and combinations)	*Suppl. 7*, 296 (1987)
Shale-oils	35, 161 (1985); *Suppl. 7*, 339 (1987)
Shikimic acid (*see also* Bracken fern)	40, 55 (1986); *Suppl. 7*, 71 (1987)
Shoe manufacture and repair (*see* Boot and shoe manufacture and repair)	
Silica (*see also* Amorphous silica; Crystalline silica)	42, 39 (1987)
Slagwool (*see* Man-made mineral fibres)	
Sodium arsenate (*see* Arsenic and arsenic compounds)	
Sodium arsenite (*see* Arsenic and arsenic compounds)	
Sodium cacodylate (*see* Arsenic and arsenic compounds)	
Sodium chlorite	52, 145 (1991)
Sodium chromate (*see* Chromium and chromium compounds)	
Sodium cyclamate (*see* Cyclamates)	
Sodium dichromate (*see* Chromium and chromium compounds)	
Sodium diethyldithiocarbamate	12, 217 (1976); *Suppl. 7*, 71 (1987)
Sodium equilin sulphate (*see* Conjugated oestrogens)	
Sodium fluoride (*see* Fluorides)	
Sodium monofluorophosphate (*see* Fluorides)	
Sodium oestrone sulphate (*see* Conjugated oestrogens)	
Sodium *ortho*-phenylphenate (*see also ortho*-Phenylphenol)	30, 329 (1983); *Suppl. 7*, 392 (1987)
Sodium saccharin (*see* Saccharin)	
Sodium selenate (*see* Selenium and selenium compounds)	

Sodium selenite (see Selenium and selenium compounds)
Sodium silicofluoride (see Fluorides)
Soots   3, 22 (1973); 35, 219 (1985);
                                                        Suppl. 7, 343 (1987)
Spironolactone   24, 259 (1980); Suppl. 7, 344 (1987)
Stannous fluoride (see Fluorides)
Steel founding (see Iron and steel founding)
Sterigmatocystin   1, 175 (1972); 10, 245 (1976);
                                                        Suppl. 7, 72 (1987)
Steroidal oestrogens (see also Oestrogens, progestins and   Suppl. 7, 280 (1987)
   combinations)
Streptozotocin   4, 221 (1974); 17, 337 (1978);
                                                        Suppl. 7, 72 (1987)
Strobane® (see Terpene polychlorinates)
Strontium chromate (see Chromium and chromium compounds)
Styrene   19, 231 (1979) (corr. 42, 258);
                                                        Suppl. 7, 345 (1987)
Styrene-acrylonitrile copolymers   19, 97 (1979); Suppl. 7, 72 (1987)
Styrene-butadiene copolymers   19, 252 (1979); Suppl. 7, 72 (1987)
Styrene oxide   11, 201 (1976); 19, 275 (1979);
                                                        36, 245 (1985); Suppl. 7, 72 (1987)
Succinic anhydride   15, 265 (1977); Suppl. 7, 72 (1987)
Sudan I   8, 225 (1975); Suppl. 7, 72 (1987)
Sudan II   8, 233 (1975); Suppl. 7, 72 (1987)
Sudan III   8, 241 (1975); Suppl. 7, 72 (1987)
Sudan Brown RR   8, 249 (1975); Suppl. 7, 72 (1987)
Sudan Red 7B   8, 253 (1975); Suppl. 7, 72 (1987)
Sulfafurazole   24, 275 (1980); Suppl. 7, 347 (1987)
Sulfallate   30, 283 (1983); Suppl. 7, 72 (1987)
Sulfamethoxazole   24, 285 (1980); Suppl. 7, 348 (1987)
Sulphisoxazole (see Sulfafurazole)
Sulphur mustard (see Mustard gas)
Sunset Yellow FCF   8, 257 (1975); Suppl. 7, 72 (1987)
Symphytine   31, 239 (1983); Suppl. 7, 72 (1987)

T

2,4,5-T (see also Chlorophenoxy herbicides; Chlorophenoxy
   herbicides, occupational exposures to)   15, 273 (1977)
Talc   42, 185 (1987); Suppl. 7, 349 (1987)
Tannic acid   10, 253 (1976) (corr. 42, 255);
                                                        Suppl. 7, 72 (1987)
Tannins (see also Tannic acid)   10, 254 (1976); Suppl. 7, 72 (1987)
TCDD (see 2,3,7,8-Tetrachlorodibenzo-para-dioxin)
TDE (see DDT)

Tea	*51*, 207 (1991)
Terpene polychlorinates	*5*, 219 (1974); *Suppl. 7*, 72 (1987)
Testosterone (*see also* Androgenic (anabolic) steroids)	*6*, 209 (1974); *21*, 519 (1979)
Testosterone oenanthate (*see* Testosterone)	
Testosterone propionate (*see* Testosterone)	
2,2',5,5'-Tetrachlorobenzidine	*27*, 141 (1982); *Suppl. 7*, 72 (1987)
2,3,7,8-Tetrachlorodibenzo-*para*-dioxin	*15*, 41 (1977); *Suppl. 7*, 350 (1987)
1,1,1,2-Tetrachloroethane	*41*, 87 (1986); *Suppl. 7*, 72 (1987)
1,1,2,2-Tetrachloroethane	*20*, 477 (1979); *Suppl. 7*, 354 (1987)
Tetrachloroethylene	*20*, 491 (1979); *Suppl. 7*, 355 (1987)
2,3,4,6-Tetrachlorophenol (*see* Chlorophenols; Chlorophenols, occupational exposures to)	
Tetrachlorvinphos	*30*, 197 (1983); *Suppl. 7*, 72 (1987)
Tetraethyllead (*see* Lead and lead compounds)	
Tetrafluoroethylene	*19*, 285 (1979); *Suppl. 7*, 72 (1987)
Tetrakis(hydroxymethyl) phosphonium salts	*48*, 95 (1990)
Tetramethyllead (*see* Lead and lead compounds)	
Textile manufacturing industry, exposures in	*48*, 215 (1990) (*corr. 51*, 483)
Theobromine	*51*, 421 (1991)
Theophylline	*51*, 391 (1991)
Thioacetamide	*7*, 77 (1974); *Suppl. 7*, 72 (1987)
4,4'-Thiodianiline	*16*, 343 (1978); *27*, 147 (1982); *Suppl. 7*, 72 (1987)
Thiotepa	*9*, 85 (1975); *Suppl. 7*, 368 (1987); *50*, 123 (1990)
Thiouracil	*7*, 85 (1974); *Suppl. 7*, 72 (1987)
Thiourea	*7*, 95 (1974); *Suppl. 7*, 72 (1987)
Thiram	*12*, 225 (1976); *Suppl. 7*, 72 (1987)
Titanium dioxide	*47*, 307 (1989)
Tobacco habits other than smoking (*see* Tobacco products, smokeless)	
Tobacco products, smokeless	*37* (1985) (*corr. 42*, 263; *52*, 513); *Suppl. 7*, 357 (1987)
Tobacco smoke	*38* (1986) (*corr. 42*, 263); *Suppl. 7*, 357 (1987)
Tobacco smoking (*see* Tobacco smoke)	
ortho-Tolidine (*see* 3,3'-Dimethylbenzidine)	
2,4-Toluene diisocyanate (*see also* Toluene diisocyanates)	*19*, 303 (1979); *39*, 287 (1986)
2,6-Toluene diisocyanate (*see also* Toluene diisocyanates)	*19*, 303 (1979); *39*, 289 (1986)
Toluene	*47*, 79 (1989)
Toluene diisocyanates	*39*, 287 (1986) (*corr. 42*, 264); *Suppl. 7*, 72 (1987)
Toluenes, α-chlorinated (*see* α-Chlorinated toluenes)	
ortho-Toluenesulphonamide (*see* Saccharin)	

ortho-Toluidine	*16*, 349 (1978); *27*, 155 (1982); *Suppl. 7*, 362 (1987)
Toxaphene	*20*, 327 (1979); *Suppl. 7*, 72 (1987)
Tremolite (*see* Asbestos)	
Treosulphan	*26*, 341 (1981); *Suppl. 7*, 363 (1987)
Triaziquone (*see* Tris(aziridinyl)-*para*-benzoquinone)	
Trichlorfon	*30*, 207 (1983); *Suppl. 7*, 73 (1987)
Trichlormethine	*9*, 229 (1975); *Suppl. 7*, 73 (1987); *50*, 143 (1990)
Trichloroacetonitrile (*see* Halogenated acetonitriles)	
1,1,1-Trichloroethane	*20*, 515 (1979); *Suppl. 7*, 73 (1987)
1,1,2-Trichloroethane	*20*, 533 (1979); *Suppl. 7*, 73 (1987); *52*, 337 (1991)
Trichloroethylene	*11*, 263 (1976); *20*, 545 (1979); *Suppl. 7*, 364 (1987)
2,4,5-Trichlorophenol (*see also* Chlorophenols; Chlorophenols occupational exposures to)	*20*, 349 (1979)
2,4,6-Trichlorophenol (*see also* Chlorophenols; Chlorophenols, occupational exposures to)	*20*, 349 (1979)
(2,4,5-Trichlorophenoxy)acetic acid (*see* 2,4,5-T)	
Trichlorotriethylamine hydrochloride (*see* Trichlormethine)	
T_2-Trichothecene	*31*, 265 (1983); *Suppl. 7*, 73 (1987)
Triethylene glycol diglycidyl ether	*11*, 209 (1976); *Suppl. 7*, 73 (1987)
4,4',6-Trimethylangelicin plus ultraviolet radiation (*see also* Angelicin and some synthetic derivatives)	*Suppl. 7*, 57 (1987)
2,4,5-Trimethylaniline	*27*, 177 (1982); *Suppl. 7*, 73 (1987)
2,4,6-Trimethylaniline	*27*, 178 (1982); *Suppl. 7*, 73 (1'987)
4,5',8-Trimethylpsoralen	*40*, 357 (1986); *Suppl. 7*, 366 (1987)
Trimustine hydrochloride (*see* Trichlormethine)	
Triphenylene	*32*, 447 (1983); *Suppl. 7*, 73 (1987)
Tris(aziridinyl)-*para*-benzoquinone	*9*, 67 (1975); *Suppl. 7*, 367 (1987)
Tris(1-aziridinyl)phosphine oxide	*9*, 75 (1975); *Suppl. 7*, 73 (1987)
Tris(1-aziridinyl)phosphine sulphide (*see* Thiotepa)	
2,4,6-Tris(1-aziridinyl)-*s*-triazine	*9*, 95 (1975); *Suppl. 7*, 73 (1987)
Tris(2-chloroethyl) phosphate	*48*, 109 (1990)
1,2,3-Tris(chloromethoxy)propane	*15*, 301 (1977); *Suppl. 7*, 73 (1987)
Tris(2,3-dibromopropyl)phosphate	*20*, 575 (1979); *Suppl. 7*, 369 (1987)
Tris(2-methyl-1-aziridinyl)phosphine oxide	*9*, 107 (1975); *Suppl. 7*, 73 (1987)
Trp-P-1	*31*, 247 (1983); *Suppl. 7*, 73 (1987)
Trp-P-2	*31*, 255 (1983); *Suppl. 7*, 73 (1987)
Trypan blue	*8*, 267 (1975); *Suppl. 7*, 73 (1987)
Tussilago farfara L. (*see* Pyrrolizidine alkaloids)	

U

Ultraviolet radiation	*40*, 379 (1986)

Underground haematite mining with exposure to radon	*1*, 29 (1972); *Suppl. 7*, 216 (1987)
Uracil mustard	*9*, 235 (1975); *Suppl. 7*, 370 (1987)
Urethane	*7*, 111 (1974); *Suppl. 7*, 73 (1987)

V

Vat Yellow 4	*48*, 161 (1990)
Vinblastine sulphate	*26*, 349 (1981) (*corr. 42*, 261); *Suppl. 7*, 371 (1987)
Vincristine sulphate	*26*, 365 (1981); *Suppl. 7*, 372 (1987)
Vinyl acetate	*19*, 341 (1979); *39*, 113 (1986); *Suppl. 7*, 73 (1987)
Vinyl bromide	*19*, 367 (1979); *39*, 133 (1986); *Suppl. 7*, 73 (1987)
Vinyl chloride	*7*, 291 (1974); *19*, 377 (1979) (*corr. 42*, 258); *Suppl. 7*, 373 (1987)
Vinyl chloride-vinyl acetate copolymers	*7*, 311 (1976); *19*, 412 (1979) (*corr. 42*, 258); *Suppl. 7*, 73 (1987)
4-Vinylcyclohexene	*11*, 277 (1976); *39*, 181 (1986); *Suppl. 7*, 73 (1987)
Vinyl fluoride	*39*, 147 (1986); *Suppl. 7*, 73 (1987)
Vinylidene chloride	*19*, 439 (1979); *39*, 195 (1986); *Suppl. 7*, 376 (1987)
Vinylidene chloride-vinyl chloride copolymers	*19*, 448 (1979) (*corr. 42*, 258); *Suppl. 7*, 73 (1987)
Vinylidene fluoride	*39*, 227 (1986); *Suppl. 7*, 73 (1987)
N-Vinyl-2-pyrrolidone	*19*, 461 (1979); *Suppl. 7*, 73 (1987)

W

Welding	*49*, 447 (1990) (*corr. 52*, 513)
Wollastonite	*42*, 145 (1987); *Suppl. 7*, 377 (1987)
Wood industries	*25* (1981); *Suppl. 7*, 378 (1987)

X

Xylene	*47*, 125 (1989)
2,4-Xylidine	*16*, 367 (1978); *Suppl. 7*, 74 (1987)
2,5-Xylidine	*16*, 377 (1978); *Suppl. 7*, 74 (1987)

Y

Yellow AB	*8*, 279 (1975); *Suppl. 7*, 74 (1987)
Yellow OB	*8*, 287 (1975); *Suppl. 7*, 74 (1987)

Z

Zearalenone	*31*, 279 (1983); *Suppl. 7*, 74 (1987)
Zectran	*12*, 237 (1976); *Suppl. 7*, 74 (1987)
Zinc beryllium silicate (*see* Beryllium and beryllium compounds)	
Zinc chromate (*see* Chromium and chromium compounds)	
Zinc chromate hydroxide (*see* Chromium and chromium compounds)	
Zinc potassium chromate (*see* Chromium and chromium compounds)	
Zinc yellow (*see* Chromium and chromium compounds)	
Zineb	*12*, 245 (1976); *Suppl. 7*, 74 (1987)
Ziram	*12*, 259 (1976); *Suppl. 7*, 74 (1987)

PUBLICATIONS OF THE INTERNATIONAL AGENCY FOR RESEARCH ON CANCER
Scientific Publications Series

(Available from Oxford University Press through local bookshops)

No. 1 Liver Cancer
1971; 176 pages (*out of print*)

No. 2 Oncogenesis and Herpesviruses
Edited by P.M. Biggs, G. de-Thé and L.N. Payne
1972; 515 pages (*out of print*)

No. 3 *N*-Nitroso Compounds: Analysis and Formation
Edited by P. Bogovski, R. Preussman and E.A. Walker
1972; 140 pages (*out of print*)

No. 4 Transplacental Carcinogenesis
Edited by L. Tomatis and U. Mohr
1973; 181 pages (*out of print*)

No. 5/6 Pathology of Tumours in Laboratory Animals, Volume 1, Tumours of the Rat
Edited by V.S. Turusov
1973/1976; 533 pages; £50.00

No. 7 Host Environment Interactions in the Etiology of Cancer in Man
Edited by R. Doll and I. Vodopija
1973; 464 pages; £32.50

No. 8 Biological Effects of Asbestos
Edited by P. Bogovski, J.C. Gilson, V. Timbrell and J.C. Wagner
1973; 346 pages (*out of print*)

No. 9 *N*-Nitroso Compounds in the Environment
Edited by P. Bogovski and E.A. Walker
1974; 243 pages; £21.00

No. 10 Chemical Carcinogenesis Essays
Edited by R. Montesano and L. Tomatis
1974; 230 pages (*out of print*)

No. 11 Oncogenesis and Herpesviruses II
Edited by G. de-Thé, M.A. Epstein and H. zur Hausen
1975; Part I: 511 pages
Part II: 403 pages; £65.00

No. 12 Screening Tests in Chemical Carcinogenesis
Edited by R. Montesano, H. Bartsch and L. Tomatis
1976; 666 pages; £45.00

No. 13 Environmental Pollution and Carcinogenic Risks
Edited by C. Rosenfeld and W. Davis
1975; 441 pages (*out of print*)

No. 14 Environmental *N*-Nitroso Compounds. Analysis and Formation
Edited by E.A. Walker, P. Bogovski and L. Griciute
1976; 512 pages; £37.50

No. 15 Cancer Incidence in Five Continents, Volume III
Edited by J.A.H. Waterhouse, C. Muir, P. Correa and J. Powell
1976; 584 pages; (*out of print*)

No. 16 Air Pollution and Cancer in Man
Edited by U. Mohr, D. Schmähl and L. Tomatis
1977; 328 pages (*out of print*)

No. 17 Directory of On-going Research in Cancer Epidemiology 1977
Edited by C.S. Muir and G. Wagner
1977; 599 pages (*out of print*)

No. 18 Environmental Carcinogens. Selected Methods of Analysis. Volume 1: Analysis of Volatile Nitrosamines in Food
Editor-in-Chief: H. Egan
1978; 212 pages (*out of print*)

No. 19 Environmental Aspects of *N*-Nitroso Compounds
Edited by E.A. Walker, M. Castegnaro, L. Griciute and R.E. Lyle
1978; 561 pages (*out of print*)

No. 20 Nasopharyngeal Carcinoma: Etiology and Control
Edited by G. de-Thé and Y. Ito
1978; 606 pages (*out of print*)

No. 21 Cancer Registration and its Techniques
Edited by R. MacLennan, C. Muir, R. Steinitz and A. Winkler
1978; 235 pages; £35.00

No. 22 Environmental Carcinogens. Selected Methods of Analysis. Volume 2: Methods for the Measurement of Vinyl Chloride in Poly(vinyl chloride), Air, Water and Foodstuffs
Editor-in-Chief: H. Egan
1978; 142 pages (*out of print*)

No. 23 Pathology of Tumours in Laboratory Animals. Volume II: Tumours of the Mouse
Editor-in-Chief: V.S. Turusov
1979; 669 pages (*out of print*)

No. 24 Oncogenesis and Herpesviruses III
Edited by G. de-Thé, W. Henle and F. Rapp
1978; Part I: 580 pages, Part II: 512 pages (*out of print*)

List of IARC Publications

No. 25 Carcinogenic Risk. Strategies for Intervention
Edited by W. Davis and C. Rosenfeld
1979; 280 pages (*out of print*)

No. 26 Directory of On-going Research in Cancer Epidemiology 1978
Edited by C.S. Muir and G. Wagner
1978; 550 pages (*out of print*)

No. 27 Molecular and Cellular Aspects of Carcinogen Screening Tests
Edited by R. Montesano, H. Bartsch and L. Tomatis
1980; 372 pages; £29.00

No. 28 Directory of On-going Research in Cancer Epidemiology 1979
Edited by C.S. Muir and G. Wagner
1979; 672 pages (*out of print*)

No. 29 Environmental Carcinogens. Selected Methods of Analysis. Volume 3: Analysis of Polycyclic Aromatic Hydrocarbons in Environmental Samples
Editor-in-Chief: H. Egan
1979; 240 pages (*out of print*)

No. 30 Biological Effects of Mineral Fibres
Editor-in-Chief: J.C. Wagner
1980; Volume 1: 494 pages; Volume 2: 513 pages; £65.00

No. 31 N-Nitroso Compounds: Analysis, Formation and Occurrence
Edited by E.A. Walker, L. Griciute, M. Castegnaro and M. Börzsönyi
1980; 835 pages (*out of print*)

No. 32 Statistical Methods in Cancer Research. Volume 1. The Analysis of Case-control Studies
By N.E. Breslow and N.E. Day
1980; 338 pages; £20.00

No. 33 Handling Chemical Carcinogens in the Laboratory
Edited by R. Montesano *et al.*
1979; 32 pages (*out of print*)

No. 34 Pathology of Tumours in Laboratory Animals. Volume III. Tumours of the Hamster
Editor-in-Chief: V.S. Turusov
1982; 461 pages; £39.00

No. 35 Directory of On-going Research in Cancer Epidemiology 1980
Edited by C.S. Muir and G. Wagner
1980; 660 pages (*out of print*)

No. 36 Cancer Mortality by Occupation and Social Class 1851-1971
Edited by W.P.D. Logan
1982; 253 pages; £22.50

No. 37 Laboratory Decontamination and Destruction of Aflatoxins B_1, B_2, G_1, G_2 in Laboratory Wastes
Edited by M. Castegnaro *et al.*
1980; 56 pages; £6.50

No. 38 Directory of On-going Research in Cancer Epidemiology 1981
Edited by C.S. Muir and G. Wagner
1981; 696 pages (*out of print*)

No. 39 Host Factors in Human Carcinogenesis
Edited by H. Bartsch and B. Armstrong
1982; 583 pages; £46.00

No. 40 Environmental Carcinogens. Selected Methods of Analysis. Volume 4: Some Aromatic Amines and Azo Dyes in the General and Industrial Environment
Edited by L. Fishbein, M. Castegnaro, I.K. O'Neill and H. Bartsch
1981; 347 pages; £29.00

No. 41 N-Nitroso Compounds: Occurrence and Biological Effects
Edited by H. Bartsch, I.K. O'Neill, M. Castegnaro and M. Okada
1982; 755 pages; £48.00

No. 42 Cancer Incidence in Five Continents, Volume IV
Edited by J. Waterhouse, C. Muir, K. Shanmugaratnam and J. Powell
1982; 811 pages (*out of print*)

No. 43 Laboratory Decontamination and Destruction of Carcinogens in Laboratory Wastes: Some N-Nitrosamines
Edited by M. Castegnaro *et al.*
1982; 73 pages; £7.50

No. 44 Environmental Carcinogens. Selected Methods of Analysis. Volume 5: Some Mycotoxins
Edited by L. Stoloff, M. Castegnaro, P. Scott, I.K. O'Neill and H. Bartsch
1983; 455 pages; £29.00

No. 45 Environmental Carcinogens. Selected Methods of Analysis. Volume 6: N-Nitroso Compounds
Edited by R. Preussmann, I.K. O'Neill, G. Eisenbrand, B. Spiegelhalder and H. Bartsch
1983; 508 pages; £29.00

No. 46 Directory of On-going Research in Cancer Epidemiology 1982
Edited by C.S. Muir and G. Wagner
1982; 722 pages (*out of print*)

No. 47 Cancer Incidence in Singapore 1968-1977
Edited by K. Shanmugaratnam, H.P. Lee and N.E. Day
1983; 171 pages (*out of print*)

No. 48 Cancer Incidence in the USSR (2nd Revised Edition)
Edited by N.P. Napalkov, G.F. Tserkovny, V.M. Merabishvili, D.M. Parkin, M. Smans and C.S. Muir
1983; 75 pages; £12.00

No. 49 Laboratory Decontamination and Destruction of Carcinogens in Laboratory Wastes: Some Polycyclic Aromatic Hydrocarbons
Edited by M. Castegnaro *et al.*
1983; 87 pages; £9.00

No. 50 Directory of On-going Research in Cancer Epidemiology 1983
Edited by C.S. Muir and G. Wagner
1983; 731 pages (*out of print*)

No. 51 Modulators of Experimental Carcinogenesis
Edited by V. Turusov and R. Montesano
1983; 307 pages; £22.50

List of IARC Publications

No. 52 Second Cancers in Relation to Radiation Treatment for Cervical Cancer: Results of a Cancer Registry Collaboration
Edited by N.E. Day and J.C. Boice, Jr
1984; 207 pages; £20.00

No. 53 Nickel in the Human Environment
Editor-in-Chief: F.W. Sunderman, Jr
1984; 529 pages; £41.00

No. 54 Laboratory Decontamination and Destruction of Carcinogens in Laboratory Wastes: Some Hydrazines
Edited by M. Castegnaro et al.
1983; 87 pages; £9.00

No. 55 Laboratory Decontamination and Destruction of Carcinogens in Laboratory Wastes: Some N-Nitrosamides
Edited by M. Castegnaro et al.
1984; 66 pages; £7.50

No. 56 Models, Mechanisms and Etiology of Tumour Promotion
Edited by M. Börzsönyi, N.E. Day, K. Lapis and H. Yamasaki
1984; 532 pages; £42.00

No. 57 N-Nitroso Compounds: Occurrence, Biological Effects and Relevance to Human Cancer
Edited by I.K. O'Neill, R.C. von Borstel, C.T. Miller, J. Long and H. Bartsch
1984; 1013 pages; £80.00

No. 58 Age-related Factors in Carcinogenesis
Edited by A. Likhachev, V. Anisimov and R. Montesano
1985; 288 pages; £20.00

No. 59 Monitoring Human Exposure to Carcinogenic and Mutagenic Agents
Edited by A. Berlin, M. Draper, K. Hemminki and H. Vainio
1984; 457 pages; £27.50

No. 60 Burkitt's Lymphoma: A Human Cancer Model
Edited by G. Lenoir, G. O'Conor and C.L.M. Olweny
1985; 484 pages; £29.00

No. 61 Laboratory Decontamination and Destruction of Carcinogens in Laboratory Wastes: Some Haloethers
Edited by M. Castegnaro et al.
1985; 55 pages; £7.50

No. 62 Directory of On-going Research in Cancer Epidemiology 1984
Edited by C.S. Muir and G. Wagner
1984; 717 pages (out of print)

No. 63 Virus-associated Cancers in Africa
Edited by A.O. Williams, G.T. O'Conor, G.B. de-Thé and C.A. Johnson
1984; 773 pages; £22.00

No. 64 Laboratory Decontamination and Destruction of Carcinogens in Laboratory Wastes: Some Aromatic Amines and 4-Nitrobiphenyl
Edited by M. Castegnaro et al.
1985; 84 pages; £6.95

No. 65 Interpretation of Negative Epidemiological Evidence for Carcinogenicity
Edited by N.J. Wald and R. Doll
1985; 232 pages; £20.00

No. 66 The Role of the Registry in Cancer Control
Edited by D.M. Parkin, G. Wagner and C.S. Muir
1985; 152 pages; £10.00

No. 67 Transformation Assay of Established Cell Lines: Mechanisms and Application
Edited by T. Kakunaga and H. Yamasaki
1985; 225 pages; £20.00

No. 68 Environmental Carcinogens. Selected Methods of Analysis. Volume 7. Some Volatile Halogenated Hydrocarbons
Edited by L. Fishbein and I.K. O'Neill
1985; 479 pages; £42.00

No. 69 Directory of On-going Research in Cancer Epidemiology 1985
Edited by C.S. Muir and G. Wagner
1985; 745 pages; £22.00

No. 70 The Role of Cyclic Nucleic Acid Adducts in Carcinogenesis and Mutagenesis
Edited by B. Singer and H. Bartsch
1986; 467 pages; £40.00

No. 71 Environmental Carcinogens. Selected Methods of Analysis. Volume 8: Some Metals: As, Be, Cd, Cr, Ni, Pb, Se Zn
Edited by I.K. O'Neill, P. Schuller and L. Fishbein
1986; 485 pages; £42.00

No. 72 Atlas of Cancer in Scotland, 1975–1980. Incidence and Epidemiological Perspective
Edited by I. Kemp, P. Boyle, M. Smans and C.S. Muir
1985; 285 pages; £35.00

No. 73 Laboratory Decontamination and Destruction of Carcinogens in Laboratory Wastes: Some Antineoplastic Agents
Edited by M. Castegnaro et al.
1985; 163 pages; £10.00

No. 74 Tobacco: A Major International Health Hazard
Edited by D. Zaridze and R. Peto
1986; 324 pages; £20.00

No. 75 Cancer Occurrence in Developing Countries
Edited by D.M. Parkin
1986; 339 pages; £20.00

No. 76 Screening for Cancer of the Uterine Cervix
Edited by M. Hakama, A.B. Miller and N.E. Day
1986; 315 pages; £25.00

List of IARC Publications

No. 77 Hexachlorobenzene: Proceedings of an International Symposium
Edited by C.R. Morris and J.R.P. Cabral
1986; 668 pages; £50.00

No. 78 Carcinogenicity of Alkylating Cytostatic Drugs
Edited by D. Schmähl and J.M. Kaldor
1986; 337 pages; £25.00

No. 79 Statistical Methods in Cancer Research. Volume III: The Design and Analysis of Long-term Animal Experiments
By J.J. Gart, D. Krewski, P.N. Lee, R.E. Tarone and J. Wahrendorf
1986; 213 pages; £20.00

No. 80 Directory of On-going Research in Cancer Epidemiology 1986
Edited by C.S. Muir and G. Wagner
1986; 805 pages; £22.00

No. 81 Environmental Carcinogens: Methods of Analysis and Exposure Measurement. Volume 9: Passive Smoking
Edited by I.K. O'Neill, K.D. Brunnemann, B. Dodet and D. Hoffmann
1987; 383 pages; £35.00

No. 82 Statistical Methods in Cancer Research. Volume II: The Design and Analysis of Cohort Studies
By N.E. Breslow and N.E. Day
1987; 404 pages; £30.00

No. 83 Long-term and Short-term Assays for Carcinogens: A Critical Appraisal
Edited by R. Montesano, H. Bartsch, H. Vainio, J. Wilbourn and H. Yamasaki
1986; 575 pages; £48.00

No. 84 The Relevance of N-Nitroso Compounds to Human Cancer: Exposure and Mechanisms
Edited by H. Bartsch, I.K. O'Neill and R. Schulte-Hermann
1987; 671 pages; £50.00

No. 85 Environmental Carcinogens: Methods of Analysis and Exposure Measurement. Volume 10: Benzene and Alkylated Benzenes
Edited by L. Fishbein and I.K. O'Neill
1988; 327 pages; £35.00

No. 86 Directory of On-going Research in Cancer Epidemiology 1987
Edited by D.M. Parkin and J. Wahrendorf
1987; 676 pages; £22.00

No. 87 International Incidence of Childhood Cancer
Edited by D.M. Parkin, C.A. Stiller, C.A. Bieber, G.J. Draper, B. Terracini and J.L. Young
1988; 401 pages; £35.00

No. 88 Cancer Incidence in Five Continents Volume V
Edited by C. Muir, J. Waterhouse, T. Mack, J. Powell and S. Whelan
1987; 1004 pages; £50.00

No. 89 Method for Detecting DNA Damaging Agents in Humans: Applications in Cancer Epidemiology and Prevention
Edited by H. Bartsch, K. Hemminki and I.K. O'Neill
1988; 518 pages; £45.00

No. 90 Non-occupational Exposure to Mineral Fibres
Edited by J. Bignon, J. Peto and R. Saracci
1989; 500 pages; £45.00

No. 91 Trends in Cancer Incidence in Singapore 1968-1982
Edited by H.P. Lee, N.E. Day and K. Shanmugaratnam
1988; 160 pages; £25.00

No. 92 Cell Differentiation, Genes and Cancer
Edited by T. Kakunaga, T. Sugimura, L. Tomatis and H. Yamasaki
1988; 204 pages; £25.00

No. 93 Directory of On-going Research in Cancer Epidemiology 1988
Edited by M. Coleman and J. Wahrendorf
1988; 662 pages (*out of print*)

No. 94 Human Papillomavirus and Cervical Cancer
Edited by N. Muñoz, F.X. Bosch and O.M. Jensen
1989; 154 pages; £19.00

No. 95 Cancer Registration: Principles and Methods
Edited by O.M. Jensen, D.M. Parkin, R. MacLennan, C.S. Muir and R. Skeet
1991; 288 pages; £28.00

No. 96 Perinatal and Multigeneration Carcinogenesis
Edited by N.P. Napalkov, J.M. Rice, L. Tomatis and H. Yamasaki
1989; 436 pages; £48.00

No. 97 Occupational Exposure to Silica and Cancer Risk
Edited by L. Simonato, A.C. Fletcher, R. Saracci and T. Thomas
1990; 124 pages; £19.00

No. 98 Cancer Incidence in Jewish Migrants to Israel, 1961-1981
Edited by R. Steinitz, D.M. Parkin, J.L. Young, C.A. Bieber and L. Katz
1989; 320 pages; £30.00

No. 99 Pathology of Tumours in Laboratory Animals, Second Edition, Volume 1, Tumours of the Rat
Edited by V.S. Turusov and U. Mohr
740 pages; £85.00

No. 100 Cancer: Causes, Occurrence and Control
Editor-in-Chief L. Tomatis
1990; 352 pages; £24.00

List of IARC Publications

No. 101 Directory of On-going Research in Cancer Epidemiology 1989/90
Edited by M. Coleman and J. Wahrendorf
1989; 818 pages; £36.00

No. 102 Patterns of Cancer in Five Continents
Edited by S.L. Whelan and D.M. Parkin
1990; 162 pages; £25.00

No. 103 Evaluating Effectiveness of Primary Prevention of Cancer
Edited by M. Hakama, V. Beral, J.W. Cullen and D.M. Parkin
1990; 250 pages; £32.00

No. 104 Complex Mixtures and Cancer Risk
Edited by H. Vainio, M. Sorsa and A.J. McMichael
1990; 442 pages; £38.00

No. 105 Relevance to Human Cancer of N-Nitroso Compounds, Tobacco Smoke and Mycotoxins
Edited by I.K. O'Neill, J. Chen and H. Bartsch
1991; 614 pages; £70.00

No. 107 Atlas of Cancer Mortality in the European Economic Community
Edited by M. Smans, C.S. Muir and P. Boyle
Publ. due 1991; approx. 230 pages; £35.00

No. 108 Environmental Carcinogens: Methods of Analysis and Exposure Measurement. Volume 11: Polychlorinated Dioxins and Dibenzofurans
Edited by C. Rappe, H.R. Buser, B. Dodet and I.K. O'Neill
Publ. due 1991; approx. 400 pages; £45.00

No. 109 Environmental Carcinogens: Methods of Analysis and Exposure Measurement. Volume 12: Indoor Air Contaminants
Edited by B. Seifert, B. Dodet and I.K. O'Neill
Publ. due 1991; approx. 400 pages

No. 110 Directory of On-going Research in Cancer Epidemiology 1991
Edited by M. Coleman and J. Wahrendorf
1991; 753 pages; £38.00

No. 111 Pathology of Tumours in Laboratory Animals, Second Edition, Volume 2, Tumours of the Mouse
Edited by V.S. Turusov and U. Mohr
Publ. due 1991; approx. 500 pages

No. 112 Autopsy in Epidemiology and Medical Research
Edited by E. Riboli and M. Delendi
1991; 288 pages; £25.00

No. 113 Laboratory Decontamination and Destruction of Carcinogens in Laboratory Wastes: Some Mycotoxins
Edited by M. Castegnaro, J. Barek, J.-M. Frémy, M. Lafontaine, M. Miraglia, E.B. Sansone and G.M. Telling
Publ. due 1991; approx. 60 pages; £11.00

No. 114 Laboratory Decontamination and Destruction of Carcinogens in Laboratory Wastes: Some Polycyclic Heterocyclic Hydrocarbons
Edited by M. Castegnaro, J. Barek, J. Jacob, U. Kirso, M. Lafontaine, E.B. Sansone, G.M. Telling and T. Vu Duc
Publ. due 1991; approx. 40 pages; £8.00

List of IARC Publications

IARC MONOGRAPHS ON THE EVALUATION OF CARCINOGENIC RISKS TO HUMANS

(Available from booksellers through the network of WHO Sales Agents)

Volume 1 Some Inorganic Substances, Chlorinated Hydrocarbons, Aromatic Amines, N-Nitroso Compounds, and Natural Products
1972; 184 pages (*out of print*)

Volume 2 Some Inorganic and Organometallic Compounds
1973; 181 pages (out of print)

Volume 3 Certain Polycyclic Aromatic Hydrocarbons and Heterocyclic Compounds
1973; 271 pages (*out of print*)

Volume 4 Some Aromatic Amines, Hydrazine and Related Substances, N-Nitroso Compounds and Miscellaneous Alkylating Agents
1974; 286 pages;
Sw. fr. 18.-/US $14.40

Volume 5 Some Organochlorine Pesticides
1974; 241 pages (*out of print*)

Volume 6 Sex Hormones
1974; 243 pages (*out of print*)

Volume 7 Some Anti-Thyroid and Related Substances, Nitrofurans and Industrial Chemicals
1974; 326 pages (*out of print*)

Volume 8 Some Aromatic Azo Compounds
1975; 375 pages;
Sw. fr. 36.-/US $28.80

Volume 9 Some Aziridines, N-, S- and O-Mustards and Selenium
1975; 268 pages;
Sw.fr. 27.-/US $21.60

Volume 10 Some Naturally Occurring Substances
1976; 353 pages (*out of print*)

Volume 11 Cadmium, Nickel, Some Epoxides, Miscellaneous Industrial Chemicals and General Considerations on Volatile Anaesthetics
1976; 306 pages (*out of print*)

Volume 12 Some Carbamates, Thiocarbamates and Carbazides
1976; 282 pages;
Sw. fr. 34.-/US $27.20

Volume 13 Some Miscellaneous Pharmaceutical Substances
1977; 255 pages;
Sw. fr. 30.-/US$ 24.00

Volume 14 Asbestos
1977; 106 pages (*out of print*)

Volume 15 Some Fumigants, The Herbicides 2,4-D and 2,4,5-T, Chlorinated Dibenzodioxins and Miscellaneous Industrial Chemicals
1977; 354 pages;
Sw. fr. 50.-/US $40.00

Volume 16 Some Aromatic Amines and Related Nitro Compounds - Hair Dyes, Colouring Agents and Miscellaneous Industrial Chemicals
1978; 400 pages;
Sw. fr. 50.-/US $40.00

Volume 17 Some N-Nitroso Compounds
1987; 365 pages;
Sw. fr. 50.-/US $40.00

Volume 18 Polychlorinated Biphenyls and Polybrominated Biphenyls
1978; 140 pages;
Sw. fr. 20.-/US $16.00

Volume 19 Some Monomers, Plastics and Synthetic Elastomers, and Acrolein
1979; 513 pages;
Sw. fr. 60.-/US $48.00

Volume 20 Some Halogenated Hydrocarbons
1979; 609 pages (*out of print*)

Volume 21 Sex Hormones (II)
1979; 583 pages;
Sw. fr. 60.-/US $48.00

Volume 22 Some Non-Nutritive Sweetening Agents
1980; 208 pages;
Sw. fr. 25.-/US $20.00

Volume 23 Some Metals and Metallic Compounds
1980; 438 pages (*out of print*)

Volume 24 Some Pharmaceutical Drugs
1980; 337 pages;
Sw. fr. 40.-/US $32.00

Volume 25 Wood, Leather and Some Associated Industries
1981; 412 pages;
Sw. fr. 60-/US $48.00

Volume 26 Some Antineoplastic and Immunosuppressive Agents
1981; 411 pages;
Sw. fr. 62.-/US $49.60

Volume 27 Some Aromatic Amines, Anthraquinones and Nitroso Compounds, and Inorganic Fluorides Used in Drinking Water and Dental Preparations
1982; 341 pages;
Sw. fr. 40.-/US $32.00

Volume 28 The Rubber Industry
1982; 486 pages;
Sw. fr. 70.-/US $56.00

Volume 29 Some Industrial Chemicals and Dyestuffs
1982; 416 pages;
Sw. fr. 60.-/US $48.00

Volume 30 Miscellaneous Pesticides
1983; 424 pages;
Sw. fr. 60.-/US $48.00

Volume 31 Some Food Additives, Feed Additives and Naturally Occurring Substances
1983; 314 pages;
Sw. fr. 60-/US $48.00

List of IARC Publications

Volume 32 Polynuclear Aromatic Compounds, Part 1: Chemical, Environmental and Experimental Data
1984; 477 pages;
Sw. fr. 60.-/US $48.00

Volume 33 Polynuclear Aromatic Compounds, Part 2: Carbon Blacks, Mineral Oils and Some Nitroarenes
1984; 245 pages;
Sw. fr. 50.-/US $40.00

Volume 34 Polynuclear Aromatic Compounds, Part 3: Industrial Exposures in Aluminium Production, Coal Gasification, Coke Production, and Iron and Steel Founding
1984; 219 pages;
Sw. fr. 48.-/US $38.40

Volume 35 Polynuclear Aromatic Compounds, Part 4: Bitumens, Coal-tars and Derived Products, Shale-oils and Soots
1985; 271 pages;
Sw. fr. 70.-/US $56.00

Volume 36 Allyl Compounds, Aldehydes, Epoxides and Peroxides
1985; 369 pages;
Sw. fr. 70.-/US $70.00

Volume 37 Tobacco Habits Other than Smoking: Betel-quid and Areca-nut Chewing; and some Related Nitrosamines
1985; 291 pages;
Sw. fr. 70.-/US $56.00

Volume 38 Tobacco Smoking
1986; 421 pages;
Sw. fr. 75.-/US $60.00

Volume 39 Some Chemicals Used in Plastics and Elastomers
1986; 403 pages;
Sw. fr. 60.-/US $48.00

Volume 40 Some Naturally Occurring and Synthetic Food Components, Furocoumarins and Ultraviolet Radiation
1986; 444 pages;
Sw. fr. 65.-/US $52.00

Volume 41 Some Halogenated Hydrocarbons and Pesticide Exposures
1986; 434 pages;
Sw. fr. 65.-/US $52.00

Volume 42 Silica and Some Silicates
1987; 289 pages;
Sw. fr. 65.-/US $52.00

Volume 43 Man-Made Mineral Fibres and Radon
1988; 300 pages;
Sw. fr. 65.-/US $52.00

Volume 44 Alcohol Drinking
1988; 416 pages;
Sw. fr. 65.-/US $52.00

Volume 45 Occupational Exposures in Petroleum Refining; Crude Oil and Major Petroleum Fuels
1989; 322 pages;
Sw. fr. 65.-/US $52.00

Volume 46 Diesel and Gasoline Engine Exhausts and Some Nitroarenes
1989; 458 pages;
Sw. fr. 65.-/US $52.00

Volume 47 Some Organic Solvents, Resin Monomers and Related Compounds, Pigments and Occupational Exposures in Paint Manufacture and Painting
1990; 536 pages;
Sw. fr. 85.-/US $68.00

Volume 48 Some Flame Retardants and Textile Chemicals, and Exposures in the Textile Manufacturing Industry
1990; 345 pages;
Sw. fr. 65.-/US $52.00

Volume 49 Chromium, Nickel and Welding
1990; 677 pages;
Sw. fr. 95.-/US$76.00

Volume 50 Pharmaceutical Drugs
1990; 415 pages;
Sw. fr. 65.-/US$52.00

Volume 51 Coffee, Tea, Mate, Methylxanthines and Methylglyoxal
1991; 513 pages;
Sw. fr. 80.-/US$64.00

Volume 52 Chlorinated Drinking-water; Chlorination By-products; Some Other Halogenated Compounds; Cobalt and Cobalt Compounds
1991; 544 pages;
Sw. fr. 80.-/US$64.00

Supplement No. 1
Chemicals and Industrial Processes Associated with Cancer in Humans (IARC Monographs, Volumes 1 to 20)
1979; 71 pages; (*out of print*)

Supplement No. 2
Long-term and Short-term Screening Assays for Carcinogens: A Critical Appraisal
1980; 426 pages;
Sw. fr. 40.-/US $32.00

Supplement No. 3
Cross Index of Synonyms and Trade Names in Volumes 1 to 26
1982; 199 pages (*out of print*)

Supplement No. 4
Chemicals, Industrial Processes and Industries Associated with Cancer in Humans (IARC Monographs, Volumes 1 to 29)
1982; 292 pages (*out of print*)

Supplement No. 5
Cross Index of Synonyms and Trade Names in Volumes 1 to 36
1985; 259 pages;
Sw. fr. 46.-/US $36.80

Supplement No. 6
Genetic and Related Effects: An Updating of Selected IARC Monographs from Volumes 1 to 42
1987; 729 pages;
Sw. fr. 80.-/US $64.00

Supplement No. 7
Overall Evaluations of Carcinogenicity: An Updating of IARC Monographs Volumes 1-42
1987; 434 pages;
Sw. fr. 65.-/US $52.00

Supplement No. 8
Cross Index of Synonyms and Trade Names in Volumes 1 to 46 of the IARC Monographs
1990; 260 pages;
Sw. fr. 60.-/US $48.00

List of IARC Publications

IARC TECHNICAL REPORTS*

No. 1 **Cancer in Costa Rica**
Edited by R. Sierra,
R. Barrantes, G. Muñoz Leiva, D.M. Parkin, C.A. Bieber and
N. Muñoz Calero
1988; 124 pages;
Sw. fr. 30.-/US $24.00

No. 2 **SEARCH: A Computer Package to Assist the Statistical Analysis of Case-control Studies**
Edited by G.J. Macfarlane,
P. Boyle and P. Maisonneuve (in press)

No. 3 **Cancer Registration in the European Economic Community**
Edited by M.P. Coleman and
E. Démaret
1988; 188 pages;
Sw. fr. 30.-/US $24.00

No. 4 **Diet, Hormones and Cancer: Methodological Issues for Prospective Studies**
Edited by E. Riboli and
R. Saracci
1988; 156 pages;
Sw. fr. 30.-/US $24.00

No. 5 **Cancer in the Philippines**
Edited by A.V. Laudico,
D. Esteban and D.M. Parkin
1989; 186 pages;
Sw. fr. 30.-/US $24.00

No. 6 **La genèse du Centre International de Recherche sur le Cancer**
Par R. Sohier et A.G.B. Sutherland
1990; 104 pages
Sw. fr. 30.-/US $24.00

No. 7 **Epidémiologie du cancer dans les pays de langue latine**
1990; 310 pages
Sw. fr. 30.-/US $24.00

No. 8 **Comparative Study of Anti-smoking Legislation in Countries of the European Economic Community**
Edited by A. Sasco
1990; c. 80 pages
Sw. fr. 30.-/US $24.00
(English and French editions available) (in press)

DIRECTORY OF AGENTS BEING TESTED FOR CARCINOGENICITY (Until Vol. 13 Information Bulletin on the Survey of Chemicals Being Tested for Carcinogenicity)*

No. 8 Edited by M.-J. Ghess,
H. Bartsch and L. Tomatis
1979; 604 pages; Sw. fr. 40.-

No. 9 Edited by M.-J. Ghess,
J.D. Wilbourn, H. Bartsch and
L. Tomatis
1981; 294 pages; Sw. fr. 41.-

No. 10 Edited by M.-J. Ghess,
J.D. Wilbourn and H. Bartsch
1982; 362 pages; Sw. fr. 42.-

No. 11 Edited by M.-J. Ghess,
J.D. Wilbourn, H. Vainio and
H. Bartsch
1984; 362 pages; Sw. fr. 50.-

No. 12 Edited by M.-J. Ghess,
J.D. Wilbourn, A. Tossavainen and
H. Vainio
1986; 385 pages; Sw. fr. 50.-

No. 13 Edited by M.-J. Ghess,
J.D. Wilbourn and A. Aitio 1988;
404 pages; Sw. fr. 43.-

No. 14 Edited by M.-J. Ghess,
J.D. Wilbourn and H. Vainio
1990; 370 pages; Sw. fr. 45.-

NON-SERIAL PUBLICATIONS †

Alcool et Cancer
By A. Tuyns (in French only)
1978; 42 pages; Fr. fr. 35.-

Cancer Morbidity and Causes of Death Among Danish Brewery Workers
By O.M. Jensen
1980; 143 pages; Fr. fr. 75.-

Directory of Computer Systems Used in Cancer Registries
By H.R. Menck and D.M. Parkin
1986; 236 pages; Fr. fr. 50.-

---

* Available from booksellers through the network of WHO sales agents.

† Available directly from IARC

www.ingramcontent.com/pod-product-compliance
Ingram Content Group UK Ltd.
Pitfield, Milton Keynes, MK11 3LW, UK
UKHW051257180426
11947UKWH00020B/1768